T0271010

STATISTICAL THERMODYNAMICS
AND
PROPERTIES OF MATTER

STATISTICAL THERMODYNAMICS
AND
PROPERTIES OF MATTER

Lucienne Couture and Robert Zitoun
Université Pierre et Marie Curie
Paris, France

Translated by

Erik Geissler
CNRS, Grenoble

CRC Press
Taylor & Francis Group
Boca Raton   London   New York

Amsteldijk 166
1st Floor
1079 LH Amsterdam
The Netherlands

British Library Cataloguing in Publication Data

Couture, Lucienne
    Statistical thermodynamics and properties of matter
    1. Thermodynamics   2. Matter – Thermal properties
    I. Title II. Zitoun, Robert
    536.7

    ISBN: 9-0569919-65 (hardcover)

Cover illustration:
Molar heat capacity at constant pressure of diamond in units of $R = \mathcal{N}k$ [A. Einstein, *Ann. Physik*, **22**, 180 (1907)]. (See Fig. 3.5)

Contents

Foreword

There is no shortage of textbooks or monographs about the several aspects of thermodynamics and its interpretation as a branch of statistical physics. The publication of yet another title - this time an English translation of the French original *Physique Statistique* (published in 1992) - merits a brief explanation. The authors, who have been teaching the subject at the University of Paris to 3rd and 4th year physics students, observed that thermodynamics was very unpopular with the students. The reason for this was that both the lectures and the textbooks presented it as an abstract discipline, overlooking its importance to practically all other scientific disciplines. The realization of this fact led first to the publication in 1989 of *Thermodynamique* by L. Couture, Ch. Chahine and R. Zitoun (Dunod, Paris) and then, in 1992 the French original of this book. By skilfully inserting into the text various types of exercises, descriptions of experiments and their results as well as nearly 300 illustrations, the authors have produced a lively textbook with none of the shortcomings of its French predecessors.

True there are many books in English covering the same topic, achieving the same aim, using somewhat similar methods, nevertheless the present book has certain novel features which will recommend it to many readers.

First and foremost, whenever experimental results or applications of theory are quoted the full reference is given, either within the text or in the legend of the figure or in the excellent bibliographies at the ends of the fifteen chapters. This encourages students to develop the habit of consulting primary literature. Another useful feature is that numerous figures present both the curves predicted by theory and the actual measurement points so that the reader may judge at a glance the degree of agreement between experiment and theory and the limits of validity of the latter. And what better way to remind the student that physics has a history than to illustrate Einstein's theory of the specific heats of solids with a curve, complete with experimental points, taken from Einstein's original publication in 1907? A truly felicitous choice.

There are three types of exercises. There are those referred to by this name (exercises) which form an integral part of the text. They are fully worked out so that the reader is almost forced to consider how a theory may be tested or applied in practice.

Exercises of 'comprehension' which follow every chapter are usually snappy questions with only the answers given but not how they are arrived at.

Finally, each chapter ends with one or two "Problems" including their complete solution. The main reason for including these in the English edition is to enable those not familiar with the French higher education examination system to get some idea of it. Written examinations, usually of four hours duration, consist of one or two essays about theory and its applications taken from a discipline, e.g. thermodynamics, optics.

The translation by E. Geissler is excellent. The clear, crisp sentences read as if their contents had been conceived in English and not translated from another language.

This book is a worthy and welcome addition to the list of distinguished titles in this field of physics, mainly by virtue of some novel approaches to presentation such as insistence on experimental illustrations, often with an historical slant. It can be recommended for study, for reference and even for browsing.

Nicholas Kurti, CBE, FRS
Emeritus Professor of Physics
Oxford, September 1998

Preface

This book is based on a degree course taught over a period of about fifteen years to BSc and Masters students in physics at the Pierre and Marie Curie University of Paris. For the sake of clarity and in order to familiarize our students with the physicist's approach, we decided to apply statistical theory as early as possible to real physical problems. The method we employ is

- to expound the theory by stages, beginning with an elementary theory that handles systems of particles without interactions; the Gibbs theory of ensembles is developed later;

- to describe different macroscopic properties of substances: relationships, measurement techniques, etc.;

- at the same time to describe the phenomena at the atomic scale;

- to choose appropriate models, then apply statistical theory and confront predictions with experimental results.

The progress of the theory can be seen through the discussion of magnetic phenomena in two chapters, the first treating ideal paramagnetism and the second ordered magnetic phases. In addition, whole chapters are devoted to special topics such as superfluidity in helium, superconductivity, etc. This approach explains the large number of figures, tables, exercises and problems involving experimental results.

This book could not have been what it is without the collaboration of Charles Chahine in our team over the last fifteen years. It is his inspiration that lies behind the presentation of most of the theoretical methods and illustrative physical examples. May this be a testimony of our friendship and recognition.

Our gratitude goes to all our colleagues who took part in the teaching programme, as well as to the students upon whom we tried our methods. We also thank Nathalie Laberrigue and Jean Louis Gorrand for their help in preparing this book. Last, we are grateful to the Laboratoire de Physique Nucléaire et des Hautes Energies (LPNHE) of the Pierre et Marie Curie and Paris VII Universities (associate laboratory of the Institut de Physique Nucléaire et de Physique des Particules (IN2P3)), as well as to the Aimé Cotton Laboratory of the Centre National de la Recherche Scientifique (CNRS).

General References

NOTATIONS AND UNITS

Quantities, Units, and Symbols, The Royal Society, London, 1975.
E.R. Cohen and P. Giacomo, *Symbols, Units, Nomenclature and Fundamental Constants in Physics*, Document I.U.P.A.P.-25 (SUNAMCO 87-1), *Physica A*, **146A**, 1 (1987).

CLASSICAL THERMODYNAMICS

L.Couture, Ch. Chahine and R. Zitoun, *Thermodynamique*, Dunod, Paris, 1989.
E.A. Guggenheim, *Thermodynamics*, North Holland Publishing Company, Amsterdam, 1957.
M.W. Zemansky, *Heat and Thermodynamics*, McGraw Hill, New York, 1962.

GENERAL STATISTICAL PHYSICS

R. Balian, *Du microscopique au macroscopique*, Ellipses, Paris, 1982.
Ch. Chahine and P. Devaux, *Thermodynamique Statistique*, Dunod, Paris, 1976.
N. Davidson, *Statistical Mechanics*, McGraw Hill, New York, 1962
B. Diu, C. Guthmann, D. Lederer and B. Roulet, *Eléments de Physique Statistique*, Hermann, Paris, 1982.
R.H. Fowler and E.A. Guggenheim, *Statistical Thermodynamics*, Cambridge University Press, Cambridge, 1949.
D. ter Haar, *Elements of Thermostatistics*, Holt, Rinehart and Winston, New York, 1966.
T.L. Hill, *Statistical Mechanics*, McGraw Hill, New York, 1956.
K. Huang, *Statistical Mechanics*, John Wiley and Sons, New York, 1963.
C. Kittel, *Elementary Statistical Physics*, Wiley, New York, 1958.
R. Kubo, *Statistical Mechanics*, North Holland, Amsterdam, 1965.
L.D. Landau and E.M. Lifschitz, *Theoretical Physics Volume 5, Statistical Physics*, Pergamon, Oxford, 1969.
J.E. Mayer and M.G. Mayer, *Statistical Mechanics*, New York, 1946.
D.A. McQuarrie, *Statistical Mechanics*, Harper and Row, New York, 1976.

A. Münster, *Statistische Thermodynamik*, Springer Verlag, Berlin, 1956.

E. Schrödinger, *Statistical Thermodynamics*, Cambridge University Press, Cambridge, 1952.

R.C. Tolman, *Statistical Mechanics*, Oxford, 1938.

QUANTUM MECHANICS

C. Cohen-Tannoudji, B. Diu and F. Laloë, *Mécanique quantique*, Hermann, Paris, 1973.

H. Eyring, J. Walter and G.E. Kimball, *Quantum Chemistry*, John Wiley and Sons, New York, 1961.

L.D. Landau and E.M. Lifschitz, *Theoretical Physics Volume 3, Quantum Mechanics*, Pergamon, Oxford, 1977.

J.M. Lévy-Leblond and F. Balibar, *Quantique (Rudiments)*, Interéditions, Paris, 1984.

A. Messiah, *Mécanique quantique*, Dunod, Paris, 1959.

SOLIDS

N.W. Ashcroft and N.D. Mermin, *Solid State Physics*, Holt, Rinehart and Winston, New York, 1976.

J.S. Blakemore, *Solid State Physics*, Saunders, Philadelpia, 1974.

R.J. Elliot and A.F. Gibson, *Solid State Physics*, Macmillan, London, 1974.

C. Kittel, *Introduction to Solid State Physics*, Wiley, New York, 1970.

J.F. Nye, *Physical Properties of Crystals*, Dunod, Clarendon Press, Oxford, 1957.

H.M. Rosenberg, *Low Temperature Solid State Physics*, Oxford, 1963.

ELECTROMAGNETISM

B.I. Bleaney and B. Bleaney, *Electricity and Magnetism*, University Press, Oxford, 1976.

J. Crangle, *The Magnetic Properties of Solids*, Edward Arnold, London, 1977.

Chapter 1

Entropy and the Boltzmann Relation

1.1 INTRODUCTION

The statistics of any system is determined by the properties of its constituent particles on a molecular scale. In the course of this book we shall meet with two families of systems that are distinguished by the fact that, in the first, the particles are mobile (the family of gases), while in the second the particles are localized. We take here as an example an ideal case of each family: a perfect (or ideal) monatomic gas, and an ideal paramagnetic substance. For these model systems, as in the majority of the cases investigated in the first part of this book, we shall neglect interactions between the particles.

Statistics relates macroscopic quantities to molecular quantities. Some of the relations used are obvious. For example the internal energy is defined as the sum of the energies of the particles in the system, while the magnetic moment is the vector sum of their magnetic moments. In contrast, entropy has no clear relationship with molecular properties. It was by studying the changes in systems that Boltzmann was able to relate entropy to the number of different microscopic states accessible to the system. The Boltzmann relation provides an interpretation on a molecular level of the origin of irreversibility and gives meaning to the Second Law. In particular, it will be seen that this relation predicts the value of the entropy at absolute zero for all systems, whether or not they obey the Third Law.

If the statistical expressions for the energy and the entropy are known, then all other thermodynamic quantities can be found. Problem 1.1 applies the elementary statistical method described in this section to a copper salt in which the magnetic particles possess only two energy levels.

1.2 IDEAL MONATOMIC GASES

1.2.1 Translational Energy Levels

There are many monatomic gases, in particular the rare gases. These are the simplest gases to study, since the energy of each atom comes only from its translational movement. In this paragraph we neglect electronic excitations since they exert an influence only at very high temperatures (§5.2.1).

1

In the model of an ideal monatomic gas, the atoms are considered to be point-like particles constrained to move inside their container, which is taken to be a rectangular box. Interactions between particles are neglected, with the result that each atom moves as if it were alone in the volume of the box. During collisions between atoms, however, interactions are necessary for thermal equilibrium to be established in the gas.

In this model the quantum state of each particle is determined by three quantum numbers m_x, m_y and m_z that are positive integers and correspond to the three degrees of translational freedom along the axes Ox, Oy and Oz of a rectangular parallelepiped of sides a, b and c. The energy of translation of a particle of mass m is then

$$\epsilon(m_x, m_y, m_z) = \frac{h^2}{8m}\left(\frac{m_x^2}{a^2} + \frac{m_y^2}{b^2} + \frac{m_z^2}{c^2}\right) \quad , \quad m_x, m_y, m_z = 1, 2, 3, \ldots \quad (1.1)$$

where h denotes Planck's constant ($h = 6.63 \times 10^{-34}$ J s). If the box is a cube, i.e. if $a = b = c = V^{1/3}$, this energy can be written

$$\epsilon(m_x, m_y, m_z) = \left(m_x^2 + m_y^2 + m_z^2\right)\epsilon^0 \quad \text{with} \quad \epsilon^0 = \frac{h^2}{8mV^{2/3}} = \frac{h^2\mathcal{N}}{8MV^{2/3}} \quad , \quad (1.2)$$

where M is the molar mass of the gas and $\mathcal{N}$ is Avogadro's number ($\mathcal{N} = 6.02 \times 10^{23}$ mol^{-1}).

It is interesting to note that in a typical case, that of helium ($M = 4.0 \times 10^{-3}$ kg mol^{-1}) in a cubic container of one litre, we have

$$\epsilon^0 = 8.3 \times 10^{-40} \text{ J} = 5.2 \times 10^{-21} \text{ eV} \quad . \quad (1.3)$$

This value cannot be measured experimentally as it is much smaller than the mean kinetic energy of a gas particle. We shall see later (§5.2.3) that, at temperature T, the latter energy is equal to $\bar{\epsilon} = 3kT/2$ ($k = R/\mathcal{N} = 1.38 \times 10^{-23}$ J K^{-1} is the Boltzmann constant and R the gas constant); i.e., for $T = 273$ K,

$$\bar{\epsilon} = 5.7 \times 10^{-21} \text{ J} = 3.5 \times 10^{-2} \text{ eV} \quad . \quad (1.4)$$

This mean energy corresponds to values of m_x, m_y and m_z of the order of 10^9. The quantum numbers involved are therefore very large and the translational energy levels closely spaced, i.e. their separation is much smaller than the mean energy of a particle. It is therefore justified to view the energy levels as forming a continuous spectrum, as is assumed in classical mechanics. This is an application of the correspondence principle according to which in the limit of large quantum numbers the behaviour of an atomic system is identical to that holding in classical mechanics. In order to count the number of states, however, it is still necessary to take account of the results of quantum mechanics.

Exercise 1.1 *Degeneracy of the translational energy levels*

Calculate the energy ϵ_i and sketch the first 15 energy levels of a particle moving in a cubic box. Determine the degeneracy g_i, the number of different states occupying

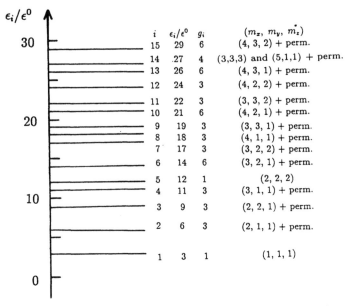

Figure 1.1: Translational energy levels of a particle in a cubic box.

the same level i. What is the reason for the degeneracy in the first four degenerate levels ($g_i > 1$)? For level 14 show that the degeneracy may have a different origin.

Solution
The energies of the first fifteen levels can be obtained from equation (1.2) by incrementing the integers m_x, m_y and m_z, starting from 1. The energies ϵ_i of the first 15 levels are given in Figure 1.1 in units of ϵ^0. The triplets (m_x, m_y, m_z) of the quantum states are also shown. The levels $i = 1$ and $i = 5$ are not degenerate ($g_i = 1$). For the first four degenerate levels ($i = 2, 3, 4$ and 6), it can be seen that the degeneracy comes from the symmetry of the model (equivalence of the three axes Ox, Oy, Oz). In general, starting from a given state, 6 different quantum states with the same energy can be constructed by permutation if the numbers m_x, m_y and m_z are all different, and 3 states if two of them are equal. Energy level 14 is the first to exhibit a degeneracy that is due not to symmetry, but this is accidental, since the three states $(5, 1, 1)$, $(1, 5, 1)$ and $(1, 1, 5)$ have the same energy as the state $(3, 3, 3)$.

Exercise 1.2 *Enumeration of the translational quantum states*
A moving particle is confined inside a rectangular box of sides a, b, and c. Each of its quantum states can be represented by a point M in a *reciprocal space*, such that $\mathbf{OM} = m_x \mathbf{e}_x/a + m_y \mathbf{e}_y/b + m_z \mathbf{e}_z/c$, where $\mathbf{e}_x$, $\mathbf{e}_y$, $\mathbf{e}_z$ are unit vectors along the three orthogonal axes.

1 Show that for every quantum state an elementary cell exists in this reciprocal space. Calculate its size.
2 In this space, calculate the size of the domain that includes all states of mo-

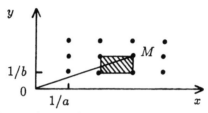

Figure 1.2: Reciprocal space in two dimensions. The point M corresponds to the quantum state $m_x = 3, m_y = 2$. The elementary cell associated with M is hatched.

mentum $p < p_0$. Hence deduce the number of quantum states having momentum less than p_0.

Solution

1 The reciprocal space corresponding to two dimensional space is shown in Figure 1.2. In three dimensions, if the different points are joined by straight lines parallel to the co-ordinate axes, a lattice is generated whose elementary cell is a rectangular parallelepiped. To each node of the lattice there corresponds a cell. The edges of this cell are of length $1/a, 1/b, 1/c$. Its size is therefore $1/abc = 1/V$, where V is the volume of the box.

2 From relation (1.1) we have

$$\mathbf{OM}^2 = \frac{m_x^2}{a^2} + \frac{m_y^2}{b^2} + \frac{m_z^2}{c^2} = \frac{8m\epsilon}{h^2} = \frac{4p^2}{h^2} \quad .$$

States with $p < p_0$ have $\mathbf{OM}^2 < 4p_0^2/h^2$ and hence are located inside a sphere of radius $2p_0/h$, in the positive octant, since m_x, m_y and m_z are positive. The volume of the octant is

$$\frac{1}{8} \frac{4\pi}{3} \left(\frac{2p_0}{h} \right)^3 = \frac{\pi}{6} \left(\frac{2p_0}{h} \right)^3 \quad .$$

The number of cells, and hence the number of quantum states, is

$$n(p_0) = \frac{\pi}{6} \left(\frac{2p_0}{h} \right)^3 \div \frac{1}{V} = \frac{\frac{4\pi}{3} p_0^3 \times V}{h^3} \quad . \tag{1.5}$$

1.2.2 Phase Space

In classical mechanics, the state of a particle is defined by its position $\mathbf{r}$ and its momentum $\mathbf{p}$. To represent this state, the notion of *phase space* for a particle is introduced. This six-dimensional space is the Cartesian product of position space (variables x, y, z) and momentum space (variables p_x, p_y, p_z). From the point of view of classical mechanics, the total microscopic state of a particle is represented in phase space by a point with co-ordinates x, y, z, p_x, p_y, p_z, and vice versa.

In the course of time the representative point moves in phase space. To illustrate this, if we restrict ourselves to the movement of a particle in a one-dimensional box, phase space possesses two dimensions (x, p_x) and, for a given motion, the

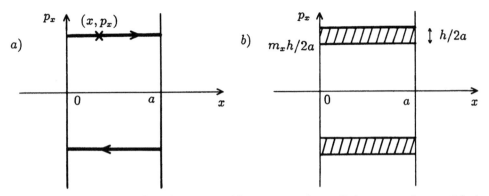

Figure 1.3: Classical a) and quantum b) representations of the state of a particle in one-dimensional space (two-dimensional phase space).

representative point of the particle moves along the two symmetrical straight-line segments shown in Figure 1.3a. After collision with a wall at $x = 0$ or $x = a$, the momentum p_x changes sign. This momentum, and the energy of translation $\epsilon = p_x^2/2m$ can take any value (continuous spectrum).

In quantum mechanics, in contrast, the states of a particle correspond to specific values of the energy and momentum. For motion in one dimension, the only possible values are

$$\epsilon = \frac{h^2}{8ma^2}m_x^2 \quad \text{and} \quad |p_x| = \frac{h}{2a}m_x \quad , \quad m_x = 1, 2, 3, \dots$$

These are associated in phase space with pairs of symmetric straight-line segments regularly spaced by $h/2a$ (Fig. 1.3b), and two neighbouring pairs enclose an area in phase space equal to $2 \times h/2a \times a = h$. This result is, moreover, independent of the problem under consideration and is related to the Heisenberg inequalities ($\Delta x \Delta p_x \geq \hbar/2$ with $\hbar = h/2\pi$). Thus, for large quantum numbers the classical description of the motion remains valid, provided a *cell* of volume h in phase space is attributed to each state.

Generalization to motion in d dimensions is immediate: to each quantum state corresponds in $2d$ dimensional phase space a cell of volume h^d. In particular, for motion in 3 dimensions this volume is equal to h^3. More precisely, a volume in phase space of $d^3\mathbf{r}d^3\mathbf{p} = dxdydzdp_xdp_ydp_z$ contains a number of quantum states equal to

$$g(\mathbf{r}, \mathbf{p})d^3\mathbf{r}d^3\mathbf{p} = \frac{d^3\mathbf{r}d^3\mathbf{p}}{h^3} \tag{1.6}$$

where $g(\mathbf{r}, \mathbf{p}) = 1/h^3$ is the density of quantum states in phase space.

Exercise 1.3 *Counting states in phase space*

1 From the previous formula calculate the number of translational quantum states, $n(p_0)$, having momentum less than p_0 (or energy less than $\epsilon_0 = p_0^2/2m$). Compare this result with that of Exercise 1.2.

2 Evaluate the number of states $g(p_0)dp$ with momentum lying between p_0 and $p_0 + dp$, and also the number of states $g(\epsilon_0)d\epsilon$ with energy between ϵ_0 and $\epsilon_0 + d\epsilon$.

Solution

1 $n(p_0)$ is obtained by integrating expression (1.6) over all possible values of $\mathbf{r}$ within the volume V and over all $\mathbf{p}$ such that $|\mathbf{p}| < p_0$. This gives

$$n(p_0) = \int \frac{d^3\mathbf{r}d^3\mathbf{p}}{h^3} = \frac{V}{h^3} \int_{|\mathbf{p}|<p_0} d^3\mathbf{p} = \frac{V}{h^3}\frac{4\pi p_0^3}{3} \quad . \tag{1.7a}$$

This result, whose meaning is clear, is identical to that found in Exercise 1.2. The same physical quantity is obtained by two different methods.

It is also possible to write

$$n(\epsilon_0) = \frac{V}{h^3}\frac{4\pi \left(2m\epsilon_0\right)^{3/2}}{3} \quad . \tag{1.7b}$$

2 On differentiating formula (1.7a) with respect to p, we get for $p = p_0$

$$dn = g(p_0)dp = \frac{V}{h^3}4\pi p_0^2 dp \quad ; \tag{1.8}$$

$g(p_0)dp$ is the number of states contained between two spheres of radius p_0 and $p_0 + dp$ in $\mathbf{p}$ space. In the same way, using (1.7b), we get

$$dn = g(\epsilon_0)d\epsilon = \frac{2\pi V}{h^3}\left(2m\right)^{3/2}\epsilon_0^{1/2}d\epsilon \quad . \tag{1.9}$$

The densities of states are increasing functions of momentum and energy.

1.2.3 Microscopic and Macroscopic Descriptions of Gases

Particles that are identical and not localized, as in a gas, are *indistinguishable*. This is a principle of quantum mechanics. In classical mechanics we may talk about the trajectory of a particle. This is no longer true in quantum mechanics, however, owing to the uncertainty principle, which makes it impossible to define the exact position of a particle and at the same time measure its momentum. Thus for a set of identical particles, we cannot follow their individual motions. It is in this respect that identical particles are indistinguishable in quantum mechanics.

It follows that, if the quantum numbers of two particles are exchanged, the same quantum state for the gas is obtained. Hence, in order to describe the *microscopic* states of a gas composed of N identical particles, it is necessary and sufficient to specify the numbers $n_1, n_2, \ldots n_j, \ldots$ of particles in the quantum states (or cells) $1, 2, \ldots j, \ldots$ of energy $\epsilon_1, \epsilon_2, \ldots \epsilon_j, \ldots$ The statistical expression for the macroscopic quantities depends on these numbers. Thus, the total number of particles in the gas is given by

$$N = \sum_j n_j \quad . \tag{1.10}$$

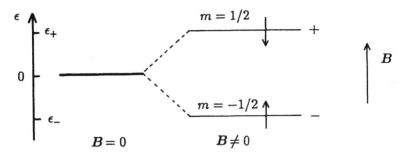

Figure 1.6: Energy levels of a spin $1/2$ in a magnetic field

where the energies ϵ_+ and ϵ_- are given by equation (1.19)

$$\epsilon_+ = \frac{1}{2}g\mu_B B \quad \text{and} \quad \epsilon_- = -\frac{1}{2}g\mu_B B \quad .$$

Therefore

$$U = \frac{1}{2}g\mu_B B\,(N_+ - N_-) = (N_+ - N_-)\,\overline{m}B \quad . \tag{1.25}$$

3 The magnetic moment $\mathcal{M}$ of the salt is given by formula (1.22) and is equal to

$$\mathcal{M} = -g\mu_B \left(\frac{1}{2}N_+ - \frac{1}{2}N_-\right) = -(N_+ - N_-)\,\overline{m} \quad . \tag{1.26}$$

Note that we again obtain the general relation

$$U = -\mathcal{M}B \quad . \tag{1.27}$$

Saturation corresponds to the maximum value of $\mathcal{M}$, with $N_- = N$ and $N_+ = 0$, i.e.

$$\mathcal{M}_S = N\overline{m} \quad .$$

The order parameter is

$$R = \frac{N_- - N_+}{N} = -\frac{U}{N\overline{m}B} \tag{1.28}$$

R is dimensionless and lies between the limits -1 and $+1$.
  4 Combining equations (1.24) and (1.28), we find

$$N_+ = \frac{N}{2}(1 - R) \quad \text{and} \quad N_- = \frac{N}{2}(1 + R) \quad . \tag{1.29}$$

We see that for this system with two levels, a knowledge of U and of N is sufficient to determine the populations N_+ and N_-. The same is true for all systems in which the particles can occupy two levels.

1.4   BASIC POSTULATES OF ELEMENTARY STATISTICAL THEORY

A closed isolated system of N particles that is in a given microscopic state at any one instant remains in that state only for a very short time, since collisions or interactions induce changes. It therefore changes its microscopic state at a very high frequency. One might think it possible to determine the changes in the system using the (classical or quantum mechanical) equations of motion. This would, however, mean solving N differential equations, which is out of the question when N is of the same order as Avogadro's number $\mathcal{N}$. Confronted with the impossibility of exactly predicting the changes in the system, Gibbs introduced the idea that all microscopic states accessible to an isolated system are explored equally in the course of time. This led him to expound the first postulate of statistical thermodynamics, which states that *all microscopic states accessible to an isolated system are equally probable.*

Although all the microscopic states are equally probable, the same is not true for the macroscopic states, whose probability is greater when they contain a large number of microscopic states. The number of microscopic states $W(N_1, \ldots N_i, \ldots)$ that contribute to a macroscopic state defined by the numbers $N_1, \ldots N_i, \ldots$ is called the *thermodynamic probability* of the macroscopic state.

To clarify these ideas we return to the example of a paramagnetic salt containing N ions of spin $J = 1/2$ in zero field (Ex. 1.4). The energy U of this system, given by (1.25), is zero whatever the value of the magnetic moment $\mathcal{M}$ (or the order parameter R). As the numbers N_+ and N_- are connected by relation (1.24), the macroscopic state of the system can be defined by a single parameter, e.g. N_+, or by the order parameter R (1.28). The thermodynamic probability W of a macroscopic state in this case is the number of microscopic states comprising N_+ spins selected out of N, i.e. the number of combinations $\mathrm{C}_N^{N_+}$

$$W = \mathrm{C}_N^{N_+} = \frac{N!}{N_+! \, (N - N_+)!} = \frac{N!}{N_+! N_-!} \quad . \tag{1.30}$$

This probability takes its maximum value when $N_+ = N_-$, i.e. when

$$N_+ = N/2 \quad . \tag{1.31}$$

This macroscopic state is the most probable. It is characterized by an equal number of ions with their magnetic moment lying in the same direction as the field ($m = -1/2$) as in the opposite direction ($m = 1/2$), i.e. the total magnetic moment $\mathcal{M}$ and the order parameter R are zero.

We shall now show that the system actually spends very little time in other macroscopic states, even in the neighbouring states. To do this, we look at the function $W(R)$, which, along with (1.29) and Stirling's formula

$$\ln x! = x \ln x - x + \ln \sqrt{2\pi x} + O\left(x^{-1}\right) \quad , \tag{1.32}$$

can be written successively as

$$W = \frac{N!}{[N(1 - R)/2]! \, [N(1 + R)/2]!}$$

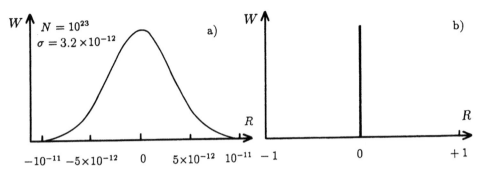

Figure 1.7: Thermodynamic probability as a function of the order parameter R for $N = 10^{23}$ at two different scales.

$$\ln W = N \ln N - \frac{N}{2}(1-R)\ln\left[\frac{N}{2}(1-R)\right] - \frac{N}{2}(1+R)\ln\left[\frac{N}{2}(1+R)\right]$$
$$+ \ln \frac{\sqrt{2\pi N}}{\pi N\sqrt{1-R^2}}$$
$$= N \ln 2 - \frac{N}{2}(1-R)\ln(1-R) - \frac{N}{2}(1+R)\ln(1+R)$$
$$- \ln\sqrt{\frac{\pi N}{2}} - \ln\sqrt{1-R^2} \quad .$$

Expanding around $R = 0$ to second order, we get

$$\ln W = N \ln 2 - (N-1)R^2/2 - \ln\sqrt{\pi N/2} \quad , \tag{1.33}$$

and hence

$$W(R) = \frac{2^N}{\sqrt{\pi N/2}} e^{-(N-1)R^2/2} \quad . \tag{1.34}$$

The function $W(R)$, shown on Figure 1.7, is a Gaussian curve ($y \propto \exp(-x^2/2\sigma^2)$) centred at $R = 0$, with mean deviation

$$\sigma = 1/\sqrt{(N-1)} \simeq 1/\sqrt{N}. \tag{1.35}$$

Its maximum value is

$$W_{max} = C_N^{N/2} = 2^N/\sqrt{\pi N/2} \quad . \tag{1.36}$$

The probability W is practically zero for $|R|$ larger than a few mean deviations. This means that if $N \sim 10^{23}$, the system spends almost all its time in states for which $|R|\lesssim 10^{-11}$ (Fig. 1.7a). Such states are macroscopically indistinguishable from the most probable state $R = 0$ (Fig. 1.7b).

Although the above discussion is addressed to a particular case, it can be generalized to other systems, provided that the number of particles N is very large. This allows us to introduce a second postulate that will be used in the elementary theory:

the equilibrium state of a closed isolated system is the most probable macroscopic state.

These postulates yield the macroscopic equilibrium state for isolated closed systems of non-interacting particles. The thermodynamic probability W is maximized subject to the constraints

$$\sum_i N_i = N \text{ (closed system)} \tag{1.37a}$$

$$\text{and } \sum_i N_i \epsilon_i = U \text{ (isolated system).} \tag{1.37b}$$

Note that the thermodynamic probability is not a true probability since it is a number greater than or equal to 1. The true probability of a macroscopic state is defined by $P_l = W_l / \Sigma W_l$. In the above example, the true probability $P(R)$ is

$$P(R) = \frac{W(R)}{\Sigma W} = \frac{W(R)}{2^N} \simeq \frac{1}{\sqrt{\pi N/2}} e^{-(N-1)R^2/2} \quad . \tag{1.38}$$

1.5 STATISTICAL DEFINITION OF ENTROPY

1.5.1 Boltzmann Relation

When a system is not at equilibrium its thermodynamic probability W is smaller than W_{max}. The system then moves towards states of higher probability until the maximum value $W = W_{max}$, corresponding to the equilibrium state, is reached.

In classical thermodynamics, the Second Law may be stated as follows: *the entropy of an isolated system remains constant or increases.* This expresses the fact that a system that is out of equilibrium evolves in time in such a way that its entropy increases until it reaches a maximum at equilibrium. A mathematical relation must therefore exist between the entropy S of a system and its thermodynamic probability W, i.e.

$$S = f(W) \quad .$$

The form of this relation can be determined by simple considerations of extensivity. Consider two independent isolated systems in equilibrium having entropies S_1 and S_2 respectively. Since it is an extensive quantity, the entropy of both systems together is $S_1 + S_2$. Now let W_1 and W_2 be the thermodynamic probabilities of each system at equilibrium. The thermodynamic probability corresponding to the two systems, which are assumed to be independent, is $W_1 \times W_2$ (combination of probabilities). We then have

$$S_1 = f(W_1) \,, \; S_2 = f(W_2) \quad \text{and} \quad S_1 + S_2 = f(W_1 \times W_2) \quad .$$

These equalities lead to the functional relation

$$f(W_1 \times W_2) = f(W_1) + f(W_2)$$

which is satisfied by the logarithmic function

$$S = k \ln W \quad . \tag{1.39}$$

This relation, called the Boltzmann relation (1872), does not depend on the nature of the systems whose entropy is S_1 and S_2. It is therefore general. The constant k, called the Boltzmann constant, is determined experimentally and is found to be $k = 1.38 \times 10^{-23}$ J K^{-1}. The example of ideal molecular gases (§5.2.3) shows that

$$k = \frac{R}{N} \quad , \tag{1.40}$$

the ratio of the gas constant R to Avogadro's number.

The concept of entropy is a macroscopic notion that is applicable to systems containing large numbers of particles. For example, equation (1.33) shows that the entropy of a system of spins $J = 1/2$ in equilibrium at zero field is given by

$$S = k \ln W(R = 0) = N k \ln 2 - k \ln \sqrt{\pi N/2} \quad . \tag{1.41a}$$

This expression for the entropy is extensive only in so far as N is very large so that the second term can be neglected. For $N = 10^{23}$, the second term is 3.7×10^{-22} J K^{-1} while the first is 0.96 J K^{-1}. In the *thermodynamic limit*, $N \to \infty$, the extensive formula is obtained

$$S = N k \ln 2 \quad . \tag{1.41b}$$

The Boltzmann relation provides a definition of the entropy of a system even if it is out of equilibrium.

1.5.2   Statistical Interpretation of the Second Law. Irreversibility.

For the sake of illustration we consider a system of N spins $J = 1/2$ in a macroscopic state characterized by $N_- = N$ and $N_+ = 0$. In this state, which is obtained by applying a sufficiently strong magnetic field of flux density B at low temperature, the total magnetic moment (1.26) is $\mathcal{M} = \mathcal{M}_S \equiv N\overline{m}$. This macroscopic state corresponds to only one microscopic state, that in which the magnetic moments all lie in the direction of the field. Its thermodynamic probability is thus $W = 1$, in accordance with equation (1.30), and the Boltzmann relation yields $S = 0$ for the entropy.

At a given instant, $t = 0$, the system is thermally isolated and the magnetic field is reduced to zero. The energy of the spin system, previously equal to $-\mathcal{M}_S B$, also falls to zero. Under the influence of crystal lattice vibrations and dipole-dipole interactions the spins start to rotate, so that the system becomes disordered and evolves towards its equilibrium state. This state, investigated in paragraph 1.4, is defined by its magnetic moment $\mathcal{M} = 0$ and its entropy $S = N k \ln 2$. During this spontaneous change, the entropy of the system increases irreversibly, in accordance with the Second Law of Thermodynamics. To understand these changes, we must take account of the idea of time duration in the phenomena occurring at a molecular scale. Let τ be the average interval of time between two consecutive rotations of a given spin. Assume furthermore that at regular time intervals τ/N, one spin, arbitrarily chosen among N, rotates. The quantities N_+ and N_- must then be considered as being functions of time. Between the instant t and the instant $t + \tau/N$,

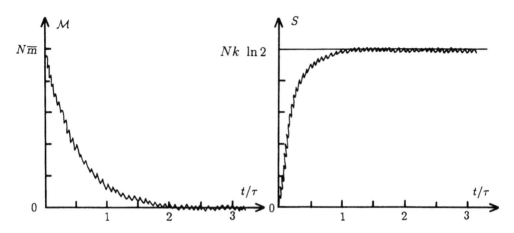

Figure 1.8: Magnetic moment and entropy of a system of N spins 1/2. The amplitude of the fluctuations is greatly exaggerated and their frequency reduced.

one spin has rotated. The probability that it is a spin belonging to N_+ (or N_-) is N_+/N (or N_-/N) and in this case N_+ decreases (or increases) by one unit. On average, therefore

$$N_+(t + \tau/N) - N_+(t) = -1 \times \frac{N_+}{N} + 1 \times \frac{N_-}{N}$$

i.e. $\quad \dfrac{dN_+}{dt} \times \dfrac{\tau}{N} = \dfrac{N_- - N_+}{N} = \dfrac{N - 2N_+}{N} \quad .$ \hfill (1.42)

This differential equation describes an average change with time of the system. With the initial condition, $N_+(0) = 0$, the solution becomes

$$N_+(t) = \frac{N}{2} \left(1 - e^{-2t/\tau}\right) \quad .$$ \hfill (1.43)

The above result shows that the system evolves with a time constant $\tau/2$, called the *relaxation time*. After an interval of a few τ, the system reaches its equilibrium state defined by $N_+ = N/2$. The relaxation times are determined experimentally, and can have a very wide range of values. They provide information about the interactions on a molecular scale.

From expression (1.43) for $N_+(t)$, the magnetic moment $\mathcal{M}(t)$ can be determined as well as the entropy $S(t)$ using equations (1.26 and 33). Figure 1.8 displays these functions. Also shown diagrammatically are the fluctuations occurring around the mean values, the above relations being only statistical. In particular, no single microscopic state is reached, since rotations continue endlessly. The fluctuations, however, which obey the probability relation (1.38), produce extremely small macroscopic effects ($R \lesssim N^{-1/2} \sim 10^{-11}$ from (1.35)).

More precise results can be extracted from this model. In particular, it is found that the system can successively explore all the possible states with an average frequency that is proportional to the true probability (1.38) of the state under consideration. The reciprocal of this frequency, called the recurrence time, is given by

$$T(R) = \frac{\tau}{N} \times \frac{1}{P(R)} \ . \tag{1.44}$$

Thus the most probable state $(R = 0)$ recurs on average in a time interval

$$T(0) = \tau \sqrt{\frac{\pi}{2N}}$$

which is very small if $N \sim 10^{23}$. The initial state $(R = 1)$ can recur after a time

$$T(1) = \frac{\tau}{N} \times 2^N = \frac{\tau}{N} \times 10^{0.3 \times N}$$

which is far greater than the age of the universe ($\sim 10^{17}$ s $\sim 10^{10}$ years). It is therefore clear that the notion of irreversibility, which is absent from the model at the molecular scale, appears on a macroscopic scale as a statistical result emanating from the very large number of particles involved. For a system containing a small number of particles ($N \sim 10$), all macroscopic states have probabilities with the same order of magnitude, and the fluctuations are as large as the quantities themselves.

1.5.3 Statistical Interpretation of the Third Law

The Third Law of classical thermodynamics (W. Nernst, 1906) can be stated as follows: *at absolute zero, the entropy of a pure substance, whether liquid or in the form of an ordered crystal, is zero.* This law lends itself directly to a statistical interpretation. According to the Boltzmann relation (1.39), zero entropy corresponds to a value of the thermodynamic probability W equal to 1. The Third Law thus corresponds to the fact that at absolute zero only one state is available to the system.

For example, a pure ordered crystalline body at 0 K occupies a single state, its quantum ground state, whatever the allotropic variety or the external constraints such as pressure or magnetic field. The entropy at 0 K is thus zero both for graphite (stable phase of carbon) and for diamond (metastable phase). Similarly, for a magnetic material, the spins are in a perfectly ordered state (ferromagnetic, ...). In these examples, the macroscopic state can be constructed in only one way ($W = 1$) and the entropy is zero.

It happens, however, that certain solids retain a residual entropy at absolute zero. This is not in contradiction with the Third Law, because such materials are not in the form of an ordered crystal. The remaining disorder corresponds to a value of W greater than 1. For example, crystals of nitrous oxide N_2O at 0 K have a residual molar entropy of $(0.57 \pm 0.05)R = 1.14 \pm 0.1$ cal K^{-1} mol^{-1}, i.e. significantly greater than the experimental uncertainty [R.W. Blue and W.F. Giauque, *J. Am. Chem. Soc.* **57**, 991 (1935)]. The Boltzmann relation provides a statistical interpretation of this phenomenon. Molecular crystals of N_2O are composed of non symmetrical linear molecules of the form N-N-O, whose dipole moment is very small. X-ray diffraction measurements of the structure show that at each site the molecules can take two opposing orientations that are related to each other by a rotation (N-N-O or O-N-N). The stable state corresponding to perfect order cannot be reached, because the molecules are hindered by their surroundings and are unable to rotate as the crystal is cooled. Since each molecule has two possible orientations, the number of possible configurations for a completely random distribution is found from the rule of combined probabilities. For a crystal containing N molecules, it is given by $W = 2^N$. Hence, using the Boltzmann relation, the entropy is

$$S = k \ln W = Nk \ln 2 \quad,$$

which yields for the molar entropy

$$s = \mathcal{N}k \ln 2 = R \ln 2 = 1.38 \text{ cal K}^{-1} \text{ mol}^{-1} \quad.$$

This result is in fairly good agreement with the experimental value. The fact that the latter is smaller shows that the assumption of a random distribution is not entirely valid, and that as the crystal cools, it tends towards the ordered state.

An important remark is in order about the notion of accessible states. The example of nitrous oxide provides an illustration. A crystalline sample of N_2O at absolute zero is in a given microscopic state and stays there, since rotation of the molecules is impossible. It would be incorrect to conclude that $W = 1$ and hence $S = 0$, since W, the number of microscopic states making up the macroscopic state of the crystal, corresponds to all possible configurations that could have been obtained on cooling the nitrous oxide to 0 K.

The notion of statistical ensembles introduced by J.W. Gibbs (1902) helps in understanding this aspect (Ch. 12). It substitutes the statistical analysis of a closed isolated system by that of an ensemble containing a very large number of systems that are all macroscopically identical to the system under consideration, and are called *microcanonical ensembles*. The thermodynamic probability W of the system in a given macroscopic state is then the number of microscopically different states ocurring in the ensemble. Returning to the case of nitrous oxide, the systems in the ensemble will give rise to 2^N possible microscopic states at absolute zero, even though the real system occupies only one of these states. Hence its molar entropy is equal to $R \ln 2$.

Exercise 1.5 *Statistical determination of the residual entropy of ice at absolute zero*

In a molecular crystal of ordinary ice I_h, each oxygen atom has four oxygen neighbours located at the vertices of an almost regular tetrahedron of which it is the centre. The H_2O molecules are connected to each other by hydrogen bonds. In Figure 1.9 it can be seen that the hydrogen atoms are located on the segments O-H $\cdots$ O, holding closer to the oxygen with which they form a covalent bond. The orientation of the molecules is not fixed, however, and a large number of configurations is possible.

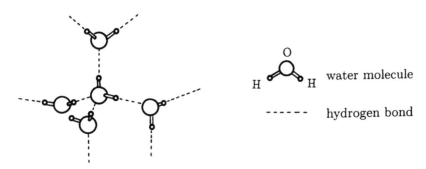

Figure 1.9: Tetrahedral surroundings of a water molecule in ordinary ice I_h. One of the possible configurations for the hydrogen atoms is shown.

1 Assuming that each hydrogen atom can occupy with equal probability the sites given by the two situations O-H $\cdots$ O or O $\cdots$ H-O on each bond between oxygens, find the number of configurations for a crystal containing N molecules.

2 Not all of the configurations considered above are suitable. Since the crystal is composed of H_2O molecules, each oxygen atom must have two close and two more distant hydrogen atom neighbours. For the immediate environment of a single oxygen, find the fraction of configurations that satisfy this constraint.

3 Hence deduce the thermodynamic probability W (number of possible configurations), neglecting correlations between neighbouring oxygens. Find the corresponding value of the molar entropy s_{stat} and compare this with the residual molar entropy of ice $s_{exp} = (0.41 \pm 0.03)R$.

Solution

1 For N molecules of H_2O there are $2N$ hydrogen atoms and hence $2N$ bonds between oxygens. As each has 2 equally probable positions, the total number of possible cases is 2^{2N}.

2 Confining attention just to the four bonds around one oxygen, the number of configurations evaluated as above is $2^4 = 16$. Of these configurations, only those with two near and two distant oxygens can be retained. This is given by the number of ways of selecting 2 objects out of 4, independent of order, i.e., $C_4^2 = 4!/(2!2!) = 6$. The proportion of physically acceptable configurations around one oxygen is thus $6/16 = 3/8$.

3 Consideration of the immediate surroundings of a single oxygen atom leaves only a fraction $3/8$ of the 16 possible cases. Neglecting correlations between neighbouring oxygens, the factor $3/8$ must be applied to each oxygen. The proportion of correct configurations is thus $(3/8)^N$. Hence, the thermodynamic probability is

$$W = 2^{2N} \times (3/8)^N = (3/2)^N \quad .$$

The entropy is thus

$$S = k \ln W = Nk \ln(3/2) \quad ,$$

i.e. the molar entropy is

$$s_{stat} = R\ln(3/2) = 0.406\,R \quad.$$

This result is in agreement with the experimental value. Correlations between orientations of neighbouring molecules, neglected in the above calculation, increase the theoretical value by merely 1% and the agreement with experiment still holds [N.H. Fletcher, *Rep. Prog. Phys.* **34**, 913 (1971)].

1.6   THERMODYNAMIC QUANTITIES AND FUNCTIONS

1.6.1   Thermodynamic functions

For each macroscopic state defined by the numbers $N_1, \ldots N_i, \ldots$, the postulates of statistical thermodynamics give the entropy as a function of N_i as

$$S = k\ln W(N_i) \quad.$$

At equilibrium, defined by the maximum of the function $W(N_i)$, subject to the constraints (1.37), the N_i take the values N_i^0. These depend on the number of particles N, the internal energy U, as well as on external parameters (volume V, flux density B, etc.) through the energy levels ϵ_i (Ch. 2). The equilibrium entropy is thus a function $S(N, U, V, B, \ldots)$. We shall see that if this function is known, all the other thermodynamic functions can be determined at equilibrium.

We first consider a pure chemical substance. To determine the temperature T, the pressure P and the chemical potential of a particle μ, we use the differential relation

$$dU = T\,dS - P\,dV + \mu\,dN \tag{1.45}$$

which expresses the First Law of thermodynamics. This relation can be rewritten

$$dS = \frac{dU}{T} + \frac{P}{T}dV - \frac{\mu}{T}dN, \tag{1.46}$$

showing that T, P and μ are obtained from the partial derivatives of $S(U, V, N)$, i.e.,

$$\frac{1}{T} = \left(\frac{\partial S}{\partial U}\right)_{V,N} \quad, \quad P = T\left(\frac{\partial S}{\partial V}\right)_{U,N} \quad \text{and} \quad \mu = -T\left(\frac{\partial S}{\partial N}\right)_{U,V} \quad. \tag{1.47}$$

The following example illustrates these relations. Consider a gas contained in a rigid isolated container, which is divided into two compartments by a fixed wall that only allows exchange of heat (Fig. 1.10). The equilibrium of the gas is determined by the condition that the total entropy $S = S_1 + S_2$ is a maximum, with the constraint $U_1 + U_2 = U = $ constant, i.e.

$$dS_1 + dS_2 = \left(\frac{\partial S_1}{\partial U_1}\right)dU_1 + \left(\frac{\partial S_2}{\partial U_2}\right)dU_2 = 0$$
$$dU_1 + dU_2 = 0 \quad.$$

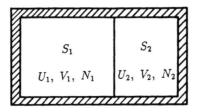

Figure 1.10: Two systems in contact

These two conditions, which govern the thermal equilibrium of the system, yield

$$\left(\frac{\partial S_1}{\partial U_1}\right)_{V_1,N_1} = \left(\frac{\partial S_2}{\partial U_2}\right)_{V_2,N_2} \quad , \tag{1.48}$$

In view of equation (1.47), this relation requires that the temperatures must be equal, $T_1 = T_2$.

By choosing either a mobile wall or one that is permeable to the gas we obtain one or other of the following relations

$$\left(\frac{\partial S_1}{\partial V_1}\right)_{U_1,N_1} = \left(\frac{\partial S_2}{\partial V_2}\right)_{U_2,N_2} \quad \text{or} \quad \left(\frac{\partial S_1}{\partial N_1}\right)_{U_1,V_1} = \left(\frac{\partial S_2}{\partial N_2}\right)_{U_2,V_2}$$

which give either the mechanical or the chemical equilibrium of the system. According to (1.47) and (1.48), these relations correspond either to equality of the pressures P_1 and P_2 or that of the chemical potentials μ_1 and μ_2 in the two compartments. Expressions (1.47) also yield the other thermodynamic quantities such as specific heat, as well as the energy functions (enthalpy H, free energy F, free enthalpy G, ...) in the desired set of variables.

The method used for a pure chemical substance can be transposed to other cases, in particular to magnetic materials which possess a magnetic work

$$d\mathcal{W} = -\mathcal{M}dB \tag{1.49}$$

that must be added to expression (1.45) to obtain the internal energy. The magnetic moment of the material is then given from the entropy by

$$\mathcal{M} = T\left(\frac{\partial S}{\partial B}\right)_{U,N,V} \quad . \tag{1.50}$$

Thus, starting from properties on the molecular scale, it can be seen how statistical thermodynamics yields the macroscopic properties of physical systems at equilibrium. Away from equilibrium, however, only N, U and S are defined. The other thermodynamic quantities (temperature, pressure, free energy, etc.) are not defined.

1.6.2   Work and Heat

Classical thermodynamics was constructed from the concepts of work and heat. We give an interpretation of these concepts starting at the molecular scale.

Elementary transformations

When a closed system of independent particles undergoes an elementary transformation starting from an equilibrium state, a variation of the energy levels $\epsilon_i \to \epsilon_i + d\epsilon_i$ may occur owing to the changes in the external parameters, as well as a modification in the population of these levels $N_i^0 \to N_i^0 + dN_i$. In the transformation, the internal energy U and the entropy S change by

$$dU = \sum_i N_i^0 d\epsilon_i + \sum_i \epsilon_i dN_i \quad \text{and} \quad dS = k \sum_i \left(\frac{\partial \ln W}{\partial N_i} \right)_0 dN_i \quad . \qquad (1.51)$$

To transform this formula, we anticipate the general relation (2.8)

$$\left(\frac{\partial \ln W}{\partial N_i} \right)_0 = \frac{\epsilon_i - \mu}{kT}$$

which gives the derivatives of $\ln W$ at equilibrium as a function of the chemical potential μ and of the temperature T of the system. Recalling that $\sum dN_i = 0$, we can write for the entropy change associated with the transformation

$$dS = \frac{\sum_i \epsilon_i dN_i}{T} \quad . \qquad (1.52)$$

In contrast to the change of internal energy (1.51), the entropy change (1.52) does not depend on the shift in the energy levels, but only on the changes in their population.

Heat alone

We consider a system of particles at equilibrium in which the external parameters are fixed, i.e. the energy levels are fixed. When this system is placed in contact with another, interactions between particles of the two systems occur at the interface, which may produce a change dN_i in the populations and a variation in the internal energy of the system being investigated. In classical thermodynamics it is said that the system has received heat by thermal contact without work being done on it ($dW = 0$). The heat dQ received by the system is then

$$dQ = dU = \sum_i \epsilon_i dN_i \quad ,$$

and the change in entropy is, from (1.52),

$$dS = \frac{dQ}{T} \quad .$$

This relation is valid, whether or not the transformation is reversible, when the system has no work done on it. In classical thermodynamics the *entropy flux* $d_e S$ due to heat exchange with the outside and the *entropy production* $d_i S$ are introduced through

$$d_e S = \frac{dQ}{T} \quad \text{and} \quad d_i S = dS - d_e S \quad . \qquad (1.53)$$

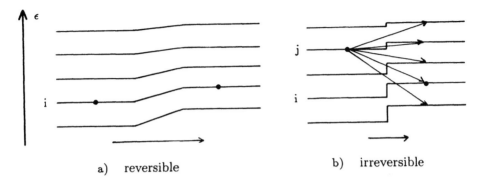

Figure 1.11: Shift in energy levels and in position of a particle during a reversible a) and irreversible b) change of external variables.

It can be seen that when the work done on the system is zero,

$$dS = d_e S \quad \text{and} \quad d_i S = 0 \quad .$$

Work alone

Now let us consider a system at equilibrium, in the absence of contact with the outside, and we vary the external parameters infinitesimally ($\epsilon_i \rightarrow \epsilon_i + d\epsilon_i$). If the transformation is performed slowly enough, the theorem of adiabatic invariants in quantum mechanics stipulates that each particle remains in the same quantum state during the transformation, which is therefore reversible (Fig. 1.11a). During this transformation in which $dN_i = 0$, the internal energy of the system varies. In classical thermodynamics, we say that the thermally isolated system ($dQ = 0$) has work done on it reversibly. The work received is equal to the change in internal energy, i.e.

$$dU = dW_{rev} = \sum_i N_i^0 d\epsilon_i \tag{1.54}$$

and the entropy change (1.52) is zero: $dS = 0$.

Denoting the various external parameters $(V, B, \dots)$ by λ_α, we can write

$$dW_{rev} = \sum_\alpha \left(\sum_i N_i^0 \frac{\partial \epsilon_i}{\partial \lambda_\alpha} \right) d\lambda_\alpha = \sum_\alpha F_\alpha d\lambda_\alpha \quad . \tag{1.55}$$

In this expression the generalized forces are introduced

$$F_\alpha = \sum_i N_i^0 \frac{\partial \epsilon_i}{\partial \lambda_\alpha} \quad . \tag{1.56a}$$

Thus the pressure P or the magnetic moment $\mathcal{M}$ of a substance can be written

$$P = -\sum_i N_i^0 \frac{\partial \epsilon_i}{\partial V} \qquad \mathcal{M} = -\sum_i N_i^0 \frac{\partial \epsilon_i}{\partial B} \quad . \tag{1.56b}$$

If the transformation is not infinitely slow (Fig. 1.11b), a particle in an initial state j can, after transformation, go into a final state i with a quantum probability P_{ji} that is subject to the condition

$$\sum_i P_{ji} = 1 \quad .$$

Variation of the external parameters in the reverse direction will return the particle to the state j only with a probability $P_{ij} = P_{ji}$, which is less than 1. The transformation is then irreversible. We can calculate the dN_i for this transformation starting from the quantum probabilities, by noting that the number of particles in the final state i is

$$N_i = \sum_j P_{ji} N_j^0 \quad .$$

We then have

$$dN_i = N_i - N_i^0 = \sum_j P_{ji} N_j^0 - \left(\sum_j P_{ji} \right) N_i^0 = \sum_j P_{ji} \left(N_j^0 - N_i^0 \right) \quad .$$

and hence

$$\sum_i \epsilon_i dN_i = \sum_{ij} P_{ji} \epsilon_i (N_j^0 - N_i^0) = \frac{1}{2} \sum_{ij} P_{ji} (\epsilon_i - \epsilon_j)(N_j^0 - N_i^0) > 0 \quad .$$

This term, which enters the formulae for dU and dS, is always positive, since, as we shall see, at equilibrium the highest energy states are less populated (§2.5.3). Hence, according to (1.51,52)

$$dU = dW = \sum N_i^0 d\epsilon_i + \sum \epsilon_i dN_i > \sum N_i^0 d\epsilon_i$$

$$\text{and} \quad dS = \sum \frac{\epsilon_i dN_i}{T} > 0 \quad .$$

In summary, when a thermally isolated system undergoes a transformation, its entropy remains constant if the transformation is reversible and increases if not. As the heat received and hence the entropy flux are zero, an irreversible transformation produces an entropy

$$d_i S = dS > 0 \quad .$$

We draw special attention to the fact that the presence of the term $\sum \epsilon_i dN_i$ in the expression for the work is always related to production of an entropy $d_i S$. This term corresponds to a change in the population of the levels due to the work. In certain transformations, it may be the only term present, as is the case in the Joule experiment where the blades supply work by turning in a fluid.

Work and heat

Lastly, we consider a system in which the external parameters are varied while thermal contact is maintained with another system. The system can absorb work and heat simultaneously. In formula (1.51) for the change in internal energy, the term $\sum N_i^0 d\epsilon_i$ comes only from the work absorbed ("reversible part"). The term $\sum \epsilon_i dN_i$, however, has contributions coming partly from the heat and partly from the work ("irreversible part"), i.e.

$$\sum_i \epsilon_i dN_i = dQ + dW \quad \text{(irreversible)} \quad .$$

This equation can be identified term by term with relation (1.53)

$$TdS = Td_eS + Td_iS \quad .$$

Since the irreversible part of the work is always positive (or zero for reversible transformations), the direct consequences of the Second Law of classical thermodynamics are recovered, i.e.,

$$d_iS \geq 0 \quad \text{or} \quad dS \geq d_eS = \frac{dQ}{T} \quad .$$

We point out that the separation into reversible and irreversible parts of the elementary work is unique. The reversible part is the work required to carry out the same variation of the external parameters (yielding another macroscopic final state), while the irreversible part makes up the difference in dW. This decomposition is, however, artificial. For example in a Joule expansion (expansion of a gas into a vacuum), the work received is zero and the reversible and irreversible parts are non zero, being equal and opposite.

The final state of the above transformations is not always an equilibrium state, although it is almost so. We take the example of a gas in a rectangular container (§1.2), the length of whose side a parallel to Ox is increased reversibly without heat transfer. Since the populations are unchanged, it can be seen from (1.1) that the contribution to the total kinetic energy of the motion along x diminishes, while that of the motion along y and z is unchanged. The distribution of velocities is thus no longer isotropic. Collisions bring the system back to equilibrium by redistributing the particles among the levels. In this rearrangement, which occurs without alteration of the levels and at constant energy, we have

$$dU = \sum_i \epsilon_i dN_i = 0 \quad .$$

By virtue of (1.52), this means that the entropy does not vary (to first order). The above results are therefore unchanged. Lastly, we point out that irreversible phenomena constitute an important branch of statistical physics. They lie outside the scope of this book.

BIBLIOGRAPHY

Quantum Mechanics
L.Landau and E. Lifschitz, *op. cit.*, Volume 3, p.82 (particle in a box).
A. Messiah, *op. cit.*, p.639 (adiabatic invariants).
Second Law and thermodynamic functions
N. Davidson, *op. cit.*, p.86.
D.A. McQuarrie, *op. cit.*
Third Law
R. Fowler and E.A. Guggenheim, *op. cit.*
F.E. Simon, *The Third Law of Thermodynamics, an Historical Survey*, Yearbook of the Physical Society, London (1956).
Historical references
L. Boltzmann, *Wien. Ber.* **66**, 275 (1872).
J. W. Gibbs, *Elementary Principles in Statistical Mechanics*, (Collected Works Vol. II) New Haven (1948).
W. Nernst, *Göttinger Nach.*, 1 (1906).

COMPREHENSION EXERCISES

1.1 A litre of helium gas is held at normal conditions of pressure and temperature ($P = 1.013 \times 10^5$ Pa and $T = 273$ K). Calculate the number of atoms in the container and the total kinetic energy [Ans.: $N = 2.7 \times 10^{22}$ atoms and $U = 152$ J].

1.2 For a two dimensional gas in a rectangular box of surface area S, determine the number of quantum states $n(p_0)$ with momentum less than p_o and the densities of states $g(p_0)$ and $g(\epsilon_0)$ [Ans.: $n(p_0) = S\pi p_0^2/h^2$, $g(p_0) = 2\pi S p_0/h^2$ and $g(\epsilon_0) = 2\pi S m/h^2$].

1.3 Discuss how the formula for the density $g(p_0)$ depends on the dimension D of geometrical space. Same question for $g(\epsilon_0)$ [Ans.: $g(p_0) \propto V_{\text{generalized}} \, p_0^{D-1}/h^D$ and $g(\epsilon_0) \propto V_{\text{generalized}} \, \epsilon_0^{(D/2)-1}/h^D$].

1.4 Determine the populations N_1 and N_2 of a system in which the particles have two energy levels ϵ_1 and ϵ_2 as a function of the total number of particles N and of the internal energy U [Ans.: $N_1 = (N\epsilon_2 - U)/(\epsilon_2 - \epsilon_1)$].

1.5 Using a programmable calculator, calculate the value of $\ln N! = \sum \ln n$ for $N = 10$ and 100. Compare with the values given by the basic Stirling formula

$$\ln n! = n \ln n - n + \frac{1}{2}\ln(2\pi n) + \frac{1}{12n}$$

and by equations (1.32 and 57) [Ans.: for $N = 10$, 15.104413 - 15.104415 - 15.096 - 13.026, respectively; for $N = 100$, 363.739375 - 363.739376 - 363.7385 - 360.517, respectively].

1.6 In a crystal of deuterated methane, each CH_3D molecule occupies a site in which the C-D bond can take 4 equivalent directions at random. Calculate its molar entropy at absolute zero [Ans.: $s = R \ln 4$].

1.7 From relations (1.2) and (1.56), show that the pressure of an ideal gas is related to its internal energy by $PV = 2U/3$.

PROBLEM 1.1 MAGNETIC PROPERTIES OF A COPPER SALT

We wish to investigate the properties of the double sulphate Cu K_2 $(SO_4)_2$, 6 H_2O in the low temperature region $(T < 300$ K$)$. The magnetic properties of this salt are due principally to the paramagnetic ions Cu^{2+}, which are localized (i.e. distinguishable) and sufficiently far apart in the crystal for their interactions to be negligible. In a uniform magnetic flux density B along the z direction, the Cu^{2+} ions possess two energy levels $\epsilon_+ = +\overline{m}B$ and $\epsilon_- = -\overline{m}B$ corresponding respectively to the two possible projections $-\overline{m}$ and $+\overline{m}$ of the dipole on the z axis.

This problem can be treated by examining a thermally isolated system of N paramagnetic ions in a magnetic field of constant induction B, i.e. having constant internal energy U.

1 N_+ and N_- are the number of ions with energies ϵ_+ and ϵ_- respectively. Use the Boltzmann relation to express the entropy S of the system in terms of N_+ and N_-.

2 Using the conditions of constraint, express N_+ and N_- in terms of N, U, B, and hence write S in terms of the same variables.

3 Show how the above formula for $S(N, U, B)$ can give the temperature T as well as the total magnetic moment $\mathcal{M}$ of the material in terms of the same variables. Hint: the elementary magnetic work is $d\mathcal{W} = -\mathcal{M}dB$.

4a Determine the temperature T of the substance and hence deduce an expression for the internal energy U in the variables N, T and B. Use the reduced Stirling formula

$$\ln n! = n \ln n - n \qquad (1.57)$$

and set $x = \overline{m}B/kT$.

4b Obtain formulae for the populations N_+ and N_- of the two levels in terms of N, T, B and calculate the ratio N_+/N_-. Comment on the limiting values of N_+ and N_- when $x \gg 1$ and $x \ll 1$.

4c Use these results to write S in terms of N, T and B. Plot S as a function of $y = 1/x$.

4d Determine c_B, the molar heat capacity at constant magnetic field, and plot the function $c_B(y)$.

5a Using the results of 2 and 3, find an expression for the magnetic moment $\mathcal{M}(N, T, B)$ of the material and show that $U = -\mathcal{M}B$.

5b Describe the variation of $\mathcal{M}$ with x and draw the function. Comment on the value of $\mathcal{M}$ for large x.

5c For the region where x is small compared with 1, show that the magnetic moment is proportional to B/T.

5d The molar magnetic susceptibility χ_M is defined as

$$\mathcal{M} = n\chi_M B/\mu_0$$

where n is the number of moles of the substance and μ_0 the permeability of vacuum. Find a formula for χ_M for the substance under consideration. Figure 1.12

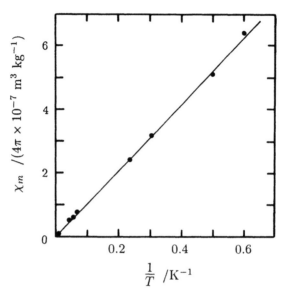

Figure 1.12: Magnetic susceptibility per unit mass χ_m in the monoclinic crystal Cu K$_2$(SO$_4$)$_2$, 6H$_2$O for a magnetic field parallel to the two-fold symmetry axis b of the crystal [J.C. Hupse, *Physica* **9**, 633 (1942)].

shows the mass susceptibility $\chi_m = \chi_M/M$ as a function of $1/T$, where $M = 441.8$ g mol^{-1} is the molar mass of the salt. Show that the experimental curve obeys the above equation. Hence derive the value of $\overline{m}$ and of the spectroscopic splitting factor $g = 2\overline{m}/\mu_B$. Compare with the value of g found directly from electron paramagnetic resonance (EPR), $g = 2.225$.

Universal constants: permeability of free space $\mu_0 = 4\pi \times 10^{-7}$ SI; Bohr magneton $\mu_B = 0.927 \times 10^{-23}$ J T^{-1}; Avogadro's number $\mathcal{N} = 6.02 \times 10^{23}$ mol^{-1}; Boltzmann constant $k = 1.38 \times 10^{-23}$ J K^{-1}.

SOLUTION

1 The thermodynamic probability W is the number of ways of selecting N_+ moments in one direction and N_- in the other, among N. It is given in (1.30). Using the Boltzmann relation, we have

$$S = k \ln W = k \ln N! - k \ln N_+! - k \ln N_-! \quad . \tag{1.58}$$

2 As in (1.24,25), the constraints (1.37) can be written

$$N = N_+ + N_- \quad \text{and} \quad U = \overline{m}B(N_+ - N_-) \quad .$$

Hence

$$N_\pm = \frac{1}{2}\left(N \pm \frac{U}{\overline{m}B}\right) \quad . \tag{1.59}$$

In this problem, as for all systems with two levels, it can be seen that the macroscopic state of the system is determined by the values of the constraints. It is not therefore necessary to seek the maximum value of W for fixed U and N, and we immediately have

$$S(N, U, B) = k \ln N! - k \ln \left(\frac{N}{2} + \frac{U}{2\overline{m}B} \right)! - k \ln \left(\frac{N}{2} - \frac{U}{2\overline{m}B} \right)! \quad . \quad (1.60)$$

3 We have $dU = TdS + dW$. Neglecting volume changes, it follows that

$$dU = TdS - \mathcal{M}dB \quad \text{or} \quad dS = \frac{1}{T}dU + \frac{\mathcal{M}}{T}dB \quad ,$$

hence

$$\frac{1}{T} = \left(\frac{\partial S}{\partial U} \right)_{B,N} \quad \text{and} \quad \frac{\mathcal{M}}{T} = \left(\frac{\partial S}{\partial B} \right)_{U,N} \quad . \quad (1.61)$$

4a Writing out in full the above expression for $1/T$ and noting that, on differentiation, Stirling's reduced formula (1.57) gives

$$\frac{d}{dn}(\ln n!) = \ln n \quad , \quad (1.62)$$

we get, from (1.60)

$$\frac{1}{kT} = \frac{1}{2\overline{m}B} \ln \frac{N\overline{m}B - U}{N\overline{m}B + U}$$

Inverting, we get

$$U = -N\overline{m}B \tanh x \quad \text{with} \quad x = \overline{m}B/kT \quad . \quad (1.63)$$

4b Substitution of the above formula for U into (1.59) yields

$$N_{\pm} = \frac{N}{2} \frac{e^{\mp x}}{\cosh x},$$

hence $N_+/N_- = e^{-2x}$. This result gives the formula for the Boltzmann distribution (§3.1), which can be written in general

$$\frac{N_i}{N_j} = e^{-(\epsilon_i - \epsilon_j)/kT} \quad .$$

At low temperature and in a strong field ($x \gg 1$), we have $N_- = N$ and $N_+ = 0$. Thus only the lower level is populated. At higher temperatures or in weak fields ($x \ll 1$), we get $N_+ = N_- = N/2$, and the two levels are equally populated.

4c With (1.58), Stirling's formula (1.57) gives

$$S/k = N \ln N - N_+ \ln N_+ - N_- \ln N_- = N_+ \ln N/N_+ + N_- \ln N/N_- \quad .$$

Using the above expressions for $N_{\pm}$, we obtain

$$S = Nk[\ln(2 \cosh x) - x \tanh x] \quad . \quad (1.64)$$

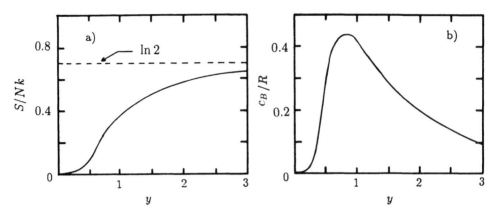

Figure 1.13: Change in a) entropy S and b) molar magnetic heat capacity c_B as a function of $y = 1/x = kT/\overline{m}B$.

When the temperature is very low and the magnetic field strong, $x \gg 1$ and with the help of the relations $2\cosh x = e^x(1 + e^{-2x})$ and $\tanh x \simeq 1 - 2e^{-2x}$, it can be seen that $S \simeq Nk(1 + 2x)e^{-2x}$ tends to zero in accordance with the Third Law. Thus the ions populate only the lower level and the macroscopic state is composed of just one microscopic state (completely ordered system).

At higher temperatures or in a weak magnetic field, $x \ll 1$. Since $\cosh x \simeq 1$ and $\tanh x \simeq x$, it can be seen that $S \simeq Nk\ln 2$. The two levels are equally populated and we recover the result (1.41). In Figure (1.13a) the variation of S/Nk with $y = 1/x = kT/\overline{m}B$ is shown. If $B = 0$, then x is identically equal to zero, and hence $S = Nk\ln 2$ at all temperatures, in contradiction with the Third Law. In reality, other phenomena appear at very low temperatures that lift the degeneracy of the ground state of the paramagnetic ions.

4d The magnetic heat capacity C_B is given by

$$C_B = \left(\frac{\partial U}{\partial T}\right)_B = T\left(\frac{\partial S}{\partial T}\right)_B \quad .$$

Using (1.63) for example, the molar value $(N \to \mathcal{N})$ is found to be

$$c_B = -\frac{\mathcal{N}\overline{m}B}{\cosh^2 x} \times \left(-\frac{\overline{m}B}{kT^2}\right) = R\frac{x^2}{\cosh^2 x} \quad . \tag{1.65}$$

The variation of c_B/R with y is shown in Figure (1.13b). It can be seen that c_B vanishes for $y \to 0$ and $y \to \infty$ and has a maximum at $y = 0.83(y = \tanh(1/y))$. The maximum value of the magnetic heat capacity c_B ($c_B^{max} = 0.44R$) is very much larger than all other heat capacities in the low temperature region.

5a From equations (1.60-62), it is found that

$$\frac{\mathcal{M}}{kT} = -\frac{U}{2\overline{m}B^2}\ln\frac{N_-}{N_+} = -\frac{U}{kTB} \quad .$$

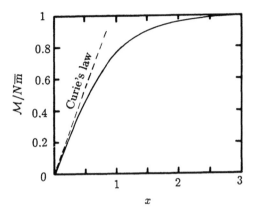

Figure 1.14: Dependence of $\mathcal{M}$ upon $x = \overline{m}B/kT$. The tangent at the origin of this curve corresponds to Curie's law.

It can be seen that $U = -\mathcal{M}B$, in accordance with (1.27), and hence

$$\mathcal{M} = -\frac{U}{B} = N\overline{m}\,\tanh x \quad . \tag{1.66}$$

5b The dependence of $\mathcal{M}/\mathcal{M}_S$ on x, where $\mathcal{M}_S = N\overline{m}$, is shown in Figure 1.14. For $x \gg 1$, i.e. at low temperature and in strong field, the magnetic moment tends towards its saturation value $\mathcal{M}_S$, at which the alignment of the moments is maximum.

5c For $x \ll 1$, $\tanh x \simeq x$, i.e.

$$\mathcal{M} \simeq N\overline{m}x = \frac{N\overline{m}^2}{k}\frac{B}{T} \quad . \tag{1.67}$$

This is Curie's Law, showing that in this region the total magnetic moment is proportional to the flux density B and inversely proportional to the temperature T.

5d With Curie's law (1.67), we find

$$\chi_M = \frac{\mu_0 \mathcal{M}}{nB} = \frac{N}{n}\frac{\mu_0 \overline{m}^2}{kT} = \mathcal{N}\frac{\mu_0 \overline{m}^2}{kT} \quad .$$

The above relation in which χ_M is proportional to $1/T$ is confirmed by the experimental data of Figure 1.12. From this figure, we find $\chi_m(\mathrm{SI}) = 13.1 \times 10^{-6}/T$. Hence

$$\overline{m}^2 = 13.1 \times 10^{-6}\frac{Mk}{\mathcal{N}\mu_0}, \quad \text{i.e.} \quad \overline{m} = 1.03 \times 10^{-23}\ \mathrm{J\ T^{-1}} \quad .$$

It follows that the spectroscopic splitting factor g is $g = 2.22$. This value is in agreement with that found by EPR ($g = 2.225$) for a magnetic field parallel to the two-fold symmetry axis b of the crystal [B. Bleaney, R.P. Penrose and B.I. Plumpton, *Proc. Roy. Soc. A* (London) **198**, 406 (1949)].

Chapter 2

The Various Statistics

2.1 INTRODUCTION

The postulates introduced earlier defined the way in which the macroscopic properties of a system at equilibrium can be predicted from a microscopic model. The equilibrium gives the maximum value of the thermodynamic probability W. In this chapter, expressions for W will be determined for the various possible cases, and the equilibrium macroscopic state (distribution relation and thermodynamic functions) will be deduced for systems of independent particles.

Two principal cases will be discerned depending on whether the particles are distinguishable or not. The calculation of the thermodynamic probabilities is different in these two cases, leading respectively to Maxwell-Boltzmann statistics or quantum statistics. Two kinds of quantum statistics exist, depending on the nature of the particles: those with half-integer spin (fermions) obey Fermi-Dirac statistics, and those with integer spin (bosons) obey Bose-Einstein statistics. We show that in the limit of low population density of the energy levels, the two types of quantum statistics converge towards the same limit, which we shall call here corrected Maxwell-Boltzmann statistics.

2.2 EQUILIBRIUM DISTRIBUTIONS

2.2.1 Equilibrium Relations

The equilibrium condition of a closed isolated thermodynamic system is expressed mathematically by the condition of maximum entropy S with respect to the variables $N_1, \ldots N_i, \ldots N_r$, that characterize its macroscopic state. We therefore try to determine the distribution $N_1^0, \ldots N_i^0, \ldots N_r^0$ that maximizes the function $S(N_i) = k \ln W(N_i)$. The variables N_i are not independent, however, since they must satisfy the conditions of a closed, isolated system

$$\sum_i N_i = N \tag{2.1}$$

33

and

$$\sum_i N_i \epsilon_i = U \ . \tag{2.2}$$

We have therefore to solve an extremum problem for a function with constrained variables. This is handled by the method of Lagrange multipliers. Two multipliers λ and λ' are introduced that are independent of the N_i, and a maximum is sought for the function

$$S(N_i) + \lambda U(N_i) + \lambda' N(N_i), \tag{2.3}$$

assuming that the variables N_i are independent. In this function, $U(N_i)$ and $N(N_i)$ are the functions of N_i that appear respectively on the left hand side of equations (2.1,2). As we shall see below, the extremum condition of the function (2.3) then gives rise to r relations (since the r derivatives with respect to N_i are zero) which, for each domain, yield the number N_i^0 as a function of the Lagrange multipliers. These multipliers, expressed as functions of U and N, are then determined from the constraints (2.1,2).

We now expose in detail the method we have just described. First replace the function (2.3) by $\overline{\Omega}(N_i)$

$$\overline{\Omega} = \frac{1}{\lambda}(S + \lambda U + \lambda' N) = U + \frac{1}{\lambda}S + \frac{\lambda'}{\lambda}N$$

which has the dimensions of energy. The coefficients $1/\lambda$ and λ'/λ have the dimensions of temperature and chemical potential respectively, and hence, by setting $\overline{T} = -1/\lambda$ and $\overline{\mu} = -\lambda'/\lambda$, $\overline{\Omega}$ can be rewritten in the form

$$\overline{\Omega}(N_i) = U(N_i) - \overline{T}S(N_i) - \overline{\mu}N(N_i) \ . \tag{2.4}$$

The multipliers $\overline{T}$ and $\overline{\mu}$ are now used instead of λ and λ'.

The r conditions that make $\overline{\Omega}$ an extremum with respect to the variables N_i, which are assumed to be independent, can be written as

$$\frac{\partial \overline{\Omega}}{\partial N_i} = 0 \ . \tag{2.5}$$

Since, according to (1.39) and (2.1,2), we have

$$\frac{\partial U}{\partial N_i} = \epsilon_i \ , \qquad \frac{\partial S}{\partial N_i} = k \frac{\partial \ln W}{\partial N_i} \quad \text{and} \quad \frac{\partial N}{\partial N_i} = 1 \ ,$$

the equilibrium conditions (2.5) become

$$\frac{\partial \overline{\Omega}}{\partial N_i} \equiv \epsilon_i - k\overline{T} \frac{\partial \ln W}{\partial N_i} - \overline{\mu} = 0$$

or

$$\frac{\partial \ln W}{\partial N_i} = \frac{\epsilon_i - \overline{\mu}}{k\overline{T}} \ . \tag{2.6}$$

These r equations yield the r equilibrium values N_i^0 in terms of the multipliers $\overline{T}$ and $\overline{\mu}$ when an explicit formula for $W(N_i)$ exists (cf. §2.3 and following).

2.2.2   Physical Interpretation of Lagrange Multipliers

What is the physical significance of the multipliers $\overline{T}$ and $\overline{\mu}$? To answer this question we compare the statistical and the classical expressions for the change in entropy of the system at equilibrium during an infinitesimal reversible transformation in which the external parameters (volume, magnetic field, etc.) remain constant (no work being done on the substance). From the Boltzmann relation and the equilibrium conditions (2.6), we have

$$dS = k\, d\ln W = k \sum_i \frac{\partial \ln W}{\partial N_i} dN_i = \sum_i \frac{\epsilon_i - \overline{\mu}}{\overline{T}} dN_i \ .$$

In such a transformation, the energy levels remain constant, i.e.

$$dU = \sum_i \epsilon_i dN_i \quad \text{and} \quad dN = \sum_i dN_i \ .$$

dS then becomes

$$dS = \frac{dU}{\overline{T}} - \frac{\overline{\mu}}{\overline{T}} dN \ .$$

Moreover, for the same transformation, classical thermodynamics gives $(dW = 0)$

$$dU = TdS + \mu dN \quad \text{or} \quad dS = \frac{dU}{T} - \frac{\mu}{T} dN \ . \tag{2.7}$$

By identifying terms, it can be seen that $\overline{T} = T$ and $\overline{\mu} = \mu$, i.e., the multipliers $\overline{T}$ and $\overline{\mu}$ can be identified with the temperature T and the chemical potential μ of a particle of the system. The equilibrium relations then become

$$\frac{\partial \ln W}{\partial N_i} = \frac{\epsilon_i - \mu}{kT} \ . \tag{2.8}$$

The extremum value of $\overline{\Omega}(N_i)$ coincides with the value of the energy function $\Omega(T, V, \mu) = U - TS - N\mu$.

This is the physical meaning of the Lagrange multipliers. The question of their determination as a function of the constraints will be discussed later in the various cases. In the following exercise, we apply the general method described above to a system of spins 1/2. The same results are found as were obtained directly in problem 1.1.

Exercise 2.1 *Equilibrium distribution for a system of spins 1/2.*
Find the equilibrium relations (2.8) for a system of spins 1/2 in a magnetic field of flux density B, whose thermodynamic probability W is given by (1.30). Hence deduce the values of the equilibrium populations N_+^0 and N_-^0 of the levels $\epsilon_+ = \overline{m}B$ and $\epsilon_- = -\overline{m}B$ in terms of the Lagrange multipliers T and μ. Write down the relations of constraint (2.1,2) for this case and state how T and μ can be obtained as a function of U and N. In particular, eliminate μ from the expressions for $N_\pm^0$,

while retaining N. Check that the result is consistent with formula (1.63), which was obtained directly.

Solution

The thermodynamic probability for the two-level system under consideration in terms of N_+ and N_- is given by

$$W = \frac{N!}{N_+!N_-!} = \frac{(N_+ + N_-)!}{N_+!N_-!} \ .$$

Using the derivative of the Stirling formula (1.62), we obtain the two relations

$$\frac{\partial \ln W}{\partial N_\pm} = \ln(N_+ + N_-) - \ln N_\pm = -\ln \frac{N_\pm}{N} \ .$$

Substituting these formulae into the equilibrium relations (2.8) we get

$$N_\pm^0 = Ne^{(\mu - \epsilon_\pm)/kT} \tag{2.9}$$

with $\epsilon_\pm = \pm \overline{m}B$. The multipliers T and μ are then obtained from the two conditions of constraint (2.1,2), which can be written here as

$$N = Ne^{\mu/kT} \left(e^{-x} + e^x\right) = 2Ne^{\mu/kT} \cosh x$$
$$U = N\overline{m}Be^{\mu/kT} \left(e^{-x} - e^x\right) = -2N\overline{m}Be^{\mu/kT} \sinh x$$

with $x = \overline{m}B/kT$. These two equations determine the two multipliers T and μ in terms of U and N. In particular, the constraint on the number of particles yields the parameter $\exp(\mu/kT) = (2 \cosh x)^{-1}$, thus allowing μ to be eliminated from the formulae for the equilibrium populations. Hence

$$N_\pm^0 = \frac{N}{2} \frac{e^{\mp x}}{\cosh x} \ .$$

This result is in agreement with that obtained by the direct method, which is applicable only to systems with two levels (Problem 1.1). To pursue the Lagrange multiplier method further, the variable T, contained in $x = \overline{m}B/kT$, should also be eliminated, leaving only U and N. However, for experimental reasons it is preferable to keep the temperature as a variable and express the internal energy as a function of T rather than the reverse. Thus, the second relation of constraint becomes

$$U = -2N\overline{m}B \times \frac{1}{2 \cosh x} \times \sinh x = -N\overline{m}B \tanh \frac{\overline{m}B}{kT}$$

in agreement with (1.63).

2.3   THERMODYNAMIC PROBABILITY : EVALUATION

We shall now evaluate the thermodynamic probability W of a macroscopic state of the system being considered, i.e. the number of different microscopic states making up this macroscopic state.

Table 2.1: Thermodynamic probability of the macroscopic distribution $N_1, \ldots N_i, \ldots N_r$.

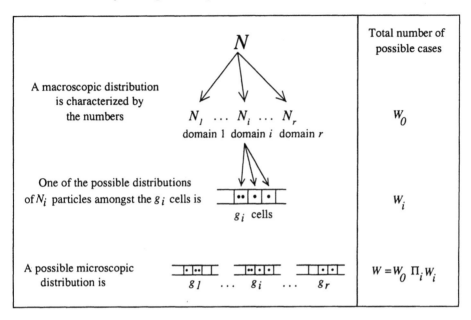

Depending on the physical system, the quantum states of a particle are either few and well separated in energy (paramagnetic substance in a magnetic field) or else infinite in number with energy intervals that are much too small to be measured (monatomic gas). Intermediate cases will also be encountered. To handle these different situations, we shall however use the same concepts as were used for gases (Table 1.1), in which a cell designates a quantum state and a domain contains a number g_i of quantum states of similar energy ϵ_i. In the case of paramagnetic materials, each energy level forms a domain, in which the number of cells is equal to the number of quantum states (or degeneracy) in the level.

A macroscopic state is then defined by the set of numbers of particles $N_1, \ldots N_i$, $\ldots N_r$ in each of the domains, having indices $1, \ldots i, \ldots r$. To find the thermodynamic probability $W(N_1, \ldots N_i, \ldots N_r)$ of this *a priori* distribution, we proceed in two steps.

First, the number of ways W_0 of obtaining the chosen distribution $N_1, \ldots N_i, \ldots N_r$ is calculated for the N particles in the domains. Then the number of ways W_i of distributing the N_i particles of the domain i in the g_i cells of this domain is evaluated. Carrying out this calculation for each of the domains, we get finally

$$W = W_0 W_1 \ldots W_i \ldots W_r = W_0 \prod_{i=1}^{r} W_i \ . \tag{2.10}$$

The method is shown diagrammatically in Table 2.1.

The explicit evaluation of W depends on the physical nature of the particles under consideration. Two categories arise depending on whether the particles are distinguishable or indistinguishable.

2.4 DISTINGUISHABLE PARTICLES : MAXWELL-BOLTZMANN STATISTICS

2.4.1 Calculation of the Thermodynamic Probability

When the particles are localized, i.e. distinguishable, the principles of Maxwell-Boltzmann statistics apply. In this case we seek an expression for the thermodynamic probability W of a macroscopic state characterized by the numbers $N_1, \ldots N_i, \ldots N_r$.

Following the method outlined in Table 2.1, we first calculate the number of ways W_0 of distributing N particles in the domains $1, \ldots i, \ldots r$ in accordance with the chosen macroscopic distribution $N_1, \ldots N_i, \ldots N_r$. Since the particles are distinguishable, this number is equal to the number of ways of taking N_1 particles among N, multiplied by the number of ways of taking N_2 among the remaining $N - N_1$, multiplied by the number of ways of taking N_3 among the remaining $N - N_1 - N_2$, etc. In terms of combinations, this gives

$$
\begin{aligned}
W_0 &= C_N^{N_1} \times C_{N-N_1}^{N_2} \cdots C_{N-N_1 \ldots -N_{r-1}}^{N_r} \\
&= \frac{N!}{N_1!(N-N_1)!} \times \frac{(N-N_1)!}{N_2!(N-N_1-N_2)!} \times \cdots \frac{(N-N_1 \ldots -N_{r-1})!}{N_r! \times 0!} \\
&= \frac{N!}{N_1!N_2!\ldots N_r!} = \frac{N!}{\prod_i N_i!} \quad .
\end{aligned}
$$

The same result can of course be obtained by observing that the $N!$ permutations of the N particles must be divided by those of the $N_1, \ldots N_i, \ldots N_r$ particles.

We must now calculate the number of ways W_i of distributing the N_i distinguishable particles in the ith domain among the g_i cells of the domain, with the knowledge that no restriction is imposed on the number of particles per cell. As each particle has g_i possibilities, we have $W_i = g_i^{N_i}$, or, finally,

$$
W_{MB} = W_0 \prod_i W_i = N! \prod_i \frac{g_i^{N_i}}{N_i!} \quad , \tag{2.11}
$$

where the subscript MB denotes Maxwell-Boltzmann statistics.

2.4.2 Equilibrium Distribution

To determine the equilibrium distribution, noting that $N = \sum N_i$, we write

$$
\ln W_{MB} = \ln \left(\sum_i N_i \right)! + \sum_i [N_i \ln g_i - \ln N_i!] \quad .
$$

Differentiating with respect to N_i and using the derivative form of the Stirling formula (1.62), we get

$$\frac{\partial \ln W_{MB}}{\partial N_i} = \ln N + \ln g_i - \ln N_i$$

and the equilibrium conditions (2.8) become

$$\ln N \frac{g_i}{N_i^0} = \frac{\epsilon_i - \mu}{kT}$$

or alternatively

$$N_i^0 = N g_i e^{(\mu - \epsilon_i)/kT}. \tag{2.12}$$

This formula gives the equilibrium distribution in terms of μ and T.

In principle, the two equations of constraint (2.1,2) allow the two multipliers T and μ to be determined as a function of U and N. This would give us the distribution and all the related thermodynamic quantities as a function of the constraints U and N, and of the external parameters (V, B, etc.) that enter the expression for the energies ϵ_i. In practice, however, although it is desirable to eliminate the multiplier μ, it is preferable to keep the temperature T as a variable. For this reason, the set of variables (T, μ) is not replaced by (U, N) but instead by (T, N).

To eliminate μ, the constraint relation (2.1) is used, $N = \sum N_i^0$, which, combined with equation (2.12), can be written

$$N = N e^{\mu/kT} \sum_i g_i e^{-\epsilon_i/kT} \quad . \tag{2.13}$$

Lastly, the constraint (2.2), $U = \sum N_i^0 \epsilon_i$, yields U as a function of N, of T and of the external parameters. The sum

$$Z = \sum_i g_i e^{-\epsilon_i/kT} \tag{2.14}$$

plays an important role in the statistical theory of Maxwell-Boltzmann. It is called the partition function of a particle, or more succinctly the *partition function*. It depends only on the temperature and the external parameters. Equation (2.13) then gives

$$e^{\mu/kT} = 1/Z \quad \text{or} \quad \mu = -kT \ln Z \quad , \tag{2.15}$$

and the distribution relation (2.12) in Maxwell-Boltzmann statistics becomes

$$N_i^0 = \frac{N}{Z} g_i e^{-\epsilon_i/kT} \quad . \tag{2.16}$$

As the factor $1/kT$ arises frequently, the following notation is used

$$\beta = 1/kT \tag{2.17}$$

This allows the distribution relation to be rewritten in the form

$$N_i^0 = \frac{N}{Z} g_i e^{-\beta \epsilon_i} \quad \text{with} \quad Z = \sum_i g_i e^{-\beta \epsilon_i} \quad . \tag{2.18}$$

2.4.3 Thermodynamic Functions

The statistical formula for the entropy at equilibrium S can be obtained from the Boltzmann relation (1.39). For the distribution at equilibrium N_i^0, we have

$$S/k = \ln W_{MB} = \ln N! + \sum_i N_i^0 \ln g_i - \sum_i \ln N_i^0! \ .$$

Use of the reduced Stirling formula (1.57) yields

$$S/k = N \ln N - \sum_i N_i^0 \ln \frac{N_i^0}{g_i} \ .$$

The distribution relation (2.16) gives the ratio N_i^0/g_i and the entropy becomes

$$S = Nk \ln N - k \sum_i N_i^0 \ln \frac{N}{Z} + \frac{1}{T} \sum_i N_i^0 \epsilon_i = Nk \ln Z + \frac{U}{T} \ . \qquad (2.19)$$

This result provides us with the very simple formula for the free energy

$$F = U - TS = -NkT \ln Z \ . \qquad (2.20)$$

Since all the thermodynamic functions can be obtained from F, this expression shows that in Maxwell-Boltzmann statistics, every problem reduces to finding the partition function Z as a function of the temperature T and of the external variables such as $V, B, \ldots$

Exercise 2.2 *Partition function and free energy of a system of spins 1/2.*
    Find the partition function Z of a particle of spin 1/2 placed in a magnetic field of flux density B. Hence deduce the free energy of a system of N spins, then its entropy and magnetization, given that

$$dF = -SdT - \mathcal{M}dB \ .$$

Solution
    This system has only two energy levels $\epsilon_+ = \overline{m}B$ and $\epsilon_- = -\overline{m}B$, of degeneracy $g_+ = g_- = 1$. The partition function Z is then

$$Z = \exp(-\beta \overline{m}B) + \exp(\beta \overline{m}B) = 2 \cosh x$$

with $x = \beta \overline{m}B = \overline{m}B/kT$. The free energy is

$$F = -NkT \ln Z = -NkT \ln(2 \cosh x) \ .$$

Hence, noting that $(\partial x/\partial T)_B = -x/T$,

$$S = -\left(\frac{\partial F}{\partial T}\right)_B = Nk \left[\ln(2 \cosh x) - x \tanh x\right]$$

$$\text{and} \quad \mathcal{M} = -\left(\frac{\partial F}{\partial B}\right)_T = N\overline{m} \tanh x \ .$$

By means of the general method we thus recover the results (1.64 and 66). These could be obtained directly in this particular case where there are only two energy levels.

2.5 INDISTINGUISHABLE PARTICLES : QUANTUM STATISTICS

2.5.1 Symmetrization Rule

Example of a two particle system

The principle of indistinguishability for identical particles (§1.2.3) has far-reaching consequences. Imagine first a system of two identical particles defined by the indices 1 and 2 with wave function denoted $\Phi(1,2)$. We introduce the permutation operator $\hat{P}_{12}$ defined for an arbitrary wave function $\varphi(1,2)$ by

$$\hat{P}_{12}\varphi(1,2) = \varphi(2,1) \ .$$

By iteration, it can be seen that

$$\hat{P}_{12}^2\varphi(1,2) = \hat{P}_{12}\varphi(2,1) = \varphi(1,2) \ ,$$

i.e. the eigenvalues of $\hat{P}_{12}$ are $+1$ and -1, corresponding to wave functions that are respectively symmetric and antisymmetric under the exchange $1 \leftrightarrow 2$. As the particles 1 and 2 are indistinguishable, the wave functions $\Phi(1,2)$ and $\Phi(2,1)$ describe the same physical reality, and thus differ only by a phase factor

$$\Phi(2,1) \equiv \hat{P}_{12}\Phi(1,2) = e^{i\alpha}\Phi(1,2) \ .$$

As the eigenvalues of $\hat{P}_{12}$ are ± 1, it can be seen that the wave functions of a system of 2 indistinguishable particles are necessarily either symmetric or antisymmetric upon exchange of the variables 1 and 2.

Bosons and fermions

The above discussion, based on the principle of indistinguishability, can be generalized to the case of N identical particles by introducing the same number of operators $\hat{P}_{ij}$ as there are pairs of particles. The wave function of the system must then be either entirely symmetric or entirely antisymmetric with respect to exchange of any pair of variables i and j. In nature, it is found that particles are divided into two categories:
- *bosons*, particles whose wave function is always symmetric, and
- *fermions*, particles whose wave function is always antisymmetric.

In quantum field theory it is shown that the set of bosons is identical to the set of particles with integer spin $0, 1, \ldots$ in units of $\hbar$, and that the set of fermions is identical to that of particles with half-integer spin $1/2, 3/2, 5/2, \ldots$ Fermions include *leptons* (electrons, muons, neutrinos, $\ldots$), which have spin $1/2$, *baryons* (protons, neutrons, $\ldots$), as well as all particles composed of an odd number of fermions, such as for example the nucleus of helium ^{3}He (2 protons + 1 neutron) or the atom ^{3}He (a ^{3}He nucleus + 2 electrons). Bosons include *gauge particles* such as the photon with spin 1, intermediate bosons W^+, W^- and Z^0, as well as *mesons* (π , K, $\ldots$) and all particles composed of an even number of fermions such as the ^{4}He nucleus (2 protons + 2 neutrons) or the helium atom ^{4}He (a ^{4}He nucleus + 2 electrons). It is presently believed that baryons and mesons are composed of more elementary particles of spin 1/2 - quarks - three for baryons and two (actually a quark and an antiquark) for mesons. Lastly, in condensed phase theory we shall encounter field quanta of boson type, such as phonons, rotons, magnons...

Exercise 2.3 *Quantum character of atoms and molecules.*
Determine the character (boson or fermion) of the atoms H $\equiv {}^1$H, D $\equiv {}^2$H, ^{3}H and of the molecules H_2, HD and D_2.

Solution
The quantum character of these particles is defined by the parity of the sum of the number of protons (N_p), neutrons (N_n) and electrons (N_e) (all fermions) that they contain. Since all particles considered in this exercise are neutral, we have $N_p = N_e$ and the sum $N_p + N_n + N_e$ has the same parity as the number of neutrons N_n. Hence H $(N_n = 0)$, ^{3}H (2), H_2 (0) and D_2 (2) are bosons, while D $(N_n = 1)$ and HD (1) are fermions.

Pauli rule

Consider a system of two non-interacting particles. The solutions of Schrödinger's equation for a single particle form a series $\Psi^{(1)}, \Psi^{(2)}, \ldots \Psi^{(i)}, \ldots$ For the particles taken together, every function $\Psi^{(i)}(1)\Psi^{(j)}(2)$ will in turn be a solution of Schrödinger's equation for two particles. But by the symmetrization rule, this form does not correspond to any physical reality. The only suitable functions are the combinations

$$\text{symmetric} \qquad (\Psi^{(i)}(1)\Psi^{(j)}(2) + \Psi^{(j)}(1)\Psi^{(i)}(2))/\sqrt{2} ,$$
$$\text{and antisymmetric} \qquad (\Psi^{(i)}(1)\Psi^{(j)}(2) - \Psi^{(j)}(1)\Psi^{(i)}(2))/\sqrt{2} .$$

In the case of two identical wave functions $(i = j)$, a symmetric wave function can be constructed having the form $\Psi^{(i)}(1)\Psi^{(i)}(2)$ that is suitable for bosons. It is impossible, however, to construct an antisymmetric wave function. As a result, two fermions cannot occupy the same quantum state. This is the Pauli rule (or the Pauli exclusion principle) (1925). This principle provides an interpretation for the periodic table of the elements, or Mendeleev table, by virtue of the fact that two electrons of an atom cannot occupy the same quantum state (same orbital state and same spin). This principle was introduced into statistical physics by E. Fermi (1926).

These considerations lead to important differences between the thermodynamic properties of boson and fermion systems, since the enumeration of the quantum states is different for the two species.

If each particle can occupy any of r possible states, then r^2 2-particle wave functions can be constructed of the form $\Psi^{(i)}(1)\Psi^{(j)}(2)$. The r functions obtained by setting $i = j$ are symmetric, but the remaining $r^2 - r = r(r-1)$ have no symmetric character. Linear combinations of these, however, yield $r(r-1)/2$ symmetric functions and $r(r-1)/2$ antisymmetric functions. The total number of symmetric functions $r(r+1)/2$ is therefore larger than that of the antisymmetric functions. It

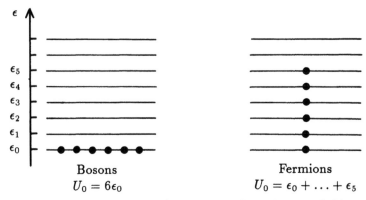

Figure 2.1: State of minimum energy for a system of 6 indistinguishable particles with non degenerate energy levels.

follows that there are more states available for a system of bosons than for a system of fermions.

An important consequence of the exclusion principle has to do with the minimum energy U_0 of a system of N particles (Fig. 2.1). For a system of bosons, this energy is equal to N times the minimum energy of one particle. For a system of fermions, however, since each state can be occupied by only one particle, U_0 is the sum of the energies of the N states of lowest energy. We note that the fermion distribution can be interpreted as a "quantum mechanical repulsion" between fermions. Likewise, we may consider bosons as having a "quantum mechanical attraction", which provides an interpretation of the phenomena of stimulated emission (§7.4), superfluidity (Chap. 8) and superconductivity (Chap.15).

The repulsion between fermions explains the very high mean kinetic energy of electrons in a metal (§10.3.3). Moreover, as the lowest quantum states are all occupied, fermions are unable to change their quantum state and are therefore insensitive to their interactions. This is the reason, for example, why the electrons in a metal, which form a very dense gas, can nonetheless be treated as independent particles (§10.3.4).

2.5.2 Fermi-Dirac Distribution

The statistics that applies to fermions bears the name of Fermi-Dirac. The thermodynamic probability is calculated from the general relation (2.10).

Since the fermions of a gas are indistinguishable particles, the number of ways W_0 of arranging N fermions in the r levels according to the desired distribution $N_1, \ldots N_i, \ldots N_r$ is equal to 1. Any N_1 may be taken (since they are indistinguishable), then $N_2, \ldots$ and lastly N_r. This value, $W_0 = 1$ is entirely different from that found in Maxwell-Boltzmann statistics for the case of distinguishable particles. In the case of fermions the number of possible arrangements W_i of the N_i particles in the g_i cells of the ith domain is found by requiring that each cell should contain at most one particle (i.e. 0 or 1). The following diagram illustrates a distribution.

It can be seen that N_i must lie in the range $0 \le N_i \le g_i$. From the g_i cells, N_i can be selected to place the particles. The number of ways W_i of choosing N_i objects among g_i different objects is

$$W_i = C_{g_i}^{N_i} = \frac{g_i!}{N_i!(g_i - N_i)!} \quad .$$

The thermodynamic probability for the fermions is thus

$$W_{FD} = W_0 \prod_i W_i = \prod_i \frac{g_i!}{N_i!(g_i - N_i)!} \tag{2.21}$$

where the subscript FD refers to Fermi-Dirac quantum statistics. To determine the equilibrium distribution from relation (2.8), we write

$$\ln W_{FD} = \sum_i [\ln g_i! - \ln N_i! - \ln(g_i - N_i)!] \quad , \tag{2.22}$$

which, combined with the derivative form of the Stirling relation (1.62), gives

$$\frac{\partial \ln W_{FD}}{\partial N_i} = -\ln N_i + \ln(g_i - N_i) = \ln \frac{g_i - N_i}{N_i} \quad .$$

Upon setting $\beta = 1/kT$, the equilibrium condition (2.8) becomes

$$\ln \frac{g_i - N_i^0}{N_i^0} = \beta(\epsilon_i - \mu) \quad .$$

Solving for N_i^0, we obtain the equilibrium distribution in Fermi-Dirac statistics

$$N_i^0 = \frac{g_i}{e^{\beta(\epsilon_i - \mu)} + 1} \quad (\beta = 1/kT) \quad . \tag{2.23}$$

Before discussing and using this formula, we first establish the distribution relation for bosons.

2.5.3   Bose-Einstein Distribution

The case of bosons is handled similarly to that of fermions. Thus we also have $W_0 = 1$, but the expressions for the W_i are different since an arbitrary number of particles can now occupy each cell. One arrangement might be represented, for example, by the following diagram:

N_i particles
g_i cells

or in a simplified representation that contains the same information:

$$\bullet \mid \mid \bullet \bullet \bullet \mid \bullet \bullet \mid --- \mid \bullet \quad \begin{matrix} N_i & \text{dots} \\ g_i\text{-}1 & \text{rods} \end{matrix} \Big\} N_i + g_i\text{-}1 \text{ objects}$$

The N_i dots and $g_i - 1$ rods make up $N_i + g_i - 1$ objects which in the first place we shall assume to be distinct. The number of ways of arranging them is equal to the number of permutations $(N_i + g_i - 1)!$. We then take account of the fact that the N_i dots are all identical by dividing by the number of permutations $N_i!$. Likewise, since the $g_i - 1$ rods are not distinct, we divide by their number of permutations $(g_i - 1)!$. Hence

$$W_i = \frac{(N_i + g_i - 1)!}{N_i!(g_i - 1)!} = C_{N_i + g_i - 1}^{N_i} \quad .$$

The thermodynamic probability for the bosons is thus

$$W_{BE} = W_0 \prod_i W_i = \prod_i \frac{(N_i + g_i - 1)!}{N_i!(g_i - 1)!} \quad . \tag{2.24}$$

The subscript BE refers to the name Bose-Einstein for the statistics that applies to bosons.

The equilibrium distribution is found as for fermions. We have

$$\ln W_{BE} = \sum_i [\ln(g_i + N_i - 1)! - \ln N_i! - \ln(g_i - 1)!] \quad , \tag{2.25}$$

which, on differentiation, gives

$$\frac{\partial \ln W_{BE}}{\partial N_i} = \ln(g_i + N_i - 1) - \ln N_i = \ln \frac{g_i + N_i - 1}{N_i} \quad .$$

If the term $g_i + N_i$ is much greater than 1, which, in the following, is always the case, then we have

$$\frac{\partial Log W_{BE}}{\partial N_i} = \ln \frac{g_i + N_i}{N_i} \quad .$$

The equilibrium condition can then be written

$$\ln \frac{g_i + N_i^0}{N_i^0} = \beta(\epsilon_i - \mu)$$

or, on solving,

$$N_i^0 = \frac{g_i}{e^{\beta(\epsilon_i - \mu)} - 1} \quad (\beta = 1/kT) \quad . \tag{2.26}$$

This equilibrium distribution relation, called the Bose-Einstein distribution, has a form that is similar to that of Fermi-Dirac, thus allowing both types of statistics to be considered at the same time. The two equilibrium distributions for quantum statistics can be written in the form

$$N_i^0 = \frac{g_i}{e^{\beta(\epsilon_i - \mu)} \pm 1} \quad \binom{FD}{BE} \tag{2.27}$$

in which, as throughout the rest of this chapter, we use the upper sign for Fermi-Dirac statistics and the lower sign for that of Bose-Einstein. We observe that the population N_i^0 of the domain i is proportional to g_i, which is what determines the size of the domain. This means that the way in which the division into domains has been performed is irrelevant. The significant quantity is the ratio

$$n_i^0 = \frac{N_i^0}{g_i} = \frac{1}{e^{\beta(\epsilon_i - \mu)} \pm 1} \quad , \tag{2.28}$$

which is the mean number of particles per cell (or per quantum state). It can be seen that this number is a decreasing function of the energy ϵ_i.

2.5.4 Thermodynamic Functions in Quantum Statistics

According to the method of Lagrange multipliers, the parameters T (or β) and μ should be determined by the constraints (2.1,2) acting on N and U. However, just as in Maxwell-Boltzmann statistics (§2.4.2), it is preferable to keep T as a variable rather than introduce the variable U. Also, even though it would be desirable to eliminate μ and keep N, the condition (2.1)

$$N = \sum_i N_i^0 = \sum_i \frac{g_i}{e^{\beta(\epsilon_i - \mu)} \pm 1} \tag{2.29}$$

fails to yield μ in an analytical form. We therefore have to keep μ as an intermediate variable in the calculation. In summary, the parameters T and μ are retained as variables,
- T, because it is preferable to the internal energy U;
- μ, because in general it cannot be replaced by N.

The energy function that is best suited to this choice of variables is the grand potential $\Omega(T, \mu, V)$, defined by

$$\Omega = U - TS - N\mu \tag{2.30}$$

the differential of which is

$$d\Omega = -SdT - Nd\mu - PdV \quad . \tag{2.31}$$

A knowledge of the function Ω in each system allows expressions to be found for the entropy S, the number of particles N and for the pressure P in terms of T, μ, V, through the relations

$$S = -\left(\frac{\partial \Omega}{\partial T}\right)_{\mu, V}, \quad N = -\left(\frac{\partial \Omega}{\partial \mu}\right)_{T, V} \quad \text{and} \quad P = -\left(\frac{\partial \Omega}{\partial V}\right)_{T, \mu} \quad . \tag{2.32}$$

The second of these equations is the implicit equation that determines μ, which, as we shall see, is the analogue of (2.29). If external parameters other than V are present, such as a magnetic field B, work terms must be added to $d\Omega$. For example $dW = -\mathcal{M}dB$, which gives, in addition to (2.32)

$$\mathcal{M}(T, \mu, V, B) = -\left(\frac{\partial\Omega}{\partial B}\right)_{T,\mu,V} . \tag{2.33}$$

To find an expression for the grand potential Ω in quantum statistics, we start by determining the entropy from the Boltzmann relation. Applying the reduced Stirling formula (1.57) to expressions (2.22) for $\ln W_{FD}$ and (2.25) for $\ln W_{BE}$, we obtain

$$\ln W_{FD} = \sum_i \left[N_i \ln \frac{g_i - N_i}{N_i} - g_i \ln \frac{g_i - N_i}{g_i} \right]$$

$$\text{and} \quad \ln W_{BE} = \sum_i \left[N_i \ln \frac{(g_i + N_i - 1)}{N_i} + (g_i - 1) \ln \frac{(g_i + N_i - 1)}{(g_i - 1)} \right] .$$

In the approximation $g_i - 1 \simeq g_i$, these two equations reduce to the single expression

$$\ln W = \sum_i \left[N_i \ln \frac{g_i \mp N_i}{N_i} \mp g_i \ln \frac{g_i \mp N_i}{g_i} \right] \quad \begin{pmatrix} FD \\ BE \end{pmatrix} .$$

Introducing the equilibrium distribution relations (2.27), we get

$$\frac{g_i \mp N_i^0}{N_i^0} = e^{\beta(\epsilon_i - \mu)} \quad \text{and} \quad \frac{g_i \mp N_i^0}{g_i} = \frac{1}{1 \pm e^{-\beta(\epsilon_i - \mu)}}$$

and hence $\quad S = k \ln W_{max} = k \sum_i \left[N_i^0 \left(\frac{\epsilon_i - \mu}{kT} \right) \pm g_i \ln(1 \pm e^{-\beta(\epsilon_i - \mu)}) \right] .$

Substitution of the constraints $N = \sum N_i$ and $U = \sum N_i \epsilon_i$ into these formulae gives

$$S = \frac{U}{T} - \frac{N\mu}{T} \pm k \sum_i g_i \ln(1 \pm e^{-\beta(\epsilon_i - \mu)}) ,$$

and hence the final expression for the grand potential $\Omega = U - TS - N\mu$ becomes

$$\Omega = \mp kT \sum_i g_i \ln(1 \pm e^{-\beta(\epsilon_i - \mu)}) \quad \begin{pmatrix} FD \\ BE \end{pmatrix} . \tag{2.34}$$

In this equation we see that Ω is an explicit function of T (with $\beta = 1/kT$) and of μ, and that the external variables $(V, B, \dots)$, which depend on the particular physical problem, enter through the ϵ_i. In later chapters we shall meet with examples of how this function is determined and how the potential μ can be eliminated and replaced by N. It can be seen that with this expression for Ω, the relation

$$N = -\left(\frac{\partial\Omega}{\partial\mu}\right)_{T,V} = \pm kT \sum_i g_i \frac{\pm\beta e^{-\beta(\epsilon_i - \mu)}}{1 \pm e^{-\beta(\epsilon_i - \mu)}}$$

reproduces the condition of constraint (2.29).

2.5.5   Corrected Maxwell-Boltzmann Statistics

Common limit in quantum statistics

When the population density $n_i^0 = N_i^0/g_i$ is much smaller than 1 for all domains i, then there are far fewer particles than cells in each domain and the number of microscopic states in which two or more particles (for the case of bosons) occupy the same cell is relatively very small and negligible. Almost all the microscopic states that are accessible to a system of bosons are also accessible to a system of fermions, i.e. $W_{FD} \simeq W_{BE}$. This argument can be seen to be quantitatively correct by making the approximation $N_i \ll g_i$ in expressions (2.21) and (2.24) for the thermodynamic probabilities W_{FD} and W_{BE}. With this approximation

$$W_{FD} = \prod_i \frac{g_i!}{N_i!(g_i - N_i)!} = \prod_i \frac{g_i(g_i - 1) \ \ldots \ (g_i - N_i + 1)}{N_i!}$$

$$\simeq \prod_i \frac{g_i g_i \ \cdots \ g_i}{N_i!} = \prod_i \frac{g_i^{N_i}}{N_i!}$$

$$\text{and} \ \ W_{BE} = \prod_i \frac{(N_i + g_i - 1)!}{N_i!(g_i - 1)!} = \prod_i \frac{(g_i + N_i - 1) \ (g_i + N_i - 2) \ \cdots \ g_i}{N_i!}$$

$$\simeq \prod_i \frac{g_i^{N_i}}{N_i!} \ .$$

This shows that when $N_i \ll g_i$ the two formulae for W tend to the same limit,

$$W_{MBc} = \prod_i \frac{g_i^{N_i}}{N_i!} \ , \tag{2.35}$$

given by the lower value in Fermi-Dirac statistics and by the upper value in Bose-Einstein statistics. This limit is similar to expression (2.11) for W_{MB}, apart from the factor $N!$ due to indistinguishability, whence the name "corrected" Maxwell-Boltzmann statistics that we assign to the limiting statistics based on expression (2.35) for W. We emphasize the fact that both types of Maxwell-Boltzmann statistics, whose formalism is closely similar, apply to systems that are completely different. Maxwell-Boltzmann statistics is used for systems of distinguishable (or localized) particles with arbitrary population density $n_i^0 = N_i^0/g_i$, while corrected Maxwell-Boltzmann statistics is used for indistinguishable (or non localized) particles, and only for low population densities ($N_i \ll g_i$).

Exercise 2.4 *Calculation of W for indistinguishable particles.*

Calculate the thermodynamic probability for a set of fermions (W_{FD}) and a set of bosons (W_{BE}) when there is only one single domain ($r = 1$) containing $g_1 = 1\,000$ cells, and for a number of particles N_1 in turn equal to 500 and 10. Also, calculate W in corrected Maxwell-Boltzmann statistics with $N_1 = 10$ and compare this result with the values of W_{FD} and W_{BE}.

Solution

We use relations (2.21, 24 and 35). For $N_1 = 500$, we have $W_{FD} = 1\,000!/(500! \times 500!) = 1.07 \times 10^{301}$ and $W_{BE} = 1\,499!/(500! \times 999!) = 2.99 \times 10^{414}$.
For $N_1 = 10$, $W_{FD} = 1\,000!/(10! \times 990!) = 2.62 \times 10^{23}$ and $W_{BE} = 1\,009!/(10! \times 999!) = 2.87 \times 10^{23}$.

It can be seen that for $N_1 = 10$ ($N_1/g_1 \ll 1$), the discrepancy between the two types of statistics is small ($\Delta W/W \sim 10\%$), in contrast to the case $N_1 = 500$ ($N_1/g_1 = 0.5$). In corrected Maxwell-Boltzmann statistics, with $N_1 = 10$, we have $W_{MBc} = 1\,000^{10}/10! = 2.76 \times 10^{23}$, which is intermediate to the two previous values. Note that for $N_1 = 10$ the values of the logarithms of W are respectively $\ln W_{FD} = 53.92$, $\ln W_{BE} = 54.01$ and $\ln W_{MBc} = 53.97$, differing only by about 10^{-3} in relative value. (The factorials, apart from 10!, were obtained with a pocket calculator using the reduced Stirling approximation.)

Distribution law and thermodynamic functions

The general method that yields the distribution relation starting from the thermo-dynamic probability can be applied here. It is, however, simpler to look for the quantum statistical expressions in the limit $N_i \ll g_i$. Since we have from (2.27)

$$\frac{N_i^0}{g_i} = \frac{1}{e^{\beta(\epsilon_i - \mu)} \pm 1} \ll 1 \quad,$$

the denominator, and hence the exponential $e^{\beta(\epsilon_i - \mu)}$, are very much larger than 1. In corrected Maxwell-Boltzmann statistics the quantum statistical distribution relations (2.27) then become

$$N_i^0 \simeq \frac{g_i}{e^{\beta(\epsilon_i - \mu)}}, \quad \text{i.e.} \quad N_i^0 = g_i e^{\beta(\mu - \epsilon_i)} \quad. \tag{2.36}$$

Similarly, since $e^{-\beta(\epsilon_i - \mu)} \ll 1$, expression (2.34) for the grand potential Ω simplifies, yielding

$$\Omega \simeq \mp kT \sum_i g_i \times (\pm e^{-\beta(\epsilon_i - \mu)}) = -kTe^{\beta\mu} Z \tag{2.37}$$

where the partition function

$$Z(T, V, B, \ldots) = \sum_i g_i e^{-\beta\epsilon_i} \quad, \tag{2.38}$$

is a function only of T and of the external parameters such as $V, B, \ldots$ through the ϵ_i.

The grand potential function $\Omega(T, \mu, V, \ldots)$ was introduced in quantum statistics because the variable μ cannot in general be eliminated. In the present case, however, the constraint $N = \sum N_i^0$ allows this operation to be made. We have

$$N = e^{\beta\mu} \sum_i g_i e^{-\beta\epsilon_i} = e^{\beta\mu} Z$$

and hence

$$e^{-\beta\mu} = \frac{Z}{N} \quad \text{or} \quad \mu = -kT \ln \frac{Z(T,V,\dots)}{N} \quad , \tag{2.39}$$

which gives the formula for μ in terms of $T, N, V, \dots$

We can then eliminate the potential μ and leave the variable N in the distribution relation, i.e.

$$N_i^0 = \frac{N}{Z} g_i e^{-\beta\epsilon_i} \quad (\beta = 1/kT) \quad . \tag{2.40}$$

This distribution is the same as that obtained for Maxwell-Boltzmann statistics (2.16). Originally, Boltzmann was not aware of the principle of indistinguishability of particles and incorrectly applied Maxwell-Boltzmann statistics to gases. The identity of the two distribution relations nevertheless allowed him to provide a successful interpretation of certain properties of gases. However, difficulties encountered with entropy led to the introduction of the concept of indistinguishability, which accounted for these difficulties.

Since we can eliminate the variable μ, in corrected Maxwell-Boltzmann statistics we use the free energy function F, the natural variables of which are T, N, V, etc. We then have

$$F = \Omega + N\mu = -kTe^{\beta\mu}Z + N\mu$$

and, substituting for μ from relation (2.39), we find

$$F = -NkT \left(\ln \frac{Z}{N} + 1 \right) \quad . \tag{2.41}$$

The free energy F is then a function only of the variables T, N, as well as of the external parameters, by virtue of the partition function Z. From the formula for F we can then deduce the entropy S, the pressure P and the magnetization $\mathcal{M}$ through

$$S = -\left(\frac{\partial F}{\partial T}\right)_{N,V,\,\dots} , \quad P = -\left(\frac{\partial F}{\partial V}\right)_{T,N,\,\dots} \quad \text{and} \quad \mathcal{M} = -\left(\frac{\partial F}{\partial B}\right)_{T,N,V,\,\dots} \tag{2.42}$$

as well as the internal energy U, from the relation $U = F + TS$. We thus see that the determination of each thermodynamic quantity requires the partition function to be defined.

The condition $N_i^0 \ll g_i$, corresponding to low occupation of all levels, is valid if and only if the degree of occupation n_0^0 of the ground level is small. If this level is chosen as the origin for the energy, the condition of validity for corrected Maxwell-Boltzmann statistics becomes

$$\alpha = n_0^0 = \frac{N}{Z} \ll 1 \quad . \tag{2.43}$$

Only after calculating the partition function of the system can we confirm that $\alpha = n_0^0$ is much smaller than 1, and hence that corrected Maxwell-Boltzmann statistics is applicable.

2.6 CONCLUSION

We have just used a general method to obtain expressions for the distribution relations and for the thermodynamic functions in each of the different statistics. In all applications in physical systems, the same procedure must be followed:

1. Definition of the model to be used at the atomic scale.
2. Determination of the energy levels and their degeneracy.
3. Choice of statistics.

For this last step (Table 2.2), if the particles are distinguishable (or localized), Maxwell-Boltzmann statistics is selected. In the opposite situation (indistinguishable particles), Fermi-Dirac statistics must be used for fermions (half-integer spin) and Bose-Einstein statistics for bosons (integer spin). In the two latter cases, with low population densities, it is simpler to use the limiting case of corrected Maxwell-Boltzmann statistics. The following chapters provide numerous examples of this procedure.

BIBLIOGRAPHY

Historical References
L. Boltzmann, *Wien. Ber.* **58**, 517 (1868).
S.N. Bose, *Z. Physik* **26**, 178 (1924).
P.A.M. Dirac, *Proc. Roy. Soc. A* (London) **112**, 661 (1926).
A. Einstein, *Berliner Ber.* 261 (1924) ; 3 (1925) ; 18 (1925).
E. Fermi, *Z. Physik* **36**, 902 (1926).
J.C. Maxwell, *Phil. Mag.* **19**, 19 (1860).
W. Pauli, *Z. Physik* **31**, 776 (1925).

COMPREHENSION EXERCISES

2.1 100 distinguishable objects are distributed in two compartments. What is the thermodynamic probability of the most probable distribution? What is the probability of the distribution $N_1 = 60$, $N_2 = 40$, compared with the most probable one? [Ans.: 1.27×10^{30} ; 0.134].

Table 2.2: Summary of the various statistics.

Statistics		Thermodynamical probability	Distribution law	Energetic function
Distinguishable	Maxwell-Boltzmann (MB)	$W_{MB} = N! \prod_i \dfrac{g_i^{N_i}}{N_i!}$	$N_i^0 = \dfrac{N}{Z} g_i \exp(-\beta \epsilon_i)$ $Z = \sum_i g_i \exp(-\beta \epsilon_i)$	$F(T, B, \ldots) =$ $-NkT \ln Z(T, B, \ldots)$
Inistinguishable	Fermi-Dirac (FD)	$W_{FD} = \prod_i \dfrac{g_i!}{N_i!(g_i - N_i)!}$	$N_i^0 = \dfrac{g_i}{\exp \beta(\epsilon_i - \mu) + 1}$	$\Omega(T, \mu, V, \ldots) =$ $-kT \sum_i g_i \ln[1 + \exp -\beta(\epsilon_i - \mu)]$
Inistinguishable	Bose-Einstein (BE)	$W_{BE} = \prod_i \dfrac{(N_i + g_i - 1)!}{N_i!(g_i - 1)!}$	$N_i^0 = \dfrac{g_i}{\exp \beta(\epsilon_i - \mu) - 1}$	$\Omega(T, \mu, V, \ldots) =$ $-kT \sum_i g_i \ln[1 - \exp -\beta(\epsilon_i - \mu)]$
Inistinguishable	corrected Maxwell-Boltzmann (MBc)	$W_{MBc} = \prod_i \dfrac{g_i^{N_i}}{N_i!}$	$N_i^0 = \dfrac{N}{Z} g_i \exp(-\beta \epsilon_i)$ $Z = \sum_i g_i \exp(-\beta \epsilon_i)$	$F(T, N, V, \ldots) =$ $-NkT \left[\ln \dfrac{Z(T, B, \ldots)}{N} + 1\right]$

Chapter 3

Maxwell-Boltzmann Statistics

3.1 THE BOLTZMANN DISTRIBUTION

Maxwell-Boltzmann statistics introduced in paragraph 2.4 applies to systems of independent particles that are localized, i.e. distinguishable. At equilibrium the population of each level or domain, namely N_i for level i, is given by the Boltzmann distribution (2.16)

$$N_i = \frac{N}{Z} g_i e^{-\beta \epsilon_i} \quad \text{or} \quad n_i = \frac{N_i}{g_i} = \frac{N}{Z} e^{-\beta \epsilon_i} \tag{3.1}$$

where we have set $\beta = 1/kT$, and where

$$Z = \sum_i g_i e^{-\beta \epsilon_i} \tag{3.2a}$$

denotes the partition function for one particle. This function depends on the temperature T only through β, and on the external parameters, for example the magnetic field B, through the energy ϵ_i of the ith level whose degeneracy is g_i. It enters as a normalization factor in such a way that the distribution relation (3.1) obeys the condition $N = \sum N_i$ (from now on we write N_i instead of N_i^0 to denote the equilibrium distribution). Observe that the sum (3.2a) is taken over the levels, and can be written as a sum over all states

$$Z = \sum_j e^{-\beta \epsilon_j}. \tag{3.2b}$$

For fixed temperature and external parameters, the population density $n_i = N_i/g_i$ of the level i is proportional to the Boltzmann factor $e^{-\beta \epsilon_i}$, and therefore the lower the energy ϵ_i the higher is this density. In particular, the ratio of the population densities of any two levels i and j is

$$\frac{n_j}{n_i} = e^{-\beta(\epsilon_j - \epsilon_i)} = e^{-\beta \Delta \epsilon} \tag{3.3}$$

and depends only on the ratio $\Delta \epsilon / kT$. If the energy difference $\Delta \epsilon$ between the levels is large compared with kT, then the population density of the lower level is

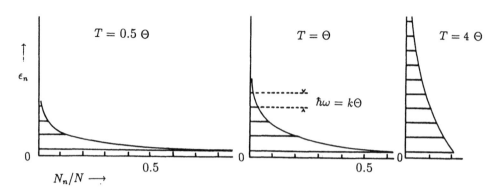

Figure 3.1: Relative population N_n/N of the energy levels in a system of N harmonic oscillators for three temperatures.

much greater than that of the upper level, while if the difference is small compared with kT then the population densities of the two levels are practically equal.

Exercise 3.1 *Distribution law for a system of harmonic oscillators.*

Consider a system of localized non interacting one-dimensional harmonic oscillators. The energy and degeneracy of the levels are

$$\epsilon_n = (n + \frac{1}{2})\hbar\omega \quad \text{and} \quad g_n = 1 \quad (n = 0, 1, 2, \ldots) \quad , \tag{3.4}$$

where ω is the natural frequency of the oscillators. Calculate the partition function Z of the system. Make a diagram of the populations of the levels for $T = \Theta/2, \Theta$ and 4Θ where $\Theta = \hbar\omega/k$ is the characteristic temperature of the spacing of the energy levels.

Solution

The partition function Z takes the form of a geometrical series and is given by

$$Z = \sum_{n=0}^{\infty} e^{-(n+1/2)\beta\hbar\omega} = e^{-\beta\hbar\omega/2} \sum_{n=0}^{\infty} (e^{-\beta\hbar\omega})^n = \frac{e^{-\beta\hbar\omega/2}}{1 - e^{-\beta\hbar\omega}} \quad . \tag{3.5}$$

The population of level n is, from (3.1)

$$N_n = \frac{N}{Z} e^{-\beta\epsilon_n} = N(1 - e^{-\beta\hbar\omega})e^{-n\beta\hbar\omega} \quad .$$

Substituting $\Theta = \hbar\omega/k$, we get

$$N_n = N(1 - e^{-\Theta/T})e^{-n\Theta/T} \quad .$$

Figure 3.1 shows the relative populations N_n/N of the levels as a function of energy, for the three temperatures considered.

3.2   THERMODYNAMIC FUNCTIONS

In paragraph 2.4.3 we saw that the free energy is given as a function of Z in terms of its natural variables by

$$F(T, N, V, B\ldots) = -NkT \ln Z(T, V, B, \ldots) \quad . \tag{3.6}$$

Differentiation of F in (3.6), together with the standard expression for the differential of F

$$dF = -SdT + \mu dN + d\mathcal{W}$$

yields the entropy S and the chemical potential μ of a single particle, i.e.

$$S = -\left(\frac{\partial F}{\partial T}\right)_{N,X} = Nk\left(\ln Z + T\frac{\partial \ln Z}{\partial T}\right) = Nk\left(\frac{\partial(T \ln Z)}{\partial T}\right)_X \tag{3.7}$$

and

$$\mu = \left(\frac{\partial F}{\partial N}\right)_{(T,X)} = -kT \ln Z \quad . \tag{3.8}$$

X stands for the set of external parameters $(V, B, \ldots)$. This formula for μ is identical to (2.15), with which we were able to eliminate the variable μ and express the result in terms of N.

The internal energy $U = F + TS$ derived from the above relations is given by

$$U = NkT^2\frac{\partial \ln Z}{\partial T} = NkT^2\frac{\partial \ln Z}{\partial \beta}\frac{d\beta}{dT} = -N\left(\frac{\partial \ln Z}{\partial \beta}\right)_X \tag{3.9}$$

on substitution of the variable $\beta = 1/kT$ for the temperature. The heat capacity of the system is then

$$C_X = \left(\frac{\partial U}{\partial T}\right)_{N,X} = \frac{(\partial U/\partial \beta)_{N,X}}{(dT/d\beta)} = Nk\beta^2\left(\frac{\partial^2 \ln Z}{\partial \beta^2}\right)_X \tag{3.10}$$

where, to make the formula more compact, β has been left as an intermediate variable instead of T.

Furthermore, since the work $d\mathcal{W}$ contains mechanical work terms $-PdV$, magnetic work $-\mathcal{M}dB, \ldots$ we have

$$P = -\left(\frac{\partial F}{\partial V}\right)_{T,N,B} = NkT\left(\frac{\partial \ln Z}{\partial V}\right)_{T,B}$$

$$\mathcal{M} = -\left(\frac{\partial F}{\partial B}\right)_{T,N,V} = NkT\left(\frac{\partial \ln Z}{\partial B}\right)_{T,V} \quad . \tag{3.11}$$

Observe that when the particles are localized the sample is usually a solid, and therefore in many cases the variable V can be omitted.

It can happen that the origin of the energies has to be shifted by an amount ϵ_0 such that the new energies are expressed in terms of the old energies through $\epsilon'_i = \epsilon_i + \epsilon_0$. The new partition function is then

$$Z' = \sum_i g_i e^{-\beta(\epsilon_i + \epsilon_0)} = e^{-\beta\epsilon_0} Z \qquad (3.12)$$

Hence, from (3.6), the new free energy is

$$F' = F + N\epsilon_0 \ . \qquad (3.13)$$

Changing the origin thus increases the free energy by an amount $N\epsilon_0$. The internal energy (3.9) increases by the same amount and the chemical potential of a single particle increases by ϵ_0. The distribution relation (3.1), however, remains unchanged. The entropy (3.7) is also unchanged, in agreement with the Boltzmann relation $S = k \ln W(N_i)$, as also is the heat capacity (3.10). It can be seen that the change in the origin affects the energy functions only.

Exercise 3.2 *Internal energy in Maxwell-Boltzmann statistics.*
   Starting from the relation $U = \sum N_i \epsilon_i$, derive formula (3.9) for the internal energy.

Solution
   We have directly

$$U \ = \ \sum N_i \epsilon_i = \frac{N}{Z} \sum g_i \epsilon_i e^{-\beta\epsilon_i} = -\frac{N}{Z} \frac{\partial}{\partial\beta} \left(\sum_i g_i e^{-\beta\epsilon_i} \right),$$

from which the required formula follows.

3.3   SEPARATING DIFFERENT ADDITIVE FORMS OF ENERGY

The energy of the particles encountered in physical problems is frequently a sum of various contributions. In the case of paramagnetic substances, for example, the energy is the sum of the vibrational energy of the ions in the lattice and that of the magnetic ions in the magnetic field. Each energy level is then defined by a sequence of quantum numbers $j, k, \dots$ relating to the various contributions grouped together under the symbol i. In a first approximation at least, the corresponding energy can be written as a sum of terms, each of which corresponds to one of the contributions

$$\epsilon_i = \epsilon_{j,k,\dots} = \epsilon_j^{(1)} + \epsilon_k^{(2)} + \dots \qquad (3.14)$$

With the assumption that the various contributions are independent, the degeneracy of level i is equal to the product of the degeneracy of each contribution, i.e.

$$g_i = g_{j,k,\dots} = g_j^{(1)} \times g_k^{(2)} \times \dots \qquad (3.15)$$

The partition function

$$Z = \sum_i g_i e^{-\beta \epsilon_i} = \sum_{j,k,\ldots} g_j^{(1)} e^{-\beta \epsilon_j^{(1)}} \times g_k^{(2)} e^{-\beta \epsilon_k^{(2)}} \times \ldots$$

then factorizes into the form

$$Z = \left(\sum_j g_j^{(1)} e^{-\beta \epsilon_j^{(1)}} \right) \times \left(\sum_k g_k^{(2)} e^{-\beta \epsilon_k^{(2)}} \right) \times \ldots$$

On introducing the partial partition functions for each contribution

$$Z^{(1)} = \sum_j g_j^{(1)} e^{-\beta \epsilon_j^{(1)}}, \quad Z^{(2)} = \sum_k g_k^{(2)} e^{-\beta \epsilon_k^{(2)}}, \quad \ldots$$

the partition function Z can be written

$$Z = Z^{(1)} \times Z^{(2)} \times \ldots \tag{3.16}$$

As a first consequence, we examine the distribution of particles among the energy levels for one of the forms of energy only (the magnetic energy of the ions, for example). The general distribution is

$$N_i = N_{j,k,\ldots} = \frac{N}{Z} g_i e^{-\beta \epsilon_i} = \frac{N}{Z^{(1)} Z^{(2)} \ldots} g_j^{(1)} g_k^{(2)} \ldots \times e^{-\beta(\epsilon_j^{(1)} + \epsilon_k^{(2)} + \cdots)}.$$

The number of particles $N_j^{(1)}$ in state j of the primary energy can be deduced from this expression, whatever the states of the other forms of energy, by summing $N_i = N_{j,k,\ldots}$ over all states of these other contributions $k, \ldots$, i.e.

$$N_j^{(1)} = \sum_{k,\ldots} N_{j,k,\ldots} = \frac{N}{Z^{(1)} Z^{(2)} \ldots} g_j^{(1)} e^{-\beta \epsilon_j^{(1)}} \times \sum_k g_k^{(2)} e^{-\beta \epsilon_k^{(2)}} \times \ldots = \frac{N}{Z^{(1)}} g_j^{(1)} e^{-\beta \epsilon_j^{(1)}}.$$

It can be seen that the partial distribution has exactly the same form as the general distribution.

A second consequence is to be found in the fact that the factorization property of the partition function makes the free energy F additive, since, from (3.6)

$$\begin{aligned} F &= -NkT \ln Z \\ &= -NkT \ln Z^{(1)} - NkT \ln Z^{(2)} - \ldots = F^{(1)} + F^{(2)} + \ldots \end{aligned} \tag{3.17}$$

As the thermodynamic functions are derived from F, they also possess the same property of additivity. Thus

$$S = S^{(1)} + S^{(2)} + \ldots \quad \text{with} \quad S^{(1)} = Nk \frac{\partial (T \ln Z^{(1)})}{\partial T}, \ldots \tag{3.18a}$$

$$U = U^{(1)} + U^{(2)} + \ldots \quad \text{with} \quad U^{(1)} = -N \frac{\partial \ln Z^{(1)}}{\partial \beta}, \ldots \tag{3.18b}$$

$$C_X = C_X^{(1)} + C_X^{(2)} + \ldots \quad \text{with} \quad C_X^{(1)} = Nk\beta^2 \frac{\partial^2 \ln Z^{(1)}}{\partial \beta^2}, \ldots \tag{3.18c}$$

Provided the energy of the particles can be expressed as a sum of independent terms, it can be seen that the various thermodynamic quantities in Maxwell-Boltzmann statistics take simple forms in which each energy contribution enters independently of the others. An important consequence is that each of these contributions can be investigated and explained separately. In any real experiment, however, only total quantities are measured. Nevertheless, in many cases the domains of temperature variation of each of these terms are separate.

3.4   TWO ENERGY LEVEL SYSTEMS

3.4.1   Distribution Relation

We apply the above formalism to the very simple case of a system of independent distinguishable particles having only two energy levels with equal degeneracy. We already encountered this case while investigating the system of spins 1/2 in a magnetic field (Problem. 1.1), and we shall come across further examples.

We choose the origin of the energies such that the lowest level has the energy $\epsilon_1 = 0$ and the highest level has energy $\epsilon_2 = \epsilon$ and we call g their degeneracy $g_1 = g_2 = g$. The partition function of this system is, from (3.2)

$$Z = g + ge^{-\beta\epsilon} = g(1 + e^{-\beta\epsilon}) \tag{3.19}$$

and the populations N_1 et N_2 of the two levels are given by (3.1), i.e.

$$N_1 = \frac{N}{1 + e^{-\beta\epsilon}} \quad \text{and} \quad N_2 = \frac{Ne^{-\beta\epsilon}}{1 + e^{-\beta\epsilon}} = \frac{N}{1 + e^{\beta\epsilon}} \ . \tag{3.20}$$

These quantities depend only on the product $\beta\epsilon = \epsilon/kT$. When the characteristic temperature $\Theta = \epsilon/k$ is introduced to describe the energy gap between these two levels, it can be seen that these quantities depend only on $\beta\epsilon = \Theta/T$, or alternatively on T/Θ.

The curves in Figure 3.2 show the ratios N_1/N, N_2/N, Z/g and $N_2/N_1 = e^{-\beta\epsilon}$ as a function of the reduced temperature T/Θ. At low temperature $(T \ll \Theta)$ where $\beta\epsilon \gg 1$, all the particles occupy the ground level $\epsilon_1 = 0$ $(N_1 = N$ and $N_2 = 0)$. The upper level $\epsilon_2 = \epsilon$ starts to become significantly populated only when $T = 0.15\ \Theta$ $(N_2/N = 1.27 \times 10^{-3})$. The point of inflection of the curves N_1/N and N_2/N is located at $T = 0.42\ \Theta$, around which the variation in population as a function of temperature is the greatest. At high temperatures $(T \gg \Theta)$, the two populations become equal, each tending to $N/2$. This value is reached to within 10% for $T = 5\Theta$.

It is thus clear that the changes in the thermodynamic functions are essentially confined to the interval $0.15\ \Theta \lesssim T \lesssim 5\ \Theta$, which illustrates the importance of the notion of characteristic temperature. Below $0.15\ \Theta$ (low temperature regime for the system), the sparsely populated upper level makes no physical contribution. The condition $\epsilon = k\Theta \gg kT$ corresponds to the fact that the upper level which is very high (compared to kT) can be neglected. For $T \gg \Theta$ (high temperature regime), the two levels are equally populated and can be considered degenerate since

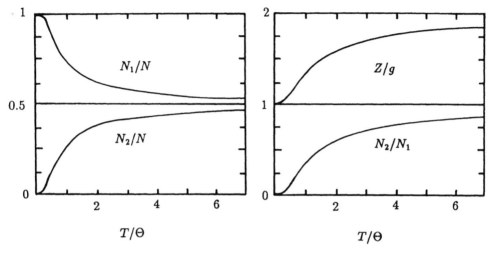

Figure 3.2: The distribution (3.20) and the partition function (3.19) for a system of two energy levels having the same degeneracy g, shown as a function of $T/\Theta = kT/\epsilon$.

their separation is much smaller than kT. Thus at low temperatures the partition function Z is equal to g, the number of quantum states in level 1, and at high temperatures it is equal to $2g$, the total number of quantum states in levels 1 and 2. At intermediate temperatures, each level contributes to Z proportionately to its population. This result can be generalized to several levels, as may be seen by inspection of equation (3.2) for Z and (3.1) for the populations N_i.

3.4.2 Thermodynamic Functions

The thermodynamic functions of the two-level system can be derived from expression (3.19) for the partition function using the relations from paragraph 3.2. Thus

$$F \;=\; -NkT \ln \left[g(1 + e^{-\beta\epsilon}) \right], \tag{3.21}$$

$$S \;=\; Nk \ln \left[g(1 + e^{-\beta\epsilon}) \right] + Nk \frac{\beta\epsilon}{1 + e^{\beta\epsilon}}, \tag{3.22}$$

$$U \;=\; \frac{N\epsilon}{1 + e^{\beta\epsilon}}, \tag{3.23}$$

$$C_X \;=\; Nk \frac{\beta^2 \epsilon^2 e^{\beta\epsilon}}{(1 + e^{\beta\epsilon})^2} = Nk \left(\frac{\beta\epsilon/2}{\cosh \beta\epsilon/2} \right)^2. \tag{3.24}$$

The entropy S is shown in Figure 3.3a. At low temperatures, the only accessible states are the g ground states, and the thermodynamic probability W given by equation (2.11) is ($N_1 = N, N_2 = 0$) $W = g^N$, and hence $S = Nk \ln g$, which is the limiting value of (3.22) as $\beta\epsilon \to \infty$. At high temperatures, it is legitimate to consider that there is only one level, of degeneracy $2g$, and W is equal to $(2g)^N$; i.e. $S = Nk \ln 2g$, the limiting value of (3.22) as $\beta\epsilon \to 0$.

The heat capacity $C_X(T/\Theta)$ is shown in Figure 3.3b. It falls to zero rapidly for

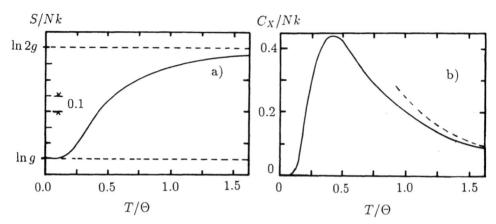

Figure 3.3: a) Entropy and b) heat capacity of a system with two energy levels as a function of $T/\Theta = kT/\epsilon$. The dashed curve shows the variation in A/T^2 of equation (3.25b).

$T \ll \Theta$, following the exponential relation

$$C_X = Nk \left(\frac{\Theta}{T}\right)^2 e^{-\Theta/T} \quad (T \ll \Theta). \tag{3.25a}$$

For $T \gg \Theta$ it varies as $1/T^2$

$$C_X = \frac{Nk}{4} \left(\frac{\Theta}{T}\right)^2 = \frac{A}{T^2} \quad (T \gg \Theta) \ . \tag{3.25b}$$

The maximum occurs at $\tanh \beta\epsilon/2 = 2/\beta\epsilon$, the numerical solution of which yields $T = 0.42\ \Theta$. Its value at the maximum is $0.44Nk$. Note that the curves obtained for a system of spins 1/2 in a magnetic field (Fig. 1.13) are a particular case of systems with two energy levels in which $\epsilon = \overline{m}B$ and $g = 1$.

3.4.3   Schottky Anomaly

Below 1 K most substances have very small heat capacities, for example, 0.75×10^{-4} J K^{-1} mol^{-1} for copper at 0.1 K. Contributions from the vibration of the ions in crystal lattices and that from the motion of conduction electrons become very small indeed (Ch. 8 and 10). However, substances such as chromium methylammonium alum Cr (CH_3NH_3) $(SO_4)_2$, $12H_2O$ display a maximum in their molar heat capacity at these temperatures that can be as much as 50 000 times the molar heat capacity of copper at the same temperature (Schottky anomaly). The increase in heat capacity of these substances when the temperature decreases can be understood in terms of the two level model proposed by W. Schottky (1922). In the absence of a magnetic field, the degeneracy of the lowest electronic level of many magnetic ions in a crystal is lifted by the crystalline field (§4.5.2). When the two lowest levels in the resulting splitting are sufficiently far away from the other levels, the system can, at low enough temperature, be considered as having only two levels. For example, the

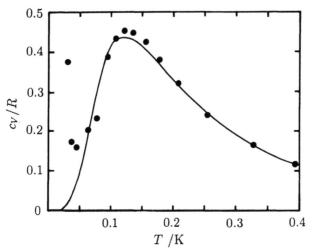

Figure 3.4: Heat capacity of chromium methylammonium alum as a function of the absolute temperature [W.E. Gardner and N. Kurti, *Proc. Roy. Soc.* A (London), **223**, 542 (1954), Table 2]. The curve is that given by equation (3.24) with $\Theta = \epsilon/k = 0.29$ K and $N \to \mathcal{N}$.

chromium ion Cr^{3+} in chromium methylammonium alum has its ground state of effective spin $J = 3/2$ split by the crystalline field into two doubly degenerate levels ($g_1 = g_2 = 2$) separated by $\epsilon = 2.11 \times 10^{-5}$ eV, that is $\Theta = \epsilon/k = 0.245$ K. These two are well separated from the other levels, the closest of which lies at 1.80 eV [B. Bleaney, *Proc. Roy. Soc.* A (London) **204**, 216 (1950) and S. Sugano and Y. Tanabe, *Ions of the Transition Elements*, A General Discussion of the Faraday Society (1958)]. Figure 3.4 shows the measured values of the molar heat capacity c_V for this alum as a function of temperature, together with the theoretical curve (3.24) for a value of $\Theta = \epsilon/k = 0.29$ K. The reasonable agreement between the spectroscopic and calorimetric measurements of Θ demonstrates the validity of Schottky's interpretation. The large low temperature heat capacity of substances that display a Schottky anomaly makes them useful as thermostats in a temperature range below that of liquid helium (Prob. 3.1 and §4.6.4).

3.5 THEOREM OF EQUIPARTITION OF ENERGY

3.5.1 Theorem

The theorem of equipartition of energy applies to systems in which all of the following conditions hold:

1. the system obeys Maxwell-Boltzmann statistics;
2. the component forms of the energy are additive, i.e.

$$\epsilon_{ijk} = \epsilon_i^{(1)} + \epsilon_{j,k,\dots} \quad ;$$

3. the number of levels in a given energy form is unlimited;

4. the temperature range is such that $T \gg \Theta$, Θ being the characteristic temperature of the spacing between two successive levels in the energy form being considered. $\epsilon_i^{(1)}$ can thus be treated as a continuous variable, which can then be expressed classically (correspondence principle in quantum mechanics);

5. classically, $\epsilon^{(1)}$ can be expressed as $\epsilon = aq^2$ or $\epsilon = bp^2$, where q (or p) is a generalized co-ordinate (or momentum), and the associated degeneracy is proportional to dq (or dp).

For such systems, the equipartition of energy theorem establishes that the contribution to the internal energy for each form of energy considered is equal to $NkT/2$, where N is the number of particles.

The partial partition function for this form of energy is given by

$$Z^{(1)} = \sum_i g_i^{(1)} e^{-\beta \epsilon_i^{(1)}} \propto \int_{-\infty}^{+\infty} dq \, e^{-\beta a q^2} ,$$

i.e., on substituting for the integral and making use of Table 6.1,

$$Z^{(1)} \propto \left(\frac{\pi}{\beta a} \right)^{1/2} \quad \text{or} \quad \ln Z^{(1)} = \frac{1}{2} \ln \frac{\pi}{a} - \frac{1}{2} \ln \beta + \text{const} ,$$

where the additive constant is independent of β. The contribution to the internal energy is then, from (3.18)

$$U^{(1)} = -N \frac{\partial \ln Z^{(1)}}{\partial \beta} = \frac{N}{2\beta} = \frac{1}{2} NkT . \tag{3.26}$$

This proves the theorem. The result is independent of the value of a, of the proportionality factor between g and dq (or dp) and of the fact that the energy appears as aq^2 or bp^2. It is therefore quite general, provided that conditions 1 to 5 above are valid.

A consequence of this theorem is that the form of energy considered contributes to the constant volume molar heat capacity ($N \to \mathcal{N}$) through a term

$$c_V^{(1)} = \left(\frac{\partial u^{(1)}}{\partial T} \right)_V = \frac{1}{2} \mathcal{N} k = \frac{1}{2} R . \tag{3.27a}$$

The corresponding numerical value is

$$c_V^{(1)} = 4.16 \text{ J K}^{-1} \text{ mol}^{-1} = 0.99 \text{ cal K}^{-1} \text{ mol}^{-1} . \tag{3.27b}$$

This simple result is employed frequently. The theorem may be stated in another way: if the energy has a quadratic form, then the mean energy of a particle is equal to

$$\epsilon^{(1)} = \frac{1}{2} kT . \tag{3.28}$$

The mean energy is defined by

$$\epsilon^{(1)} = \sum \epsilon_i^{(1)} \times \frac{N_i}{N} = \frac{U^{(1)}}{N} .$$

Table 3.1: Molar heat capacity at constant volume for various metals at room temperature [A. Eucken, *Lehrbuch der chemischen Physik*, Leipzig (1944)].

	Na	Al	K	Fe	Cu	Ag	Pt
c_V /J K^{-1} mol^{-1}	26.0	22.8	25.8	24.8	23.7	24.2	24.5

3.5.2 Dulong-Petit Law

Before the development of quantum mechanics, heat capacities of the chemical elements in the solid state were calculated by assuming that the atoms vibrate around their equilibrium position like independent harmonic oscillators in three dimensions, each oscillator of mass m vibrating at angular frequency ω. The energy of each oscillator is then the sum of three terms corresponding to the three directions of space

$$\epsilon = \epsilon_x + \epsilon_y + \epsilon_z \quad .$$

Each of these terms is a sum of kinetic energy and potential energy, with the form

$$\epsilon_x = \frac{p_x^2}{2m} + \frac{1}{2}m\omega^2 x^2 \quad . \tag{3.29}$$

The theorem of equipartition of energy can be applied to these expressions. The energy is thus the sum of 6 independent terms that are quadratic in x, y, z, p_x, p_y and p_z, respectively, and the number of quantum states in a cell of phase space (degeneracy) is given by (1.6)

$$g_i \rightarrow \frac{d^3\mathbf{r}\,d^3\mathbf{p}}{h^3} \quad . \tag{3.30}$$

This fulfils condition 5 for applicability of the theorem. It follows that the molar heat capacity of solid chemical elements is, by (3.27)

$$c_V = 6 \times \frac{1}{2}R = 3R = 24.94 \text{ J K}^{-1} \text{ mol}^{-1} \quad . \tag{3.31}$$

This relation is found experimentally to be valid at ordinary temperatures for the majority of solid elements (Dulong-Petit law, 1819). Table 3.1, which gives the molar heat capacities at constant volume for various metals at room temperature, illustrates this point.

It will be observed that in this model the molar heat capacity is independent of the mass of the atoms and their vibration frequency. The Dulong-Petit law is, however, only an approximation, as it offers no explanation for the observed dependence of c_V on T (§3.6).

For solid chemical compounds, the number of atoms per mole is a multiple of Avogadro's number $\mathcal{N}$ ($2\mathcal{N}$ for NaCl, $3\mathcal{N}$ for PbI$_2$, ...). The molar heat capacity of these substances is then equal to $3R$ multiplied by the number of atoms in a

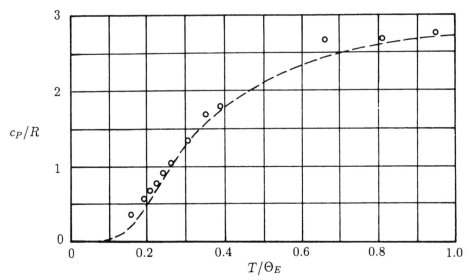

Figure 3.5: Molar heat capacity at constant pressure $(c_P \simeq c_V)$ of diamond in units of $R = \mathcal{N}k$. The curve shown is the function (3.34) of the variable $T/\Theta_E = 1/x$, where $\Theta_E = \hbar\omega/k = 1320$ K [A. Einstein, *Ann. Physik* **22**, 180 (1907)].

molecule. This relation holds fairly well for ionic crystals ($c_V = 48.32$ J K^{-1} mol^{-1} for NaCl, for example).

It is important to note that the constant R, the so-called gas constant, also plays a role in solid state physics. This is a result of the universal character of Boltzmann's constant k.

3.6   EINSTEIN MODEL

Although the Dulong-Petit law was fairly successful in interpreting heat capacities, it fails to explain why they decrease at low temperatures and tend to zero at absolute zero (Fig. 3.5). The first satisfactory explanation for these properties was provided by A. Einstein (1907), through quantization of the energy of $3N$ independent harmonic oscillators in one dimension, each vibrating with the same angular frequency ω. The energy levels of an oscillator having angular frequency ω is given from quantum mechanics by

$$\epsilon_n = (n + 1/2)\hbar\omega \quad \text{and} \quad g_n = 1 \ .$$

This yields for the partition function (3.5) the result already found in exercise 3.1

$$Z = \frac{e^{-\beta\hbar\omega/2}}{1 - e^{-\beta\hbar\omega}} \ .$$

The free energy, internal energy and heat capacity in the Einstein model are obtained from (3.6,9,10) as follows,

$$F = 3N \left[kT \ln(1 - e^{-\beta\hbar\omega}) + \frac{\hbar\omega}{2} \right] , \tag{3.32}$$

$$U = 3N\hbar\omega \left[\frac{1}{e^{\beta\hbar\omega} - 1} + \frac{1}{2} \right] , \tag{3.33}$$

$$C_V = 3Nk\beta^2 \left(\frac{\partial^2 \ln Z}{\partial \beta^2} \right)_V = 3NkE(\beta\hbar\omega) \tag{3.34a}$$

where the Einstein function $E(x)$ is

$$E(x) = \frac{(x/2)^2}{(\sinh x/2)^2} = \frac{x^2 e^x}{(e^x - 1)^2} . \tag{3.34b}$$

In this model the Dulong-Petit law is retrieved for $T \gtrsim \Theta_E = \hbar\omega/k$, since $E(x) \sim 1$ for $x \lesssim 1$. At low temperatures the heat capacity decreases with T in the same way as equation (3.25) for a two-level system, vanishing at $T = 0$. Einstein made a quantitative comparison of the heat capacity obtained in this way with that of diamond by taking the value $\Theta_E = 1320$ K (Fig. 3.5). The general agreement is satisfactory, although, as can be discerned in the figure, the model shows a large discrepancy at low temperatures $(T \lesssim 0.2\,\Theta_E)$. The Einstein model is in fact only a rough approximation, since the atoms of a solid cannot vibrate independently of each other. Other more realistic models will be described in chapter 8.

3.7   NEGATIVE TEMPERATURES

We return to the simple case of a system of N independent particles with two energy levels separated by ϵ and of the same degeneracy (§3.4). At thermal equilibrium, equation (3.3),

$$\frac{N_2}{N_1} = e^{-\beta\epsilon} = e^{-\Theta/T} , \tag{3.35}$$

describes the relation between the populations of the levels and the temperature. Conversely, this relation can be used to determine the temperature from the ratio of the populations, through

$$T = \Theta \, \frac{1}{\ln N_1/N_2} . \tag{3.36}$$

We have seen that at low temperatures $(T \ll \Theta)$ only the lower level is populated, $N_2/N_1 \simeq 0$. As the temperature increases the upper level becomes occupied at the expense of the lower level, until the limit $N_2 = N_1 = N/2$ is reached at very high temperatures $(T \gg \Theta)$. The populations are then equal. The different values of the temperature can be chosen through contact with a thermostat.

However, if the system is thermally isolated it is possible to establish a *population inversion*, in which the second level becomes more populated than the first $(N_2 > N_1)$. This transformation calls for an increase in energy which can be introduced

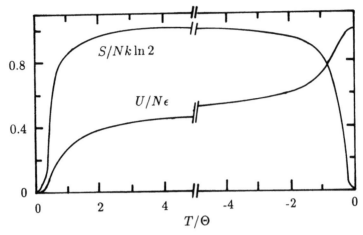

Figure 3.6: Reduced energy and entropy of a system having two levels separated by $\epsilon = k\Theta$ and degeneracy $g = 1$, for positive and negative temperatures.

into the system by absorption of photons of energy $\epsilon = h\nu$. If we still accept equation (3.36) as the definition of T, a negative value for the temperature is obtained. The first system to be brought to a negative temperature was a system of nuclear spins, those of Li^+ ions in a crystal of lithium fluoride Li F [E.M. Purcell and R.V. Pound, *Phys. Rev.* **81**, 279 (1951)]. For this substance thermal equilibrium between the spin system and the rest of the lattice is reached in a time τ_1 of the order of a few minutes to a few hours, while equilibrium among the spins is reached in a time of the order of $\tau_2 \simeq 10^{-5}$ s. During the interval τ_1, the spin system can be considered to be in internal thermal equilibrium, but isolated from the outside world (the lattice) for a time long enough for its properties to be investigated at negative temperatures. The energy and the entropy of a two-level system (3.22,23) are shown as a function of T/Θ in Figure 3.6.   We note that a system with a population inversion, i.e. having a negative temperature, has a higher energy than in a positive temperature state. It should therefore be considered as being "hotter". The negative temperature domain is located beyond infinite temperature, rather than below absolute zero, which we know to be impossible. When placed in contact with a system at positive temperature, a system at negative temperature will transfer heat to it. The notion of negative temperatures can be applied to systems of spins that have more than two levels in a magnetic field. We observe, however, that only systems possessing a finite number of levels can be put in a state of negative temperature. Investigations into the heat exchange between two systems of spins at different temperatures (spins of the nuclei of $^7Li^+$ and $^{19}F^-$ in Li F) demonstrated the validity of extending calorimetry to the domain of negative temperatures. [A. Abragam and W.G. Proctor, *Phys. Rev.* **109**, 1441 (1958)]

Access to negative temperatures, which calls for complex experimental techniques, thus opens up a new field for physical investigations. For example, for the $^{19}F^-$ spins in Ca F_2, with a magnetic field H_0 parallel to the twofold axes [110] of the cubic crystal, an ordered antiferromagnetic structure is predicted at positive temperatures and ferromagnetic at negative temperatures [A. Abragam and M. Goldman, *Nuclear Magnetism: Order and Disorder*, Clarendon Press, Oxford, 1982]. It should be added that negative temperatures can be obtained also in systems of electronic spins [A. Abragam and B. Bleaney, *Electron Paramagnetic Resonance of Transition Ions*, Oxford University Press, Oxford, 1970].

BIBLIOGRAPHY

Historical References
P. Dulong and A. Petit, *Ann. Chim.* (Paris) **10**, 395 (1819).
A. Einstein, *Ann. Physik* **22**, 180 (1907).
F.A. Lindemann, *Physik. Z.* **11**, 609 (1910).
W. Schottky, *Physik. Z.* **23**, 448 (1922).

COMPREHENSION EXERCISES

3.1 Calculate the ratios of population densities of two levels separated by $\Delta\epsilon = k\Theta$ for $T = \Theta/2, \Theta$ and 2Θ [Ans.: $n_2/n_1 = 0.14$; 0.37 and 0.61].

3.2 Same exercise for $T = \Theta/10$ and $T = 10\ \Theta$ [Ans.: $n_2/n_1 = 4.5 \times 10^{-5}$ and 0.90].

3.3 The two lowest levels of the ion Tb^{3+} in terbium ethylsulphate are separated by $\delta = 1/\lambda = 0.39$ cm^{-1}. Calculate the wavelength, frequency and energy of the photons that excite a transition between these levels. Calculate the characteristic temperature of the gap between these levels. What is the range of wavelengths of the corresponding electromagnetic waves? [Ans.: $\lambda = 2.56$ cm; $\nu = 1.17 \times 10^{10}$ s^{-1}; $\epsilon = 4.84 \times 10^{-5}$eV; $\Theta = 0.56$ K; centimetre waves].

PROBLEM 3.1 PROPERTIES OF PARAMAGNETIC SALTS

1 Theoretical Model
We consider a crystal containing N magnetic ions having no interactions and with two non degenerate levels. In a magnetic field of flux density B the energy is

$$\epsilon_- = -\overline{m}B \quad \text{and} \quad \epsilon_+ = \epsilon + \overline{m}B \ ;$$

ϵ is the energy splitting of the levels in zero field and $\overline{m}$ is a magnetic moment.

1.1 Write down the partition function for an ion. Set $\epsilon' = \epsilon + 2\overline{m}B$. Hence deduce the contribution of the ions to the free energy of the substance.

1.2 Find a formula for the molar heat capacity c_V of the system of ions in a magnetic field and in zero field. Draw the curve $c_V(T)$. Give the limiting form in the "high" temperature region ($kT \gg \epsilon$).

1.3 Determine the magnetic moment $\mathcal{M}$ of the substance. Under what circumstances can ϵ be neglected? Show that $\mathcal{M}$ then depends only on $x = \overline{m}B/kT$ and draw the curve $\mathcal{M}(x)$.

1.4 Find an expression for the magnetic susceptibility χ_M defined here by

$$\chi_M = \frac{\mu_0}{n}\frac{\partial \mathcal{M}}{\partial B}(B = 0)$$

where n is the number of moles of the substance and μ_0 the magnetic permeability of empty space. For a given temperature range (to be defined) show that the sample obeys the Curie law $\chi_M = C_M/T$. Find a formula for the Curie constant C_M.

2 Properties of Terbium Ethyl Sulphate

The electronic ground state of terbium ions Tb^{3+} in a crystal of terbium ethyl sulphate Tb $(C_2H_5 SO_4)_3, 9H_2O$ (TbES) is a doublet that displays a small splitting $\delta = 0.39$ cm^{-1}. This quantity, in wave numbers, is the reciprocal of the wavelength λ of the radiation governing the transition between two levels. It is related to the energy separation by $\epsilon = h\nu = hc/\lambda = hc\delta$. The next electronic level in this salt occurs at $\Delta = 88$ cm^{-1}.

2.1 For $T = 4$ K calculate the ratio of population densities between the excited level at 88 cm^{-1} and the ground state in zero field. Hence deduce that this excited level and those beyond it have no effect in the temperature range below $T = 4$ K.

2.2 For the temperature range 2 K to 4 K in zero field, show that the contribution to the molar heat capacity of the crystal due to the doublet of the terbium ion has the form $c_V^S = A/T^2$ (Schottky anomaly) and determine the value of A.

2.3 From electronic paramagnetic resonance the value of the spectroscopic splitting factor of the doublet in a magnetic field is found to be $g = 17.82$. This means that the projection of the magnetic moment of the ions on the field can take the two values $\mu_z = \pm\overline{m} = \pm g\mu_B/2$, where μ_B is the Bohr magneton. Calculate the value of the Curie constant C_M.

2.4 Between 2 and 4 K, the molar heat capacity of terbium ethyl sulphate takes the form $1.791/T^2 + 0.00649T^3$ J K^{-1} mol^{-1}, and its Curie constant C_M is equal to 4.78×10^{-4} K m^3 mol^{-1}. Does the above model account for these experimental results?

Universal constants: Avogadro's number $\mathcal{N} = 6.02 \times 10^{23}$ mol^{-1}, Boltzmann constant $k = 1.38 \times 10^{-23}$ J K^{-1}, Bohr magneton $\mu_B = 0.927 \times 10^{-23}$ J T^{-1}, magnetic permeability of vacuum $\mu_0 = 4\pi \times 10^{-7}$ SI, ratio $hc/k = 1.44$ cm K.

Reference: M.T. Hirvonen et al., Phys. Rev. B **11**, 4652 (1975).

SOLUTION 1 Theoretical Model

1.1 From (3.2), the partition function for an ion is

$$Z = e^{-\beta\epsilon_-} + e^{-\beta\epsilon_+} = e^{\beta\overline{m}B}\left[1 + e^{-\beta(\epsilon+2\overline{m}B)}\right] .$$

Denoting the spacing between the two levels by $\epsilon' = \epsilon + 2\overline{m}B$, we can write the partition function as

$$Z = e^{\beta\overline{m}B}(1 + e^{-\beta\epsilon'}) .$$

This expression contains a term of the form (3.19) characteristic of two level systems, and the factor $\exp(\beta\overline{m}B)$ corresponding to a non zero energy ground state (3.12).

The contribution of the ions to the free energy of the sample is then, from (3.6)

$$F = -NkT \ln Z = -NkT\ln\left[1 + e^{-\beta(\epsilon+2\overline{m}B)}\right] - N\overline{m}B .$$

1.2 The heat capacity C_B in a magnetic field is given by (3.10) with

$$\ln Z = \beta \overline{m} B + \ln(1 + e^{-\beta \epsilon'})$$

and its molar value is $(N \rightarrow \mathcal{N} = R/k)$

$$c_B = R \frac{\beta^2 \epsilon'^2 e^{\beta \epsilon'}}{(1 + e^{\beta \epsilon'})^2} \ .$$

In zero field, $c_B = c_V$ and $\epsilon' = \epsilon$. This expression is the same as (3.24). The heat capacity is shown in Figure 3.3b. When $T \gg \epsilon/k = \Theta$, then $\beta \epsilon \ll 1$ and the exponentials can be replaced by 1 in the formula for c_V, yielding the limiting form

$$c_V = R \left(\frac{\epsilon}{2kT} \right)^2 = \frac{R}{4} \left(\frac{\Theta}{T} \right)^2 \ .$$

This is identical with the $1/T^2$ behaviour already seen in (3.25b).

1.3 The magnetic moment of the substance is obtained from equation (3.11). Substitution for Z yields

$$\mathcal{M} = N \overline{m} - 2N \overline{m} \frac{e^{-\beta \epsilon'}}{1 + e^{-\beta \epsilon'}} = N \overline{m} \left[\tanh \beta \left(\overline{m} B + \frac{\epsilon}{2} \right) \right] \ .$$

When $\overline{m} B \gg \epsilon$, the magnetic moment takes the form

$$\mathcal{M} = N \overline{m} \tanh x \quad \text{with} \quad x = \overline{m} B / kT \ ,$$

which is the same result as exercise 2.2. The function $\mathcal{M}(x)$ is displayed in Figure 1.14.

1.4 Substitution of the general formula for $\mathcal{M}$ yields

$$\chi_M = \mu_0 \mathcal{N} \frac{\overline{m}^2}{kT} \times \frac{1}{\cosh^2(\beta \epsilon'/2)} \ .$$

When $kT \gg \epsilon$, the susceptibility takes the form of a Curie Law

$$\chi_M = \frac{C_M}{T} \quad \text{with} \quad C_M = \mu_0 \mathcal{N} \frac{\overline{m}^2}{k} \ .$$

2 Properties of Terbium Ethyl Sulphate

2.1 From the Maxwell-Boltzmann distribution, the ratio of population densities in the two levels given in (3.3) is

$$\frac{n \ (88 \ \text{cm}^{-1})}{n \ (0 \ \text{cm}^{-1})} = e^{-hc\Delta/kT} \ .$$

For $T = 4$ K, this ratio is 1.8×10^{-14}. This number decreases as T decreases. Below 4 K, therefore, the excited levels at 88 cm^{-1} and beyond have no effect.

2.2 The characteristic temperature Θ of the doublet is

$$\Theta = \frac{\epsilon}{k} = \frac{hc}{k}\delta = 0.56 \text{ K} \quad .$$

In the temperature range of interest, in the approximation $T \gg \Theta$, the limiting relation for question 1.2 gives $c_V^S = A/T^2$ with

$$A = \frac{R}{4}\Theta^2 = 0.65 \text{ J K mol}^{-1} \quad .$$

2.3 For a TbES crystal, the Curie constant in question 1.4 becomes

$$C_M = \mu_0 \mathcal{N}\frac{g^2\mu_B^2}{4k} = 3.74 \times 10^{-4} \text{ K m}^3 \text{ mol}^{-1} \quad .$$

2.4 The heat capacity is the sum of several contributions. The term in T^3 is due to vibrations of the lattice (Ch. 8). The Schottky anomaly contributes to the term in $1/T^2$, but accounts for only $0.65/1.791 \simeq 35$ %. Magnetic interactions between ions, which are neglected in the model, are responsible for the remaining 65 % of the term in $1/T^2$. As for the magnetic susceptibility, the non interacting model accounts for about 80 % of the measured Curie constant.

The model used is not entirely satisfactory, since interactions between Tb^{3+} ions, which are separated by approximately 8 Å, are not negligible in this temperature range. In particular, the dipolar interaction energy $\epsilon \sim (\mu_0/4\pi) \times \overline{m}^2/r^3$ is about 10^{-24} J, which corresponds to a characteristic temperature of 0.1 K, and is comparable with Θ.

PROBLEM 3.2 MODEL OF SOLIDS

In one model of solids, it is assumed that each of the N identical atoms of a crystal, oscillating about its equilibrium position, is equivalent to 3 independent one-dimensional oscillators of mass m subject to a potential $\epsilon_p(x)$.

1 Dulong-Petit Law
First it is assumed that $\epsilon_p(x) = m\omega^2 x^2/2$ (harmonic oscillator).

1.1a What type of statistics is needed for this problem? For this type of statistics write down the formula for the number of particles N_i at equilibrium in the discrete levels of energy ϵ_i and degeneracy g_i.

1.1b Hence deduce the expression for the number of one-dimensional oscillators $d^2N_{x,p}$ whose position and momentum lie between x and $x + dx$ and between p and $p + dp$ respectively. It is recalled that the number of quantum states in a cell of phase space (x,p) is $g_i \to dx\, dp/h$, where h is Planck's constant.

1.2a Find the formula for the partition function Z of a harmonic oscillator.

1.2b Specify the temperature range in which the classical partition function is a good approximation for the partition function of a quantum harmonic oscillator (3.5)

$$Z = \frac{e^{-\beta\hbar\omega/2}}{1 - e^{-\beta\hbar\omega}} \quad , \quad \beta = 1/kT \quad .$$

Table 3.2: Properties of various metals

	Mg	Al	Cu	Zn	Ag	Pb
M /g mol^{-1}	24.3	27.0	63.5	65.4	107.9	207.2
T_M /K	924	933	1356	692	1234	601
Θ_E /K	~ 340	400	320	230	220	92
d /Å	3.19	2.86	2.55	2.66	2.88	3.49

1.3 Derive the formula for the internal energy U of the crystal and hence deduce the molar heat capacity at constant volume c_V of the solid. Compare with the experimental Dulong and Petit relation ($c_V \simeq 25$ J K^{-1} mol^{-1}).

2 Melting Temperature

Lindemann (1910) suggested that solids melt when the mean amplitude of vibration of their constituent atoms reaches a certain fraction of the interatomic distance.

2.1 Calculate the number of oscillators dN_x whose position lies between x and $x + dx$.

2.2a Calculate the mean values $< x >$ and $< x^2 >$ of the position of the oscillator.

2.2b Hence deduce the mean potential energy $< \epsilon_p >$ of an oscillator and the mean vibration amplitude x_0, defined by $< \epsilon_p > = m\omega^2 x_0^2/2$.

2.3 Table 3.2 lists the atomic mass M, melting point T_M, Einstein temperature $\Theta_E = \hbar\omega/k$ and the distance d between two neighbouring atoms in the crystal for various metals.

2.3a Calculate the mean amplitude of vibration at the melting point for these elements.

2.3b For what fraction of the interatomic distance d does melting occur in the different metals? Comment on this result.

3 Anharmonicity

At high temperatures, the heat capacity of solids increases linearly with temperature instead of tending to the limit predicted by the Dulong and Petit law. In copper, for example, $c_V/(J$ K^{-1} mol$^{-1}) = 23.2 + 2.9 \times 10^{-3}$ (T /K) [from R.E. Pawel and E.E. Stanbury, *J. Phys. Chem. Solids* **26**, 607 (1965)]. One possible cause of this increase could be anharmonicity of the vibrations, i.e. the fact that the potential energy ϵ_p of the oscillator has the form $\epsilon_p = m\omega^2 x^2/2 - fx^4$, where f is a constant.

3.1 Write down the partition function Z using an integral that contains only x.

3.2 Calculate Z, retaining only the contribution of lowest order from the anharmonic term.

3.3 Hence find the formula for c_V in this approximation and compare the theoretical result with the relation found experimentally for copper.

Formulae: The following integrals I_n are given

$$I_0 = \frac{1}{2}\left(\frac{\pi}{\alpha}\right)^{1/2}, \quad I_{n+2} = \frac{n+1}{2\alpha}I_n, \quad \text{with} \quad I_n(\alpha) = \int_0^\infty x^n e^{-\alpha x^2} dx \quad .$$

Universal constants: gas constant $R = 8.31$ J K^{-1} mol^{-1}; Avogadro's number $\mathcal{N} = 6.02 \times 10^{23}$ mol^{-1}, reduced Planck constant $\hbar = 1.05 \times 10^{-34}$ J s.

Reference: J.R. Partington, *An Advanced Treatise on Physical Chemistry*, vol 3, 364, Longmans, London (1957).

SOLUTION 1 Dulong-Petit Law

1.1a The atoms are localized and assumed to be independent. Maxwell-Boltzmann statistics can then be applied. In this type of statistics the distribution in the levels is given by (3.1).

1.1b To each of the N atoms correspond 3 independent oscillators (i.e. $3N$ altogether). Each oscillator possesses the classical energy (3.29)

$$\epsilon = \frac{p^2}{2m} + \frac{1}{2}m\omega^2 x^2$$

and, in continuous variables, the distribution relation (3.1) becomes

$$N_i \to d^2 N_{x,p} = \frac{3N}{Z}e^{-\beta\epsilon}\frac{dx\ dp}{h} \quad .$$

1.2a The partition function Z given in (3.2) can here be written

$$Z = \frac{1}{h}\int_{-\infty}^{+\infty} dp \int_{-\infty}^{+\infty} dx \exp\left[-\beta\left(\frac{p^2}{2m} + \frac{1}{2}m\omega^2 x^2\right)\right] \quad . \tag{3.37}$$

This integral factorizes into a product of two integrals over p and x respectively of a Gaussian function of type I_0. This yields

$$Z = \frac{1}{h}(2\pi mkT)^{1/2} \times \left(\frac{2\pi kT}{m\omega^2}\right)^{1/2} = \frac{kT}{\hbar\omega} \quad . \tag{3.38}$$

1.2b When $kT \gg \hbar\omega$, the quantum partition function (3.5) is equivalent to

$$Z = \frac{1}{1 - (1 - \beta\hbar\omega)} = \frac{kT}{\hbar\omega} \quad .$$

It can be seen that the classical limit is reached when the energy kT is much larger than the energy spacing $\hbar\omega$ of the quantum levels. This result is general.

1.3 The formula for U was obtained in (3.9). For $3N$ oscillators we find

$$U = 3NkT^2 \times \frac{1}{T} = 3NkT, \quad \text{and therefore} \quad c_V = 3\mathcal{N}k = 3R \quad .$$

This value, $c_V = 24.94$ J K^{-1} mol^{-1}, is in good agreement with the Dulong-Petit law.

2 Melting Temperature

2.1 The number dN_x is found by summing $d^2N_{x,p}$ over all values of p, i.e.

$$dN_x = \int_{-\infty}^{+\infty} d^2N_{x,p} = \frac{3N}{Z} \exp\left(-\frac{\beta m\omega^2 x^2}{2}\right) \frac{dx}{h} \times \int_{-\infty}^{+\infty} \exp\left(-\frac{\beta p^2}{2m}\right) dp \ .$$

Insertion of the formula for Z and the Gaussian integral yields

$$dN_x = 3N \left(\frac{m\omega^2}{2\pi kT}\right)^{1/2} \exp\left(-\alpha x^2\right) dx \quad \text{with} \quad \alpha = \beta\frac{m\omega^2}{2} = \frac{m\omega^2}{2kT} \ .$$

2.2a Since the function dN_x is even owing to the symmetry of the potential, we have $< x >= 0$. In addition,

$$< x^2 >= \frac{\int x^2 dN_x}{\int dN_x} = \frac{\int x^2 \exp(-\alpha x^2)dx}{\int \exp(-\alpha x^2)dx} = \frac{I_2(\alpha)}{I_0(\alpha)}$$

Evaluation of I_2/I_0 gives

$$< x^2 >= \frac{1}{2\alpha} = \frac{kT}{m\omega^2} \ .$$

2.2b The mean value of the potential energy $\epsilon_p = m\omega^2 x^2/2$ is

$$< \epsilon_p >= \frac{1}{2}m\omega^2 < x^2 >= \frac{1}{2}kT \ .$$

The above result is consistent with the theorem of equipartition of energy, which applies to this form of energy. On comparison, it can be seen that

$$x_0 = \sqrt{< x^2 >} = \sqrt{\frac{kT}{m\omega^2}} \ .$$

2.3a Introduction of the molar mass $M = Nm$ and the Einstein temperature into the above formula gives

$$x_0 = \frac{N\hbar}{\Theta_E}\sqrt{\frac{T}{MR}} = \frac{6.96 \times 10^{-10}}{\Theta_E \ /K}\sqrt{\frac{T \ /K}{M \ /(\text{g mol}^{-1})}} \ .$$

The values of x_0 at the melting temperature are listed in Table 3.3.

2.3b The ratio x_0/d is given in Table 3.3. It can be observed that it is of the order of 3.7 % for the set of metals considered, in accordance with the suggestion of Lindemann. This correspondence provides an estimate for the value of Θ_E on the basis of the melting point for other substances, through the relation

$$\Theta_E(\text{K}) = \frac{188}{d \ /10^{-10}\text{m}}\sqrt{\frac{T/\text{K}}{M \ /(\text{g mol}^{-1})}} \ .$$

Table 3.3: Mean atomic vibration amplitude in metals

	Mg	Al	Cu	Zn	Ag	Pb
x_0 /10^{-12} m	12.6	10.2	10.1	9.84	10.7	12.9
$\frac{x_0}{d}$ /10^{-2}	4.0	3.6	3.9	3.7	3.7	3.7

3 Effect of Anharmonicity

3.1 The general expression for Z is obtained from (3.37)

$$
\begin{aligned}
Z &= \frac{1}{h} \int_{-\infty}^{+\infty} dp \int_{-\infty}^{+\infty} dx \exp\left[-\beta\left(\frac{p^2}{2m} + \frac{1}{2}m\omega^2 x^2 - fx^4\right)\right] \\
&= \frac{1}{h}(2\pi mkT)^{1/2} \int_{-\infty}^{+\infty} dx \exp\left[-\beta\left(\frac{1}{2}m\omega^2 x^2 - fx^4\right)\right] \quad .
\end{aligned}
$$

3.2 Approximation of $\exp(\beta f x^4)$ by $1 + \beta f x^4$ yields

$$
\begin{aligned}
Z &= \frac{1}{h}(2\pi mkT)^{1/2} \int_{-\infty}^{+\infty} e^{-\alpha x^2}(1 + \beta f x^4) dx \\
&= \frac{2}{h}(2\pi mkT)^{1/2}\left[I_0(\alpha) + \beta f I_4(\alpha)\right] \quad .
\end{aligned}
$$

Evaluation of I_0 and I_4 gives for the partition function in this approximation

$$
Z = \frac{kT}{\hbar\omega}\left[1 + \frac{3fkT}{m^2\omega^4}\right] \quad .
$$

It can be seen that the partition function (3.38) of the harmonic oscillator appears as a factor, with a correction term that is linear in T.

3.3 From (3.9) the internal energy is

$$
U = 3NkT + 9N\frac{f(kT)^2}{m^2\omega^4} \quad ,
$$

from which the formula for the molar heat capacity at constant volume is obtained

$$
c_V = 3R\left[1 + 6\frac{fkT}{m^2\omega^4}\right] \quad . \tag{3.39}
$$

The anharmonicity considered here thus leads to a heat capacity that varies linearly with T and is an increasing function if $f > 0$. This relation has the same form as that observed experimentally for copper. However, other phenomena can also give rise to a linear contribution in T, for example the heat capacity of the electrons in a metal (Ch. 10). In addition, part of the increase in c_V comes from the fact that the asymptotic value of the Dulong-Petit law is not fully reached in the temperature range considered.

It will be noticed that the expansion of ϵ_p contained no odd terms of the type gx^3. This is justified if the oscillators possess a centre of symmetry. For other situations, a term in x^3 must be included, giving a contribution to c_V equal to

$$3R \times \frac{15g^2 kT}{(m\omega^2)^3} \quad , \tag{3.40}$$

where the exponential in x^3 must be expanded up to the term in x^6.

Chapter 4

Ideal Paramagnetism

4.1 INTRODUCTION

The magnetic properties of materials have been the subject of many investigations, not only because of their theoretical interest but also on account of their numerous applications, particularly in the area of low temperatures.

Materials fall into two categories depending on whether their constituent parts (atoms, ions or molecules) possess a magnetic moment or not. They are accordingly either magnetic or diamagnetic, respectively.

Diamagnetic substances never have a magnetic moment in zero field. When they are placed in a magnetic field the movement of the electrons in the atoms is perturbed, and these in turn acquire a very small orbital magnetic moment that is proportional to the field, but in the opposite direction. We shall not investigate the properties of these substances.

Magnetic materials owe their properties to the existence of a magnetic moment in some of the atoms. They can have various phases:

• a *paramagnetic* phase in which the magnetization is zero in zero field, the atomic dipoles being randomly oriented. Under the influence of a magnetic field these dipoles give the substance a magnetization that, in small fields, is proportional to and in the same direction as the induction.

• "ordered" phases in which the atomic dipoles adopt ordered orientations on a molecular scale in zero field. Depending on the prevailing type of order, phases occur that may be ferromagnetic, antiferromagnetic, ferrimagnetic, etc. (Ch.13).

The ordered arrangement of the dipoles is a result of their interactions, and it disappears when the temperature reaches a value such that kT is of the order of the interaction energy. Magnetic materials in general thus have a paramagnetic phase at higher temperature and one or more ordered phases at lower temperature. In this chapter, we shall investigate magnetic substances in their paramagnetic phase in a temperature range where the dipole interaction energy is much less than kT, and so can be neglected. The corresponding model is that of an ideal paramagnetic substance. Various cases in which the interactions can no longer be neglected will be investigated in chapter 13.

4.2   EXPERIMENTS ON MAGNETIC MATERIALS

4.2.1   Magnetic Equation of State

In order to determine all the thermodynamic properties of a magnetic substance, in addition to knowing its equation of state and a heat capacity in zero field, it is sufficient to know its magnetic equation of state $\mathcal{M} = \mathcal{M}(N, T, P, \mathbf{H})$ or $\mathcal{M} = \mathcal{M}(N, T, P, \mathbf{B})$. The magnetic moment $\mathcal{M}$ is related to the magnetization $\mathbf{M}$ through

$$\mathbf{M} = \frac{\mathcal{M}}{V} \ . \tag{4.1}$$

In magnetostatics, the magnetization appears in the formula

$$\mathbf{B} = \mu_0 (\mathbf{H} + \mathbf{M}) \tag{4.2a}$$

that relates the magnetic flux density or induction $\mathbf{B}$ to the magnetic field strength $\mathbf{H}$. The situations examined in this chapter are those in which the magnetization $\mathbf{M}$ of the material is much smaller than $\mathbf{H}$. For this reason the vectors $\mathbf{B}$ and $\mathbf{H}$ are assumed to be proportional,

$$\mathbf{B} = \mu_0 \mathbf{H} \ . \tag{4.2b}$$

4.2.2   Measurement of the Magnetic Moment

Measurements of magnetic moments are made using two main methods. The first is based on the relation

$$\mathbf{F} = (\mathcal{M} \cdot \nabla)\mathbf{B} \tag{4.3}$$

which defines the force $\mathbf{F}$ acting on a small piece of the substance in an inhomogeneous magnetic field. In the method of Gouy and of Faraday (Fig. 4.1a), the sample S hangs by a thread from a precision balance D and is located at the centre of a superconducting magnet SM. This generates a uniform vertical field B_0 which magnetizes the sample. Two symmetrically placed coils C_1 and C_2 carrying the same current in opposite directions generate an additional field that is zero at S and has a vertical gradient. Measurement of the vertical force $\mathbf{F}$ produced by the field gradient yields the magnetic moment of the material. Typical values of the gradient and the sensitivity of this balance are respectively 10 T m^{-1} and 10^{-8} N (the weight of one microgramme), thus permitting measurements of magnetic moments of the order of 10^{-6} A m^2.

The second method relies on the change in the flux in a probe coil generated by movements of the magnetized sample. Figure 4.1b shows a magnetometer of this type. The sample S, in the uniform field of an electromagnet M, is set into vibration by a transducer T. A sinusoidal variation of the flux is produced in two coils C_1 and C_2 wound in opposition and arranged symmetrically on either side of the sample. The induced electromotive force is proportional to the magnetic moment of the sample. Magnetic moments can be measured by this method with a precision of 10^{-7} A m^2. [S. Foner, *Rev. Sci. Instrum.* **30**, 548 (1959)].

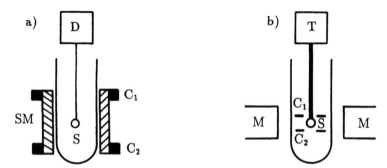

Figure 4.1: Diagrams of magnetometers. a) Method of Gouy and of Faraday: S = sample, D = precision balance, SM = superconducting magnet, C_1 and C_2 = coils generating the magnetic field gradient. b) Vibrating sample method: S = sample, M = electromagnet, T = transducer, C_1 and C_2 = probe coils.

4.2.3   Magnetic Susceptibilities

Definition

Experiments show that the magnetic moment of materials is a decreasing function of the temperature and an increasing function of the magnetic field. At low magnetic fields the magnetization is proportional to the field,

$$M = \frac{\mathcal{M}}{V} = \chi H = \chi \frac{B}{\mu_0} \quad , \tag{4.4}$$

where χ is the magnetic susceptibility, assumed to be isotropic. As the majority of experiments are carried out in the range of validity of equation (4.4), χ is an important characteristic property of the substance. It is dimensionless.

In addition to χ, the mass and molar susceptibilities χ_m and χ_M are defined from the total magnetic moment by the relations

$$\mathcal{M} = \chi V H = \chi_m m H = \chi_M n H \tag{4.5}$$

where V, m and n are respectively the volume, mass and number of moles of the substance considered. It follows that

$$\chi_m = \frac{\chi}{\rho} \quad \text{and} \quad \chi_M = v\chi = M\chi_m \tag{4.6}$$

where ρ, v and M are the density, molar volume and molar mass of the substance. The units of these susceptibilities are respectively $m^3\ kg^{-1}$ and $m^3\ mol^{-1}$.

Experimental results are often expressed in the cgs emu system. Conversion to the international system is as follows:

$$
\begin{aligned}
&\text{for } \chi &:& \quad 1 \text{ cgs emu} = 4\pi \text{ SI (without dimensions)} \\
&\text{for } \chi_m &:& \quad 1 \text{ cgs emu (or cm}^3\ g^{-1}) = 4\pi \times 10^{-3} \text{ SI (or m}^3\ kg^{-1}) \\
&\text{for } \chi_M &:& \quad 1 \text{ cgs emu (or cm}^3\ mol^{-1}) = 4\pi \times 10^{-6} \text{ SI or (m}^3\ mol^{-1}) \ .
\end{aligned} \tag{4.7}
$$

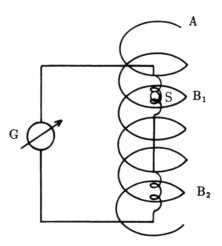

Figure 4.2: Method for measuring susceptibilities. A = field coil, B_1 and B_2 = opposed windings acting as probe, S = paramagnetic sample, G = galvanometer.

Note that for strong magnetic fields and sufficiently low temperatures, the linear relation (4.4) no longer holds and the magnetization tends to a limiting value, the so-called saturation value, at which the atomic dipoles are oriented to the maximum extent in the direction of the field.

Measurement of susceptibilities

Relative measurements of magnetic susceptibilities can be made using simple electromagnetic methods based on the phenomenon of induction. We describe one of these methods (Fig. 4.2). A first coil A carrying a low frequency alternating current generates an alternating magnetic field. Two identical coils B_1 and B_2 are placed in series in such a way that their windings are opposed. The electromotive force induced in the circuit B_1 and B_2 by the coil A is then zero. If a paramagnetic sample S is placed inside one of the coils, B_1, it develops a magnetic moment that contributes to the flux crossing B_1. The electromotive force of induction $e = -d\phi/dt$, which is proportional to $dM/dt = (dM/dB) \times (dB/dt) = (V\chi/\mu_0) \times (dB/dt)$, has an amplitude that is proportional to $V\chi$. The proportionality constant is determined using a standard sample. In modern practice, a superconducting coil is used and the electromotive force is measured by means of a SQUID [E.J. Cukauskas, D.A. Vincent and B.S. Deaver Jr., *Rev. Sci. Instrum.* **45**, 1 (1974)].

   Note that the flux density B in equation (4.4) is that prevailing inside the sample. This is different from the flux density B_0 in the absence of the sample. When the sample is an ellipsoid (a particular form of which is a sphere), the flux density B inside the sample is homogeneous, and the *demagnetizing field* $B_d = B_0 - B$ is proportional to M. In this case the proportionality constant depends only on the shape of the ellipsoid ($B_d = \mu_0 M/3$ for a sphere). If $\chi \ll 1$, the demagnetizing field is negligible and B and B_0 are interchangeable. In the contrary case, in particular at low temperatures, this field must be taken into account in order to determine χ.

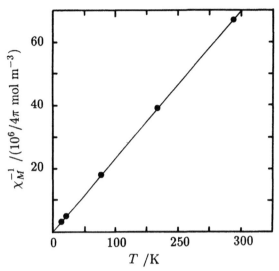

Figure 4.3: Curie law for manganese ammonium sulphate Mn $(NH_4)_2$ $(SO_4)_2$, $6H_2O$ [L.C. Jackson and H. Kamerlingh Onnes, *Proc. Roy. Soc. A* (London) **104**, 671 (1923)].

4.2.4   Curie Law

The temperature dependence of the magnetic susceptibility of a large number of materials was investigated by Pierre Curie (1895) who established the relation that bears his name

$$\chi = \frac{C}{T} \ ,$$

(4.8)

where C is the so-called Curie constant (4.8). This law is illustrated in Figure 4.3 for manganese ammonium sulphate and in Figure 1.12 for copper potassium sulphate. For these two cases the Curie constant C_M, defined by $\chi_M = C_M/T$, respectively takes the values $C_M = 5.50 \times 10^{-5}$ K m^3 mol^{-1} ($C_M = 4.38$ cgs emu) and $C_M = 5.99 \times 10^{-6}$ K m^3 mol^{-1} ($C_M = 0.476$ cgs emu). The order of magnitude found for $C_M \sim 10^{-5}$ K m^3 mol^{-1} ($C_M \sim 1$ cgs emu) is typical (Table 4.4).

The Curie law is in fact the limiting case of the more general Curie-Weiss relation

$$\chi = \frac{C}{T - \Theta}$$

(4.9)

when $T \gg |\Theta|$. The constant Θ, which can be positive or negative, is called the Weiss constant, and is such that $k|\Theta|$ is comparable in magnitude with the interaction energy between the magnetic ions. In this chapter we are interested in substances in which the interaction energy is much smaller than kT and which therefore obey Curie's law. Other terms must be added to this magnetic susceptibility that come from diamagnetism ($\chi_M \sim -10^{-10}$ m^3 mol^{-1}) and Van Vleck paramagnetism ($\chi_M \sim 10^{-11}$ m^3 mol^{-1}). The latter terms are very much smaller than that of paramagnetism and will be neglected in what follows.

In crystals, magnetic phenomena are anisotropic. The linear relation between the two vector quantities $\mathbf{M}$ and $\mathbf{H}$ is then written in a tensor form

$$\mathbf{M} = [\, \chi \,]\mathbf{H} \quad \text{or} \quad M_i = \chi_{ij} H_j \tag{4.10}$$

where, following Einstein's convention, we have omitted $\sum_j$. The magnetic susceptibility is then a tensor quantity, $[\, \chi \,]$ or χ_{ij}, a symmetric tensor of rank 2. Expressed in terms of its principal axes, it is defined by its three principal values χ_1, χ_2, χ_3. For crystals with an axis of symmetry of order higher than 2 (uniaxial crystals, §8.2.3), we have $\chi_1 = \chi_2 \equiv \chi_\perp$, $\chi_3 \equiv \chi_\parallel$, while for a cubic crystal $\chi_1 = \chi_2 = \chi_3 \equiv \chi$. In the latter case, the magnetic properties are isotropic and the simple proportionality relation (4.4) applies. When the crystalline substance is in powder form the magnetic properties are also isotropic and what is measured is

$$\chi = \frac{\chi_1 + \chi_2 + \chi_3}{3} \ . \tag{4.11}$$

4.3   VECTOR MODEL OF THE ATOM

To understand the magnetic properties of materials, a knowledge is required of the electronic states of the atoms, molecules and ions. This paragraph summarizes the principal results of atomic quantum theory. The electronic state of an atom containing Z electrons is obtained by taking Z single electron states, and then by making the appropriate couplings.

4.3.1   Quantum States of an Electron in the Atom

The state of an electron in an atom is defined by five quantum numbers n, l, m_l, s and m_s. The principal quantum number n is characteristic of the layer to which the electron belongs. It is a positive integer, which, by convention, is denoted by a capital letter.

$$
\begin{array}{cccccc}
n & 1 & 2 & 3 & 4 & \ldots \\
\text{layer} & K & L & M & N & \ldots
\end{array} \tag{4.12}
$$

The orbital quantum number l defines the modulus of the angular momentum $\boldsymbol{\sigma}_l$ of the electron in its orbital motion around the nucleus through the relation

$$|\, \boldsymbol{\sigma}_l \,| = [l(l+1)]^{1/2} \hbar \ . \tag{4.13}$$

The number l takes positive integer or zero values and defines the sub-layer to which the electron belongs, with the correspondence

$$
\begin{array}{ccccccc}
l & 0 & 1 & 2 & 3 & 4 & \ldots \\
\text{sub} - \text{layer} \ s & & p & d & f & g & \ldots
\end{array} \tag{4.14}
$$

In a given layer n, the orbital quantum number l is limited by the condition $l \leq n$.

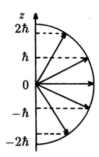

Figure 4.4: Diagram of the quantization of angular momentum σ_l for the case $l = 2$. The radius of the circle is $[l(l+1)]^{1/2}\hbar = \sqrt{6}\,\hbar$. The projections of σ_l on Oz are $m_l\hbar$ with $|m_l| \leq l$, i.e. $m_l = -2, -1, 0, 1, 2$.

The magnetic quantum number m_l defines the projection of the angular momentum σ_l on a quantization axis Oz in such a way that

$$(\sigma_l)_z = m_l\hbar \ . \tag{4.15}$$

In this expression m_l can take the $2l + 1$ values $m_l = -l, -l + 1, \ldots, l - 1, l$ (Fig. 4.4). Last, the quantum numbers s and m_s describe the intrinsic angular momentum (spin) of the electron σ_s such that

$$|\sigma_s| = [s(s+1)]^{1/2}\hbar \quad \text{and} \quad (\sigma_s)_z = m_s\hbar \ . \tag{4.16a}$$

For an electron, the value of the spin quantum number s is $1/2$ and the number m_s can take the values $m_s = \pm 1/2$. Hence

$$|\sigma_s| = \frac{\sqrt{3}}{2}\hbar \quad \text{and} \quad (\sigma_s)_z = \pm\frac{1}{2}\hbar \ . \tag{4.16b}$$

In quantum mechanics, an electronic state is represented by the ket $|n\,l\,m_l\,m_s >$. In spectroscopy, a notation is used to define the sub-layer, such as $3p$, where the number is the principal quantum number of the layer and the letter designates the sub-layer in the notation of (4.14). The numbers m_l and m_s are not specified. In the same sub-layer, therefore, there are $(2l+1)(2s+1) = 2(2l+1)$ different states. The multiplicity (or number of states) of the layers and sub-layers is shown in Table 4.1.

In the case of atoms with a single electron (hydrogen atoms or hydrogenoid ions), the energy of the electron in a quantum state depends only on the principal quantum number n in accordance with the Rydberg formula

$$\epsilon_n = -R_\infty \frac{Z^2}{n^2} \ , \tag{4.17}$$

where $R_\infty = 13.6$ eV is the Rydberg constant and Z is the atomic number of the ion.

For atoms with several electrons, the energy of each electron cannot be evaluated independently owing to their interactions. Nevertheless, as a first approximation

Table 4.1: Multiplicity of electronic states

Layer	K	L		M			N			
Sub-layer	1s	2s	2p	3s	3p	3d	4s	4p	4d	4f
Multiplicity	2	2	6	2	6	10	2	6	10	14
	2	8		18			32			

(the so-called central field approximation), each electron is subjected to an effective central field produced by the nucleus and the average effect of the other electrons. In this approximation the state of the atom can be described by all the states occupied by the electrons, which are assumed to be independent. The energy of an electron in a quantum state then depends on the values of n and l in this state. The theory is complex, but an important result is that for fixed n the energy $\epsilon_{n,l}$ increases with l.

The ground state of an atom corresponds to a distribution of its electrons in the single electronic states of lowest energy, bearing in mind the Pauli exclusion principle. The electronic configuration of the atom is defined by the set of numbers n_i and l_i for each electron and it is designated by a spectroscopic notation. For the sodium atom $(Z = 11)$ for example, we write

$$1s^2 \quad 2s^2 \quad 2p^6 \quad 3s$$

indicating that there are 2 electrons in the sub-layers $1s$ and $2s$, 6 in the sub-layer $2p$ and a single electron in sub-layer $3s$. In a full sub-layer all orientations of the orbital angular momentum and of the spin are occupied and the total angular momentum is zero. This means that full sub-layers are fairly inert and it is customary to omit them from the notation. For this reason the electronic configuration of sodium is denoted by $3s$.

During an excitation (e.g. absorption of a photon) an atom or an ion can occupy an excited configuration. The excitation energy of these configurations is large and we shall not have to include them in our investigation into the magnetic properties of materials.

Exercise 4.1 *Configuration degeneracy*

The praseodymium ion Pr^{3+} of the rare earth group has 56 electrons. The configuration of its electronic ground state is

$$4s^2 \quad 4p^6 \quad 4d^{10} \quad 4f^2 \quad 5s^2 \quad 5p^6 \,, \tag{4.18}$$

where the full layers K, L, M are omitted, or $4f^2$ where full sub-layers are also omitted. Find the number of different states of Pr^{3+} belonging to this configuration.

Solution

The two equivalent electrons of the incomplete sub-layer $4f$ $(n = 4, l = 3)$ can occupy the $2(2l+1) = 14$ states of this sub-layer. Given the Pauli exclusion principle and the indistinguishability of the electrons, the number of different states of Pr^{3+}

is equal to $C^2_{14} = 14 \times 13/2 = 91$. In the central field approximation these states have the same energy and are mutually degenerate. Typically, the energy difference between two configurations of the rare earths is 10 eV or 10^5 cm^{-1}.

4.3.2   Quantum States of Atoms

In the central field approximation, the state of an atom is determined by the single electron states that are occupied, and all states having the same configuration are mutually degenerate, i.e. they possess the same energy. Account should, however, be taken of non central forces, which are of two kinds. The first is Coulomb repulsion between the electrons. The second, which is weaker, is spin-orbit coupling, and is magnetic in origin.

Electrostatic coupling

Electrostatic coupling due to Coulomb repulsion between electrons can be treated as a perturbation of the central potential. It produces partial lifting of the degeneracy in all states of a given configuration. These states then recombine to form others that are defined by the values of the total orbital angular momentum and of the total spin of the atom

$$\sigma_L = \sum_i \sigma_l(i) \quad , \quad \sigma_S = \sum_i \sigma_s(i) \quad , \tag{4.19}$$

(index i refers to the electrons) such that

$$\begin{aligned} |\,\sigma_L\,| &= [L(L+1)]^{1/2}\hbar, \quad (\sigma_L)_z = M_L\hbar \\ |\,\sigma_S\,| &= [S(S+1)]^{1/2}\hbar, \quad (\sigma_S)_z = M_S\hbar \ . \end{aligned} \tag{4.20}$$

The new quantum states are represented by kets $|LM_LSM_S >$. The energy of the states depends only on the orbital angular momentum of the atom, given by L, and on the spin angular momentum defined by S. It is independent of M_L and M_S. Quantum states having the same L and S are therefore degenerate in energy and constitute a term with degeneracy $(2L+1)(2S+1)$. In spectroscopy, terms are denoted by a capital letter corresponding to the value of L , i.e.

$$\begin{array}{c|cccccccc} L & 0 & 1 & 2 & 3 & 4 & 5 & \dots \\ [\,L\,] & S & P & D & F & G & H & \dots \end{array} \tag{4.21}$$

and a left superscript equal to $(2S+1)$. For example, ^3H includes states of the atom such as $L = 5$ and $S = 1$. The degeneracy of this term is $(2L+1)(2S+1) = 33$.

To find all the terms belonging to the same configuration, all the vector sums must be performed for the vectors $\sigma_l(i)$ and $\sigma_s(i)$ of the electrons, taking into account the Pauli exclusion principle. The problem is in general complex. For example, in the case of Pr^{3+} ions, which have two $4f$ electrons ($l = 3$ and $s = 1/2$), we have

$$0 \leq L \leq 6 \quad \text{and} \quad 0 \leq S \leq 1$$

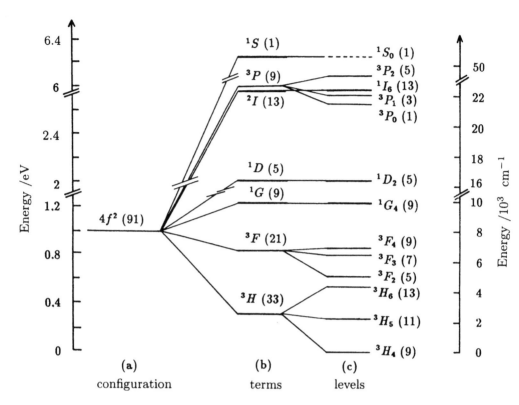

Figure 4.5: Atomic energy levels of the praseodymium ion Pr^{3+}. The three columns correspond to successive approximations. a) central field approximation; b) term splitting due to Coulomb repulsion between electrons; c) level splitting due to spin-orbit coupling. The numbers in brackets indicate the degeneracy. [*Atomic Energy Levels, The Rare Earth Elements*, p.120, Natl. Bur. Stand. NSRDS-NBS 60 (1978)].

and, owing to the Pauli exclusion principle, the only allowed terms are

$$^1S \quad ^3P \quad ^1D \quad ^3F \quad ^1G \quad ^3H \quad ^1I . \tag{4.22}$$

Exercise 4.2 *Degeneracy of the terms*

Determine the degeneracy of the seven terms in the $4f^2$ configuration of the Pr^{3+} ion.

Solution

Since the degeneracy of a term is $(2L+1)(2S+1)$, the seven terms (4.22) contain 1, 9, 5, 21, 9, 33 and 13 quantum states, respectively. The total number of states is thus equal to 91, which is the multiplicity of the $4f^2$ configuration (Fig. 4.5). Typically the difference between the terms for the Pr^{3+} ion is 1 eV or 10^4 cm^{-1}.

Spin-orbit coupling

When relativistic effects are taken into account, a further term must be added to the Hamiltonian

$$H_{SO} = \sum_i \zeta(r_i) \ \boldsymbol{\sigma}_s(i) \cdot \boldsymbol{\sigma}_l(i) \ .$$

This term is called spin-orbit coupling. Treated as a perturbation in addition to the central potential and the electrostatic coupling, it partially raises the degeneracy among the states of a given term. The new quantum states are represented by kets $|LSJM_J>$, where J and M_J are the quantum numbers of the total angular momentum

$$\boldsymbol{\sigma}_J = \boldsymbol{\sigma}_L + \boldsymbol{\sigma}_S \tag{4.23}$$

such that

$$| \ \boldsymbol{\sigma}_J \ | = [J(J+1)]^{1/2}\hbar \quad \text{and} \quad (\boldsymbol{\sigma}_J)_z = M_J\hbar \ . \tag{4.24}$$

The energy of these new states is independent of M_J, and hence they yield levels with degeneracy $2J+1$. Levels are defined by the three numbers L, S and J, and are denoted in the spectroscopic term notation by a right subscript that indicates the value of J. For example, a $^3H_4$ level contains states of the atom such that $L = 5, S = 1$ and $J = 4$. The degeneracy of this level is $2J + 1 = 9$.

To find all levels belonging to a given term, the sum of the two vectors $\boldsymbol{\sigma}_L$ and $\boldsymbol{\sigma}_S$ is performed, giving

$$|L - S| \leq J \leq L + S \tag{4.25}$$

The number of levels in the term is $2S + 1$ if $S \leq L$, and $2L + 1$ if $S \geq L$. This number is called the multiplicity of the term.

Exercise 4.3 *Splitting of terms into levels*
Determine the notations of all levels belonging to the $4f^2$ configuration of the Pr^{3+} ion.

Solution
The terms of the $4f^2$ configuration are given in (4.22). These correspond to the values $S = 0$ or $S = 1$. The multiplicity of the terms is therefore 1 or 3. Figure 4.5 shows the notations of the various levels with the J values that satisfy (4.25). Typical splittings between levels in the same term are of the order of 0.1 eV, or 10^3 cm^{-1}.

Ground state. Hund's rule

The states of the lowest level of an atom (or ion) are particularly important since this is the level that is generally occupied in the temperature range below a few hundred Kelvin. Here we shall state Hund's rule which, together with the Pauli exclusion principle, determines the ground state.

i) The spin quantum number S is equal to the maximum value of $M_S = \sum_i m_s(i)$;

ii) for this spin state, the orbital quantum number L is equal to the maximum value of $M_L = \sum_i m_l(i)$;

iii) for these spin and orbital angular momentum states, the quantum number J takes the value $|L - S|$ (J minimum) if the layer is less than half full and $L + S$ (J maximum) otherwise.

Exercise 4.4 *Ground states of the ions Pr^{3+} and Tm^{3+}.*

Determine the ground states of praseodymium Pr^{3+} and thulium Tm^{3+} ions, whose ground configurations are respectively $4f^2$ and $4f^{12}$.

Solution

The diagram below summarizes the discussion.

m_l	3	2	1	0	-1	-2	-3
Pr^{3+}	↑	↑					
Tm^{3+}	↑↓	↑↓	↑↓	↑↓	↑↓	↑	↑

The electronic sub-layer $4f (l = 3)$ contains $(2l + 1)(2s + 1) = 14$ single electron states. Applying Hund's rule to the Pr^{3+} ion, first all the m_s should be taken equal to $+1/2$(↑), i.e. $S = 1$, then the values of m_l are taken to be 3 and 2, i.e. $L = 5$. Since the sub-layer is less than half full, $J = L - S = 4$. The ground state of the Pr^{3+} ion is therefore $^3H_4$ (Fig. 4.5).

For the Tm^{3+} ion, 7 electrons have $m_s = +1/2$(↑) and 5 have $m_s = -1/2$(↓), i.e. $S = 1$. The greatest value of $M_L = \sum_i m_l(i)$ is found as shown in the diagram and is equal to 5, i.e. $L = 5$. Since the sub-layer is more than half full, it follows that $J = L + S = 6$. The ground state of the Tm^{3+} ion is thus $^3H_6$ (Table 4.2).

4.3.3 Magnetic Moment of the Atom

When a magnetic field of flux density **B** is applied to a free atom, its interaction with the electrons can be described by the Hamiltonian

$$\hat{H} = -\mu_B(\mathbf{L} + 2\mathbf{S}).\mathbf{B} \ , \tag{4.26}$$

where the dimensionless quantities

$$\mathbf{L} = \frac{\sigma_L}{\hbar} \quad \text{and} \quad \mathbf{S} = \frac{\sigma_S}{\hbar}$$

are used, and μ_B is the Bohr magneton

$$\mu_B = \frac{e\hbar}{2m_e} = 0.927 \times 10^{-23} \text{ A m}^2 = 5.79 \times 10^{-5} \text{ eV T}^{-1} \ . \tag{4.27}$$

Here m_e is the mass of the electron. The first term of the sum is the interaction of the orbital motion of the electrons with the magnetic field, and the second is that between the spins and the field. Note the factor 2 in front of **S**.

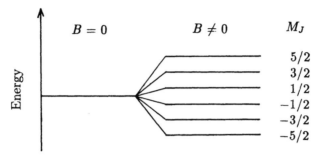

Figure 4.6: Zeeman effect for a $J = 5/2$ level.

The interaction (4.26) between atom and magnetic field completely lifts the degeneracy of the atomic levels (Zeeman effect). For a level with quantum numbers L, S, J, the $2J + 1$ states acquire an additional energy

$$\epsilon(M_J) = M_J g \mu_B B \quad (M_J = -J, -J + 1, \ldots, J) \ . \tag{4.28}$$

Here g is the Landé factor of the level, with

$$g = \frac{3}{2} + \frac{S(S+1) - L(L+1)}{2J(J+1)} \ . \tag{4.29}$$

The Zeeman sub-levels are shown diagrammatically in Figure 4.6, where it can be seen that they have a regular spacing of $\Delta\epsilon = g\mu_B B$. In a magnetic field of 1 Tesla and for $g = 2$, this spacing is equal to $\Delta\epsilon = 1.85 \times 10^{-23}$ J$= 1.16 \times 10^{-4}$ eV, i.e., $\delta = \Delta\epsilon/hc = 0.934$ cm^{-1} or $\Theta = \Delta\epsilon/k = 1.34$ K.

The above results can be explained by attributing a magnetic moment to the atom

$$\boldsymbol{\mu} = -g\mu_B \mathbf{J} \quad \text{with} \quad \mathbf{J} = \frac{\boldsymbol{\sigma}_J}{\hbar} \tag{4.30}$$

such that

$$\mu = |\ \boldsymbol{\mu}\ | = [J(J+1)]^{1/2} g\mu_B \quad \text{and} \quad \mu_z = (\ \boldsymbol{\mu}\ )_z = -M_J g\mu_B \ . \tag{4.31}$$

Equation (4.28) for the energy of the levels gives the interaction of the magnetic moment with the field

$$\epsilon = -\boldsymbol{\mu} \cdot \mathbf{B} \ . \tag{4.32}$$

A more complete theory of atoms in a magnetic field has, in addition to the Hamiltonian $\hat{H}$ which is linear in B, an extra term that is quadratic in B, and which gives rise to Larmor diamagnetism. Moreover, application of perturbation theory produces a further term that comes from Van Vleck paramagnetism. Both these terms are of the order of 10^{-10} eV at $B = 1\ T$ and are negligible for atoms or ions in which $J \neq 0$. The latter are magnetic. Those for which $J = 0$ are called diamagnetic.

Exercise 4.5
Magnetic moment of the Pr^{3+} *ion*
    Determine, in units of μ_B, the magnitude of the magnetic moment of the Pr^{3+} ion in its ground state $^3H_4$.

Solution
    In this case, $L = 5, S = 1$ and $J = 4$. The Landé factor (4.29) is then $g = 4/5$ and, from (4.31),

$$\frac{|\ \mu\ |}{\mu_B} = \sqrt{20}\ \frac{4}{5} = 3.58 \ .$$

4.4 BRILLOUIN MODEL OF PARAMAGNETISM

4.4.1 Description of the Model

To describe an ideal paramagnetic substance we take as a model a system containing N identical magnetic ions fixed in a solid. We neglect the interactions among the ions or with the other ions in the solid. The sample is thus composed of a set of N paramagnetic ions characterized by the quantum number J and the Landé g factor of their ground state. When a magnetic field of flux density B is applied, this level splits into $2J + 1$ Zeeman sub-levels that are non degenerate in energy

$$\epsilon_m = mg\mu_B B \quad (m = -J, \ldots, +J) \tag{4.33}$$

(henceforth we write m instead of M_J (4.28)).

4.4.2 Distribution Relation

Since the ions are localized and hence distinguishable, Maxwell-Boltzmann statistics applies. The number of ions N_m in the state m is given by the Boltzmann relation (3.1)

$$\frac{N_m}{N} = \frac{e^{-\beta\epsilon_m}}{Z} \tag{4.34}$$

where Z is the partition function (3.2). Introduction of the dimensionless variable

$$x = Jg\ \frac{\mu_B B}{kT} \quad , \tag{4.35}$$

yields for the partition function in the Brillouin model

$$Z = \sum_{m=-J}^{J} e^{-\beta\epsilon_m} = \sum_{m=-J}^{J} e^{-mx/J} = e^x + e^{(J-1)x/J} + \ldots + e^{-x} \ . \tag{4.36}$$

The various terms in Z make up a geometrical series, the sum of which is

$$Z = e^x\ \frac{1 - e^{-(2J+1)x/J}}{1 - e^{-x/J}} = \frac{e^{(2J+1)x/2J} - e^{-(2J+1)x/2J}}{e^{x/2J} - e^{-x/2J}} \quad ,$$

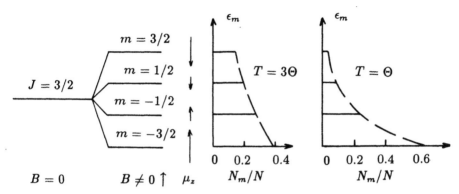

Figure 4.7: Relative populations N_m/N (4.34) of the energy levels of a system of ions with quantum number $J = 3/2$ in a magnetic field of flux density B, for two temperatures. We have set $\Theta = g\mu_B B/k$.

i.e.

$$Z = \frac{\sinh \frac{2J+1}{2J} x}{\sinh \frac{1}{2J} x} \quad . \tag{4.37}$$

Figure 4.7 shows the relative populations N_m/N (4.34) of the levels for two temperatures. It can be seen that lowering the temperature causes the upper levels to depopulate in favour of the lower ones, and orients the magnetic moments along the direction of the field. In the limit $T = 0$, only the lowest level is populated, the magnetic moment is maximum and the entropy is zero. The characteristic temperature of the level spacing, $\Theta = g\mu_B B/k$, separates the high temperature region $(T \gg \Theta)$ from that of low temperatures $(T \ll \Theta)$.

4.4.3 Thermodynamic Functions

The thermodynamic quantities, entropy, internal energy, magnetic moment, etc., can be obtained from the free energy (3.6)

$$F = -NkT \ln Z \quad ,$$

through the differential relation

$$dF = -SdT - MdB \quad .$$

In classical thermodynamics this form corresponds to the free energy of the substance, provided the self-energy of the field is excluded. The thermodynamic quantities are expressed as functions of Z in equations (3.7, 9 and 11). Substitution of expression (4.37) for Z yields for the Brillouin model

$$S/Nk = \ln \sinh \left(\frac{2J+1}{2J} x \right) - \ln \sinh \left(\frac{1}{2J} x \right) - x B_J(x) \quad , \tag{4.38}$$

$$U = F + TS = -N Jg\mu_B B B_J(x) \quad , \tag{4.39}$$

$$\text{and} \quad M = N Jg\mu_B B_J(x) \quad , \tag{4.40}$$

where

$$B_J(x) = \frac{d \ln Z}{dx} = \frac{2J+1}{2J} \coth\left(\frac{2J+1}{2J}x\right) - \frac{1}{2J}\coth\left(\frac{1}{2J}x\right) \qquad (4.41)$$

is called the Brillouin function for the number J. In the following paragraphs we shall examine in turn the magnetic moment and the entropy. The internal energy is related to M through $U = -MB$.

Exercise 4.6 *Brillouin function for $J = 1/2$*
From the general relations (4.37 and 41) for Z and B_J, rederive the particular forms that were obtained directly for $J = 1/2$ in exercise 2.2.

Solution
The general expressions give for $J = 1/2$

$$Z = \frac{\sinh 2x}{\sinh x} \quad \text{and} \quad B_{1/2}(x) = 2\coth 2x - \coth x \quad .$$

Using the relations

$$\sinh 2x = 2\sinh x \cosh x \quad \text{and} \quad \cosh 2x = \cosh^2 x + \sinh^2 x \quad ,$$

we get on substitution

$$Z = 2\cosh x \quad \text{and} \quad B_{1/2}(x) = \tanh x \quad . \qquad (4.42)$$

We thus retrieve the results of exercise 2.2, with $\overline{m} = Jg\mu_B = g\mu_B/2$.

4.4.4   Magnetic Moment

The total magnetic moment of a substance containing N magnetic ions is given by the *Brillouin relation* (4.40)

$$M = NJg\mu_B B_J(x) \quad \text{with} \quad x = Jg\frac{\mu_B B}{kT} \quad . \qquad (4.43)$$

For one mole of magnetic ions (i.e. not necessarily a mole of the substance) this becomes

$$M_M = \mathcal{N}Jg\mu_B B_J(x) \quad . \qquad (4.44)$$

The molar magnetic moment $M_M(B,T)$, which is proportional to the Brillouin function, depends only on the ratio $B/T = \mu_0 H/T$. This property defines an ideal paramagnetic material.

Brillouin function

The Brillouin functions $B_J(x)$ defined in (4.41) are displayed in Figure 4.8 for several values of J. The $B_J(x)$ are increasing functions of x and, for fixed x, $B_J(x) > B_{J'}(x)$ if $J < J'$. From the relations

$$\coth u = 1 + 2e^{-2u} + \cdots \quad (u \gg 1) \quad \text{and} \quad \coth u = \frac{1}{u} + \frac{u}{3} - \frac{u^3}{45} + \cdots \quad (u \ll 1) \quad ,$$

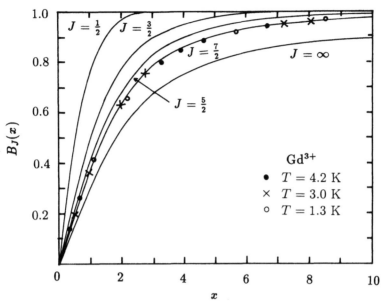

Figure 4.8: Brillouin functions for different values of J and the Langevin function corresponding to $J = \infty$. The experimental points are for gadolinium sulphate $Gd_2(SO_4)_3$, $8H_2O$ [W.E. Henry, *Phys. Rev.* **88**, 559 (1952)]. The horizontal axis is $x = M_{Ms}B/\mathcal{N}kT = Jg\mu_B/kT$, while the vertical axis is $y = M_M/M_{Ms} = B_J(x)$, where M_{Ms} is the molar magnetic moment at saturation (4.47).

we get

$$B_J(x) = 1 - \frac{1}{J}\,\exp(-x/J) + \cdots \quad (x \gg 1)$$

$$\text{and} \quad B_J(x) = \frac{J+1}{3J}x - \frac{(2J+1)^4 - 1}{45(2J)^4}x^3 + \cdots \quad (x \ll 1) \ . \qquad (4.45)$$

The functions B_J thus have a horizontal asymptote with ordinate $y = 1$ and an extended linear region in the neighbourhood of $x = 0$.

In the framework of classical mechanics P. Langevin (1905) had investigated a model that was similar to that of L. Brillouin (1927). The Langevin function

$$L(x) = \lim_{J \to \infty} B_J(x) = \coth x - \frac{1}{x} \qquad (4.46)$$

was used instead of that of L. Brillouin. This function, obtained by integrating over unquantized orientations, is no longer used in magnetism, but remains a good approximation in investigations of dielectric properties (Ex. 5.4).

Saturation moment

When $\mu_B B \gg kT(x \gg 1)$, the molar magnetic moment tends to the limiting value

$$M_{Ms} = \mathcal{N}Jg\mu_B \ . \qquad (4.47)$$

This saturation phenomenon can be interpreted at the atomic scale by the "alignment" of the magnetic moments along the direction of the field. More precisely, all the ions occupy the state in which the value of the projection of the moment along the field is maximum, $m = -J$ and $(\mu_z)_{max} = Jg\mu_B$, and therefore $M_{Ms} = \mathcal{N}(\mu_z)_{max}$. This condition is reached at low temperatures $T \lesssim 1$ K and in high fields $B \gtrsim 1$ T. Saturation corresponds to a minimum in the molar energy $u_s = -M_{Ms}B$.

Curie's law

When $\mu_B B \ll kT (x \ll 1)$, it can be seen from (4.45) that the molar magnetic moment takes the form

$$M_M = \mathcal{N}\frac{J(J+1)g^2\mu_B^2}{3k}\frac{B}{T} \quad . \tag{4.48}$$

The magnetic susceptibility χ_M, defined in (4.4,5) then becomes

$$\chi_M = M_M\frac{\mu_0}{B} = \mu_0\mathcal{N}\frac{J(J+1)g^2\mu_B^2}{3kT} \quad . \tag{4.49}$$

Curie's law (4.8) $\chi_M = C_M/T$ is obeyed with

$$C_M = \mu_0\mathcal{N}\frac{J(J+1)g^2\mu_B^2}{3k} \quad . \tag{4.50}$$

Since the magnitude μ of the magnetic moment of an ion is given by (4.31), the Curie constant can be written in the form

$$C_M = \mu_0\mathcal{N}\frac{\mu^2}{3k} \quad . \tag{4.51}$$

Standard experiments using paramagnetic substances are generally conducted in the region where Curie's law applies. The transformations for the units of the Curie constant are the same as those for susceptibilities (4.7).

Experimental discussion

The Brillouin relation has been shown to be valid for several substances where the magnetic moments behave as if they are free down to temperatures of about 1 K. The experimental results for gadolinium sulphate are displayed in Figure 4.8. The magnetic moment per mole of Gd^{3+} ions has a saturation value $M_{Ms} = 39.1$ A m^2 mol^{-1} [W.E. Henry, *Phys. Rev.* **88**, 559 (1952)]. From this result the product Jg can be determined, where

$$Jg = \frac{M_{Ms}}{\mathcal{N}\mu_B} = 7.0 \quad . \tag{4.52}$$

The numerical value of x (4.43) is then

$$x = Jg\frac{\mu_B B}{kT} = 0.59 \times 10^{-5}\frac{H\,/(A\,m^{-1})}{T\,/K} \quad .$$

The experimental values of M_M/M_{Ms} as a function of x lie on the curve $B_J(x)$ for $J = 7/2$, which, from (4.52), yields $g = 2.0$. These results are in agreement with the spectroscopic data, i.e., the ground state of the Gd^{3+} ion in the $4f^7$ configuration is $^8S_{7/2}(L = 0, S = 7/2, J = 7/2)$, and its Landé factor (4.29) is therefore $g = 2$.

In cases where the magnetic moment can be measured up to saturation, the above method provides a means of determining the effective values of J and g, which we denote by J_{eff} and g_{eff}. Since magnetic saturation experiments are difficult to perform, experimental data are generally obtained for the Curie law region. Measurement of the Curie constant then yields only the product

$$p_{eff} = [J_{eff}(J_{eff} + 1)]^{1/2} g_{eff} \quad , \tag{4.53}$$

the so-called effective paramagnetic Bohr magneton number. As will be seen in paragraph 4.5, p_{eff} is often different from the theoretical value for the free ion, on account of the crystal field.

4.4.5 Entropy

In the Brillouin model, the entropy of a substance containing N magnetic ions is given by (4.38)

$$S/Nk = \ln \sinh\left(\frac{2J+1}{2J}x\right) - \ln \sinh\left(\frac{1}{2J}x\right) - xB_J(x) \; .$$

This is a decreasing function of $x = Jg\mu_B B/kT$.

The limiting value of the entropy in zero field ($B \to 0$) or at high temperature is

$$\lim_{x \to 0} S = Nk \ln(2J + 1) \quad . \tag{4.54}$$

This corresponds to the fact that in zero field the $2J + 1$ levels are degenerate, and the thermodynamic probability W is $(2J + 1)^N$. The Boltzmann relation, $S = k \ln W$, then yields expression (4.54).

When the temperature tends to zero in a non zero magnetic field, the entropy (4.38) tends to zero. This reflects the fact that the lowest level, which is non degenerate, is the only one to be populated (Fig. 4.7). The entropy function will be used in the discussion of adiabatic demagnetization (§4.6).

The Brillouin model applies to localized atoms. Nevertheless, the Brillouin relation for the magnetic moment is valid for the few cases of gases whose molecules have a permanent magnetic moment, such as O_2, NO and NO_2.

4.5 PROPERTIES OF PARAMAGNETIC SOLIDS

4.5.1 Transition Element Ions

In ionic solids, most of the ions have full electron layers (Na^+, Cl^-, O^{2-}, etc.) and therefore have no permanent moment. The majority of solids are therefore diamagnetic. The only ions that give rise to paramagnetic phenomena belong to the transition series, in which a d or f sub-layer is incomplete. Three series correspond to the filling of the d sub-layers, i.e. that of iron (3d), palladium (4d) and platinum (5d). Two series correspond to the f sub-layers, i.e. the rare-earth group (4f) and the actinides (5f). The salts that are most frequently used for their magnetic properties are those containing ions from the iron or the rare-earth groups. We shall examine these more in detail, and take into account their interactions with the crystal.

4.5.2   Crystal Field

When the ions in a crystal are sufficiently far apart, their magnetic interactions can be neglected down to quite low temperatures. Interactions with the diamagnetic ions of their immediate surroundings in the crystal, however, cannot be overlooked.

The magnetic ion is in fact surrounded by a crystal field of electrostatic origin. The $2J + 1$ different states of a given atomic level correspond to different electronic charge distributions and are perturbed differently by the crystal field. This results in a lifting of the degeneracy, and the splitting is more pronounced if the symmetry of the crystal field is low. H.A. Kramers (1930), however, showed that an ion with an odd number of electrons (half-integer spin quantum number) has levels that remain at least doubly degenerate, states with opposing magnetic moments being associated in pairs. This degeneracy, related to time reversal symmetry $(t \rightarrow -t)$, can be lifted only by a magnetic field acting on the ion. Singlet (non degenerate) levels can thus exist only for ions having an even number of electrons.

For each level of the ion in the crystal the notion of an effective spin $\tilde{S}$ is introduced, chosen in such a way that $2\tilde{S} + 1$ is equal to the degeneracy of the level in zero field. At the same time a spectroscopic splitting factor $\tilde{g}$ is also introduced such that in the presence of a field the energies of the Zeeman sub-levels are, by analogy with (4.30 and 32),

$$\epsilon_m = -\, \boldsymbol{\mu}\, .\mathbf{B} = \tilde{g}\mu_B \tilde{\mathbf{S}} \cdot \mathbf{B} = \tilde{m}\tilde{g}\mu_B B \ , \tag{4.55}$$

where $\tilde{m} = -\tilde{S}, \ldots, \tilde{S}$. For example, for a *Kramers doublet*, we set $\tilde{S} = 1/2$ and $\tilde{m} = \pm 1/2$. It is important to notice that the effective spin $\tilde{S}$ and the factor $\tilde{g}$ are not generally equal to the angular momentum and the Landé factor of the ground state of the free ion. We shall see applications of these concepts later on.

A large number of spectroscopic investigations have been directed to the study of the energy levels of magnetic ions in crystals, and in particular to the measurement of $\tilde{S}$ and $\tilde{g}$. In general, a magnetic field of flux density B is applied to the crystal, splitting the energy levels into $2\tilde{S} + 1$ sub-levels that are separated by $\Delta\epsilon = \tilde{g}\mu_B B$. When an electromagnetic wave passes through the material, resonant absorption occurs at frequencies $h\nu = \Delta\epsilon$. Measurement of the energy of the absorbed or emitted photons thus allows $\tilde{g}$ and $\tilde{S}$ to be determined. In optical spectroscopy, the range explored extends from far infrared ($\simeq 10 \text{ cm}^{-1}$) to ultraviolet ($\simeq 50\ 000$ cm^{-1}). In electron paramagnetic resonance (EPR), only the range between zero and a few cm^{-1} is covered, but with a higher precision.

It should be noted that the Zeeman effect is not in general isotropic, i.e. ϵ_m varies with the orientation of B. Hence $\tilde{g}$ should be considered as a symmetric tensor of order 2, such that

$$\epsilon_m = \mu_B \tilde{S}[\, \tilde{g}\,]\mathbf{B} \quad \text{or} \quad \epsilon_m = \mu_B \tilde{g}_{ij} \tilde{S}_i B_j \; . \tag{4.56}$$

When this tensor is expressed in its principal axes its three principal values are $\tilde{g}_{xx}, \tilde{g}_{yy}$ and $\tilde{g}_{zz}$. If the surroundings of the ion have a symmetry axis of order higher than 2, then $\tilde{g}_{xx} = \tilde{g}_{yy} = \tilde{g}_\perp$ and $\tilde{g}_{zz} = \tilde{g}_\parallel$. If the surroundings have cubic symmetry, then $\tilde{g}_{xx} = \tilde{g}_{yy} = \tilde{g}_{zz} = \tilde{g}$.

4.5.3   Rare Earth Ions

Rare earth ions are relatively simple in that the incomplete $4f$ layer is "protected" from the influence of the crystal field by the complete $5s$ and $5p$ sub-layers lying outside. The influence of the crystal field, which is weaker than the spin-orbit interaction, then enters as a perturbation of the L, S, J levels in the free ion. For a level J, the result is a lifting of the degeneracy of degree $2J + 1$, into sub-levels whose total splitting is of the order of 100 cm^{-1}, i.e. about 100 K.

For example, according to Hund's rule, the ground state of the free terbium ion $Tb^{3+}(4f^8)$ is $^7F_6$, with the first excited level $^7F_5$ lying at about 2 000 cm^{-1}. In the ethylsulphate crystal Tb $(C_2H_5SO_4)_3$, 9 H$_2$O (Pb. 3.1), the Tb^{3+} ion has a threefold symmetry axis and the crystal field splits its ground state into 9 singlet or doublet sub-levels with energy 0 cm^{-1}, 88 cm^{-1}, ... 131 cm^{-1}. The lowest sub-level is a doublet that exhibits a very slight splitting, 0.387 cm^{-1} (due to higher orders in the perturbation). If an effective spin $\tilde{S} = 1/2$ is attributed to this doublet, paramagnetic resonance yields for the value of the spectroscopic splitting factor $\tilde{g} = 17.82$. To determine the origin of this doublet, the splitting $\Delta\epsilon = \tilde{g}\mu_B B$ between the two Zeeman levels (4.55) is identified with the theoretical splitting $\Delta\epsilon = \Delta m g \mu_B B$ between the two Zeeman levels of the free ion (4.28), where $g = 1.5$ is the Landé factor (4.29) of the $^7F_6$ level of the free ion. This yields $\Delta m = \tilde{g}/g = 11.88$. This result, close to 12, indicates that the lowest doublet contains the two states $|\pm m >= |\pm 6 >$. Crystal field theory confirms that these two states must form a doublet. Observe that in this uniaxial crystal, $\tilde{g}$ is anisotropic with $\tilde{g}_\parallel = 17.82$ and $\tilde{g}_\perp \simeq 0$.

At ordinary temperatures, susceptibility measurements provide a determination of the Curie constant in rare earth salts, and hence yield the corresponding effective paramagnetic Bohr magneton number. The latter are given in Table 4.2. At 300 K, the crystal field splitting of the sub-levels corresponding to the ground state of the free ion is much smaller than kT, and the ion behaves as if it were free. From (4.53)

$$p_{th} = [J(J+1)]^{1/2} g \; .$$

Table 4.1 compares the experimental and theoretical values of p. The agreement is satisfactory except for Sm^{3+} and Eu^{3+} ions, whose free ions have excited electronic levels close to the ground state.

When kT is of the same order as the crystal field splitting of the sub-levels, the interpretation becomes complicated because the effect of the magnetic field has to

Table 4.2: Effective paramagnetic Bohr magneton number for rare earth ions at ordinary temperatures. n is the number of electrons in the 4f sub-layer. The values of p_{exp} are those given by R. Kubo and T. Nagamiya, Solid State Physics, McGraw-Hill (1969). Promethium Pm has no stable isotope.

Ion	n	S	L	J	Ground state	g	$p_{th} = [J(J+1)]^{1/2}g$	p_{exp}
Ce^{3+}	1	1/2	3	5/2	$^2F_{5/2}$	6/7	2.54	2.4
Pr^{3+}	2	1	5	4	$^3H_4$	4/5	3.58	3.5
Nd^{3+}	3	3/2	6	9/2	$^4I_{9/2}$	8/11	3.62	3.5
Pm^{3+}	4	2	6	4	$^5I_4$	3/5	2.68	-
Sm^{3+}	5	5/2	5	5/2	$^6H_{5/2}$	2/7	0.84	1.5
Eu^{3+}	6	3	3	0	$^7F_0$	-	0	3.4
Gd^{3+}	7	7/2	0	7/2	$^8S_{7/2}$	2	7.94	8.0
Tb^{3+}	8	3	3	6	$^7F_6$	3/2	9.72	9.5
Dy^{3+}	9	5/2	5	15/2	$^6H_{15/2}$	4/3	10.63	10.6
Ho^{3+}	10	2	6	8	$^5I_8$	5/4	10.60	10.4
Er^{3+}	11	3/2	6	15/2	$^4I_{15/2}$	6/5	9.59	9.5
Tm^{3+}	12	1	5	6	$^3H_6$	7/6	7.57	7.3
Yb^{3+}	13	1/2	3	7/2	$^2F_{7/2}$	8/7	4.54	4.5

be considered for several sub-levels. Curie's law is no longer valid. When kT is smaller than the splitting of the two lowest sub-levels, only the lowest sub-level needs to be taken into account, and the value for p is then equal to $[\tilde{S}(\tilde{S}+1)]^{1/2}\tilde{g}$ (Problem 3.1).

4.5.4   Ions of the Iron Group

For ions of the transition series, the incomplete $3d, 4d$ or $5d$ electronic layer is outermost and the crystal field produces a stronger perturbation than that due to spin-orbit coupling (§4.3.2). The effect of this field on the ionic terms should therefore be treated before that of spin-orbit coupling.

The crystal field lifts the degeneracy connected with the orbital moment of the term, giving splittings of about 10^4 cm^{-1}. In the resulting levels, the projection of the angular momentum on an axis is no longer constant, and its mean value is zero for singlets. The same is true for the magnetic moment associated with the orbital motion. The orbital angular momentum is said to be "quenched" (i.e. extinguished). The spin, which can orient freely, then provides the only contribution to the magnetic moment of the ion. The spin degeneracy, equal to $2S+1$, is lifted by the spin-orbit coupling, which then yields only very small splittings ($\simeq 0.1$ cm^{-1}). If the level is not a singlet, the orbital angular momentum is no longer completely quenched, and spin-orbit coupling is stronger.

As an example we take the chromium ion Cr^{3+} ($3d^3$). From Hund's rule, the ground state in the free ion is $^4F_{3/2}$ and comes from the term ^4F. In chromium cesium alum, the Cr^{3+} ion has threefold symmetry and the crystal field lifts the

Table 4.3: Effective paramagnetic Bohr magneton number for iron group ions. n is the number of electrons in the $3d$ sub-layer. The experimental values are from R. Kuboand T. Nagamiya,*Solid State Physics*, McGraw-Hill (1969).

Ion	n	S	L	J	Ground state	$p_{th} = 2[S(S+1)]^{1/2}$	p_{exp}
Ti^{3+}	1	1/2	2	3/2	$^2D_{3/2}$	1.73	
V^{4+}	1	1/2	2	3/2	$^2D_{3/2}$	1.73	1.8
V^{3+}	2	1	3	2	$^3F_2$	2.83	2.8
V^{2+}	3	3/2	3	3/2	$^4F_{3/2}$	3.87	3.8
Cr^{3+}	3	3/2	3	3/2	$^4F_{3/2}$	3.87	3.7
Mn^{4+}	3	3/2	3	3/2	$^4F_{3/2}$	3.87	4.0
Cr^{2+}	4	2	2	0	$^5D_0$	4.90	4.8
Mn^{3+}	4	2	2	0	$^5D_0$	4.90	5.0
Mn^{2+}	5	5/2	0	5/2	$^6S_{5/2}$	5.92	5.9
Fe^{3+}	5	5/2	0	5/2	$^6S_{5/2}$	5.92	5.9
Fe^{2+}	6	2	2	4	$^5D_4$	4.90	5.4
Co^{2+}	7	3/2	3	9/2	$^4F_{9/2}$	3.87	4.8
Ni^{2+}	8	1	3	4	$^3F_4$	2.83	3.2
Cu^{2+}	9	1/2	2	5/2	$^2D_{5/2}$	1.73	1.9

orbital degeneracy, the lowest level being an orbital singlet. The orbital angular momentum is quenched and the lowest level has a spin degeneracy of $2S + 1 = 4$. Paramagnetic resonance measurements show that the lowest level is indeed a quadruplet ($\tilde{S} = 3/2$) and that the spectroscopic splitting factor $\tilde{g} = 1.98$ is very close to 2, i.e., the value expected for a spin alone. The small difference between these values comes from the residual spin-orbit interaction, which also causes the lowest quadruplet to split into two Kramers doublets $| \pm 3/2 >$ and $| \pm 1/2 >$, separated by $\delta = 0.133$ cm^{-1}.

At ordinary temperatures measurement of the susceptibility of salts of the iron group yields the value of the Curie constant and hence the effective paramagnetic Bohr magneton number. Table 4.3 compares the experimental values p_{exp} of this number and the theoretical values for the spin alone

$$p_{th} = [S(S+1)]^{1/2}g = 2[S(S+1)]^{1/2} \ .$$

The agreement is satisfactory except for Co^{2+}, the lowest level of which is not an orbital singlet.

For ions having no orbital degeneracy in the ground state ($L = 0$) or whose orbital angular momentum is quenched in the crystal, Curie's law holds down to very low temperatures (~ 1 K). This is the case for Cr^{3+}, Mn^{2+} and Fe^{3+} ions.

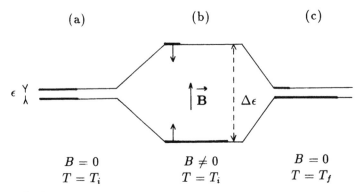

Figure 4.9: Outline of the principle of adiabatic demagnetization. (a) → (b) isothermal magnetization, (b) → (c) adiabatic demagnetization. The thick horizontal lines represent the populations of the energy levels.

4.6 ELECTRONIC ADIABATIC DEMAGNETIZATION

4.6.1 Introduction

The method of electronic adiabatic demagnetization was suggested independently by P. Debye (1926) and W.F. Giauque (1927). In view of the experimental difficulties the first experimental results started to appear only in 1933 (W.F. Giauque et al., 1933, W.J. de Haas et al., 1933, N. Kurti et al., 1934). This method, which uses liquid helium at about 1 K as the first cooling stage, routinely yields temperatures of the order of 10^{-2} K, and even down to 10^{-3} K.

4.6.2 Description of the Method

We consider the simplified case of a substance containing non-interacting magnetic ions whose ground state in the crystal is a doublet ($\tilde{S} = 1/2$) that is well separated from the first excited level. We assume that the two components of this doublet have an energy splitting $\epsilon = hc\delta = k\Theta$ with $\Theta \ll 1$ K ($\delta \ll 1$ cm^{-1}). In the temperature region considered, the energy, entropy and heat capacity are those of the system of magnetic ions, since the crystal lattice contribution is negligible.

In the initial state the specimen is placed at zero magnetic field in a bath of liquid helium under reduced pressure at temperature $T_i \simeq 1$ K. In this state the ratio of the populations of the two components of the doublet is (3.3)

$$\frac{N_2}{N_1} = \exp\left(-\frac{\epsilon}{kT_i}\right) = \exp\left(-\frac{\Theta}{T_i}\right). \tag{4.57}$$

The populations are thus practically equal, since $T_i \gg \Theta$ (Fig. 4.9a) and the magnetization is negligible. The molar entropy of the ions is (4.54) $s = R\ln 2$.

In the first step (ab), a magnetic flux density B of the order of 1 T is established, while the sample is kept in contact with the bath. The magnetic field splits the two components by $\Delta\epsilon = g\mu_B B \sim 10^{-23}$ J $\sim 10^{-4}$ eV, which corresponds to a characteristic temperature $\Delta\epsilon/k \sim 1$ K very much greater than Θ (Fig. 4.9b). The populations of the levels then readjust to give

$$\frac{N_2}{N_1} \simeq \exp\left(-\frac{\Delta\epsilon}{kT_i}\right) = \exp\left(-\frac{g\mu_B B}{kT_i}\right). \tag{4.58}$$

The ground state therefore has the greater population and the sample develops magnetization along the direction of the field. During this isothermal transformation the energy and the entropy decrease, and heat is liberated.

In the second stage (bc) the field is reduced to $B = 0$, after the sample has been isolated from the helium bath. The same energy levels as in the initial state a are present (Fig. 4.9c), but these now have the populations of state b, since in the adiabatic transformation (bc) the populations do not change (§1.6.2). The final temperature of the system of spins T_f can then be determined by the Boltzmann relation

$$\frac{N_2}{N_1} = \exp\left(-\frac{\epsilon}{kT_f}\right) = \exp\left(-\frac{\Theta}{T_f}\right). \tag{4.59}$$

Comparison with (4.58) shows that T_f is given by

$$T_f = T_i \frac{k\Theta}{g\mu_B B}. \tag{4.60}$$

With the typical values $B = 1$ T, $g = 2$, $\Theta = 0.01$ K and $T_i = 1$ K we obtain $T_f = 7.4 \times 10^{-3}$ K.

Adiabatic demagnetization has very important applications, since in addition to the fact that the temperature of paramagnetic samples can be reduced to $T \simeq \Theta \simeq 10^{-2}$ K, these substances can be used as thermostats for cooling other substances. Indeed, near $T = \Theta$, a sample like that just described has a heat capacity with a Schottky anomaly (§3.4.3) the molar value of which is of the order of R. This is much larger than that of other materials at the same temperature. Adiabatic demagnetization thus opens up a new temperature range.

The reader will observe that, during the adiabatic transformation, the populations of the levels, and hence the magnetization of the sample, remain constant. The term demagnetization, widely accepted for describing this transformation, is thus incorrect.

4.6.3   Entropy Diagram

The above discussion can be illustrated by referring to the entropy diagram showing the molar entropy s as a function of the temperature T for different values of the magnetic induction B. Keeping to the two level model discussed above, the molar entropy is (3.22)

$$\frac{s}{R} = \ln(1 + e^{-y}) + \frac{y}{1 + e^y}, \quad \text{with} \quad y = \frac{\epsilon + g\mu_B B}{kT}. \tag{4.61}$$

Figure 4.10 is an entropy diagram in which are shown two curves calculated from ·(4.61) with $B = 0$ and $B \neq 0$. The isothermal magnetization (ab) appears in this

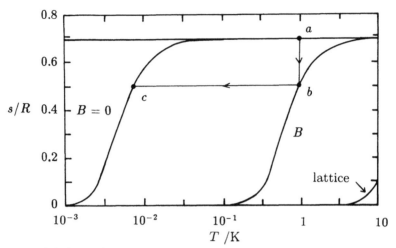

Figure 4.10: Adiabatic demagnetization in an entropy diagram. The two curves show the molar entropy (4.61) at $B = 0$ and $B = 1$ T. The values taken for the parameters are $\Theta = \epsilon/k = 10^{-2}$ K, $J = 1/2$ and $g = 2$. With $T_i = 1$ K, a final temperature $T_f = 7.4 \times 10^{-3}$ K is reached. The entropy of the lattice, also shown, is negligible below 3 K.

figure as a vertical line, since the entropy decreases at constant temperature. The adiabatic demagnetization (bc), represented by a horizontal line, corresponds to the sample being cooled at constant entropy.

The above discussion involving a simple model can be generalized to real substances in which the entropy curves have the same shape. In particular, in the temperature range covered, the approximate expression for the molar entropy (4.61) in zero field, namely

$$\frac{s}{R} \simeq \ln 2 - \frac{y^2}{8} = \ln 2 - \frac{\Theta^2}{8T^2}, \qquad (4.62)$$

remains valid, provided that $\ln 2$ is replaced by $\ln(2\tilde{S} + 1)$. The parameter Θ is obtained experimentally by measuring the heat capacity. This gives (cf. 3.25)

$$c_V = T\frac{ds}{dT} = \frac{R}{4}\frac{\Theta^2}{T^2} = \frac{A}{T^2} . \qquad (4.63)$$

The effective parameter Θ found from these measurements includes the contributions from the level splitting in the crystal field as well as those from magnetic interactions between the ions and from the hyperfine structure.

Similarly, in non-zero field, expression (4.61) for the molar entropy

$$\frac{s}{R} \simeq \ln 2 - \frac{g^2\mu_B^2 B^2}{8k^2T^2} \qquad (4.64)$$

generalizes to

$$\frac{s}{R} \simeq \ln(2\tilde{S} + 1) - \frac{C_M}{2\mu_0 R}\frac{B^2}{T^2} , \qquad (4.65)$$

where C_M is the Curie constant (4.50) per mole of ions.

Table 4.4: Properties of salts used in adiabatic demagnetization. C_M is the Curie constant per mole of magnetic ions, A is the constant in relation (4.63). T_f/T_i is the ratio of final to initial temperatures in an adiabatic demagnetization, evaluated approximately from equation (4.84), with $B_i = 1$ T [A.H. Cooke, *Prog. Low Temp. Phys.* Vol. I, p. 224, North-Holland 1955]. For Cr^{3+} ions, see [E. Ambler and R. P. Hudson, *J. Chem. Phys.* **27**, 378 (1957)].

Ion	Substance	$\rho/$ 10^3 kg m^{-3}	$M/$ 10^{-3} kg mol^{-1}
Mn^{2+}	Mn $(NH_4)_2$ $(SO_4)_2$, $6H_2O$	1.83	391
Fe^{3+}	Fe (NH_4) $(SO_4)_2$, $12H_2O$	1.70	482
Gd^{3+}	Gd_2 $(SO_4)_3$, $8H_2O$	3.01	2×373
Cr^{3+}	Cr $(NH_3 CH_3)$ $(SO_4)_2$, $12H_2O$	1.66	492
Cu^{2+}	Cu K_2 $(SO_4)_2$, $6H_2O$	2.22	442
Ce^{3+}	Ce_2 Mg_3 $(NO_3)_{12}$, $24H_2O$	2.0	2×765

Ion	$\tilde{S}$	$\tilde{g}$	C_M $/(4\pi.10^{-6}$ K m^3 mol^{-1})	A/R	T_f/T_i $(B_i = 1$ T$)$
Mn^{2+}	5/2	2.00	4.38	3.4×10^{-2}	8.0×10^{-2}
Fe^{3+}	5/2	2.00	4.37	1.3×10^{-2}	5.0×10^{-2}
Gd^{3+}	7/2	1.99	7.82	3.7×10^{-1}	1.9×10^{-1}
Cr^{3+}	3/2	1.98	1.87	1.9×10^{-2}	9.2×10^{-2}
Cu^{2+}	1/2	2.45 (∥) 2.14 ($\perp$)	0.445_{powder}	6.0×10^{-4}	3.3×10^{-2}
Ce^{3+}	1/2	0.25 (∥) 1.84 ($\perp$)	0.318 ($\perp$)	7.5×10^{-6}	4.4×10^{-3}

4.6.4 Choice of Paramagnetic Substance

The choice of substance to be used in an adiabatic demagnetization experiment is very important. The entropy diagram of Figure 4.10 shows clearly that a reduction in the initial temperature T_i, or an increase in the initial induction B, produces only minor changes in the final temperature T_f, since the entropy of the sample in zero field varies strongly in the neighbourhood of Θ, and hence $T_f \sim \Theta$. To obtain the lowest temperatures it is therefore necessary to use a material whose value of Θ or of $A = R\Theta^2/4$ is as small as possible. Furthermore, a large Curie constant C_M is advantageous since it reduces the value of the entropy at the point b (Fig. 4.10).

In general the materials used are double hydrate salts, such as alums, in which the magnetic ions are well separated and interact only weakly. Dilute magnetic crystals are also used, consisting of a diamagnetic substance doped with paramagnetic ions. The magnetic ions generally chosen are manganese Mn^{2+}, iron Fe^{3+} and gadolinium Gd^{3+}, since, with a half-full electronic layer, their orbital moment is zero (Table 4.2,3). Chromium Cr^{3+} , whose orbital moment is quenched, is also used, as well as copper Cu^{2+} and cerium Ce^{3+} whose ground state is a Kramers doublet that is split only by the magnetic interactions. Table 4.4 lists the principal properties of the salts most commonly used in adiabatic demagnetization.

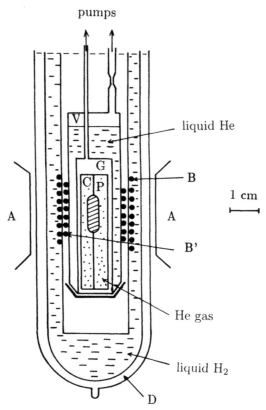

pumps

liquid He

B

1 cm

A

A

B'

He gas

liquid H$_2$

D

Figure 4.11: Cryostat for adiabatic demagnetization. D = Dewar vessel, V = helium jacket, C = cell containing the paramagnetic sample P, A = magnet, G = space containing the exchange gas, B and B' = mutual inductance coils acting as thermometer [A.H. Cooke and R.A. Hull, *Proc. Roy. Soc. A* (London) **181**, 83 (1942)].

4.6.5 Adiabatic Demagnetization Experiments

Figure 4.11 shows an example of a cryostat. A Dewar vessel D filled with hydrogen acts as a thermal shield for a metal container V containing liquid helium. The vapour pressure of the helium is reduced by a powerful pump which maintains the temperature close to 1 K. The paramagnetic sample P is suspended by threads inside the cell C which contains gaseous helium, thus allowing heat to be exchanged. An electromagnet A (or superconducting coil) generates a field that magnetizes the sample. This magnetization is carried out isothermally, with the exchange gas in the free volume G providing thermal contact between the sample and the liquid helium.

To perform the adiabatic demagnetization, the cell C is isolated from the liquid helium by pumping the exchange gas out of G, then the cryostat is removed from the field of the electromagnet. The final temperature reached is found by measuring the susceptibility (§4.2.3) by means of coils B and B'. The sample P cools to a temperature of about 10 mK. Since its molar heat capacity at these temperatures is very large (~ 0.1 R), the sample can be used for cooling and investigating other substances in this temperature range.

4.7 NUCLEAR ADIABATIC DEMAGNETIZATION

4.7.1 Principle

The principle of nuclear adiabatic demagnetization or cooling was proposed independently by C.J. Gorter (1934) and by N. Kurti and F.E. Simon (1935). To reach the temperature range below one millikelvin, these authors suggested applying the adiabatic demagnetization method to the magnetic moments of the nuclei, rather than to the electronic magnetic moments of atoms.

Qualitatively, electronic and nuclear adiabatic demagnetization are similar, and the previous formalism remains valid. There is, however, a significant quantitative difference in that the magnetic moments of nuclei are of the order of magnitude of the nuclear magneton

$$\mu_N = \frac{e\hbar}{2m_p} = 5.051 \times 10^{-27} \text{ J T}^{-1} \tag{4.66}$$

the value of which is $m_p/m_e = 1836$ times smaller than that of the Bohr magneton μ_B (m_p is the mass of the proton). In particular, identical values of the entropy can be obtained for values of B/T approximately 1000 times greater. The initial conditions for nuclear cooling are thus more difficult to establish, as they require lower initial temperatures ($T_i \sim 10$ mK) and stronger magnetic fields ($B_i \sim 5$ T).

As we saw previously, the final temperature depends on the entropy of the system of nuclear spins in zero field. The change in entropy comes from Schottky anomalies resulting from magnetic dipole interactions between the nuclei. The latter are described by an internal magnetic flux density

$$b \sim \frac{\mu_0}{4\pi} \frac{|\mu|}{r^3} \tag{4.67}$$

the value of which is approximately 10^{-3} T. Thus the final temperatures that can be reached by this method are, by analogy with (4.60),

$$T_f = T_i \frac{b}{B_i} \sim 10^{-6} \text{ K} . \tag{4.68}$$

The microkelvin temperature range thus becomes accessible. The first experiment on nuclear magnetic cooling, performed by N. Kurti et al.(1956), reached a temperature of 20 μK in a system of nuclear spins.

4.7.2 Example of Copper

Copper is a substance that has frequently been used for nuclear cooling [W.J. Huyscamp and O.V. Lounasmaa, *op. cit.*]. Its two stable isotopes have nuclear spin $J = 3/2$ and mean spectroscopic splitting factor $g = 1.5$. Since the distance between the nuclei of the nearest neighbour atoms is $r = 1.63$ Å, the internal magnetic

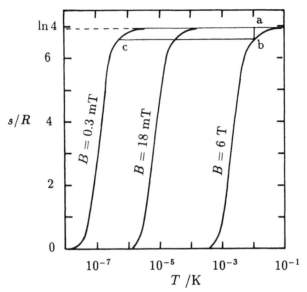

Figure 4.12: Entropy diagram for nuclear adiabatic demagnetization in copper ($J = 3/2$, $g = 1.5$). Starting from $T_i = 10$ mK (point b) and $B_i = 6$ T, the spin system can reach a final temperature $T_f = 0.5$ μK when the external field is removed (point c). It is, however, advantageous to interrupt the demagnetization at $B = 18$ mT (see text).

field b, calculated from (4.67), is 3.4×10^{-4} T, which is close to the experimental value $b = 3 \times 10^{-4}$ T.

Figure 4.12 shows the entropy (4.38) as a function of temperature for the spin system in an external field of flux density $B_i = 6$ T and for a field $B = b$ (zero external field). Isothermal magnetization at temperature $T_i = 10$ mK up to the magnetic field B_i is indicated by the segment ab. In these conditions it can be seen that the relative reduction in entropy is very small. The adiabatic demagnetization represented by the segment bc, however, generates a temperature $T_f = 0.5$ μK in the spin system.

The above result assumes that the spin system is decoupled from the conduction electrons and the lattice. In reality heat exchange occurs between nuclei and electrons with a time constant τ_1 (called the "spin-lattice" relaxation time) defined by

$$\frac{d}{dt}\left(\frac{1}{T_N}\right) = -\frac{1}{\tau_1}\left(\frac{1}{T_N} - \frac{1}{T_e}\right) \tag{4.69}$$

where T_e is the temperature of the system of electrons and lattice, and T_N is the instantaneous temperature of the spin system. This relaxation time is proportional to $1/T_e$

$$\tau_1 T_e = \kappa = 0.4 \text{ s K} \quad , \tag{4.70}$$

since the electrons involved in the exchange lie close to the Fermi level and their number is proportional to T_e (§10.2). κ is called the Korringa constant [J. Korringa, *Physica* **16**, 601 (1950)].

Under the influence of this exchange alone, the final equilibrium temperature would be very close to T_f, since the heat capacity of the lattice and the electrons is very small. The inevitable heat leaks $\dot{Q}$ must, however, be taken into account. This leads to a regime in which the nuclei have a temperature $T_N \simeq T_f$, while the electrons and the lattice have a higher temperature T_e. The latter is found by equating the heat flow from the leaks to that entering the nuclei

$$\dot{Q} = C_N \dot{T}_N \ ,$$

where C_N is the heat capacity of the nuclear spin system. Substituting (4.69,70) into this relation yields

$$T_e = T_N \left[1 + \frac{\dot{Q}\kappa}{C_N T_N^2} \right] \ . \tag{4.71}$$

Taking the numerical value $C_N/R = 0.14$ and a heat leak $\dot{Q} = 10^{-9}$ W mol^{-1}, we get (with $T_N \simeq T_f$) $T_e = 700 \ \mu$K. This is the important temperature if the intention is to cool another substance. Note that it can be maintained for only a few minutes.

The lowest possible temperature T_e is not obtained with the lowest value of T_N, but with the temperature

$$T_N = \left[\frac{\dot{Q}\kappa}{C_N} \right]^{1/2} = 30 \ \mu\text{K} \tag{4.72}$$

which minimizes expression (4.71) for T_e (here we have taken the value of κ for a field $B > 10^{-2}$ T, $\kappa = 1.1$ s K). We then have $T_e = 60 \ \mu$K. To achieve this temperature T_N, demagnetization should be arrested when the flux density reaches the value

$$B = B_i \frac{T_N}{T_i} = 1.8 \times 10^{-2} \text{ T} \ .$$

The minimum value of T_e is a compromise between obtaining a sufficiently low value of T_N and maintaining a large (negative) energy in the spin system. It follows that the temperature T_e obtained in this way can be conserved for a long time (several hours).

4.7.3 Experimental Developments

The above discussion shows that the lowest temperatures are obtained for materials having a large heat capacity, i.e. with a high nuclear Curie constant, and a small Korringa constant κ (4.70). Furthermore, in order to cool other materials, the substance should have a high thermal conductivity, and, in addition, have excellent mechanical properties. These conditions exist only in metals or alloys that are not superconducting. Copper, an optimum case, is the material that is used most frequently. At present it is common to have two nuclear cooling stages, for example using the alloy PrNi$_5$ for the first and copper for the second.

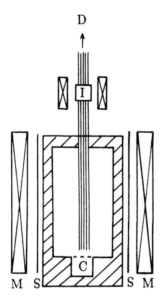

Figure 4.13: Outline of the final stage of a nuclear refrigerator. D = dilution refrigerator stage, I = superconducting heat switch, M and S = magnets. The cell is filled with copper shavings immersed in helium. The lower space C provides an experimental compartment.

Figure 4.13 shows a sketch of an experimental cell that uses copper shavings as the refrigerant for investigations into ^{3}He-^{4}He mixtures [D.I. Bradley et al., *J. Low Temp. Phys.* **57**, 359 (1984)]. The cell is placed in thermal contact with the mixing chamber D of a dilution refrigerator (§9.7) by means of a superconducting thermal switch I (§15.2.6). This device brings the cell to an initial temperature of about 15 mK. The superconducting magnet M generates an external field of flux density $B = 7$ T, while thermal contact with the dilution refrigerator is maintained. Thermal contact is then broken and the field from magnet M is removed, leaving only the weak final field $B = 0.35$ T produced by the auxiliary solenoid S. In this way a temperature of 0.85 mK is reached in the cell containing the copper shavings immersed in the ^{3}He. With a heat leak of 5 nW, the cell can remain below 3 mK for 6 days.

4.8   MEASUREMENT OF TEMPERATURES BELOW 1 K

Below 1 K, an absolute temperature scale cannot be established on the basis of the properties of gases. Intermediate scales of "magnetic temperatures" T^* are used. These temperatures are defined from susceptibility measurements through the relation

$$\chi = \frac{C}{T^*}$$

This is an extrapolation of Curie's law to a temperature range where, for a given substance, it may not be valid. In the region where Curie's law is obeyed, the

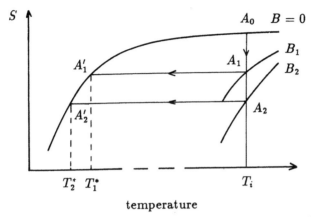

temperature

Figure 4.14: Determination of the absolute temperature by isothermal magnetizations A_0A_i and adiabatic demagnetizations A_iA_i'. The energy required to go from A_2' to A_1', measured separately, yields $T = (\partial U/dS)_{B=0}$.

magnetic temperature T^* and the absolute temperature T are equal. Below this range, only a thermodynamic method, based on the relation $T = (\partial U/\partial S)_B$, can establish the correspondence between T and T^*.

To do this, the relation $S(T^*, B = 0)$ is determined by a series of adiabatic demagnetizations for various values of the initial field B, starting from the same initial temperature T_i (Fig. 4.14). At the initial temperature, the absolute scale is known and the function $S(T_i, B)$ is determined by calorimetry. Note that if the substance obeys Curie's law, then expression (4.65) can be used. During an adiabatic demagnetization the entropy remains constant, i.e. $S(A_1') = S(A_1)$, and, since the final magnetic temperature T^* is determined, it follows that $S(T^*, B = 0) = S(T_i, B)$.

Next, the function $U(T^*, B = 0)$ is found as follows. Starting from the lowest temperature reached, the specimen is progressively heated, while the successive increments of energy are measured and the corresponding magnetic temperatures charted. Hence, at every point the quantity

$$T = \left(\frac{\partial U}{\partial S}\right)_B = \frac{dU/dT^*}{dS/dT^*}$$

can be deduced. Figure 4.15 shows the $T(T^*)$ dependence for chromium methylammonium alum in the range 0.025 to 0.5 K. Determining the function $T(T^*)$ is very tedious and is valid only for one given substance. Nonetheless, if this material is used as a thermometer, it provides a measurement of the absolute temperature. It is important to realize that at very low temperatures it is not easy to ensure uniform heating of the sample owing to the low thermal conductivity and poor efficiency of thermal contacts. This difficulty has been overcome by irradiating the sample homogeneously with γ ray beams.

This type of thermometer can be used down to temperatures of the order of 1 mK. Below this temperature, other methods must be introduced, such as anisotropic γ ray emission of oriented nuclei, nuclear magnetic resonance, nuclear magnetization,

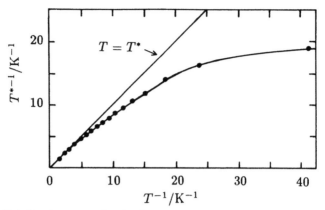

Figure 4.15: Relation between the absolute temperature T and the magnetic temperature T^* for chromium methylammonium alum [E. Ambler and R.P. Hudson, *J. Chem. Phys.* **27**, 378 (1957)].

spin-lattice relaxation, . . .

4.9   HISTORICAL ACCOUNT OF LOW TEMPERATURES

Exploration of the realm of low temperatures first began with the liquefaction of the so-called permanent gases, oxygen $O_2(T_B = 90.2$ K), nitrogen N_2 $(T_B = 77.3$ K), hydrogen H_2 $(T_B = 20.4$ K) and helium ^{4}He $(T_B = 4.2$ K). Then, immediately after liquefying helium, H. Kamerlingh Onnes (1908) reached a temperature of 1.5 K (Fig. 4.16) by reducing its vapour pressure. Nowadays, with very powerful pumps, 0.8 K can be reached with helium ^{4}He, and 0.3 K with the isotope ^{3}He (§9.7.1). These values, long considered to be the extreme limit of what could be obtained by chemical substances, have nevertheless been overtaken by dilution refrigerators (§9.7.3), which reached the value of 60 mK in 1965 . At present, temperatures down to 2mK can be produced with this kind of refrigerator.

Previously, temperatures between 1 K and 1 mK could be attained by electronic adiabatic demagnetization (§4.6). After the first such cooling to 0.25 K carried out by W.F. Giauque and D. P. MacDougall in 1933, a temperature of 18 mK was soon reached. In 1956, temperatures of about 1 mK were obtained, which is the lower limit of the range that can attainable by this method.

The temperature range $10^{-3} - 10^{-6}$ K was subsequently explored by means of nuclear adiabatic demagnetization (§4.7). After the pioneering work of F.E. Simon, N. Kurti and their collaborators in 1956 (20 μK for the spin system), this method proved itself to be operational and at present holds the record for low temperatures: 20 nK for the spin system of copper [M.T. Huiku et al. *AIP Conf. Proc.* (USA) **103**, 441 (1983)] and 13 μK for copper in its entirety [G.R. Pickett et al., *Nature* **302**, 695 (1983)]. The latter authors also cooled pure helium ^{3}He to 125 μK by means of copper. Comparable results (215 μK for ^{3}He - ^{4}He mixtures) have also been obtained [F. Pobell et al., *AIP Conf. Proc.* (USA) **103**, 192 (1983)].

This continuous quest for low temperatures has been marked along its way by

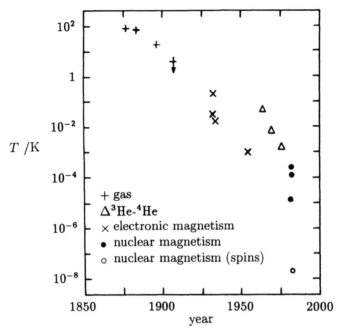

Figure 4.16: Lowest temperatures reached as a function of time [S. Balibar, *La Recherche* **15**, 1004 (1984)].

discoveries of phenomena of great importance for fundamental physics and technology: superconductivity, superfluidity in helium ^{4}He, then in ^{3}He, magnetism and anisotropy of liquid ^{3}He, the Kondo effect (resistance minimum in magnetic alloys), nuclear magnetic ordering, angular distribution of α, β and γ ray emissions from oriented nuclei, parity violation (asymmetric β emission).

No new method has yet been proposed for cooling to even lower temperatures. A new breakthrough can occur only through the discovery of new phenomena in the microkelvin range.

BIBLIOGRAPHY

A. Abragam and B. Bleaney, *Electron Paramagnetic Resonance of Transition Ions*, Oxford University Press, Oxford (1970).

B.I. Bleaney and B. Bleaney, *op. cit.*, p. 432.

A.H. Cooke, Paramagnetic Crystals in Use for Low Temperature Research, *Prog. Low Temp. Phys.* **1**, 224 (1955).

J. Crangle, *The Magnetic Properties of Solids*, Edward Arnold (1977).

G. Herzberg, *Atomic Spectra and Atomic Structure*, Dover (1944).

W.J. Huyscamp and O.V. Lounasmaa, Ultralow temperatures - how and why, *Rep. Prog. Phys.* **36**, 423 (1973).

D. de Klerk, Adiabatic demagnetization, *Handbuch der Physik* **15**, 38 (1956).

D. de Klerk and M.J. Steenland, Adiabatic demagnetization, *Prog. Low Temp. Phys.* **1**, 273 (1955).

A.H. Morrish, *The Physical Principles of Magnetism*, Wiley (1965).

H. M. Rosenberg, *op. cit.*, p. 285.

J.H. Van Vleck, *Theory of Electric and Magnetic Susceptibilities*, Oxford (1952).

M.W. Zemansky, *op. cit.*, p. 442.

Historical References

L. Brillouin, *J. Phys.* (Paris) **8**, 74 (1927).

P. Curie, *Ann. Chim. Phys.* (Paris) (7) **5**, 289 (1895).

P. Debye, *Ann. Physik* **81**, 1154 (1926).

W.F. Giauque, *J. Am. Chem. Soc.* **49**, 1870 (1927).

W. F. Giauque and D.P. Mac Dougall, *Phys. Rev.* **43**, 768 (1933).

C. J. Gorter, *Phys. Z.* **35** 923 (1934).

W.J. de Haas, E.C. Wiersma and H.A. Kramers, *Nature* **131**, 719 (1933).

H. Kamerlingh Onnes, *Leiden Comm.* **108**; *Proc. Sect. Sci. K. ned. Akad. Wet.* **11**, 168 (1908).

H. A. Kramers, *Proc. Amsterdam Acad.* **33**, 959 (1930).

N. Kurti and F.E. Simon, *Nature*, **133**, 907 (1934); *Proc. Roy. Soc. A* (London) **149**, 152 (1935).

N. Kurti et al. *Nature* **178**, 450 (1956).

P. Langevin, *J. Phys.* (Paris) **4**, 678 (1905); *Ann. Chim. Phys.* (Paris) **5**, 70 (1905).

COMPREHENSION EXERCISES

4.1 The magnetic susceptibility per unit mass of a substance having molar mass $M = 392$ g mol^{-1} is $\chi_m = 1.12 \times 10^{-2}$ cgs emu. What is its molar magnetic susceptibility χ_M in the international system of units (SI)? [Ans.: $\chi_M = 5.50 \times 10^{-5}$ m^3 mol^{-1}].

4.2 The Curie constant of iron ammonium sulphate is $C_M = 5.49 \times 10^{-5}$ K m^3 mol^{-1} and its molar mass is $M = 482$ g mol^{-1}. Calculate the magnetic moment of 3 grams of the substance placed in a magnetic field of flux density $B = 0.5$ T at $T = 77$ K [Ans.: $\mathcal{M} = 1.76 \times 10^{-3}$ A m^2].

4.3 Calculate the angle θ that the angular momentum σ_J makes with the z axis in the state $M_J = J$. Give the numerical value when $J = 1/2$ and $J = \infty$ [Ans. : $\theta = 55°$ and $0°$].

4.4 The Curie constant C_M per mole of Gd^{3+} ions in gadolinium sulphate is 9.83×10^{-5} K m^3 mol^{-1}. Hence derive the effective paramagnetic Bohr magneton number for the Gd^{3+} ion. The saturation magnetic moment per mole of Gd^{3+} ions is $M_{Ms} = 3.91$ A m^2 mol^{-1}. Calculate the values of J_{eff} and g_{eff}. Compare with the values of J and g in Table 4.2 [Ans.: $p_{\text{eff}} = 7.91$, $J_{\text{eff}} = 3.6$, $g_{\text{eff}} = 1.94$].

4.5 Same exercise for the Fe^{3+} ion in iron ammonium sulphate, the Curie constant of which is $C_M = 5.49 \times 10^{-5}$ K m^3 mol^{-1} and the molar saturation moment is $M_{Ms} = 27.9$ A m^2 mol^{-1}. Compare with the values in Table 4.3 [Ans.: $p_{\text{eff}} = 5.91$, $J_{\text{eff}} = 2.5$, $g_{\text{eff}} = 2$].

PROBLEM 4.1 ELECTRONIC ADIABATIC DEMAGNETIZATION

Here we investigate adiabatic demagnetization in chromium methylammonium alum $Cr(CH_3NH_3)$ $(SO_4)_2$, $12\,H_2O$.

1 In Zero Field

In this crystal, only the chromium Cr^{3+} ions are magnetic. Their electronic ground state has spin angular momentum $J = 3/2$, the orbital angular momentum being quenched. The crystal field splits this level into two Kramers doublets separated by an energy gap ϵ of characteristic temperature $\Theta = \epsilon/k = 0.275$ K.

1.1 Write down the expression for the populations of the levels in Maxwell-Boltzmann statistics and define the partition function Z. Calculate the population ratio of the two doubly degenerate levels of the Cr^{3+} ion at $T = 2.4$ K.

1.2 What is the partition function for the Cr^{3+} ion? Take the lowest level as the origin for the energy.

1.3 The free energy F of a system of N particles is $F = -NkT \ln Z$. Hence derive formulae for the entropy S, internal energy U and molar heat capacity c_V of the system of Cr^{3+} ions.

1.4a Inspect the function $S(T)$ and give its limiting form as $T \gg \Theta$. What is the meaning of the limiting value of S as $T \to \infty$?

1.4b In this same range, show that c_V takes the form A/T^2. Calculate the numerical value of A/R. Substituting A for Θ, derive the formula for S.

1.5a Find the limit of the function S as $T \to 0$. Is this result in agreement with the Third Law of thermodynamics?

1.5b The magnetic dipolar interaction energy between the chromium ions is approximately

$$\epsilon_d \simeq \frac{\mu_0 \mu^2}{4\pi r^3}$$

where $\mu = \sqrt{15}\,\mu_B$ is the magnetic moment of the ion, and $r = 8.7$ Å the distance between two nearest neighbour Cr^{3+} ions. In what temperature range is this interaction no longer negligible? What effect can this interaction have on the entropy? Discuss the experimental results of Figure 4.17.

2 In a Magnetic Field

When a uniform magnetic field of flux density B is applied to the crystal, the electronic level of the Cr^{3+} ions splits into four levels of energy $\epsilon_m = \epsilon_m^0 + mg\mu_B B$, where the magnetic quantum number m takes the values $-3/2, -1/2, 1/2, 3/2$, and where $\epsilon_m^0 = 0$ for $m = \pm 3/2$ and $\epsilon_m^0 = \epsilon$ for $m = \pm 1/2$. The spectroscopic splitting factor g may be taken to be 2.

2.1 Write the expression for the partition function Z. Verify that for $B = 0$ this gives the same result as in question 1.2. Set $x = Jg\mu_B B/kT$. The curve of S obtained from this expression is shown in Figure 4.17 for $B = 1.5$ T.

2.2 Give the limiting form for Z as $T \to 0$. Hence derive the limiting forms for the free energy F and S. Does the entropy vanish at absolute zero?

2.3a Express Z as a finite series expansion for $T \to \infty$ up to the term $1/T^3$, and

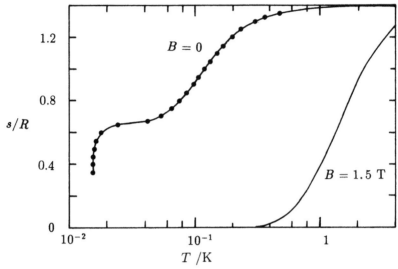

Figure 4.17: Entropy diagram of chromium methylammonium alum. The curve for $B = 0$ is experimental. The curve for $B = 1.5$ T is from theory (see text) [E. Ambler and R.P. Hudson, *J. Chem. Phys.* **27**, 378 (1957)].

hence determine the corresponding expression for F.

2.3b Using the approximate formula for F, deduce the molar magnetic moment M_M of the material. Show that Curie's law $M_M = C_M B / \mu_0 T$ is obeyed and find an expression for the Curie constant C_M. Calculate the numerical value of this constant and compare with the experimental result 2.35×10^{-5} m^3 K mol^{-1}.

2.3c Determine likewise the approximate formula for the molar entropy s and express it as a function of the constants A (defined in 1.4b) and C_M.

3 Adiabatic Demagnetization

A sample containing $n = 0.22$ moles of alum is cooled to $T_i = 1.2$ K by thermal contact with a heat bath of liquid helium under reduced pressure.

3.1 A magnetic field of flux density B is then applied, while thermal contact is maintained. The transformation can be considered as being isothermal and reversible. Use the approximate expression for the molar entropy from question 2.3c to find the heat released by the sample. Hence deduce the mass of helium lost by evaporation when the final field is $B = 0.2$ T, as well as the corresponding volume of gas under normal conditions . For the latent heat per unit mass of helium at 1.2 K , take $L = 22$ J g^{-1}. At the molecular level, what is the physical reason for the release of heat?

3.2 The sample is then thermally isolated and the external field removed. Determine the final temperature T_f as a function of T_i and of B and calculate its value for $B = 0.2$ T. Figure 4.18 shows the experimental values of $(T_i/T_f)^2$ as a function of B^2 . In which range are these values in agreement with the answers to this question? Give a physical explanation for the cooling.

3.3 Figure 4.17 shows the entropy diagram of the alum. The curve for $B = 0$

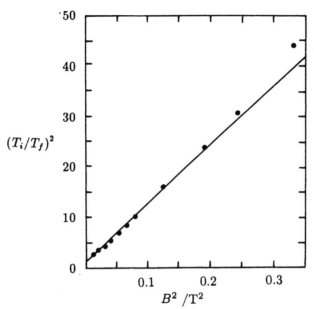

Figure 4.18: Temperature reached by chromium methylammonium alum under adiabatic demagnetization in a weak field [D. de Klerk and R.P. Hudson, *Phys. Rev.* **91**, 278 (1953)].

was obtained experimentally by calorimetric measurements. The curve for $B = 1.5$ T is calculated theoretically from the partition function of 2.1. Use this diagram to determine the final temperature obtained in an adiabatic demagnetization with starting temperature $T_i = 1.2$ K and induction $B = 1.5$ T.

Numerical constants: magnetic permeability of free space $\mu_0 = 4\pi \times 10^{-7}$ SI, Bohr magneton $\mu_B = 0.927 \times 10^{-23}$ J T^{-1}, Avogadro's number $\mathcal{N} = 6.02 \times 10^{23}$ mol^{-1}, gas constant $R = 8.31$ J K^{-1} mol^{-1}.

SOLUTION
1 In Zero Field

1.1 The population of a level i is

$$N_i = \frac{N}{Z} g_i e^{-\beta \epsilon_i} \quad \text{with} \quad Z = \sum g_i e^{-\beta \epsilon_i}.$$

The ratio of populations of the levels is

$$\frac{N_2}{N_1} = \frac{g_2 e^{-\beta \epsilon_2}}{g_1 e^{-\beta \epsilon_1}} = e^{-\beta(\epsilon_2 - \epsilon_1)} = e^{-\epsilon/kT} = e^{-\Theta/T},$$

since the degeneracy of the levels is $g_1 = g_2 = 2$. At $T = 2.4$ K, we have $N_2/N_1 = 0.89$ and the two levels are almost equally populated.

1.2 and **1.3** As the ions in the system have two energy levels, the reader is referred to paragraph 3.4. In particular it is found that ($g_1 = g_2 = 2$)

$$\frac{S}{Nk} = \ln\left[2(1 + e^{-\beta \epsilon})\right] + \frac{\beta \epsilon}{1 + e^{\beta \epsilon}} \quad \text{and} \quad \frac{c_V}{R} = \left[\frac{\beta \epsilon/2}{\cosh \beta \epsilon/2}\right]^2. \tag{4.73}$$

1.4a The function $S(T)$, displayed in Figure 3.3, fits the experimental data of Figure 4.17 above about 0.1 K. For $T \gg \Theta$, i.e. $\beta\epsilon \ll 1$, we have

$$\frac{S}{Nk} \simeq \ln\left[4\left(1 - \frac{\beta\epsilon}{2} + \frac{\beta^2\epsilon^2}{4}\right)\right] + \frac{\beta\epsilon}{2 + \beta\epsilon} \simeq \ln 4 - \frac{\beta^2\epsilon^2}{8} = \ln 4 - \frac{\Theta^2}{8T^2} \quad .$$

This expansion is accurate to 1% down to $T = \Theta$, i.e., in this case, $T \simeq 0.3$ K.

At high temperatures $(T \gg \Theta)$, the limiting value of the entropy is $S = Nk\ln 4$. This value corresponds to the Boltzmann relation $S = k\ln W$ for four equally populated levels, for which $W = 4^N$.

1.4b The function $c_V(T)$ is displayed in Figure 3.3. Its limiting form at high temperatures (3.25b) is

$$\frac{c_V}{R} = \frac{T}{R}\frac{dS}{dT} = \frac{\beta^2\epsilon^2}{4} = \frac{\Theta^2}{4T^2} \quad .$$

This heat capacity is $c_V = A/T^2$ with

$$\frac{A}{R} = \frac{\Theta^2}{4} \quad . \tag{4.74}$$

Numerically, $A/R = 1.89 \times 10^{-2}$ K^2. The molar entropy can then be written in the form

$$s(B = 0, T) = R\ln 4 - \frac{A}{2T^2} \quad . \tag{4.75}$$

1.5a From (4.73), the limit of S for $T = 0$ is $S = Nk\ln 2$. This result is in disagreement with the Third Law of Thermodynamics which states that the entropy is zero at absolute zero.

1.5b The interaction energy between ions is of order of magnitude $\epsilon_d \simeq 1.96 \times 10^{-25}$ J, and the corresponding characteristic temperature is $\Theta_d = \epsilon_d/k \simeq 1.4 \times 10^{-2}$ K. When T is of the order of Θ_d, the interactions between ions can no longer be neglected. They lift the degeneracy of the ground state, and the entropy is zero for $T \ll \Theta_d$. In Figure 4.17 a sudden drop in entropy can be seen near 1.6×10^{-2} K, which is close to Θ_d.

2 In a Magnetic Field

2.1 The partition function is

$$Z = \sum_m e^{-\beta\epsilon_m} = 2(\cosh 3\beta\mu_B B + e^{-\beta\epsilon}\cosh\beta\mu_B B)$$

$$= 2(\cosh x + e^{-\beta\epsilon}\cosh x/3) \quad \text{with} \quad x = 3\mu_B B/kT \quad . \tag{4.76}$$

When $B = 0$, (3.19) is recovered. The formula for the entropy, derived from (4.76) by applying the general relation (3.7), is quite complicated, and will not be detailed here. For its applications, however, the curve of $S(T)$ at $B = 1.5$ T is displayed graphically (Fig. 4.17).

2.2 When $T \to 0$, $x \to \infty$ and $\beta\epsilon \to \infty$, i.e.

$$Z \simeq e^x + e^{x/3 - \beta\epsilon} = e^x \left[1 + e^{-(2x/3 + \beta\epsilon)} \right]$$

The free energy then becomes

$$F = -NkT \ln Z \simeq -3N\mu_B B - NkT e^{-(2x/3 + \beta\epsilon)}$$

and the entropy is

$$S = -\left(\frac{\partial F}{\partial T} \right)_B \simeq Nk(1 + 2x/3 + \beta\epsilon) e^{-(2x/3 + \beta\epsilon)} \ .$$

This entropy vanishes at $T = 0$ in accordance with the Third Law, because only the lowest level ($m = -3/2$) is populated.

2.3a When $T \to \infty$, $x \to 0$ and $\beta\epsilon \to 0$. Restricting ourselves to the terms in x^2 and in $(\beta\epsilon)^2$, we get for Z

$$Z \simeq 2 \left[1 + \frac{x^2}{2} + \left(1 - \beta\epsilon + \frac{\beta^2\epsilon^2}{2} \right) \left(1 + \frac{x^2}{18} \right) \right]$$

$$\simeq 4 \left[1 - \frac{\beta\epsilon}{2} + \frac{\beta^2\epsilon^2}{4} + \frac{5}{18} x^2 \right] = 4 \left[1 - \frac{\Theta}{2T} + \frac{\Theta^2}{4T^2} + \frac{5}{2} \frac{\mu_B^2 B^2}{k^2 T^2} \right] .$$

Since $\ln(1 + u) \simeq u - u^2/2$, the limiting form for the free energy can be written

$$F = -NkT \ln Z \simeq -NkT \ln 4 + \frac{N}{2} k\Theta - \frac{Nk}{T} \left(\frac{\Theta^2}{8} + \frac{5}{2} \frac{\mu_B^2 B^2}{k^2} \right). \quad (4.77)$$

2.3b The total magnetic moment of the material is

$$\mathcal{M} = -\left(\frac{\partial F}{\partial B} \right)_T \simeq \frac{Nk}{T} \times 5 \frac{\mu_B^2}{k^2} B \ ;$$

the molar magnetic moment is then given by ($N \to \mathcal{N}$)

$$M_M = 5 \frac{\mathcal{N}\mu_B^2}{k} \frac{B}{T} \ . \quad (4.78)$$

The substance thus obeys Curie's law with

$$C_M = 5\mathcal{N} \frac{\mu_0 \mu_B^2}{k} \ . \quad (4.79)$$

This formula is identical to (4.51) for $\mu = [J(J+1)]^{1/2} g\mu_B = \sqrt{15}\mu_B$. The numerical value of C_M is $C_M = 2.36 \times 10^{-5}$ SI, in good agreement with the experimental value.

2.3c The entropy at high temperatures is derived from the approximate formula for F. We have

$$S = -\left(\frac{\partial F}{\partial T} \right)_B = Nk \left[\ln 4 - \frac{1}{T^2} \left(\frac{\Theta^2}{8} + \frac{5}{2} \frac{\mu_B^2 B^2}{k^2} \right) \right]. \quad (4.80)$$

Substitution of the constant A (4.74) and the Curie constant C_M (4.79) yields finally for the molar entropy

$$s(B,T) = R \ln 4 - \frac{A}{2T^2} - \frac{C_M}{2\mu_0} \frac{B^2}{T^2} \quad . \tag{4.81}$$

When $B = 0$, this approximate formula reduces to expression (4.75), which was found in 1.4. It is valid as long as s/R is greater than 1.25. In particular, for $T = 1.2$ K, it holds for $B \lesssim 0.4$ T.

Remark: In the general case where $\epsilon_m = \epsilon_m^0 - (\mu_z)_m B$, in place of Curie's law (4.78) and formula (4.81), a similar calculation gives

$$M_M = \frac{C_M B}{\mu_0 T} - \frac{D}{T} \quad \text{and} \quad s = R \ln Z_0 - \frac{A}{2T^2} + D\frac{B}{T^2} - \frac{C_M}{2\mu_0} \frac{B^2}{T^2} \tag{4.82}$$

where Z_0 is the number of states involved and

$$A = \frac{\mathcal{N}}{k} \left[\frac{\sum \epsilon_m^2}{Z_0} - \left(\frac{\sum \epsilon_m}{Z_0} \right)^2 \right], \quad D = \frac{\mathcal{N}}{Z_0 k} \sum (\mu_z)_m \epsilon_m$$

$$\text{and } C_M = \frac{\mathcal{N}\mu_0}{Z_0 k} \sum (\mu_z)_m^2 \quad . \tag{4.83}$$

Curie's law is no longer obeyed if $D \neq 0$ (cf. Problem 3.1). However, if, as here, the levels of the ion are Kramers doublets, D is zero and both the Curie law (4.78) and expression (4.81) for the entropy remain valid.

3 Adiabatic Demagnetization

3.1 The isothermal transformation changes the substance from state a ($B = 0, T = T_i$) to state b (B, T_i) (Fig.4.10), and thus, according to (4.81), the molar entropy changes by

$$\Delta s = s_b - s_a = -\frac{C_M}{2\mu_0} \frac{B^2}{T_i^2}.$$

Since the transformation is isothermal and the entropy change is negative, the sample releases a quantity of heat

$$|Q| = nT_i|\Delta s| = n\frac{C_M}{2\mu_0} \frac{B^2}{T_i}.$$

For $B = 0.2$ T, $|Q| = 6.9 \times 10^{-2}$ J. This heat evaporates $|Q|/L = 3.1 \times 10^{-3}$ g of helium, i.e. about 20 cm^3 of helium gas at normal temperature and pressure.

During the isothermal magnetization, the magnetic field causes the energy levels to split. The particles then fall to the lowest levels, thereby releasing heat (Fig. 4.9).

3.2 During adiabatic demagnetization the system moves from point b (B, T_i) to c ($B = 0, T_f$) (Fig. 4.10), while the entropy stays constant, i.e. $s_b = s_c$. Expression (4.81) for s yields

$$\frac{A}{2T_i^2} + \frac{C_M}{2\mu_0} \frac{B^2}{T_i^2} = \frac{A}{2T_f^2}, \quad \text{and hence} \quad T_f = T_i \left(1 + \frac{C_M}{\mu_0 A} B^2 \right)^{-1/2} . \tag{4.84}$$

For $B = 0.2$ T, we find $T_f = 0.50$ K. Formula (4.84) can be written

$$\left(\frac{T_i}{T_f}\right)^2 = 1 + \lambda B^2, \quad \text{with} \quad \lambda = \frac{C_M}{\mu_0 A} = 119 \text{ T}^{-2}.$$

This theoretical result is shown by the straight line on Figure 4.18. The experimental points deviate from this line for $B^2 \simeq 0.15$ T^2, i.e., $B \simeq 0.4$ T, in agreement with the result of 2.3c.

During the isentropic transformation, the populations of the levels remain constant. Since the levels come closer together, the temperature of this distribution falls.

3.3 From the point a $(B = 0, T_i = 1.2$ K) in Figure 4.17, a line parallel to the vertical axis is drawn to represent the isothermal transformation. This cuts the curve $B = 1.5$ T at b with $s/R = 0.52$. The isentropic transformation corresponds to a line parallel to the horizontal axis going through b and cutting the curve $B = 0$ at c having the same entropy as b and at temperature $T_f \simeq 1.6 \times 10^{-2}$ K.

PROBLEM 4.2 FERROMAGNETISM OF EUROPIUM OXIDE EuO BY THE WEISS MOLECULAR FIELD METHOD

In europium oxide EuO, the magnetic ions Eu^{2+} have a total angular momentum $J = 7/2$, due only to the electron spins, and a spectroscopic splitting factor $g = 2$. This oxide becomes ferromagnetic below $T_c = 77$ K with a saturation magnetic moment per mole of $M_{Ms} = 38.5$ A m^2 mol^{-1}. Above this temperature it obeys the Curie-Weiss relation

$$M_M = \frac{C_M}{T - T_c} \frac{B}{\mu_0},$$

with a Curie constant C_M equal to 9.90×10^{-5} K m^3 mol^{-1}.

To interpret these experimental results, the model of Weiss can be used. This model replaces interactions between magnetic ions by a fictitious magnetic field B_i (called the molecular field) which is proportional to the magnetic moment M_M, $B_i = q M_M$, where q is a constant.

We recall that the projection of the magnetic moment of an ion along the direction of the magnetic field can take the values $(\mu_z)_m = -mg\mu_B$, where μ_B is the Bohr magneton and m the magnetic quantum number $(m = -J, -J+1, \ldots, J)$.

1a Give the formula for the magnetic energy levels ϵ_m of the ions. These ions are assumed to be non interacting and subjected to a magnetic flux density B'.

1b Write the formula for the populations N_m of these levels as a function of the total number of ions N and of the partition function Z.

1c Show that the various terms of Z form a geometrical series and express it in terms of hyperbolic functions. Set $x = Jg\mu_B B'/kT$.

2 Give the formula for the magnetic moment $\mathcal{M}$ as a function of the level populations N_m, and show that

$$\mathcal{M} = N Jg\mu_B \frac{d \ln Z}{dx} = N Jg\mu_B B_J(x) .$$

Table 4.5: Temperature dependence of molar magnetization in europeum oxyde EuO

T /K	5	10	20	30	40	50	68.6	72.3
M_M /A m^2 mol^{-1}	38.5	38.5	37.1	35.5	33.3	30.2	15.0	10.1

3a Write out in full the Brillouin function $B_J(x) = d \ln Z/dx$ and find its limiting value for $x = \infty$.

3b Expand $B_J(x)$ to order 3 around $x = 0$, using the relation $\coth u = 1/u + u/3 - u^3/45 + \ldots$

3c Sketch the Brillouin function for the case of europium ions $Eu^{2+}(J = 7/2)$.

4 Write the formula for the molar magnetic moment M_M and calculate its saturation value M_{Ms} for europium oxide EuO. Compare with the experimental value.

5 To describe the interactions between ions, each ion is assumed to be subjected to a magnetic field of flux density $B' = B + qM_M$, where B is the flux density in the substance and qM_M is the Weiss molecular field.

5a Use the results of the previous question to show that the magnetic equation of state $M_M(B,T)$ is determined in implicit form.

5b Show that the reduced magnetization $R = M_M/M_{Ms}$ is equal to the ordinate of the intersection of $B_J(x)$ with a straight line. Define the parameters of this straight line in terms of B and T.

6a Use the expansion of $B_J(x)$ to order 1 in x (question 3b) to derive explicitly the approximate equation of state $M_M(B,T)$ at small fields and at high temperature.

6b Hence show that the material obeys the Curie-Weiss law in the range $x \ll 1$. Determine the formulae for the Curie constant C_M and the temperature T_c. Sketch the function $1/\chi_M(T)$ (the magnetic susceptibility χ_M is defined by $\chi_M = \mu_0 M_M/B$).

6c Calculate the value of the Curie constant C_M for europium oxide, and compare with the experimental result. Show that by measuring C_M and T_c the molecular field coefficient q can be determined. Calculate its value and hence deduce the saturation value of the molecular field. Comment on the result.

7a Show that in zero field $(B = 0)$, spontaneous magnetization occurs if the temperature T is below T_c.

7b Sketch the curve of the reduced magnetization R as a function of T/T_c. Find an approximate analytical formula for R in the neighbourhood of $T = T_c$.

7c Experimental measurements of the magnetic moment M_M of the oxide EuO as a function of T are listed in Table 4.5. Plot these experimental points on the previous graph and comment.

Physical constants: Magnetic permeability of free space $\mu_0 = 4\pi \times 10^{-7}$ SI; Bohr magneton $\mu_B = 0.927 \times 10^{-23}$ J T^{-1}, Avogadro's number $\mathcal{N} = 6.02 \times 10^{23}$ mol^{-1}, Boltzmann constant $k = 1.38 \times 10^{-23}$ J K^{-1}.

Reference: B.T. Matthias, R.M. Bozorth and J.H. Van Vleck, *Phys. Rev. Lett.* **7**, 160 (1961).

SOLUTION

1a We have $\epsilon_m = -(\mu_z)_m B' = m g \mu_B B'$.

1b The populations are given by the Maxwell-Boltzmann distribution

$$N_m = \frac{N}{Z} e^{-\beta \epsilon_m} \quad \text{with} \quad Z = \sum_{m=-J}^{J} e^{-\beta \epsilon_m} \ .$$

It can be seen that $\sum N_m = N$.

1c The calculation of Z was performed in paragraph 4.4.2. The result is (4.37)

$$Z = \frac{\sinh \frac{2J+1}{2J} x}{\sinh \frac{1}{2J} x} \ .$$

2 The magnetic moment is

$$\mathcal{M} = \sum_m N_m \times (\mu_z)_m = \frac{N}{Z} g \mu_B \sum_m -m e^{-mx/J} \ .$$

Apart from a factor J, the above sum is the derivative of the partition function Z (4.37). This gives the required relation.

3a The function $B_J(x)$ is given in (4.41). It is

$$B_J(x) = \frac{2J+1}{2J} \coth\left(\frac{2J+1}{2J} x\right) - \frac{1}{2J} \coth\left(\frac{x}{2J}\right) .$$

When $u \to \infty, \coth u \to 1$, which yields $\lim_{x \to \infty} B_J(x) = 1$.

3b The expansion (4.45) of the Brillouin function for x close to 0 is

$$
\begin{aligned}
B_J(x) &\simeq \left(\frac{1}{x} + \frac{(2J+1)^2}{(2J)^2}\frac{x}{3} - \frac{(2J+1)^4}{(2J)^4}\frac{x^3}{45}\right) - \left(\frac{1}{x} + \frac{1}{(2J)^2}\frac{x}{3} - \frac{1}{(2J)^4}\frac{x^3}{45}\right) \\
&= \frac{J+1}{J}\frac{x}{3} - \frac{(2J+1)^4 - 1}{(2J)^4}\frac{x^3}{45} \ .
\end{aligned}
$$

3c The Brillouin function $B_{7/2}(x)$ is shown in Figure 4.19.

4 We have

$$M_M = \mathcal{N} J g \mu_B B_J(x) \ .$$

The saturation moment, obtained for $x \to \infty$, is $M_{Ms} = \mathcal{N} J g \mu_B$. The numerical value of M_{Ms} is 39.1 A m^2 mol^{-1}. This is very close to the experimental value.

5a On expressing x in terms of B' and T, we get

$$M_M = M_{Ms} B_J \left(J g \mu_B \frac{B + q M_M}{kT}\right) ,$$

which defines M_M as an implicit function of B and of T.

5b Eliminating M_M, and keeping the variable $x = \beta J g \mu_B (B + q M_M)$, the above equation yields

$$R = \frac{M_M}{M_{Ms}} = \frac{kT}{q M_{Ms} J g \mu_B} x - \frac{B}{q M_{Ms}} = B_J(x) \ . \tag{4.85}$$

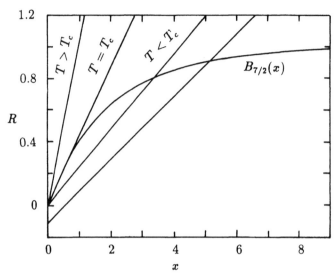

Figure 4.19: Graphic solution of equation (4.85). The curve is the Brillouin function $B_J(x)$ for $J = 7/2$.

This equation is solved graphically by looking for the intersection of the straight line of slope $kT/qM_{Ms}Jg\mu_B$, cutting the ordinate axis at $-B/qM_{Ms}$, with the curve $B_J(x)$ (Fig. 4.19). The ordinate of the point of intersection is equal to the reduced magnetization $R = M_M/M_{Ms}$.

6a Expansion to the first power in x yields

$$R = \frac{J+1}{J}\frac{x}{3} = \frac{J+1}{J}\frac{Jg\mu_B}{3}\frac{B+qM_M}{kT} \quad \text{i.e.} \quad M_M = \frac{B}{\frac{3kT}{(J+1)g\mu_B M_{Ms}} - q} \quad .$$

6b Identification with the Curie-Weiss law gives

$$C_M = \mu_0\frac{(J+1)g\mu_B M_{Ms}}{3k} = \mathcal{N}\frac{\mu_0 J(J+1)g^2\mu_B^2}{3k} \quad \text{and} \quad T_c = \frac{qC_M}{\mu_0} \quad . \tag{4.86}$$

The magnetic susceptibility χ_M is

$$\chi_M = \frac{\mu_0 M_M}{B} = \frac{C_M}{T - T_c} \quad .$$

The function $1/\chi_M(T)$ is a straight line that intersects the temperature axis at $T = T_c$. This point is the transition between the paramagnetic phase and the ferromagnetic phase, at which the susceptibility becomes infinite. The results of experimental measurements are plotted in Figure 4.20a. They are in agreement with the Curie-Weiss relation.

6c The Curie constant C_M of europium oxide, calculated from (4.86), is 9.89×10^{-5} K m^3 mol^{-1}. This lies close to the experimental value (9.90×10^{-5} K m^3 mol^{-1}). The molecular field constant q is obtained from (4.87) with

$$q = \frac{\mu_0 T_c}{C_M} = 0.98 \text{ SI} \quad .$$

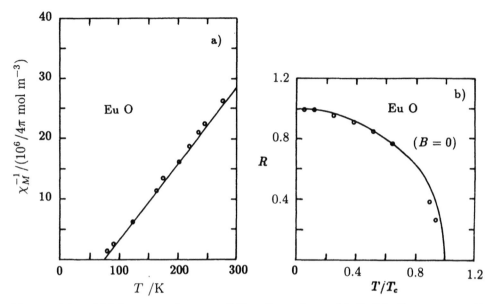

Figure 4.20: a) Molar magnetic susceptibility of europium oxide EuO in the paramagnetic state $(T > T_c)$. The straight line is the Curie-Weiss relation with $T_c = 77$ K and $C_M = 9.90 \times 10^{-5}$ K m^3 mol^{-1}. b) Reduced magnetization for the same oxide in the ferromagnetic state $(T < T_c)$. The curve is the molecular field theory result with $J = 7/2$ [B.T. Matthias et al., *Phys. Rev. Letters* **7**, 160 (1961)].

At saturation $(M_M = M_{Ms})$, the molecular field is equal to

$$B_{is} = q \times M_{Ms} = 37.6 \text{ T} .$$

This result shows that the saturation value of the molecular field is very high. As long as the moment M_M is greater than about $0.1 M_{Ms}$, the external field $(B \lesssim 1$ T) is negligible compared to the molecular field.

7a When the magnetic field is zero, the straight line in the graphic solution (4.85) goes through the origin and has slope

$$\frac{kT}{qM_{Ms}Jg\mu_B} = \frac{J+1}{3J} \times \frac{T}{T_c} .$$

The number of solutions of equation (4.85) depends on the value of this slope compared to that of $B_J(x)$ at the origin, which, from question 3b, is equal to $(J+1)/3J$. In all cases, a solution $(x = 0, R = 0)$ exists. A second solution exists when the slope of the straight line is smaller than that of B_J at the origin, i.e., when $T < T_c$ (Fig. 4.19). This solution corresponds to spontaneous magnetization in the stable ferromagnetic state below T_c.

7b If straight lines of varying slopes are drawn in Figure 4.19, the curve of Figure 4.20b is generated. As $T \to 0$, the slope of the straight line tends to zero and $R \to 1$. The magnetic moments of the Eu^{2+} ions are then completely ordered by their mutual interactions. As $T \to T_c$, $R \to 0$. More precisely, using the expansion in question 3b, the solution of (4.85) yields (for $J = 7/2$)

$$\frac{3}{7}x - \frac{13}{7^3}x^3 = \frac{3}{7}\frac{T}{T_c}x$$

the solution of which is (in addition to $x = 0$)

$$x = \left[\frac{147}{13}\left(1 - \frac{T}{T_c}\right)\right]^{1/2} .$$

Lastly, the reduced magnetization for $T \simeq T_c$ is

$$R = \frac{3}{7}\frac{T}{T_c}x = \sqrt{\frac{27}{13}\frac{T}{T_c}}\left(1 - \frac{T}{T_c}\right)^{1/2} \simeq \sqrt{\frac{27}{13}}\left(1 - \frac{T}{T_c}\right)^{1/2} . \qquad (4.88)$$

This expression contains the factor $(1 - T/T_c)^\beta$, where the critical exponent β is equal to $1/2$. This value is characteristic of the molecular field approximation (§13.4.2).

7c. Insertion of the experimental value of M_{Ms} yields the experimental points shown in Figure 4.20b. Molecular field theory thus gives a good approximation to the experimental results for europium oxide (Fig. 4.20).

Chapter 5

Corrected Maxwell-Boltzmann Statistics. Ideal Gases

5.1 DISTRIBUTION LAW AND THERMODYNAMIC FUNCTIONS

Corrected Maxwell-Boltzmann statistics (MB_c) introduced in paragraph 2.5.5 applies to systems composed of independent indistinguishable particles in the limiting case of weakly populated energy levels ($N_i \ll g_i$). This type of statistics is the limit for both Bose-Einstein and Fermi-Dirac quantum statistics. Its formalism is very similar to that of Maxwell-Boltzmann statistics (MB). Both these types of statistics are characterized by the thermodynamic probabilities (2.11,35)

$$W_{MBc} = \prod_i \frac{g_i^{N_i}}{N_i!} \quad \text{and} \quad W_{MB} = N! \prod_i \frac{g_i^{N_i}}{N_i!} \tag{5.1}$$

which differ only in the factor $N!$. This factor, which defines the number of mutual permutations of N distinguishable particles, arises in Maxwell-Boltzmann statistics, but not in corrected Maxwell-Boltzmann statistics, on account of the postulate of indistinguishability (§1.2.)

5.1.1 Distribution Relation

The distribution relation found in (2.40) is

$$N_i = \frac{N}{Z} g_i e^{-\beta \epsilon_i} \quad (\beta = 1/kT) \tag{5.2}$$

where the partition function (or sum of states) for one particle,

$$Z = \sum_i g_i e^{-\beta \epsilon_i} = \sum_j e^{-\beta \epsilon_j} \, , \tag{5.3}$$

is a sum over the energy levels (subscript i) or over the quantum states (subscript j) of a particle. This distribution is identical to that of Maxwell-Boltzmann statistics. Its properties are therefore the same:

• for a given temperature the ratio of population densities $n_i = N_i/g_i$ of two energy

levels depends only on their energy difference (§3.1);
• changing the origin of the energies does not change the distribution (§3.2);
• if the various forms of energy are additive, they contribute to the partition function as multiplying factors (§3.3);
• the theorem of equipartition of energy (§3.5) is valid under the same conditions.

5.1.2   Thermodynamic Functions

In paragraph 2.5.5 we saw that the expression for the free energy (2.41) in terms of Z is

$$F = -NkT \left(\ln \frac{Z}{N} + 1 \right). \tag{5.4}$$

As in paragraph 3.2, the differential of F,

$$dF = -SdT - PdV + \mu dN ,$$

yields the various thermodynamic quantities

$$S = -\left(\frac{\partial F}{\partial T} \right)_{V,N} = Nk \left[\ln \frac{Z}{N} + 1 + T \left(\frac{\partial \ln Z}{\partial T} \right)_V \right] \tag{5.5}$$

$$P = -\left(\frac{\partial F}{\partial V} \right)_{T,N} = NkT \left(\frac{\partial \ln Z}{\partial V} \right)_T \tag{5.6}$$

$$\mu = \left(\frac{\partial F}{\partial N} \right)_{T,V} = -kT \ln \frac{Z}{N} \tag{5.7}$$

$$U = F + TS = NkT^2 \left(\frac{\partial \ln Z}{\partial T} \right)_V = -N \left(\frac{\partial \ln Z}{\partial \beta} \right)_V \tag{5.8}$$

$$C_V = \left(\frac{\partial U}{\partial T} \right)_{V,N} = Nk \frac{\partial}{\partial T} \left[T^2 \frac{\partial \ln Z}{\partial T} \right]_V = Nk\beta^2 \left(\frac{\partial^2 \ln Z}{\partial \beta^2} \right)_V. \tag{5.9a}$$

Note that if the partition function cannot be expressed as a simple analytical expression, the specific heat can be found by writing (5.9a) in the form

$$C_V = Nk\beta^2 \left[\frac{Z''}{Z} - \frac{Z'^2}{Z^2} \right] ,$$

where $Z' = \dfrac{\partial Z}{\partial \beta} = -\sum_i g_i \epsilon_i e^{-\beta \epsilon_i}$ and $Z'' = \dfrac{\partial^2 Z}{\partial \beta^2} = \sum_i g_i \epsilon_i^2 e^{-\beta \epsilon_i}$ (5.9b)

The calculation is then performed numerically, the series being truncated according to the precision desired.

When an electric field $\mathcal{E}$ or a magnetic flux density B is present, dF contains terms $d\mathcal{W} = -\mathcal{P}d\mathcal{E}$ or $d\mathcal{W} = -\mathcal{M}dB$. The electric and magnetic dipole moments

$\mathcal{P}$ and $\mathcal{M}$ are then given by

$$\mathcal{P} = -\left(\frac{\partial F}{\partial \mathcal{E}}\right)_{T,V,N} = NkT\left(\frac{\partial \ln Z}{\partial \mathcal{E}}\right)_{T,V} \tag{5.10}$$

$$\mathcal{M} = -\left(\frac{\partial F}{\partial B}\right)_{T,V,N} = NkT\left(\frac{\partial \ln Z}{\partial B}\right)_{T,V} . \tag{5.11}$$

Exercise 5.1 *Consequences of indistinguishability*
   Compare the above expressions for the thermodynamic functions with those of Maxwell-Boltzmann statistics (§3.2). Comment.

Solution
   The entropy is

$$S_{MBc} = S_{MB} - Nk \ln N + Nk = S_{MB} - k \ln N! . \tag{5.12}$$

The extra term $-k \ln N!$ is a direct consequence of applying the Boltzmann relation $S = k \ln W$ to equations (5.1). No difference, however, arises in the expressions for the internal energy U, the heat capacity C_V, the pressure P, or the magnetic moment $\mathcal{M}$. For the free energy and the chemical potential, we have

$$F_{MBc} = F_{MB} + kT \ln N!$$
$$\mu_{MBc} = \mu_{MB} + kT \ln N .$$

The extra term comes directly from S.

5.2   IDEAL MONATOMIC GASES

In an ideal (or perfect) monatomic gas (rare gases, metal vapours, etc.), the energy of each atom is a sum of energies, i.e. that of translation, that of the electrons, and that of the nucleus,

$$\epsilon_i = \epsilon_k^{(t)} + \epsilon_l^{(e)} + \epsilon_m^{(n)} . \tag{5.13a}$$

The degeneracy of the level $i \equiv (k,l,m)$ of the atom is

$$g_i = g_k^{(t)} \times g_l^{(e)} \times g_m^{(n)} . \tag{5.13b}$$

It follows (§3.3) that the partition function is a product of three factors

$$Z = Z^{(t)} \times Z^{(e)} \times Z^{(n)} \tag{5.13c}$$

respectively called the translational, electronic and nuclear partition functions. The free energy (5.4) is then the sum of three terms

$$F = F^{(t)} + F^{(e)} + F^{(n)}$$
$$= -NkT\left(\ln \frac{Z^{(t)}}{N} + 1\right) - NkT \ln Z^{(e)} - NkT \ln Z^{(n)} . \tag{5.14}$$

The term $kT \ln N!$ is grouped with the translation term as it is the source of the indistinguishability of the particles.

5.2.1   Electronic and Nuclear Partition Functions

In the great majority of cases, the first excited electronic level of the atoms lies several electron volts above the ground state: 11.5 eV for argon, 1.85 eV for lithium, 4.67 eV for mercury ... With the choice $\epsilon_0^{(e)} = 0$, the electronic partition function is just given by its first term, i.e.

$$Z^{(e)} = g_0^{(e)} + g_1^{(e)} e^{-\beta \epsilon_1^{(e)}} + \ldots \simeq g_0^{(e)} ,$$

since the term $\beta \epsilon_1^{(e)} = \epsilon_1^{(e)}/kT = \Theta_e/T$ is much greater than 1 at normal temperatures. Indeed, for $\epsilon_1^{(e)} \sim 1$ eV, $\Theta_e \sim 10^4$ K. Thus in general the electronic partition function reduces to its first term $g_0^{(e)} = 2J + 1$, where J is the quantum number for the electronic angular momentum of the atom in its electronic ground state. The electronic partition function is thus

$$Z^{(e)} = g_0^{(e)} = 2J + 1 , \tag{5.15a}$$

i.e., $Z^{(e)} = 1$, 2 and 1, respectively, for argon ($J = 0$), lithium ($J = 1/2$) and mercury ($J = 0$).

The nuclear partition function likewise reduces to a constant

$$Z^{(n)} = g_0^{(n)} = 2I + 1 \tag{5.15b}$$

because the first excited nuclear level always lies about 1 Mev (10^6 eV) above the ground state. For example, in lithium ^{7}Li this separation is 0.5 Mev, the degeneracy of its nuclear ground level being $g_0^{(n)} = 4$ (nuclear spin $I = 3/2$). As the nuclear state does not change in a thermodynamic or chemical transformation, $Z^{(n)}$ is usually omitted.

For an ideal monatomic gas, therefore

$$Z = (2J + 1)Z^{(t)} . \tag{5.16}$$

5.2.2   Translational Partition Function

The translational partition function $Z^{(t)}$ can be calculated in two ways. In the first, we look at the quantum states of a particle of mass m in a cubic box of volume V. The energy levels (1.2) are

$$\epsilon_{n_x,n_y,n_z} = \epsilon_0(n_x^2 + n_y^2 + n_z^2) \quad \text{with } \epsilon_0 = \frac{h^2}{8mV^{2/3}} , \tag{5.17}$$

where n_x, n_y, n_z are positive integers. The partition function $Z^{(t)}$ is then

$$
\begin{aligned}
Z^{(t)} &= \sum_{n_x,n_y,n_z=1}^{\infty} \exp\left(-\beta \epsilon_{n_x,n_y,n_z}\right) \\
&= \sum_{n_x=1}^{\infty} \exp(-\beta \epsilon_0 n_x^2) \sum_{n_y=1}^{\infty} \exp(-\beta \epsilon_0 n_y^2) \sum_{n_z=1}^{\infty} \exp(-\beta \epsilon_0 n_z^2) \\
&= \left[\sum_{n=1}^{\infty} \exp(-\beta \epsilon_0 n^2)\right]^3 .
\end{aligned}
$$

The sum cannot be performed in simple analytical terms. However, since $\epsilon_0/k \sim 10^{-16}$ K (1.3), the terms are almost independent of n and the sum can be replaced by the Gaussian integral

$$\sum_{n=1}^{\infty} \exp(-\beta\epsilon_0 n^2) \rightarrow \int_0^{\infty} \exp(-\beta\epsilon_0 n^2)dn = \frac{1}{2}\left(\frac{\pi}{\beta\epsilon_0}\right)^{1/2} .$$

Substitution of expression (5.17) for ϵ_0 gives for the translational partition function

$$Z^{(t)} = \frac{V}{h^3}(2\pi mkT)^{3/2} . \tag{5.18}$$

The second method of calculating $Z^{(t)}$ uses the classical limit

$$\epsilon_i \rightarrow \frac{p^2}{2m} \quad \text{and} \quad g_i \rightarrow \frac{d^3r d^3\mathbf{p}}{h^3} . \tag{5.19}$$

This yields

$$
\begin{aligned}
Z^{(t)} &= \frac{1}{h^3}\int d^3\mathbf{r}\int d^3\mathbf{p}\exp\left[-\frac{\beta}{2m}(p_x^2 + p_y^2 + p_z^2)\right] \\
&= \frac{V}{h^3}\left[\int_{-\infty}^{+\infty} dp_x \exp\left(-\frac{\beta}{2m}p_x^2\right)\right]^3 .
\end{aligned}
$$

Since the Gaussian integral is equal to $(2\pi m/\beta)^{1/2}$, we again find expression (5.18) for the translational partition function $Z^{(t)}$.

It should be noted that $Z^{(t)}$ is an extensive quantity, and therefore Z/N is intensive. It follows that the function F (5.14) is extensive. We may write

$$Z^{(t)} = \frac{V}{\Lambda^3} \quad \text{with} \quad \Lambda = \frac{h}{(2\pi mkT)^{1/2}} \tag{5.20}$$

where Λ is called the de Broglie thermal wavelength. This wavelength is that of a particle of momentum $p = (2\pi mkT)^{1/2}$ and energy $\epsilon = p^2/2m = \pi kT$, which is approximately equal to the mean kinetic energy (5.24) of an atom.

5.2.3   Thermodynamic Functions in the Variables T and V

Substitution of the expressions (5.15 and 18) for $Z^{(e)}$ and $Z^{(t)}$ into equation (5.14) gives for the free energy

$$F = -NkT\left[\ln\left(\frac{V}{N}\frac{(2\pi mkT)^{3/2}}{h^3}\right) + 1 + \ln(2J+1)\right] . \tag{5.21}$$

This yields for the equation of state

$$P = -\left(\frac{\partial F}{\partial V}\right)_{T,N} = \frac{NkT}{V} = \frac{\mathcal{N}kT}{v} , \tag{5.22a}$$

where v is the molar volume and $\mathcal{N}$ is Avogadro's number. This equation is independent of the nature of the gas and is identical to the experimental equation of state for ideal gases

$$P = \frac{RT}{v} \; . \tag{5.22b}$$

This shows that the Boltzmann constant k is related to the macroscopic quantity R (gas constant) through

$$\mathcal{N}k = R \; .$$

The internal energy U can be calculated from relation (5.8)

$$U = NkT^2 \left(\frac{\partial \ln Z}{\partial T} \right)_V = \frac{3}{2} NkT \; . \tag{5.23}$$

This energy is independent of both the nature and the volume of the gas. Since the equipartition of energy theorem applies to all three degrees of translational freedom of the atoms, $U = 3 \times (NkT/2)$. The mean kinetic energy of an atom is equal to

$$\bar{\epsilon} = \frac{\sum N_i \epsilon_i}{\sum N_i} = \frac{U}{N} = \frac{3}{2} kT \; . \tag{5.24}$$

The molar heat capacity at constant volume is then given by

$$c_V = \left(\frac{\partial U}{\partial T} \right)_V = \frac{3}{2} \mathcal{N}k = \frac{3}{2} R \; , \tag{5.25}$$

in complete agreement with the experimental results for monatomic gases at atmospheric pressure. Finally, the entropy and the chemical potential are equal to

$$S = -\left(\frac{\partial F}{\partial T} \right)_{V,N} = Nk \left[\ln \left(\frac{V}{N} \frac{(2\pi mkT)^{3/2}}{h^3} \right) + \frac{5}{2} + \ln(2J+1) \right] \tag{5.26}$$

and $\mu = -kT \ln \dfrac{Z}{N} = -kT \left[\ln \left(\dfrac{V}{N} \dfrac{(2\pi mkT)^{3/2}}{h^3} \right) + \ln(2J+1) \right] \; . \tag{5.27}$

5.2.4   Range of Validity of the Model

The ideal gas model we have been considering is limited on the one hand by the use of corrected Maxwell-Boltzmann statistics, and, on the other, by the neglect of interactions between atoms.

Corrected Maxwell-Boltzmann statistics is applicable under the condition $N_i/g_i \ll 1$, which is equivalent to the condition (2.43) $\alpha = N/Z \ll 1$. We write the latter condition in terms of the variables T and P that define the experimental constraints. To do this, the variable V is replaced by P in the equation for Z (5.16 and 18), by means of the equation of state (5.22), i.e.

$$Z = (2J+1) \frac{NkT}{P} \frac{(2\pi mkT)^{3/2}}{h^3} = (2J+1) M^{3/2} \frac{NT^{5/2}}{P} e^{i_0} \; , \tag{5.28}$$

where M is the molar mass and i_0 is a universal constant defined by

$$i_0 = \ln \frac{(2\pi)^{3/2} k^{5/2}}{N^{3/2} h^3} = 18.223 \text{ SI} .$$ (5.29)

The condition for corrected Maxwell-Boltzmann statistics to apply is then

$$\alpha = \frac{N}{Z} = e^{-i_0} \frac{M^{-3/2}}{2J+1} P T^{-5/2} \ll 1 .$$ (5.30)

In the case of argon ($M = 39.95 \times 10^{-3}$ kg mol^{-1}) at normal temperature ($T = 273$ K) and pressure ($P = 1$ atm $= 1.013 \times 10^5$ Pa), we have $\alpha = 1.26 \times 10^{-7}$, i.e. condition (5.30) is amply satisfied. Note that α is a product of three factors: a constant, a second factor that depends on the nature of the gas, and a factor that depends on the constraints applied to the gas. It can be seen that α is large for light gases at high pressure and low temperature. Before the ratio α approaches unity, however, gases condense. The only exception to this rule is the most unfavourable case, helium, for which

$$\alpha(^4\text{He}) = 0.13 \quad \text{and} \quad \alpha(^3\text{He}) = 0.41$$

at their normal boiling point (Table 9.1). A more rigorous approach to these situations would require the use of the appropriate quantum statistics, namely Bose-Einstein (§7.5) and Fermi-Dirac (§10.3), respectively.

As far as interactions are concerned, these are negligible insofar as the mean kinetic energy $\bar{e} = 3kT/2$ is much larger than the mean interaction energy, the value of which can be estimated for a distance equal to the mean interatomic distance $\bar{r} \sim (V/N)^{1/3}$. Since the interaction at large distances has a van der Waals behaviour in Ar^{-6}, the condition for interactions to be negligible is

$$A \left(\frac{N}{V}\right)^2 \ll \frac{3}{2} kT .$$ (5.31)

For rare gases, A is of the order of 10^{-77} SI, which in normal conditions, gives a mean interaction energy of the order of 10^{-8} eV, compared with the mean kinetic energy of order 10^{-2} eV. As the critical point is approached, the interaction energy ceases to be negligible compared to the kinetic energy. In other situations the ideal gas model gives a good approximation, even in the vicinity of the boiling point. It should be emphasized that liquefaction occurs when the liquid phase becomes more stable than the gas phase, for which the interactions may still be neglected.

In summary, the ideal gas model is valid for pressures sufficiently far below the critical pressure, and, except for the case of helium, for temperatures down to the boiling point of the liquid.

5.2.5   Thermodynamic Functions in the Variables T and P

The equations for the entropy and chemical potential for one particle in terms of the variables T and P are found from (5.26 and 27) by substituting kT/P for V/N, in accordance with the equation of state (5.22). Moreover, introduction of the molar mass, $M = \mathcal{N}m$, yields

$$S = Nk \left[-\ln P + \frac{5}{2} \ln T + \ln(2J + 1) + \frac{3}{2} \ln M + i_0 + \frac{5}{2} \right] \quad (5.32)$$

$$\mu = kT \left[\ln P - \frac{5}{2} \ln T - \ln(2J + 1) - \frac{3}{2} \ln M - i_0 \right] . \quad (5.33)$$

The equation for the entropy is called the Sackur-Tetrode relation. It involves the universal constant i_0 (5.29). The value for the entropy predicted by the Sackur-Tetrode formula (5.32) has been compared with the experimental value for a large number of gases. The excellent agreement found provides a check both of the Boltzmann relation $S = k \ln W$ and of the Third Law. We recall that an experimental measurement of the entropy of a gas at pressure P_0 requires evaluation of the successive changes in entropy of the substance for each temperature increase, using integrals of the form

$$S(T_2, P_0) - S(T_1, P_0) = \int_{T_1}^{T_2} \frac{C_p(T, P_0)}{T} dT ,$$

and, at a phase transition, using

$$\Delta S = \frac{L(T_0)}{T_0} .$$

These evaluations must be made from absolute zero up to the temperature under consideration.

The free enthalpy and the enthalpy functions G and H are also employed in terms of T and P, as well as the heat capacity at constant pressure C_P. They are given by

$$G = F + PV = NkT \left[\ln P - \frac{5}{2} \ln T - \ln(2J + 1) - \frac{3}{2} \ln M - i_0 \right] \quad (5.34)$$

$$H = U + PV = \frac{5}{2} NkT \quad (5.35)$$

$$C_P = \left(\frac{\partial H}{\partial T} \right)_P = \frac{5}{2} Nk \text{ or } c_P = \frac{5}{2} R . \quad (5.36)$$

It can be seen here that $G = N\mu$ (since the substance is pure), that the enthalpy and internal energy depend only on the temperature, and that the heat capacity is constant. For a monatomic gas $\gamma = C_P/C_V = 5/3$, in good agreement with experiment (Table 6.3). Finally, we note that, to use the value of i_0 given by (5.29), P, T and M must be expressed in the international system of units.

Exercise 5.2 *Experimental verification of the Sackur-Tetrode relation*
    The molar entropy of argon at $T = 300$ K and $P = 1$ atm, measured experimentally by calorimetry, is equal to $s = 154.85$ J K^{-1} mol^{-1} [J. Hilsenrath et al., *Tables*

of *Thermodynamic and Transport Properties*, Pergamon Press, 1960]. Compare this result with the theoretical value of the entropy given by the Sackur-Tetrode relation. The molar mass of argon is $M = 39.95$ g mol^{-1}.

Solution

We employ the international system of units ($P = 1.013 \times 10^5$ Pa, $M = 39.95 \times 10^{-3}$ kg mol^{-1}), in which $i_0 = 18.223$ SI. Since rare gases have full electronic layers, the total angular momentum of the ground state is zero. Setting $R = 8.3145$ J K^{-1} mol^{-1}, we find $s = 154.87$ J K^{-1} mol^{-1}. The agreement is excellent.

5.2.6   Paradox of the Electronic Partition Function

It was shown in paragraph 5.2.1 that the second term of the electronic partition function is in general negligible. As the number of electronic levels is infinite, however, the matter deserves fuller discussion.

Consider the hypothetical case of a gas composed of hydrogen atoms. For these atoms, the electronic energy levels and the associated degeneracy are

$$\epsilon_n = R_H \left(1 - \frac{1}{n^2}\right) \quad \text{and} \quad g_n = 2n^2$$

where $R_H = 13.6$ eV is the Rydberg energy, and the level $n = 1$ is taken as the origin for the energy. The electronic partition function is then

$$Z^{(e)} = 2 \sum_{n=1}^{\infty} n^2 e^{-\beta \epsilon_n} = 2 + 2 \sum_{n=2}^{\infty} n^2 e^{-\beta \epsilon_n} .$$

In the limit $n \to \infty$, the general term of the series varies as n^2, and the series diverges. This constitutes a failure of the theory, a result which can be transposed to all gases, since they possess an infinite number of levels of the above type (Rydberg states).

The solution of this paradox comes from the fact that in the Rydberg states, the atomic radius varies as $r_0 n^2$ with $r_0 \sim 1$ Å. Hence, for a given volume V of the containing vessel, n is necessarily limited to a value n_{max}. Thus, for $V = 1$ m^3, n cannot be greater than $n_{max} \sim 10^5$. We then have

$$
\begin{aligned}
Z^{(e)} \quad &= \quad 2 + 2 \sum_{n=2}^{n_{max}} n^2 e^{-\beta \epsilon_n} \\
&< \quad 2 + 2e^{-\beta \epsilon_2} \sum_{n=2}^{n_{max}} n^2 = 2 + 2e^{-3\beta R_H/4} \left[\frac{n_{max}(n_{max}+1)(2n_{max}+1)}{6} - 1\right] .
\end{aligned}
$$

For $T \sim 300$ K and $n_{max} = 10^5$, numerical calculation gives

$$Z^{(e)} < 2(1 + 2 \times 10^{-157}) .$$

It can thus be seen that truncation of the sum at $n_{max} = 10^5$ produces no significant change in the value of $Z^{(e)}$, calculated by taking into account only the first level ($Z^{(e)} = 2$) [S.J. Strickler, *J. Chem. Educ.* **43**, 364 (1966)].

5.3   ASYMMETRIC DIATOMIC IDEAL GASES

5.3.1   Rotational and Vibrational Energy Levels

Consider a diatomic molecule composed of two atoms A and B. In the inertial frame of reference of the centre of mass, the Hamiltonian is

$$H = \frac{\mathbf{p}^2}{2\mu} + V(r) = \frac{p_r^2}{2\mu} + \frac{\mathbf{L}^2}{2\mu r^2} + V(r) \tag{5.37}$$

where $\mu = m_A m_B/(m_A + m_B)$ is the reduced mass, p_r is the radial component of the momentum and $\mathbf{L}$ is the angular momentum. The potential energy of interaction $V(r)$ of the two atoms at distance r has a minimum at the equilibrium distance r_e. The interaction is attractive for $r > r_e$ and repulsive for $r < r_e$.

When the amplitude of vibration of the molecule is small ($r \simeq r_e$), $V(r)$ can be approximated by a parabola and the Hamiltonian can be written in the form

$$H \simeq \frac{\mathbf{L}^2}{2\mu r_e^2} + \frac{p_r^2}{2\mu} + \frac{1}{2}\mu\omega^2(r - r_e)^2 + V(r_e) , \tag{5.38}$$

with $\omega = [(V''(r_e)/\mu]^{1/2}$. In this form the energy of the molecule can be considered as a sum of the energies of a rigid rotator with moment of inertia $I_e = \mu r_e^2$ and that of a harmonic oscillator of angular frequency ω, plus a constant term. In this approximation, the quantum mechanical energy levels are given by

$$\epsilon_{K,n} = \frac{\hbar^2}{2I_e}K(K+1) + \left(n + \frac{1}{2}\right)\hbar\omega + V(r_e) .$$

where K and n are two positive or zero integers, called respectively the rotational and vibrational quantum numbers. The degeneracy of a level (K, n) is equal to $2K+1$, corresponding to the $2K+1$ different values of the quantum number m ($m = -K, \ldots, K$), i.e. to the $2K+1$ projections of the angular momentum $\mathbf{L}$ along a quantization axis. The energy of the lowest level is $\epsilon_{0,0} = \hbar\omega/2 + V(r_e)$, which is the binding energy of the molecule. With this energy as the origin, the energy levels of the molecule are

$$\epsilon_{K,n} = \frac{\hbar^2}{2I_e}K(K+1) + n\hbar\omega = hc[B_e K(K+1) + \omega_e n]. \tag{5.39}$$

Here the spectroscopic notations have been introduced

$$B_e = \frac{\hbar}{4\pi c I_e} \quad \text{and} \quad \omega_e = \frac{\omega}{2\pi c} . \tag{5.40}$$

Figure 5.1 shows a diagram of the energy levels (5.39). Since ω_e is always much larger than B_e (for carbon monoxide CO, $\omega_e = 2170$ cm^{-1} and $B_e = 1.931$ cm^{-1}), the levels appear as a stack of groups of rotational levels

$$\epsilon_K^{(r)} = \frac{\hbar^2}{2I_e}K(K+1) = hcB_e K(K+1) \quad \text{with} \quad g_K^{(r)} = 2K+1 \tag{5.41}$$

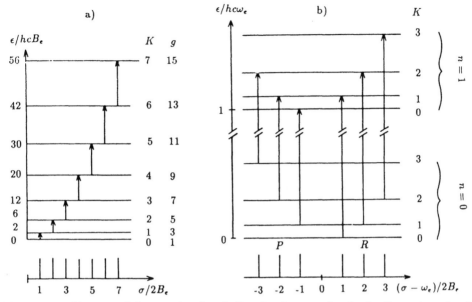

Figure 5.1: Diagram of the rotational and vibrational energy levels of a diatomic molecule (5.39). a) Pure rotational levels ($n = 0$) and associated degeneracy $g = 2K + 1$. b) Vibration-rotation levels for $n = 0$ and 1, up to $K = 3$. The arrows represent the allowed transitions for photon absorption. The absorption line spectra are shown diagrammatically below the corresponding transitions, in a scale of wave-numbers $\sigma = \Delta\epsilon/hc$.

on each vibrational level

$$\epsilon_n^{(v)} = n\hbar\omega = hc\omega_e n \quad \text{with} \quad g_n^{(v)} = 1 . \tag{5.42}$$

Note that, in spectroscopy, energies are written as $\epsilon = hc\sigma$, where $\sigma = 1/\lambda$ is the number of waves per unit length, usually measured in cm^{-1}.

At a higher level of approximation, correction terms must be introduced into (5.38) (*vibration-rotation coupling*) to take account of
- anharmonicity of the potential, i.e. $-x_e\omega_e n(n + 1)$;
- deformation of the molecule due to the centrifugal force, i.e. $-D_e K^2(K + 1)^2$, with $D_e = 4B_e^3/\omega_e^2$;

- variation of the inertial term during the vibration, i.e. replacement of B_e by $B_n = B_e - \alpha_e(n + \frac{1}{2})$.

The expression for the energy then becomes

$$\frac{\epsilon_{K,n}}{hc} = B_n K(K + 1) + \omega_e n - D_e K^2(K + 1)^2 - x_e\omega_e n(n + 1) . \tag{5.43}$$

For carbon monoxide CO, the values of the correction coefficients are $\alpha_e = 1.75 \times 10^{-2}$ cm^{-1}, $D_e = 6.10 \times 10^{-6}$ cm^{-1}, $x_e = 6.13 \times 10^{-3}$ [A.W. Mantz et al., *J. Mol. Spectrosc.* **35**, 325 (1970)].

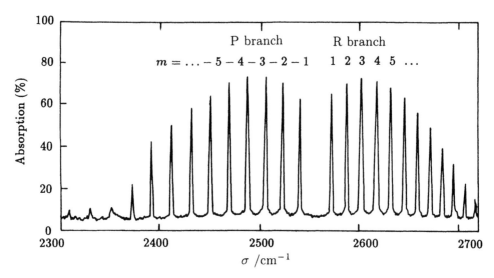

Figure 5.2: Infrared vibration-rotation absorption spectrum in HBr gas for the transition $n = 0 \rightarrow n = 1$ [G.M. Barrow, *Introduction to Molecular Spectroscopy*, Mc Graw-Hill, 1962]. The absorption lines denoted by the number m have wave numbers equal to σ_m $/\text{cm}^{-1} = 2559.25 + 16.72m - 0.226m(m+1)$. These numbers correspond to the constants $\omega_e = 2649.67 \text{ cm}^{-1}, B_e = 8.473 \text{ cm}^{-1}, x_e = 1.706 \times 10^{-2}, \alpha_e = 0.226 \text{ cm}^{-1}$ [G. Herzberg I, *op. cit.*].

5.3.2 Spectroscopy

In spectroscopy precise measurements of molecular parameters are made by investigating the transitions between levels.

Pure rotational spectra, corresponding to the selection rules $\Delta n = 0, \Delta K = \pm 1$, are generally obtained by absorption ($\Delta K = +1$) of electromagnetic waves (Fig. 5.1a). For the majority of molecules they lie in the millimetre wavelength range, and in the far infrared for molecules containing a hydrogen atom.

Vibration-rotation spectra obey the selection rules $\Delta n = 1, \Delta K = \pm 1$ (Fig 5.1b) and are obtained by absorption in the near infrared range. For each vibrational transition ($n \rightarrow n + 1$), two series of lines are observed, belonging to the branch $R(\Delta K = +1)$ and the branch $P(\Delta K = -1)$ (Fig. 5.2), lying on either side of the pure vibrational frequency (corresponding to the forbidden transition $\Delta n = 1, \Delta K = 0$). In the approximation of (5.39), these lines are equidistant, but as K increases, vibration-rotation coupling brings the lines of the R branch closer together while separating those of the P branch. The intensity of the lines is proportional to the population (5.58) of the initial level. Note that these transitions are also present in electronic transitions in the visible and ultraviolet range.

Raman spectra, obtained by scattering of a monochromatic light beam by the gas, detect transitions that obey the selection rules $\Delta K = 0, \pm 2$. Two branches, O and S, are observed lying on each side either of the Rayleigh line (pure scattering $\Delta n = 0, \Delta K = 0$) or of the pure vibrational Raman line $Q(\Delta n = 1, \Delta K = 0)$.

Note that the absorption spectra of rotation and vibration-rotation occur only for diatomic molecules that carry a permanent electric dipole moment. They therefore cannot be observed for symmetric molecules. This rule does not apply to Raman scattering, which thus provides a means of investigating diatomic molecules whether they are symmetric or not.

Exercise 5.3 *Investigation of a vibration-rotation spectrum*
　　Figure 5.2 shows a vibration-rotation spectrum for the HBr molecule, for which the molecular constants are given. Using expression (5.43), calculate the position of the lines corresponding to the transitions $K = 2 \rightarrow K' = 3$ and $K = 4 \rightarrow K' = 3$. Identify these lines in the figure and find the correspondence between the quantum numbers K and m.

Solution
　　Using the values $B_0 = B_e - \alpha_e/2 = 8.360$ cm^{-1} and $B_1 = B_e - 3\alpha_e/2 = 8.134$ cm^{-1}, we find

$$\frac{\epsilon_{3,1} - \epsilon_{2,0}}{hc} = 2656.66 - 50.16 = 2606.5 \text{ cm}^{-1}$$

$$\frac{\epsilon_{3,1} - \epsilon_{4,0}}{hc} = 2656.66 - 167.20 = 2489.5 \text{ cm}^{-1} \ .$$

These two wave numbers are those of the lines denoted $m = 3$ and $m = -4$. For the branch R, m is equal to the number K' of the final level of the transition, while, in branch P, m is the negative of the number K of the initial level.

5.3.3　Additivity of the Thermodynamic Functions

The energy ϵ of a diatomic molecule is the sum of its translational energy and its energy in the centre of inertia reference frame. The translational energy $\epsilon^{(t)}$ is the same as for a monatomic gas (§5.2.2). The energy of the molecule in its reference frame can, to a good approximation, be taken as the sum of its rotational energy $\epsilon^{(r)}$ (5.41), its vibrational energy $\epsilon^{(v)}$ (5.42) and its electronic energy $\epsilon^{(e)}$. Thus

$$\epsilon = \epsilon^{(t)} + \epsilon^{(r)} + \epsilon^{(v)} + \epsilon^{(e)} \ . \tag{5.44}$$

It follows (§3.3) that the partition function Z is the product of the partial partition functions

$$Z = Z^{(t)} Z^{(r)} Z^{(v)} Z^{(e)} \ . \tag{5.45}$$

$Z^{(t)}$ is given by the same expression (5.18) as for a monatomic gas, where $m = m_A + m_B$ is now the total mass of the molecule. The three functions $Z^{(r)}$, $Z^{(v)}$ and $Z^{(e)}$, to be investigated in the following paragraphs, depend only on the temperature, and are independent of the volume.

　　It will be noted that here we have omitted the nuclear energy. This is because only the lowest nuclear level is populated and the nuclear partition function $Z^{(n)}$ reduces to a constant that is equal to the degeneracy of that level, $g^{(n)} = (2I_1 + 1)(2I_2 + 1)$, where I_1 and I_2 are the spins of the two nuclei. As in the case of monatomic molecules (§5.2.1), $Z^{(n)} = g^{(n)}$ is generally omitted from the partition function.

As in (5.14), it follows from the factorization of the partition function Z that the free energy (5.4) is the sum of four terms

$$F = F^{(t)} + F^{(r)} + F^{(v)} + F^{(e)} . \tag{5.46a}$$

Hence the internal energy, entropy, heat capacity and the chemical potential will also consist of 4 terms. For the heat capacity, we have

$$C_V = C_V^{(t)} + C_V^{(r)} + C_V^{(v)} + C_V^{(e)} . \tag{5.46b}$$

The pressure (5.6), however, obtained by differentiating F with respect to V, contains only one term, that due to translation. The expression for the free energy of translation is, as in (5.14)

$$
\begin{aligned}
F^{(t)} &= -NkT \left(\ln \frac{Z^{(t)}}{N} + 1 \right) \\
&= -NkT \left[\ln \left(\frac{V}{N} \frac{(2\pi m kT)^{3/2}}{h^3} \right) + 1 \right] . \tag{5.47}
\end{aligned}
$$

We therefore have

$$P = - \left(\frac{\partial F}{\partial V} \right)_{T,N} = - \left(\frac{\partial F^{(t)}}{\partial V} \right)_{T,N} = \frac{NkT}{V} .$$

In other words, diatomic gases also obey the equation of state for ideal gases.

As in the case of monatomic gases (§5.2.3) we get

$$U^{(t)} = \frac{3}{2} NkT \quad \text{and} \quad c_V^{(t)} = \frac{3}{2} R . \tag{5.48}$$

It can be seen that the thermodynamic contribution from translation is the same as for a monatomic gas.

Since the partition functions other than $Z^{(t)}$ do not depend on V, the total partition function Z and the free energy F take the form

$$Z(V,T) = V f(T) \tag{5.49a}$$

$$\text{and } F(T,V,N) = -NkT \ln \frac{V}{N} + Nk\phi(T) . \tag{5.49b}$$

These expressions are valid for all ideal molecular gases.

5.3.4   Rotational Partition Function

High temperatures

From expression (5.41) for the rotational energy levels, the rotational partition function $Z^{(r)}$ is

$$Z^{(r)}(T) = \sum_{K=0}^{\infty} (2K+1) e^{-K(K+1)\Theta_r/T} \tag{5.50}$$

Table 5.1: Physical constants of several asymmetric diatomic gases. Molar mass M, normal boiling point T_B, characteristic rotational, vibrational and electronic temperatures Θ_r, Θ_v and Θ_e, and degeneracy of the electronic ground state $g_0^{(e)}$. The degeneracy of the first excited level of NO is $g_1^{(e)} = 2$ and the next excited level occurs at 63 300 K. Except for HD and NO, $\Theta_r \sim 10$ K $\ll \Theta_v \sim 10^3$ K$\ll \Theta_e \sim 10^5$ K and $\Theta_r \ll T_B \ll \Theta_v$. The characteristic temperatures were obtained from spectroscopic data [G. Herzberg I, *op. cit.*].

Molecule	M /(g mol^{-1})	T_B /K	Θ_r /K	Θ_v /K	$g_0^{(e)}$	Θ_e /K
HD	3.02	22.1	65.7	5 492	1	132 000
CO	28.01	81.7	2.78	3 122	1	93 600
NO	30.01	121	2.45	2 740	2	174
HCl	36.47	188	15.2	4 302	1	63 300
HBr	80.91	206	12.2	3 812	1	50 400

where the characteristic rotational temperature Θ_r is defined by

$$\Theta_r = \frac{\hbar^2}{2I_e k} = \frac{hc}{k}B_e \ . \tag{5.51}$$

The values of Θ_r for several gases are listed in Table 5.1. It can be seen that in general they are much lower than the boiling temperatures at atmospheric pressure T_B. For these gases the condition $T \gg \Theta_r$ holds, and the sum (5.50) defining $Z^{(r)}$ can therefore be replaced by the integral

$$Z^{(r)} = \int_0^\infty (2K + 1)e^{-K(K+1)\Theta_r/T} dK \ .$$

Setting $u = K(K + 1)$, we obtain

$$Z^{(r)} = \int_0^\infty e^{-u\Theta_r/T} du = \frac{T}{\Theta_r} \ . \tag{5.52}$$

The free energy, internal energy and rotational heat capacity then become

$$F^{(r)} = -NkT \ln Z^{(r)} = -NkT \ln \left(\frac{T}{\Theta_r}\right) , \tag{5.53}$$

$$U^{(r)} = NkT^2 \frac{d \ln Z^{(r)}}{dT} = NkT , \tag{5.54}$$

$$c_V^{(r)} = Nk = R \ . \tag{5.55}$$

The last result is in agreement with experiment.

These results correspond to the limit for which it is possible to use the classical expression for the energy of rotation

$$\epsilon^{(r)} = \frac{1}{2}I_e(\dot\theta^2 + \sin^2\theta \ \dot\phi^2) = \frac{1}{2I_e}\left(p_\theta^2 + \frac{p_\phi^2}{\sin^2\theta}\right) \tag{5.56}$$

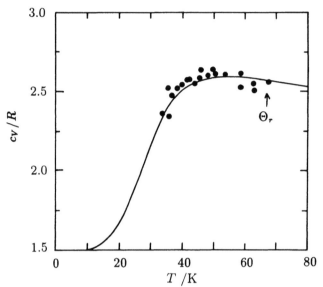

Figure 5.3: Molar heat capacity c_V of hydrogen HD expressed as a fraction of R. The continuous line is the theoretical curve for $\Theta_r = 65.7$ K. The points are the experimental data [K. Clusius and E. Bartolomé, *Z. Elektrochem.* **40**, 524 (1934)].

where $\dot\theta$ and $\dot\phi$ are the rotational angular velocities of the molecules and p_θ and p_ϕ are the associated momenta. The partition function in the classical limit is then

$$
\begin{aligned}
Z^{(r)} &= \int \frac{d\theta \; dp_\theta}{h} \frac{d\phi \; dp_\phi}{h} \exp(-\beta \epsilon^{(r)}) \\
&= \frac{1}{h^2} \int_0^{2\pi} d\phi \int_{-\infty}^{+\infty} dp_\theta \exp\left(\frac{-\beta p_\theta^2}{2I_e}\right) \times \int_0^{\pi} d\theta \int_{-\infty}^{+\infty} dp_\phi \exp\left(-\frac{\beta p_\phi^2}{2I_e \sin^2\theta}\right) \\
&= \frac{1}{h^2} 2\pi \times (2\pi I_e kT)^{1/2} \times \int_0^{\pi} d\theta (2\pi I_e kT \sin^2\theta)^{1/2} \\
&= \frac{8\pi^2 I_e kT}{h^2} = \frac{T}{\Theta_r} \; .
\end{aligned}
\tag{5.57}
$$

Note that a diatomic molecule has two degrees of rotational freedom (θ and ϕ) and that equation (5.54) for the internal energy corresponds to the application of a generalized theorem of equipartition of energy.

The HD molecule

The HD molecule is remarkable in that at atmospheric pressure the gaseous state persists at temperatures $T \lesssim \Theta_r$. Replacement of the sum by an integral in expression (5.50) for $Z^{(r)}$, which is acceptable for $T \gtrsim 2\,\Theta_r$, is no longer valid for $T < 2\Theta_r$. The heat capacity is thus calculated numerically from formulae (5.9b) by truncating the series (5.50) for $Z^{(r)}$. Truncation at $K = 4$ provides a satisfactory approximation for $c_V^{(r)}$, and is shown in Figure 5.3.

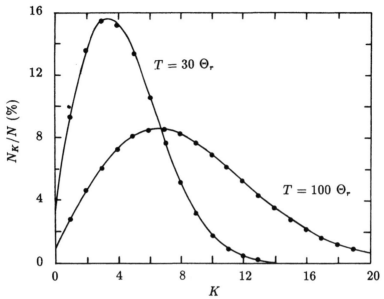

Figure 5.4: Population of the rotational levels of an asymmetric diatomic molecule for two values of the temperature (Θ_r is the characteristic temperature for rotation of the molecule).

The rotational heat capacity displays a maximum at $T = 0.81\Theta_r$ and tends to zero at absolute zero in conformity with the Third Law. The theory is in good agreement with the experimental observations for HD gas (Fig 5.3).

Distribution within the rotational levels

In the approximation we have adopted, the rotational energy of a molecule enters the total energy as an additive term. The distribution within the rotational levels is then given by (cf. §3.3)

$$N_K = \frac{N}{Z^{(r)}}(2K + 1)e^{-K(K+1)\Theta_r/T} . \tag{5.58}$$

Figure 5.4 shows this distribution as a function of the rotational quantum number K. N_K is the product of the degeneracy, which increases with K, and the Boltzmann factor, which decreases. N_K therefore exhibits a maximum. If K is taken to be a continuous variable, this maximum is located at $K_M = (T/2\Theta_r)^{1/2} - 1/2$. For example, in HBr ($\Theta_r = 12.2$ K) at $T = 300$ K, the maximum occurs at $K_M = 3$.

Since the intensity of the absorption lines in the vibration-rotation spectra (Fig 5.2) is proportional to the population of the initial level of the transition, these spectra yield a picture of the distribution. From the position of the maximum, which depends on T/Θ_r, measurements can be made of the temperature of planetary atmospheres or of interstellar clouds.

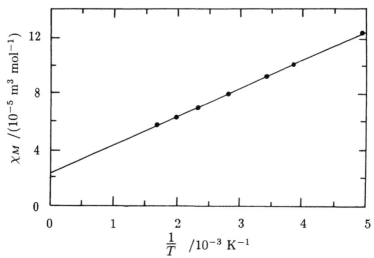

Figure 5.5: Molar dielectric susceptibility $\chi_M = \chi v$ for HCl gas. The straight line is given by $\chi_M /(m^3 \, mol^{-1}) = 2.33 \times 10^{-5} + 2.01 \times 10^{-2} \times (T /K)^{-1}$ [C.T. Zahn, *Phys. Rev.* **24**, 400 (1924)].

Exercise 5.4 *Dielectric susceptibility of hydrogen chloride gas HCl*

The HCl molecule has a permanent dipole moment of magnitude $p = 1.11$ D (1 D = 1 debye = 3.336×10^{-30} C m = 10^{-18} cgs esu) [E.W. Kaiser, *J. Chem. Phys.* **53**, 1686 (1970)]. When an electric field $\mathcal{E}$ is present, an orientational energy term $-\mathbf{p} \cdot \mathcal{E}$ must be included in the energy of the molecule.

1 In the classical limit $T \gg \Theta_r$, what is the partial partition function $Z^{(rp)}$ for rotation and polarization of the gas?

2 What is the contribution $F^{(p)}$ to the free energy from the dipole moment? Hence show that in an electric field the gas displays a dipole moment $\mathcal{P}$. Find an expression for this (Langevin equation, 1905).

3 Show that in a weak field, $\mathcal{P} = \epsilon_0 V \chi \mathcal{E}$ (linear dielectric), where $\epsilon_0 = (\mu_0 c^2)^{-1} = 8.854 \times 10^{-12}$ SI is the permittivity of free space and χ is the dielectric susceptibility of the substance. Experimental measurements of the molar susceptibility $\chi_M = \chi v$ where v is the molar volume, are shown in Figure 5.5 for HCl gas. Does the Langevin theory explain these results?

Solution

1 In the classical limit, the energy of rotation and orientation in an electric field is (cf. 5.56)

$$\epsilon^{(rp)} = \frac{1}{2I_e} \left(p_\theta^2 + \frac{p_\phi^2}{\sin^2 \theta} \right) - p\mathcal{E} \cos \theta .$$

The partition function is then

$$Z^{(rp)} = \int \frac{d\theta \, dp_\theta}{h} \int \frac{d\phi \, dp_\phi}{h} \exp(-\beta \epsilon^{(rp)}) .$$

As in (5.57) for $Z^{(r)}$, integration over p_θ, ϕ and p_ϕ yields

$$Z^{(rp)} = \frac{4\pi^2 I_e kT}{h^2} \int_0^\pi d\theta \sin\theta \exp(\beta p\mathcal{E} \cos\theta) \ .$$

Changing the variable $u = \beta p\mathcal{E} \cos\theta$, and integrating gives

$$Z^{(rp)} = \frac{8\pi^2 I_e kT}{h^2} \frac{\sinh \beta p\mathcal{E}}{\beta p\mathcal{E}} = Z^{(r)} Z^{(p)} \quad \text{with} \quad Z^{(p)} = \frac{\sinh \beta p\mathcal{E}}{\beta p\mathcal{E}} \ . \tag{5.59}$$

It can be seen that the partial partition function $Z^{(p)}$ tends to 1 as $\mathcal{E}$ tends to zero.

2 The contribution to the free energy is $F^{(p)} = -NkT \ln Z^{(p)}$. The dipole moment of the gas is then

$$\mathcal{P} = -\left(\frac{\partial F}{\partial \mathcal{E}}\right)_{T,V,N} = Np\, L(x) \quad \text{where} \quad x = \frac{p\mathcal{E}}{kT} \ , \tag{5.60}$$

and where $L(x)$ is the Langevin function (§4.4.4)

$$L(x) = \coth x - \frac{1}{x} \ . \tag{5.61}$$

3 In a weak field ($x \ll 1$), the Langevin function takes the limiting form $L(x) = x/3$ and the dipole moment of the gas becomes

$$\mathcal{P} = \epsilon_0 V \chi \mathcal{E} \quad \text{where} \quad \chi = \frac{N}{V} \frac{p^2}{3\epsilon_0 kT} \ . \tag{5.62}$$

The susceptibility is proportional to the density of particles in the gas and, like the Curie law, varies as $1/T$. For strong fields ($\mathcal{E} \simeq 10^6$ V/m) and for temperatures around 100 K, x is of the order of 10^{-3} and the Langevin equation still remains within its linear region. For hydrogen chloride gas the molar susceptibility is given in the Langevin model by

$$\chi_M = \chi v = \frac{\mathcal{N} p^2}{3\epsilon_0 kT} \ ;$$

with $p = 1.11$ D, $\chi_M = 2.26 \times 10^{-2}/T$ in SI units. This term is in good agreement with the measured $1/T$ dependence of the susceptibility. An additional constant term arises owing to the deformation of the molecule in an electric field; the polarizability α of a molecule, which is the ratio of the induced dipole moment to the product $\epsilon_0 \mathcal{E}$, is equal to $\alpha = \chi_M(T = \infty)/\mathcal{N}$. According to Figure 5.5, $\alpha = 3.87 \times 10^{-29}$ m^3 for HCl. Conversely, the term in $1/T$ in the susceptibility χ_M yields the value of the dipole moment of the molecule. For HCl, this would give $p = 1.05$ D, in good agreement with the value $p = 1.11$ D obtained from electric dipole resonance in molecular beams.

5.3.5 Vibrational Partition Function

From expression (5.42) for the vibrational energy levels, the vibrational partition function can be written

$$Z^{(v)}(T) = \sum_{n=0}^{\infty} e^{-n\Theta_v/T} \; , \tag{5.63}$$

where the characteristic temperature of vibration Θ_v is defined by

$$\Theta_v = \frac{\hbar\omega}{k} = \frac{hc}{k}\omega_e \; . \tag{5.64}$$

The values of Θ_v for several gases are listed in Table 5.1. It can be seen that in general they are of the order of several thousand degrees. At ordinary temperatures $T \ll \Theta_v$.

The partition function (5.63) is the sum of a geometrical series, and is equal to

$$Z^{(v)} = \frac{1}{1 - e^{-\Theta_v/T}} \; , \tag{5.65}$$

The vibrational free energy, the internal energy and the molar heat capacity are then given by

$$\begin{aligned}
F^{(v)} &= -NkT \ln Z^{(v)} = NkT \ln(1 - e^{-\Theta_v/T}) \; , & (5.66) \\
U^{(v)} &= NkT^2 \frac{d\ln Z^{(v)}}{dT} = \frac{Nk\Theta_v}{e^{\Theta_v/T} - 1} \; , & (5.67) \\
c_V^{(v)} &= R\left(\frac{\Theta_v}{T}\right)^2 \frac{e^{\Theta_v/T}}{(e^{\Theta_v/T} - 1)^2} \; . & (5.68)
\end{aligned}$$

These results are similar to those found for the Einstein model for solids (§3.6); in particular the heat capacity is equal to $R\,E(\Theta_v/T)$, where $E(x)$ is the Einstein function (3.34). As $T \to \infty$, the vibrational heat capacity tends to R, in accordance with the theorem of equipartition of energy for a set of one-dimensional harmonic oscillators (§3.5). When $T \lesssim 0.1\,\Theta_v$, i.e. in the vicinity of room temperature, this heat capacity is negligible and the heat capacity of the gas, $c_V = 5R/2$, is the sum of the translational and the rotational heat capacities (5.48) and (5.55). The curve of $5R/2 + c_V^{(v)}$ is shown in Figure 5.6. It is in satisfactory agreement with the experimental measurements for various gases in the temperature range 500 to 2000 K.

The relative population of the vibrational level with quantum number n is

$$\frac{N_n}{N} = (1 - e^{-\Theta_v/T})e^{-n\Theta_v/T} \; . \tag{5.69}$$

At room temperature, $T \ll \Theta_v$, and most of the molecules are therefore in the vibrational ground state.

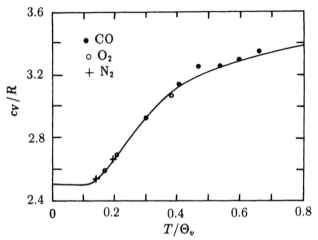

Figure 5.6: Molar heat capacity of various diatomic gases. The curve is the sum of the translational heat capacity ($3R/2$), the rotational heat capacity (R) and the vibrational heat capacity given by (5.68) [G.G. Sherratt and E. Griffiths, *Proc. Roy. Soc. A* (London) **147**, 292 (1934); A. Eucken und O. Mücke, *Z. Phys. Chem.* (Leipzig) B **18**, 167 (1932)].

5.3.6   Electronic Partition Function

As in the case of monatomic gases (§5.2.1), the first excited electronic level is in general very high (Table 5.1), and $Z^{(e)}$ reduces to

$$Z^{(e)} = g_0^{(e)} \ . \tag{5.70}$$

Thus in general the excited electronic levels make no contribution to the heat capacity.

A notable exception is the molecule of nitrous oxide, NO, the first excited state of which has $\Theta_e = 174$ K, while the other excited levels lie much higher. Since the ground state and the first excited state both have a degeneracy of 2, the electronic partition function is

$$Z^{(e)} = 2\left(1 + e^{-\Theta_e/T}\right) \ . \tag{5.71}$$

This partition function is that of a two-level system (3.19) and yields the contributions to the thermodynamic functions given in (3.21-24). In particular, the molar electronic heat capacity is

$$c_V^{(e)} = R\left(\frac{\Theta_e}{T}\right)^2 \frac{e^{\Theta_e/T}}{\left(e^{\Theta_e/T} + 1\right)^2} \ . \tag{5.72}$$

Above the normal boiling point of NO ($T_B = 121$ K) and up to about 300 K, the total molar heat capacity of the gas is $c_V = 5R/2 + c_V^{(e)}$. This relation is in good agreement with the experimental results (Fig. 5.7). Above 500 K, $c_V^{(e)}$ becomes negligible, but vibrations must be taken into account, and hence $c_V = 5R/2 + c_V^{(v)}$. Between 300 and 500 K, both contributions $c_V^{(v)}$ and $c_V^{(e)}$ must be added to $5R/2$.

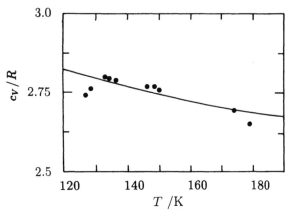

Figure 5.7: Molar heat capacity of NO gas. The curve is the sum of the heat capacities of translation ($3R/2$) and rotation (R) and of the electronic heat capacity given by (5.72) with $\Theta_e = 174$ K. [A. Eucken and L. d'Or, *Nachr. Akad. Wiss. Göttingen*, Math.-Phys. Kl., 2, 107 (1932)].

When the electronic ground state has angular momentum, this couples with the rotational angular momentum. However, for $T \gg \Theta_r$, the value of the partition function is not modified.

5.4 SYMMETRIC DIATOMIC IDEAL GASES

5.4.1 Symmetry of the Molecular Wave Function

When a molecule is composed of two identical atoms (homonuclear molecule), its wave function is governed by the principle of symmetry with respect to exchange of the two nuclei (§2.5.1): *the wave function of the molecule must either be symmetric or antisymmetric under simultaneous exchange of the nuclear co-ordinates* r_i *and the projections of the spins* m_i, *according to whether the nuclei are bosons or fermions.* That is, neglecting the electron co-ordinates,

$$\psi(\mathbf{r}_2, m_2, \mathbf{r}_1, m_1) = \pm \psi(\mathbf{r}_1, m_1, \mathbf{r}_2, m_2) \quad . \tag{5.73}$$

If couplings, such as rotation-vibration coupling, are neglected, the wave function ψ factorizes into the form

$$\psi(\mathbf{r}_1, m_1, \mathbf{r}_2, m_2) = \psi^{(t)}(\,\mathbf{R}\,)\; \psi^{(r)}(\theta, \phi)\; \psi^{(v)}(r)\; \psi^{(n)}(m_1, m_2) \tag{5.74}$$

where $\mathbf{R} = (\mathbf{r}_1 + \mathbf{r}_2)/2$ defines the centre of mass, and r, θ and ϕ are the spherical co-ordinates of the separation vector $\mathbf{r} = \mathbf{r}_2 - \mathbf{r}_1$. The wave functions $\psi^{(t)}, \psi^{(r)}, \psi^{(v)}$ and $\psi^{(n)}$ describe the state of translation, rotation, vibration and nuclear spin of the molecule. Under exchange of the two nuclei, the centre of mass $\mathbf{R}$ of the molecule, and hence $\psi^{(t)}$, remain invariant, while the distance r between the two nuclei, and therefore $\psi^{(v)}$, are also invariant. Furthermore, θ changes into $\pi - \theta$, ϕ into $\phi + \pi$, and m_1 and m_2 interchange. It is therefore clear that only the product $\psi^{(r)}\, \psi^{(n)}$ is subject to the principle of symmetrization.

Table 5.2: Symmetry conditions for the wave functions $\psi^{(n)}$ and $\psi^{(r)}$ of a homonuclear molecule. The product $\psi^{(r)}\psi^{(n)}$ must be symmetric (ss or aa) for bosons, and antisymmetric (as or sa) for fermions. The degeneracy connected with the spin I of the nuclei is shown.

$\psi^{(n)} \downarrow \quad \psi^{(r)} \rightarrow$	K even s	K odd a	Nuclear degeneracy
ortho s	boson ss	fermion sa	$(2I+1)(I+1)$
para a	fermion as	boson aa	$(2I+1)I$

When the molecule is in a quantum rotational state characterized by the two quantum numbers K and $m(m = -K, \ldots, K)$, the wave function $\psi^{(r)}$ is the spherical harmonic $Y_K^m(\theta, \phi)$ obeying the relation

$$Y_K^m(\pi - \theta, \phi + \pi) = -(1)^K Y_K^m(\theta, \phi) \ . \tag{5.75}$$

The rotational wave function is thus either symmetric (s) or antisymmetric (a) according to whether the rotational quantum number K is even or odd.

The nuclear wave function $\psi^{(n)}$ should also be symmetric (s) or antisymmetric (a) with respect to exchange of the two nuclei. For a value I of the nuclear spin, the wave function $\psi^{(n)}$ is constructed from the $\rho = 2I + 1$ wave functions $\sigma_p(p = -I, \ldots, I)$ of each nucleus in the form of products $\sigma_p(1)\sigma_q(2)$. Among these ρ^2 products, ρ are symmetric, those with $p = q$. Out of the remaining $\rho(\rho - 1)$ products, $\rho(\rho - 1)/2$ symmetric combinations and the same number of antisymmetric combinations

$$\psi^{(n)} = \frac{1}{\sqrt{2}}[\sigma_p(1)\sigma_q(2) \pm \sigma_q(1)\sigma_p(2)] \ , \tag{5.76}$$

can be formed.

There are therefore $\rho(\rho - 1)/2 = (2I + 1)I$ wave functions, of which $\psi^{(n)}$ are antisymmetric and $\rho(\rho + 1)/2 = (2I + 1)(I + 1)$ are symmetric. These different wave functions are nuclear states that can be considered as being degenerate in energy, since the nuclear spins couple very weakly with each other or with the electrons (the hyperfine coupling having a characteristic temperature $\Theta_n \sim 10^{-6}$ K). However, since the coupling between the nuclear spin and the electron is weak, it induces few transitions between the different nuclear states. It follows that a molecule remains in a given nuclear state for a long time. In the symmetric or antisymmetric state molecules are respectively designated by the name ortho or para.

Since the product $\psi^{(r)}\psi^{(n)}$ must be symmetric for bosons and antisymmetric for fermions, ortho and para molecules can occur only in even or odd rotational states respectively (Table 5.2). For the case of oxygen O_2, whose nuclear spin is $I = 0$ (boson), it can thus be seen that only one nuclear state exists (ortho-oxygen) and the molecules can occupy only even rotational levels. Hydrogen H_2 ($I = 1/2$, fermion), has 3 symmetric (orthohydrogen) and a single antisymmetric nuclear state (parahydrogen), which occupy only odd and even rotational levels

Table 5.3: Physical constants of some homonuclear diatomic gases. The notations are the same as in Table 5.1, and I designates the nuclear spin [G. Herzberg I, *op. cit.*].

Molecule	M /(g mol^{-1})	T_B /K	Θ_r /K	I	Θ_v /K	$g_0^{(e)}$	Θ_e /K
^{1}H$_2$	2.02	20.4	87.5	1/2	6 324	1	132 000
^{2}D$_2$	4.03	~ 20	43.8	1	4 487	1	132 000
^{14}N$_2$	28.01	77.3	2.89	1	3 395	1	99 700
^{16}O$_2$	31.99	90.1	2.08	0	2 274	3	11 400
^{35}Cl$_2$	69.94	239	0.351	3/2	813	1	26 300

respectively. Deuterium D_2 ($I = 1$, boson), has 6 nuclear states for ortho-deuterium and 3 for para-deuterium, respectively occupying even and odd rotational states.

5.4.2   Partition Function

As in the case of asymmetric diatomic molecules, the energy of symmetric diatomic molecules can be written as

$$\epsilon = \epsilon^{(t)} + \epsilon^{(r)} + \epsilon^{(v)} + \epsilon^{(e)} \quad ,$$

where the different forms of the energy are expressed in the same way as before. Table 5.3 lists the characteristic temperatures of a few gases. Owing to the symmetry of the molecule, the rotational level diagram splits into two series (Fig. 5.8), one containing levels with even K, the other with odd K. We therefore write the partition function in the form

$$Z = Z^{(t)} Z^{(v)} Z^{(e)} Z^{(rn)} \tag{5.77}$$

where the partition functions $Z^{(t)}, Z^{(v)}$ and $Z^{(e)}$ are the same as for asymmetric molecules and where the partition function $Z^{(rn)}$, which accounts for the nuclear and the rotational states, is for bosons

$$Z^{(rn)} = (2I + 1)(I + 1)Z_{\text{even}} + (2I + 1)I Z_{\text{odd}} \tag{5.78a}$$

and for fermions

$$Z^{(rn)} = (2I + 1)I Z_{\text{even}} + (2I + 1)(I + 1)Z_{\text{odd}} \quad , \tag{5.78b}$$

$$\text{where} \quad Z_{\text{even(odd)}} = \sum_{K \text{even(odd)}} (2K + 1)e^{-K(K+1)\Theta_r/T} \quad . \tag{5.79}$$

At high temperatures ($T \gg \Theta_r$), as in paragraph 5.3.4, replacing the sum by an integral yields

$$Z_{\text{even}} \simeq Z_{\text{odd}} \simeq \frac{1}{2}\sum_K (2K + 1)e^{-K(K+1)\Theta_r/T} \simeq \frac{1}{2}\frac{T}{\Theta_r} \quad . \tag{5.80}$$

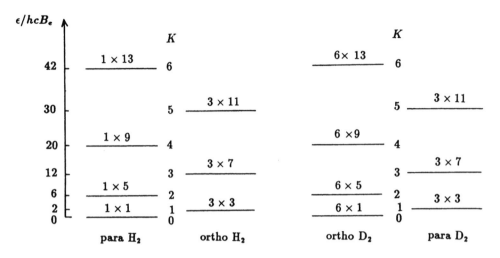

Figure 5.8: Rotational energy levels for hydrogen $H_2(I = 1/2$, fermion) and deuterium $D_2(I = 1$, boson). The rotational energy is $\epsilon_K^{(r)} = hcB_eK(K+1)$. The degeneracy of each level is the product of the nuclear degeneracy (Table 5.2) and the rotational degeneracy $2K+1$.

In both cases (bosons or fermions) we find

$$Z^{(rn)} = (2I + 1)^2\,\frac{1}{2}\frac{T}{\Theta_r}\;, \tag{5.81}$$

which is to be compared with the case of asymmetric molecules where $Z^{(rn)} = Z^{(r)}Z^{(n)} = (2I_1 + 1)(2I_2 + 1)T/\Theta_r$. Thus symmetric molecules can be regarded as a particular case of asymmetric molecules, provided a factor $1/\sigma$ is included in the partition function, where $\sigma = 2$ for diatomic molecules. From a classical standpoint, the symmetry factor or number σ stems from the fact that, as the nuclei are identical, any two opposing directions of the molecule are equivalent, and the range of integration of the angular variables is therefore reduced to a half space $(0 \le \theta \le \pi/2; 0 \le \phi < 2\pi)$.

The general expression for the partition function of symmetric molecules can then be written

$$Z = \frac{1}{\sigma}Z^{(t)}Z^{(r)}Z^{(v)}Z^{(e)}\;, \tag{5.82}$$

where $Z^{(r)}$ is calculated as for asymmetric molecules, and where, as before, the nuclear degeneracy factor $Z^{(n)}$ is omitted. The free energy of a symmetric diatomic gas is thus the same as that of an asymmetric diatomic gas (5.46), apart form the extra term $NkT\ln\sigma$. In the equation for the entropy, this gives rise to an additional term $-Nk\ln\sigma = -Nk\ln 2$. The internal energy and the heat capacity, however, are not affected.

5.4.3 Distribution within the Rotational Energy Levels

The Boltzmann relation predicts the population of the various rotational levels at
equilibrium. We first examine the case in which the nuclei are fermions. The number
of molecules occupying a rotational level with even K is

$$N_K = \frac{N}{Z^{(rn)}}(2I+1)I \times (2K+1)e^{-K(K+1)\Theta_r/T} \tag{5.83a}$$

and with odd K

$$N_K = \frac{N}{Z^{(rn)}}(2I+1)(I+1) \times (2K+1)e^{-K(K+1)\Theta_r/T} . \tag{5.83b}$$

The total number of para and ortho molecules is therefore

$$N_{\text{para}} = \frac{N}{Z^{(rn)}}(2I+1)I Z_{\text{even}} \quad \text{and} \quad N_{\text{ortho}} = \frac{N}{Z^{(rn)}}(2I+1)(I+1)Z_{\text{odd}} . \tag{5.84}$$

At high temperatures, equations (5.80,81) give

$$\frac{N_{\text{para}}}{N} = \frac{I}{2I+1} \quad \text{and} \quad \frac{N_{\text{ortho}}}{N} = \frac{I+1}{2I+1} . \tag{5.85}$$

These equations are also valid for bosons. Hence, for nuclei with zero spin, molecules
exist only in the ortho state. For spins $I = 1/2$, the ratio $N_{\text{ortho}} : N_{\text{para}} = 3 : 1$,
while for $I = 1$, $N_{\text{ortho}} : N_{\text{para}} = 2 : 1$, etc.

5.4.4 Hydrogen

If a homonuclear diatomic gas is left for a long time at room temperature ($T \gg \Theta_r$),
it is in thermodynamic equilibrium and the ortho and para molecules are in the
ratios given by (5.85). If the temperature is reduced the ratio of ortho and para
remains fixed, since transitions between the two states (reversal of a nuclear spin)
seldom occur. The gas is then in a state of metastable equilibrium in which it
behaves like a mixture of two different chemical species in the proportion (5.85).
For this mixture one should therefore write

$$F = \frac{I}{2I+1}F_{\text{para}} + \frac{I+1}{2I+1}F_{\text{ortho}} , \tag{5.86}$$

rather than $F = -NkT \ln Z$ with Z as in (5.77), since this relation would be valid
only if the gas were at thermodynamic equilibrium at the given temperature.

These considerations give rise to special behaviour only in the case of hydrogen
and deuterium H_2 and D_2, since these are the only gases that can be investigated
in the temperature range $T \lesssim \Theta_r$. In this range, only the vibrational and electronic
ground levels are occupied. For H_2 (fermions, $I = 1/2$) and D_2 (bosons, $I = 1$), we
have

$$\begin{aligned}
H_2 &: F_{\text{para}} = F_{\text{even}} \quad \text{and} \quad F_{\text{ortho}} = F_{\text{odd}} \\
D_2 &: F_{\text{para}} = F_{\text{odd}} \quad \text{and} \quad F_{\text{ortho}} = F_{\text{even}}
\end{aligned} \tag{5.87}$$

with the definitions

$$F_{\text{even(odd)}} = -NkT\left(\ln\frac{Z^{(t)}}{N} + 1\right) - NkT\ln Z_{\text{even(odd)}} \tag{5.88}$$

where $Z^{(t)}$ is the partition function (5.18) and Z_{even} and Z_{odd} are sums (5.79) over the even and odd rotational states. The free energies of metastable hydrogen and deuterium are then

$$\text{H}_2 \quad : \quad F = \frac{1}{4}F_{\text{even}} + \frac{3}{4}F_{\text{odd}}$$

$$\text{D}_2 \quad : \quad F = \frac{2}{3}F_{\text{even}} + \frac{1}{3}F_{\text{odd}} \tag{5.89}$$

The corresponding molar heat capacities are

$$\text{H}_2 \quad : \quad c_V = \frac{3}{2}R + \frac{1}{4}c_V \text{ even} + \frac{3}{4}c_V \text{ odd}$$

$$\text{D}_2 \quad : \quad c_V = \frac{3}{2}R + \frac{2}{3}c_V \text{ even} + \frac{1}{3}c_V \text{ odd} \quad , \tag{5.90}$$

with

$$c_V \text{ even} = R\beta^2\frac{d^2\ln Z_{\text{even}}}{d\beta^2} \quad \text{and} \quad c_V \text{ odd} = R\beta^2\frac{d^2\ln Z_{\text{odd}}}{d\beta^2} \quad . \tag{5.91}$$

As it is not possible to find general analytical expressions for the rotational partition functions, the numerical method applied previously is used in the case of the HD molecule (§5.3.4). The curves obtained in this way for the heat capacities (5.90) of hydrogen and deuterium, displayed in Figure 5.9, are in excellent agreement with the experimental data and confirm the validity of the model of a metastable mixture of ortho and para gases.

When hydrogen gas is cooled from room temperature to $T \simeq T_B = 20$ K, all the parahydrogen molecules (25 %) are in the rotational level $K = 0$ and the orthohydrogen (75 %) in the level $K = 1$. In the resulting liquid, interactions between molecules are stronger than in the gas, inducing reversals of the nuclear spins. This pushes the gas towards its stable thermodynamic state, which at this temperature is composed of pure parahydrogen in the level $K = 0$. The energy released in the transition $K = 1 \to K = 0$ evaporates the hydrogen, a process that may take about one week. To avoid too rapid evaporation, hydrogen gas is prepared in a state of stable equilibrium at low temperatures (parahydrogen) by contact with a chemical catalyser (dissociation and recombination of the molecules) or a paramagnetic substance (interaction between electron and nuclear spins) inside the liquefier itself.

The heat capacity of metastable hydrogen between its boiling point and 3000 K, displayed in Figure 5.10, reveals the plateaux corresponding to the classical values $c_V = 3R/2$ (translation only), $c_V = 5R/2$ (translation and rotation), and also the intermediate ranges where the rotational and vibrational levels become progressively populated. The continuous curve, given by

$$c_V = \frac{3}{2}R + \frac{1}{4}c_V \text{ even} + \frac{3}{4}c_V \text{ odd} + c_V^{(v)} \quad , \tag{5.92}$$

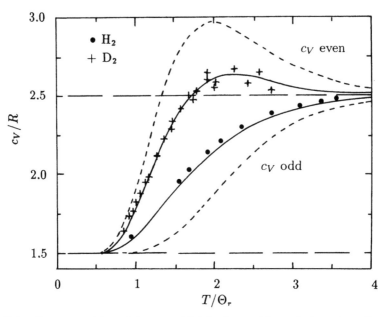

Figure 5.9: Heat capacities of (metastable) hydrogen H_2 and deuterium D_2. The full curves are the theoretical values (5.90) and the dashed curves are equations (5.91). For the experimental points, we have taken Θ_r (H_2) = 87.5 K and Θ_r (D_2) = 43.8 K. [R.E. Cornish and E.D. Eastman, *J. Am. Chem. Soc.* **50**, 627 (1928); K. Clusius and E. Bartholomé, *Z. Elektrochem.* **40**, 524 (1934)].

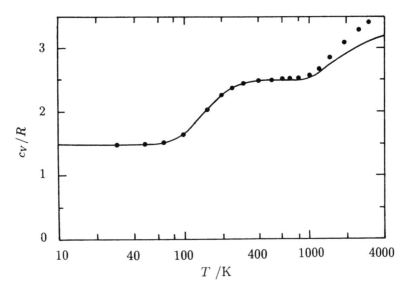

Figure 5.10: Molar heat capacity of metastable hydrogen. The curve is given by equation (5.92) [*AIP Handbook* (1963), Mc Graw-Hill].

where $c_V^{(v)}$ is the vibrational heat capacity (5.68), deviates from the experimental values for $T > 1000$ K. The discrepancy is due to anharmonicity in the vibration and to coupling with the rotation (§5.3.1). This effect is especially sensitive in the case of hydrogen owing to the small atomic mass. With the more complete expression (5.43) for the molecular energy, the statistical theory yields better agreement with experiment. In practice, heat capacities of gases at high temperatures are determined by spectroscopic methods, which are far more precise than calorimetry.

5.5   IDEAL POLYATOMIC GASES

5.5.1   Linear Molecules

A linear molecule comprising p atoms possesses $3p$ degrees of freedom, 3 for translation and 2 for rotation. Of the $3p - 5$ remaining degrees of vibrational freedom, $p - 1$ pertain to normal modes of vibration along the molecular axis, and the other $2(p - 2)$ to modes of vibration perpendicular to this axis. Owing to the axial symmetry of the molecule these modes are degenerate in pairs. For the CO_2 molecule ($p = 3$), for example, there are 4 modes of vibration, two longitudinal, and two transverse, which are mutually degenerate (Fig. 5.11). Note that the frequencies of the transverse modes, which involve bending of the molecule, are lower than those of the longitudinal modes.

To a first approximation, the energy of a molecule is equal to the sum of its energies of translation, rotation and vibration

$$\epsilon = \epsilon^{(t)} + \epsilon^{(r)} + \epsilon^{(v)}. \tag{5.93}$$

In this expression the interactions between rotation and vibration are neglected and the electronic and nuclear energies have been omitted. The translational energy has the general form $\epsilon^{(t)} = p^2/2m$, while that of rotation has the same form as for diatomic molecules (5.41,56), and the energy of vibration is the sum of the energies (5.42) of the $3p - 5$ harmonic oscillators corresponding to the normal modes of vibration. The partition function of one molecule is then

$$Z = \frac{1}{\sigma} Z^{(t)} Z^{(r)} \prod_i Z_i^{(v)} \ . \tag{5.94}$$

The translational partition function $Z^{(t)}$ is the same as that of a monatomic gas (5.18). The rotational partition function $Z^{(r)} = T/\Theta_r$ is the same as for a diatomic gas and the symmetry number σ enters in the same way (σ is equal to 2 if the molecule has a centre of symmetry and 1 otherwise). The vibrational partition function is the product of $3p - 5$ harmonic oscillator partition functions (5.65), each corresponding to a normal mode of vibration.

From the partition function Z, all the thermodynamic functions are recovered. In particular, these gases obey the ideal gas equation of state, $Pv = RT$, and their molar heat capacity is

$$c_V = \frac{5}{2} R + R \sum_i \left(\frac{\Theta_{vi}}{T} \right)^2 \frac{e^{\Theta_{vi}/T}}{\left(e^{\Theta_{vi}/T} - 1 \right)^2} \ . \tag{5.95}$$

Table 5.4: Physical constants of some linear polyatomic gases. Molar mass M in g mol^{-1}, symmetry number σ, characteristic temperatures of rotation Θ_r and vibration Θ_{vi}. The numbers in brackets indicate the degeneracy of the vibrational mode. The characteristic temperatures are obtained from spectroscopic data [G. Herzberg II, *op. cit.*].

Molecule	M	σ	Θ_r /K	Θ_{vi} /K						
HCN	27	1	2.13	1020	(2)	3010		4770		
DCN	28	1	1.74	820	(2)	2740		3780		
N$_2$O	44	1	0.602	850	(2)	1850		3200		
CO$_2$	44	2	0.560	960	(2)	2000		3380		
CS$_2$	76	2	0.157	570	(2)	950		2190		
C$_2$HD	27	1	1.43	750	(2)	980	(2)	2660	3720	4800
C$_2$H$_2$	26	2	1.69	880	(2)	1050	(2)	2840	4730	4850

In this equation $5R/2$ is the sum of the classical heat capacities of translation ($3R/2$) and rotation (R) and the Θ_{vi} are the characteristic temperatures of vibration for the $3p-5$ vibration modes (Table 5.4). Figure 5.12 compares the theoretical and experimental values of heat capacity for carbon dioxide CO$_2$.

5.5.2   Non-linear Molecules

Like linear molecules, non-linear molecules with p atoms possess $3p$ degrees of freedom, three of which are related to translation. The orientation in space of such a molecule is defined by three Euler angles θ, ϕ and ψ, and accordingly it has 3 degrees of rotational freedom. There are therefore $3p-6$ degrees of freedom of vibration, with some of the vibrational modes perhaps being degenerate, depending on the symmetry of the molecule. For example, water H$_2$O has 3 non degenerate modes of vibration and ammonia NH$_3$ has 6 modes, two of which are not degenerate, and 4 are degenerate in two pairs of two.

The formula for the energy of a polyatomic molecule has the form (5.93) and the energies of translation $\epsilon^{(t)}$ and vibration $\epsilon^{(v)}$ are the same as those already encountered, but the rotational energy $\epsilon^{(r)}$ is different. In the range of existence of gases, i.e. above their boiling point, this energy can be written in the classical form for a rigid body

$$\epsilon^{(r)} = \frac{1}{2I_1 \sin^2 \theta} [(p_\phi - p_\psi \cos \theta) \cos \psi - p_\theta \sin \theta \sin \psi]^2$$
$$+ \frac{1}{2I_2 \sin^2 \theta} [(p_\phi - p_\psi \cos \theta) \sin \psi + p_\theta \sin \theta \cos \psi]^2 + \frac{1}{2I_3} p_\psi^2 , \quad (5.96)$$

where p_θ, p_ϕ, p_ψ are the momenta associated with the angles θ, ϕ and ψ, and I_1, I_2 and I_3 are the principal moments of inertia of the molecule (setting $\psi = 0$, $p_\psi = 0$ and $I_1 = I_2$, reduces to the case of the linear molecule (5.56)).

The partition function (5.94) is the same as for a linear molecule that has for its

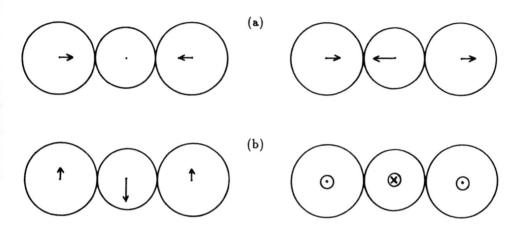

Figure 5.11: Normal modes of vibration of the CO_2 molecule. The movements of the nuclei are indicated diagrammatically by arrows. The wave numbers of the longitudinal normal modes (a) are respectively $\sigma_1 = 1388$ cm^{-1} and $\sigma_3 = 2349$ cm^{-1}. The transverse modes (b) are mutually degenerate and their wave number is equal to $\sigma_2 = 667$ cm^{-1}.

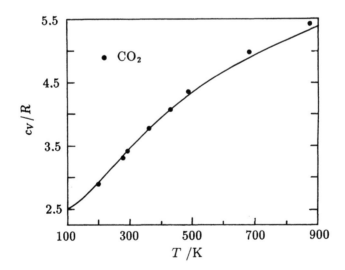

Figure 5.12: Molar heat capacity of carbon dioxide CO_2. The curve is equation (5.95) with the vibration temperatures listed in Table 5.4 [W.Heuse, *Ann. Physik* **59**, 86 (1919); A. Eucken and K. von Lüde, *Z. Phys. Chem.*, (Leipzig) B **5**, 413 (1929); A. Eucken and O. Mücke, *Z. Phys. Chem.*, (Leipzig) B **18**, 184 (1932)].

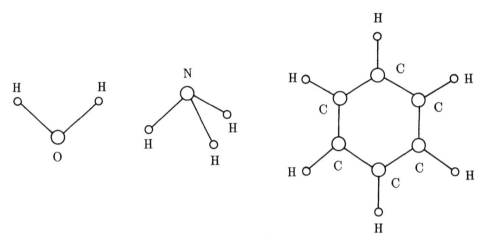

Figure 5.13: Configuration of water H_2O, ammonia NH_3 and benzene C_6H_6 molecules, with symmetry numbers $\sigma = 2, 3$ and 12 respectively.

rotational partition function (cf. 5.57)

$$Z^{(r)} = \int \frac{d\theta dp_\theta}{h} \frac{d\phi dp_\phi}{h} \frac{d\psi dp_\psi}{h} \exp(-\beta \epsilon^{(r)}) \ . \tag{5.97}$$

Integration over p_θ, p_ϕ and p_ψ (in that order) is tedious and yields

$$
\begin{aligned}
Z^{(r)} &= \frac{(2\pi kT)^{3/2}(I_1 I_2 I_3)^{1/2}}{h^3} \int_0^\pi d\theta \sin\theta \int_0^{2\pi} d\phi \int_0^{2\pi} d\psi \\
&= \pi^{1/2} \left(\frac{T^3}{\Theta_{r1}\Theta_{r2}\Theta_{r3}} \right)^{1/2} \text{ with } \Theta_{ri} = \frac{h^2}{8\pi^2 I_i k} \ .
\end{aligned} \tag{5.98}
$$

For three degrees of rotational freedom, this function is proportional to $T^{3/2}$, whereas for a linear molecule possessing two degrees of rotational freedom, it varies as T (5.52).

For an asymmetric molecule the symmetry factor σ is equal to 1. For a symmetric molecule, it is equal to the number of different but equivalent ways of orienting the molecule in space. This factor, of quantum mechanical origin, comes from the identity relation between atoms that are related to each other by a direct symmetry operation (rotation). Thus, for example, $\sigma = 2$ for water H_2O (Fig. 5.13), $\sigma = 3$ for ammonia NH_3, and $\sigma = 12$ for benzene C_6H_6.

A knowledge of the partition function from the characteristic temperatures (Table 5.5) allows the thermodynamic functions to be determined. In particular, the molar heat capacity, which can be compared directly with experiment, is given by

$$c_V = c_V^{(t)} + c_V^{(r)} + \Sigma_i c_{Vi}^{(v)} \ . \tag{5.99}$$

The translational heat capacity adopts its classical value $3R/2$, and the vibrational heat capacities are given by equation (5.68). The rotational heat capacity obtained from (5.9a) is equal to

$$c_V^{(r)} = \frac{3}{2}R \ . \tag{5.100}$$

The molar heat capacity found in this way for a gas of non-linear polyatomic molecules is

$$c_V = 3R + R\sum_i \left(\frac{\Theta_{vi}}{T}\right)^2 \frac{e^{\Theta_{vi}/T}}{\left(e^{\Theta_{vi}/T} - 1\right)^2} \ . \tag{5.101}$$

5.5.3   Hindered Rotation

In some molecules, particularly some organic molecules, certain degrees of freedom define the orientation of a group of atoms around a bond, such as the CH_3 groups in ethane C_2H_6 (configuration $H_3C\text{-}CH_3$). The interactions between one group and another give rise to a potential energy for internal rotation with a period $2\pi/3$. At low temperatures, the group oscillates around one of its equilibrium positions (torsion or libration oscillations) and the discussion of the paragraph above applies. At higher temperatures, however, the group rotates around the bond in a hindered motion. At very high temperatures, the rotation can be considered as being free. In the partition function, the factor corresponding to this degree of freedom transforms from a term of the type $Z^{(v)}$ (5.65) into one of the form

$$Z_{int}^{(r)} = \int \frac{d\phi \, dp_\phi}{h} \exp(-\beta\epsilon^{(r)}) \quad \text{where} \quad \epsilon^{(r)} = \frac{p_\phi^2}{2I} \ ,$$

$$\text{i.e.,} \quad Z_{int}^{(r)} = \left(\pi\frac{T}{\Theta_r}\right)^{1/2} \quad \text{with} \quad \Theta_r = \frac{\hbar^2}{2Ik} \tag{5.102}$$

(divided, if necessary, by the symmetry factor σ_{int} of the group). The contribution from hindered rotation to the heat capacity (Fig. 5.14) appears at low temperatures in the same form (5.68) as for a harmonic oscillator, and at high temperatures as

$$c_V^{(r)}{}_{int} = \frac{R}{2} \tag{5.103}$$

corresponding to the limiting case (5.102) of the partition function for internal rotation. In the intermediate temperature range, an exact calculation must be performed to take into account the characteristics of each molecule. For ethane, the characteristic temperature of the torsion oscillation is $\Theta_v \simeq 400$ K, while that of rotation is $\Theta_r = 7.6$ K with $\sigma_{int} = 3$ [G. Herzberg II, *op. cit.*].

5.6   IDEAL MIXTURES OF IDEAL GASES

5.6.1   Thermodynamic Functions in the Variables T, V, N_i

An ideal mixture of ideal gases is a model in which the interaction energy is neglected between molecules of the same gas as well as between those of different species. This

Table 5.5: Physical constants of some non-linear polyatomic gases. Cf. caption of Table 5.4.

Gas	$M/$ g mol^{-1}	σ	$\Theta_{r1}/$ K	$\Theta_{r2}/$ K	$\Theta_{r3}/$ K	Θ_{vi} / K
HDO	19	1				2020 ; 3910 ; $\simeq$ 5200
H$_2$O	18	2	39.3	21.0	13.7	2290 ; 5250 ; 5400
NH$_3$	17	3	14.3	9.1	9.1	1370 ; 2340 (2) ; 4800 ; 4910 (2)
CH$_4$	16	12	7.56	7.56	7.56	1880 (3) ; 2200 (2) ; 4190 ; 4350 (3)

model, like that of an ideal gas, is valid only for a mixture in which the total pressure is not too high.

Consider a vessel containing a mixture of p gases with N_i molecules of each gas i ($i = 1, 2, \ldots, p$). The energy of the mixture is

$$F(T, V, N_1, N_2, \ldots, N_p) = \sum_{i=1}^{p} F_i(T, V, N_i) \tag{5.104}$$

where F_i is obtained from the partition function Z_i by means of (5.4)

$$F_i(T, V, N_i) = -N_i kT \left[\ln \frac{Z_i}{N_i} + 1 \right] ; \tag{5.105}$$

that is, the interaction energies between the different molecules are negligible and each gas behaves microscopically as if it were alone in the container. The extensive thermodynamic quantities of the mixture are then obtained by summing the corresponding quantities for each gas on the assumption that it alone occupies the volume V at temperature T.

The following extensive variables can be obtained, either from the expression for F, or directly (§5.1.2).

$$S = \sum_i S_i \quad \text{with} \quad S_i = N_i k \left[\ln \frac{Z_i}{N_i} + 1 + T \left(\frac{\partial \ln Z_i}{\partial T} \right)_V \right] \tag{5.106}$$

$$U = F + TS = \sum_i U_i \quad \text{where} \quad U_i = N_i kT^2 \left(\frac{\partial \ln Z_i}{\partial T} \right)_V \tag{5.107}$$

$$C_V = \sum_i C_{Vi} \quad \text{with} \quad C_{Vi} = \left(\frac{\partial U_i}{\partial T} \right)_{V N_i} = N_i k\beta^2 \left(\frac{\partial^2 \ln Z_i}{\partial \beta^2} \right)_V .$$

It can be seen that the internal energy and the heat capacity of the mixture depend only on T, as in ideal gases. The pressure is obtained by differentiation,

$$P = -\left(\frac{\partial F}{\partial V} \right)_{T,N_i} = \sum_i \left[N_i kT \left(\frac{\partial \ln Z_i}{\partial V} \right)_T \right] .$$

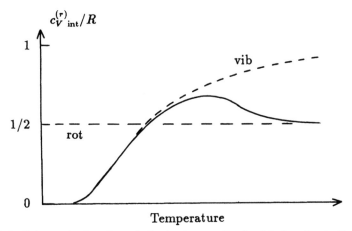

Figure 5.14: Schematic drawing of the heat capacity for hindered rotation. The heat capacities for pure vibration and pure rotation are shown as dashed lines.

In (5.49a) it was seen that for each gas

$$Z_i = V f_i(T) \tag{5.108}$$

$$\text{from which} \quad P = \sum_i \frac{N_i kT}{V} = \frac{NkT}{V} \quad ; \tag{5.109}$$

Thus, on substituting the total number of molecules $N = \sum N_i$, the same equation of state is obtained as for a pure ideal gas. It is useful to introduce the *partial pressure* P_i of the gas i , which is defined as the pressure that the gas would exert if it occupied the volume V alone, i.e.,

$$P_i = \frac{N_i kT}{V} \quad . \tag{5.110}$$

Equation (5.109) shows that the pressure of a mixture of ideal gases is the sum of the partial pressures of the constituents *(Dalton's law)*

$$P = \sum_i P_i \quad . \tag{5.111}$$

The chemical potential μ_i for one molecule of the gas i is

$$\mu_i = \left(\frac{\partial F}{\partial N_i} \right)_{T,V,N_{j \neq i}} = \left(\frac{\partial F_i}{\partial N_i} \right)_{T,V} = -kT \ln \frac{Z_i}{N_i} \quad . \tag{5.112}$$

This is the chemical potential (5.7) that the gas i would possess if it occupied the volume V on its own.

5.6.2   Thermodynamic Functions in the Variables T, P, N_i

In most applications, it is preferable to use the intensive variable P and the free enthalpy function $G = F + PV$ rather than the volume and the free energy. The equations (5.104,109) for F and P yield

$$G = - \sum_i N_i kT \ln \frac{Z_i}{N_i} = \sum_i N_i \mu_i \quad . \tag{5.113}$$

Starting from (5.108,109), we now express Z_i and μ_i in terms of the variables T, P and N_i . We obtain

$$Z_i = \frac{NkT}{P} f_i(T) = \frac{N_i kT}{P_i} f_i(T)$$

$$\text{and} \quad \mu_i = -kT \ln \left[\frac{N}{N_i} \frac{kT}{P} f_i(T) \right] = -kT \ln \left[\frac{kT}{P_i} f_i(T) \right] \quad . \tag{5.114}$$

Defining the molar fraction of the gas i in the mixture as

$$X_i = \frac{N_i}{N} = \frac{P_i}{P} \qquad (\sum_i X_i = 1) \quad , \tag{5.115}$$

we get

$$\mu_i = -kT \ln \left[\frac{kT}{P} f_i(T) \right] + kT \ln X_i$$

$$= g_i(T, P) + kT \ln X_i \quad , \tag{5.116}$$

where $g_i(T, P)$ is the free enthalpy per molecule (or chemical potential) of the pure gas i in the mixture at temperature T and pressure P. It follows that G is

$$G = - \sum_i N_i kT \ln \left[\frac{kT}{P} f_i(T) \right] + \sum_i N_i kT \ln X_i \quad . \tag{5.117}$$

From the relation

$$dG = -SdT + VdP + \sum_i \mu_i dN_i \quad , \tag{5.118}$$

the various thermodynamic quantities can be obtained in terms of the variables T, P, N_i.

The enthalpy function of the mixture $H = U + PV$ can be found directly from equations (5.107 and 109) for U and P.

$$H = \sum_i N_i kT^2 \left(\frac{\partial \ln Z_i}{\partial T} \right)_V + NkT \quad .$$

Substituting (5.108) for Z_i then gives

$$H = \sum_i N_i h_i \quad \text{with} \quad h_i = kT^2 \left[\frac{d}{dT} \ln f_i(T) + \frac{1}{T} \right] \quad , \tag{5.119}$$

where h_i is the partial enthalpy per molecule of gas i in the mixture.

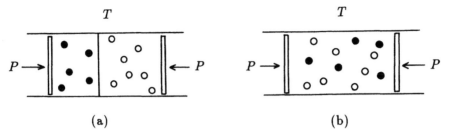

Figure 5.15: Mixing of two gases.

5.6.3 Mixing Entropy. Gibbs's Paradox

Consider a system composed of two ideal gases at the same temperature and pressure, contained in two compartments that are separated by a partition (Fig. 5.15a). The partition is withdrawn and the two gases mix by diffusion at constant temperature and pressure (Fig. 5.15b). The change in total volume ΔV and in internal energy ΔU between the initial and final state is zero. The initial volume

$$V_1 + V_2 = N_1 \frac{kT}{P} + N_2 \frac{kT}{P} = (N_1 + N_2) \frac{kT}{P}$$

is, according to (5.109), equal to the final volume, and since the temperature does not change, the internal energy of the system (5.107) does not change either. It follows that the system absorbs neither work from the pressure forces nor heat from the thermostat that maintains the temperature constant.

To calculate the change in entropy ΔS, we note that $\Delta G = \Delta U + P\Delta V - T\Delta S = -T\Delta S$. Since, from (5.113,116), we have

$$\Delta G = \sum_i N_i kT \ln X_i \ ,$$

it follows that the change in entropy is equal to

$$\Delta S = -(N_1 k \ln X_1 + N_2 k \ln X_2) \ . \tag{5.120}$$

This difference, called the entropy of mixing, is positive, since the molar fractions X_1 and X_2 are smaller than 1. In the special case where $N_1 = N_2 = N/2$, we have $X_1 = X_2 = 1/2$ and the entropy of mixing is equal to

$$\Delta S = N k \ln 2 \ .$$

This increase in entropy corresponds to the fact that the transformation is adiabatic and irreversible.

A paradox, called Gibbs's paradox, arises when the two gases are of the same kind. In fact, the transformation in this case is no longer irreversible, since replacing the partition restores the initial state. In other words, on account of extensivity, the change in entropy is zero, and not positive as in (5.120). The solution to this paradox stems from the fact that the molecules of the gas in the two compartments are identical, and thus indistinguishable from each other. It is therefore no longer possible to write $S = S_1 + S_2$ after the partition is withdrawn, since the thermodynamic probability W of one fraction of the gas cannot be calculated without taking into account the other fraction. In this case $W \neq W_1.W_2$. Note that for a mixture of isotopes, (H_2 and HD for example), the molecules are distinguishable and the entropy of mixing does indeed exist. Since the isotopic ratios remain constant in all chemical transformations, however, this entropy of mixing is generally omitted.

5.7   CHEMICAL EQUILIBRIUM IN IDEAL GASES

5.7.1   Law of Mass Action

As an example we consider the chemical reaction that occurs spontaneously at high temperature

$$2\,H_2 + O_2 \to 2\,H_2O \ . \tag{5.121}$$

Generalizing, we write a chemical reaction as

$$\sum_i \nu_i A_i \to \sum_j \nu'_j A'_j \ , \tag{5.122}$$

where A_i, A'_j are the chemical formulae of the substances to the left (reagents) and to the right (products) of the arrow and ν_i, ν'_j are their stoichiometric coefficients.

For reactions involving gases only, the gaseous mixture of the substances can, as a first approximation, be assumed to be ideal. Starting from an initial out-of-equilibrium mixture with arbitrary ratios defined by N_i^0 and $N_j'^0$, the reaction shifts the system towards an equilibrium state. The elementary changes in the number of molecules are governed by

$$-\frac{dN_i}{\nu_i} = \frac{dN'_j}{\nu'_j} = d\lambda \ . \tag{5.123}$$

The parameter λ characterizes the degree of completion of the reaction. On integration, this relation becomes

$$N_i = N_i^0 - \nu_i\lambda \quad \text{and} \quad N'_j = N_j'^0 + \nu'_j\lambda \tag{5.124}$$

where $\lambda = 0$ corresponds to the initial state of the reaction.

The equilibrium condition of a chemical system held at constant temperature T and pressure P is that its free enthalpy $G(T, P, N_i)$ should be a minimum. From (5.118), this condition can be stated as

$$\sum_i \mu_i dN_i + \sum_j \mu'_j dN'_j = 0 \ ,$$

or, using (5.123),

$$\sum_i \nu_i \mu_i = \sum_j \nu'_j \mu'_j \ . \tag{5.125}$$

This relation is called the *law of mass action*.

5.7.2 Determining the Equilibrium Constant K_p

Substitution of expression (5.114) for the chemical potentials into the law of mass action gives

$$\sum_i \nu_i \ln \left[\frac{kT}{P_i} f_i(T) \right] = \sum_j \nu'_j \ln \left[\frac{kT}{P'_j} f'_j(T) \right]$$

$$\text{or} \quad \prod_i \left(\frac{kT f_i(T)}{P_i} \right)^{\nu_i} = \prod_j \left(\frac{kT f'_j(T)}{P'_j} \right)^{\nu'_j} \quad ,$$

where P_i and P'_j are the partial pressures of the constituents at equilibrium, or alternatively,

$$\left[\frac{\prod_j P'^{\nu'_j}_j}{\prod_i P^{\nu_i}_i} \right]_{\acute{e}q.} = \frac{\prod_j [kT f'_j(T)]^{\nu'_j}}{\prod_i [kT f_i(T)]^{\nu_i}} \equiv K_p(T) . \tag{5.126}$$

This remarkable result shows that for any initial reagent/product ratio, the ratio of the partial pressures at equilibrium, raised to the power of their stoichiometric coefficients, is a function of temperature alone, called the *equilibrium constant* $K_p(T)$. Furthermore it states that K_p can be found from a knowledge of the partition functions and hence of the functions $f_i(T)$ for the different molecules present. These functions are obtained from the energy levels of the molecules.

Care must be taken with this relation. The partition functions must not be evaluated, as is usually done, by taking the ground state of each type of molecule as the origin for the energy, but rather by considering a common origin. Z_i should then be replaced by $Z_i \exp(-\beta \epsilon_i^0)$, which means also that $f_i(T)$ must be replaced by $f_i(T) \exp(-\beta \epsilon_i^0)$, where ϵ_i^0 is the ground state energy of molecule i evaluated with respect to the common origin. We then have

$$K_p(T) = \frac{\prod_j [f'_j(T)]^{\nu'_j}}{\prod_i [f_i(T)]^{\nu_i}} (kT)^{\Delta\nu} e^{-\Delta\epsilon^0/kT} \tag{5.127}$$

$$\text{with} \quad \Delta\nu = \sum_j \nu'_j - \sum_i \nu_i \quad \text{and} \quad \Delta\epsilon^0 = \sum_j \nu'_j \epsilon'^0_j - \sum_i \nu_i \epsilon^0_i . \tag{5.128}$$

In practice, the quantity

$$D_0 = -\Delta\epsilon^0 , \tag{5.129}$$

is tabulated for each type of molecule, where $\Delta\epsilon^0$ corresponds to the synthesis reaction starting from free atoms ($\Delta\epsilon^0 < 0$ for a stable molecule). Such tables allow $\Delta\epsilon^0$ to be calculated for any reaction.

5.7.3   Displacement of Equilibrium with Temperature

In order to select the optimum temperature to carry out a reaction, it is of interest to know in what direction the equilibrium will be shifted by a change in temperature. If K_p is an increasing function of T, the reaction proceeds as the temperature is raised, i.e., the reagents appearing on the left hand side of (5.122) transform into products on the right hand side. The temperature dependence of K_p can be found by inspection of the function $\ln K_p(T)$, where

$$\ln K_p = \sum_j \nu'_j \ln f'_j(T) - \sum_i \nu_i \ln f_i(T) + \Delta\nu \ln(kT) - \Delta\epsilon^0/kT \ .$$

Differentiating $\ln K_p$ with respect to T gives

$$
\begin{aligned}
\frac{d}{dT}\ln K_p &= \sum_j \nu'_j \frac{d}{dT}\ln f'_j(T) - \sum_i \nu_i \frac{d}{dT}\ln f_i(T) + \frac{\Delta\nu}{T} + \frac{\Delta\epsilon^0}{kT^2} \\
&= \sum_j \nu'_j \left[\frac{d}{dT}\ln f'_j(T) + \frac{1}{T} + \frac{\epsilon'^0_j}{kT^2} \right] \\
&\quad - \sum_i \nu_i \left[\frac{d}{dT}\ln f_i(T) + \frac{1}{T} + \frac{\epsilon^0_i}{kT^2} \right] .
\end{aligned}
$$

This can be written simply

$$\frac{d}{dT}\ln K_p = \frac{\sum_j \nu'_j h'_j - \sum_i \nu_i h_i}{kT^2},$$

using expression (5.119) for the partial molecular enthalpies and the common origin for the energy $h_i \to h_i + \epsilon^0_i$. On multiplying the numerator and denominator of the right hand side by Avogadro's number $\mathcal{N}$ we get

$$\frac{d}{dT}\ln K_p = \frac{\Delta H}{RT^2} \quad \text{with} \quad \Delta H = \sum_j \nu'_j \overline{h}'_j - \sum_i \nu_i \overline{h}_i , \tag{5.130}$$

where $\overline{h}_i, \overline{h}'_j$ are the partial molar enthalpies of the substances i and j. This result is *van't Hoff's law*, which states that, when the temperature is raised,

• the equilibrium shifts to the right if ΔH is positive (endothermic reaction),
• the equilibrium shifts to the left if ΔH is negative (exothermic reaction).

The quantity ΔH (measured in J mol^{-1}) is numerically equal to the change in enthalpy (measured in J) in a reaction that completely transforms an initial mixture consisting only of reagents with respective molar numbers ν_i into a final mixture consisting only of products with respective molar numbers ν'_j. The negative of ΔH, ($Q = -\Delta H$), is normally called the heat of reaction, so that Q is positive for an exothermic reaction.

In practice therefore, it is advantageous to work at high temperatures to obtain the reaction products if the reaction is endothermic, and at low temperatures if it is exothermic. In the latter case, since the reaction rate decreases with decreasing temperature (Pb. 6.1), catalysers are used to operate at the optimum temperature. It should be added that increasing the total pressure shifts the equilibrium to the right if $\Delta \nu$ is negative, and to the left if $\Delta \nu$ is positive (*Le Chatelier's law*).

BIBLIOGRAPHY

G.M. Barrow, *Introduction to Molecular Spectroscopy*, McGraw-Hill, (1962).
G. Herzberg, *Molecular Spectra and Molecular Structure, I: Spectra of Diatomic Molecules*, Van Nostrand (1950).
G. Herzberg, *Molecular Spectra and Molecular Structure, II: Infrared and Raman Spectra of Polyatomic Molecules*, Van Nostrand (1945).
K.P. Huber and G. Herzberg, *Molecular Spectra and Molecular Structure, IV: Constants of Diatomic Molecules*, Van Nostrand Reinhold (1979).
Historical References
P. Langevin, *Ann. Chim. Phys.* (Paris) **5**, 70, (1905).
O. Sackur, *Ann. Physik* **36**, 958 (1911) ; **40**, 67 (1913).
H. Tetrode, *Ann. Physik* **38**, 434 (1912) ; **39**, 255 (1912).

COMPREHENSION EXERCISES

5.1 Evaluate the number of atoms in a litre of argon ($M = 39.9$ g mol^{-1}) at 300 K and at a pressure of one atmosphere. Compare their mean separation D with the de Broglie wavelength Λ and their diameter $d = 3.76$ Å [Ans.: $N = 2.45 \times 10^{22}$; $D = 34.4$ Å$\sim 10d$; $\Lambda = 0.16$ Å$= 4.7 \times 10^{-3}D$].

5.2 Calculate the average kinetic energy $\bar{\epsilon}$ of an argon atom in the gas at 300 K. Hence deduce the mean square velocity u of an atom [Ans.: $\bar{\epsilon} = 3.9 \times 10^{-2}$ eV; $u = 430$ m s^{-1}].

5.3 Calculate the ratio of the populations in the first excited electronic state (11.55 eV, $J = 2$) and in the ground state for argon ($J = 0$) at $T = 300$ K [Ans.: $N_1/N_0 = 5 \times 10^{-194}$].

5.4 Explore the properties of an ideal monatomic gas in two dimensions of area A (§5.2). In particular, find the partition function, equation of state, internal energy, specific heat at constant area, and the entropy [Ans.: $Z^{(t)} = (A\, 2\pi m k T)/h^2$].

5.5 Show that the moment of inertia of a diatomic molecule has the form $I_e = \mu r_e^2$. Calculate r_e for carbon monoxide CO ($B_e = 1.931$ cm^{-1}) [Ans.: $r_e = 1.13$ Å].

5.6 Starting from equation (5.43) for the vibration-rotation energy levels and using the molecular constants for HBr given in the caption of Figure 5.2, derive the formula in the caption that gives the position of the lines.

5.7 The rotation-vibration spectrum of HBr gas (Fig. 5.2), measured at higher resolution, exhibits a splitting of each line corresponding to the two isotopes ^{79}Br and ^{81}Br. Analysis of the spectrum yields the following values for the molecular constants: for H^{79}Br, $\omega_e = 2649.855$ cm^{-1} and $B_e = 8.4678$ cm^{-1}; for H^{81}Br, $\omega_e = 2649.450$ cm^{-1} and $B_e = 8.4652$ cm^{-1} [E.K. Plyler, *J. Res. Natl. Bur. Stand.* A **64**, 377 (1960)]. Explain why the separation of the two components of each line is approximately 0.4 cm^{-1}, whatever the value of m. Using the reduced mass μ, explain the difference between the values of ω_e and B_e for the two molecules (hint:

find why r_e has the same value) [Ans.: $\delta B_e/B_e = 2\delta\omega_e/\omega_e = -\delta\mu/\mu = 3.1 \times 10^{-4}$].

5.8 For the Raman spectrum of a diatomic molecule, show that the pure rotation band displays scattering lines whose difference in wave number from the exciting line is given by $\Delta\sigma = \pm B_e(4m + 6)$. Make a drawing like that of Figure 5.1.

5.9 Using Figure 5.2, estimate approximately the characteristic temperatures of vibration Θ_v and rotation Θ_r for hydrogen bromide gas, as well as the temperature T of the gas in the experiment [Ans.: $T \sim 300$ K].

5.10 At normal temperature, what is the most populated rotational level in CO ($\Theta_r = 2.8$ K) and for HD ($\Theta_r = 66$ K) [Ans.: $K = 7; K = 1$].

5.11 Calculate the fraction of HD gas molecules in the rotational levels $K \leq 2$ at $T = 300$ K and $T = 40$ K ($\Theta_r = 65.7$ K). For each value of T use the appropriate expression for $Z^{(r)}$ [Ans.: $K = 0 : 0.22$ and 0.90 ; $K = 1 : 0.42$ and 0.10 ; $K = 2 : 0.29$ and 2×10^{-4}].

5.12 Calculate the fraction of molecules of HBr gas in the first excited vibrational level at $T = 300$ K and $T = 1000$ K ($\Theta_v = 3800$ K) [Ans.: 3.2×10^{-6} and 2.3×10^{-2}].

5.13 Construct the three nuclear spin wave functions of orthohydrogen, as well as that of parahydrogen.

5.14 From the expression for the interaction between two magnetic dipoles, estimate the order of magnitude of the difference in nuclear energy between the ortho and para states of the H_2 molecule (equilibrium separation of the nuclei $r_e \simeq 1$ Å) [Ans.: about 10^{-6} K].

5.15 Using the Boltzmann relation linking the entropy to the thermodynamic probability W, explain why the entropy of a symmetric diatomic gas contains an extra term $-Nk\ln 2$.

5.16 What is the composition in ortho- and para-hydrogen gas that is thermodynamically stable at $T = \Theta_r$? [Ans.: 71% and 29%].

5.17 A mole of hydrogen at normal temperature is liquefied ($T_B = 20.4$ K). Calculate the energy liberated during the transformation from the metastable state (para: 25 %, ortho: 75 %) into the stable state. Compare this with the latent heat of vaporization ($\Theta_r = 87.5$ K and $L = 900$ J mol^{-1}) [Ans.: 1100 J mol^{-1}].

5.18 Show that the Raman scattering lines of a band in a symmetric diatomic molecule display an alternating intensity pattern due to the existence of the para and ortho molecular species.

5.19 Verify that the symmetry number σ is respectively equal to 2, 3, 2, 2 and 4 for the following molecules: N_2, CH_3Cl, CH_2Cl_2, C_6H_5Cl, C_2H_4.

5.20 Calculate the molar heat capacity c_V of methane CH_4 at 873 K and compare the result with the experimental value $c_P = 16.8$ calories K^{-1} mol^{-1} (Table 5.5). Discuss the difference (cf. Fig. 5.10) [Ans.: $c_P = c_V + R = 15.9$ cal K^{-1} mol^{-1}].

5.21 For temperatures around 300 K, give the approximate values of the molar heat capacity c_V and the ratio $\gamma = c_P/c_V$ for the following gases: He, HCl, N_2, CO_2, H_2O, C_2H_2, CH_3Cl [Ans.: $c_V/R = 3/2, 5/2, 5/2, 5/2, 3, 3, 3; \gamma = 5/3, 7/5, 7/5, 7/5, 4/3, 4/3, 4/3$].

5.22 Use an extensivity argument to show that the partition function $Z(T,V)$ of an ideal gas particle takes the form $Z = Vf(T)$. Hence deduce that the equation of state of the gas is $Pv = RT$.

5.23 At 400K, the chemical reaction for synthesizing ammonia, $N_2 + 3\,H_2 \rightarrow 2\,NH_3$, yields a heat of reaction $Q = 25.5$ kcal mol^{-1}. Indicate the sign of the changes in temperature and pressure that favour the reaction [Ans.: decreasing T, increasing P].

PROBLEM 5.1 SUBLIMATION CURVE OF A SOLID

Using atomic data, we shall now establish the sublimation curve $P(T)$ of a solid and compare this with experimental measurements for zinc. We recall that the equilibrium conditions between two phases (1) and (2) are $T_1 = T_2$ and $\mu_1 = \mu_2$, where μ_i is the chemical potential of a particle in phase i.

1 Chemical Potential of an Ideal Monatomic Gas
For the vapour phase of zinc, we use the model of an ideal monatomic gas obeying corrected Maxwell-Boltzmann statistics. The free energy of the gas is then

$$F = -NkT \left(\ln \frac{Z}{N} + 1 \right)$$

where Z is the partition function of one particle and N is the number of particles.
1.1 Recall the general definition of the partition function Z.
1.2 Derive an expression for the partition function $Z \equiv Z^{(t)}$, remembering that the energy of an atom of the gas is purely translational ($\epsilon = p^2/2m$).
1.3 Write the general expression for the derivative of the free energy in the variables T, V, N and hence derive the pressure P and the chemical potential μ for the gas as a function of Z.
1.4 Write P and μ in terms of the variables T, V, N.
1.5 Hence deduce the equation for μ in terms of T and P. This expression contains the universal constant

$$i_0 = \ln \frac{(2\pi)^{3/2} k^{5/2}}{N^{3/2} h^3} = 18.22 \text{ SI} ,$$

and the molar mass M of the gas.

2 Chemical Potential of a Solid in the Einstein Model
For the solid phase we adopt the Einstein model in which the solid is assumed to contain a set of N' localized atoms that vibrate around their equilibrium positions like independent three-dimensional harmonic oscillators having the same constant frequency ν. The energy of vibration of an atom is then

$$\epsilon_{n_1,n_2,n_3} = (n_1 + n_2 + n_3)h\nu ,$$

with positive or zero integer n_i. In this model, where Maxwell-Boltzmann statistics is used, the free energy of the solid takes the form

$$F' = -N'l_0 - N'kT \ln Z' ,$$

where Z' is the partition function for an atom and $N'l_0$ is the cohesive energy of the solid (potential energy plus zero point vibrational energy, $3h\nu/2$ per oscillator).

2.1 Show that Z' factorizes into a product of three identical functions $Z^{(v)}$, and determine Z'.

2.2 Hence derive the function $F'(T, N')$ and the chemical potential $\mu'(T)$. Express the results in terms of the characteristic temperature $\Theta = h\nu/k$.

3 Sublimation Curve of Zinc

3.1 State the conditions of equilibrium between the gas and the solid phase, and derive the equation of the sublimation curve in the form $\ln P = f(T)$.

3.2 For the limiting case $T \gg \Theta$, show that this equation can be written in the form

$$\ln P = A + B \ln T + \frac{C}{T} + O\left(\frac{1}{T^2}\right) .$$

3.3 Calculate the values of the parameters A, B and C for zinc , for which $M = 65.38$ g mol^{-1}, $Nl_0 = 1.30 \times 10^5$ J mol^{-1} and $\Theta \simeq 240$ K. Compare with the following formula, which fits the experimental data in the temperature range 500 to 600 K,

$$\ln[P \ /\text{Pa}] = 30.09 - 0.5 \ln[T \ /\text{K}] - \frac{1.64 \times 10^4}{T \ /\text{K}} .$$

The gas constant is $R = 8.315$ J K^{-1} mol^{-1}.

References: A.C. Egerton, *Phil. Mag. S.6*, **33**, 33 (1917) and R.W. Ditchburn and J.C. Gilmour, *Rev. Mod. Phys.* **13**, 310 (1941).

SOLUTION

1 Chemical Potential of an Ideal Monatomic Gas

Ideal monatomic gases are treated in detail in paragraph 5.2. The chemical potential, expressed in the variables T and P, is

$$\begin{aligned}
\mu &= -kT \ln\left[\frac{kT}{P}\frac{(2\pi mkT)^{3/2}}{h^3}\right] \\
&= -kT\left[\frac{5}{2}\ln T - \ln P + i_0 + \frac{3}{2}\ln M\right] .
\end{aligned}$$

2 Chemical Potential of a Solid in the Einstein Model

2.1 Since the energy of an atom is the sum of three independent terms, the partition function factorizes into a product of three identical terms. We have

$$Z' = \left[Z^{(v)}\right]^3 = \left[\sum_{n=0}^{\infty} e^{-\beta n h\nu}\right]^3 = \left[\frac{1}{1 - e^{-\beta h\nu}}\right]^3 .$$

2.2 Hence, setting $\beta h\nu = \Theta/T$, we find

$$F' = -N'l_0 + 3N'kT \ln\left(1 - e^{-\Theta/T}\right)$$

$$\text{and} \quad \mu' = \left(\frac{\partial F'}{\partial N'}\right)_T = -l_0 + 3kT \ln\left(1 - e^{-\Theta/T}\right) \quad .$$

3 Sublimation Curve of Zinc

3.1 The equality of temperatures and chemical potentials in the two phases yields the relation $\mu(T, P) = \mu'(T)$ which, written out in full, gives

$$\ln P = i_0 + \frac{3}{2}\ln M + \frac{5}{2}\ln T - \frac{l_0}{kT} + 3\ln\left(1 - e^{-\Theta/T}\right) \quad . \tag{5.131}$$

3.2 In the limiting case $T \gg \Theta$, expansion in series gives

$$\ln\left(1 - e^{\Theta/T}\right) = \ln\left[\frac{\Theta}{T}\left(1 - \frac{\Theta}{2T} + \cdots\right)\right] = \ln\Theta - \ln T - \frac{\Theta}{2T} + \cdots$$

The sublimation curve (5.131) is thus given by

$$\ln P = i_0 + \frac{3}{2}\ln M + 3\ln\Theta - \frac{1}{2}\ln T - \frac{l_0 + 3k\Theta/2}{kT} + O\left(\frac{1}{T^2}\right) \quad .$$

This equation has the desired form.

3.3 The numerical values of the parameters A, B and C are

$$A = i_0 + \frac{3}{2}\ln M + 3\ln\Theta = 30.6 \text{ SI}$$

$$B = -0.5$$

$$C = -\frac{l_0 + 3k\Theta/2}{k} = -\frac{\mathcal{N}l_0}{R} - \frac{3\Theta}{2} = -1.60 \times 10^4 \text{ K} \quad .$$

The coefficients found are very close (to within 2 %) to those obtained experimentally. It will, however, be noted that for $T = 550$ K, for example, the theoretical formula predicts a vapour pressure $P = 1.9 \times 10^{-1}$ Pa, whereas the experimental formula gives $P = 5.6 \times 10^{-2}$ Pa, that is, roughly three times smaller. This discrepancy comes partly from the fact that the formula for $\ln P$ occurs in the form of a difference in which A and C/T almost compensate, thereby producing a loss in precision, and partly from the fact that P is found from $\ln P$ by taking an exponential.

PROBLEM 5.2 THERMODYNAMICS OF NITROGEN GAS N_2

1 Comparison between Spectroscopic and Calorimetric Entropy

We wish to compare the value of the entropy of nitrogen gas at its normal boiling point $T_B = 77.3$ K, found by calorimetry, with that predicted theoretically in the ideal gas model from the molecular parameters measured by spectroscopy. The nitrogen molecule N_2 is symmetric and is composed of two identical atoms. Neglect its nuclear spin I.

1.1a Write the differential of the free energy F for the gas in the variables T, V, N and show that if F is known, then the entropy S and the pressure P of the gas can be determined.

1.1b In corrected Maxwell-Boltzmann statistics, the free energy is

$$F = -NkT \left(\ln \frac{Z}{N} + 1 \right),$$

where Z is the partition function for one particle. Hence deduce the entropy and the pressure as a function of Z.

1.2a Under certain approximations, it may be assumed that the energy of each level of a nitrogen molecule is the sum of the energies of translation, rotation and vibration and that the associated degeneracy is the product of the corresponding degeneracies. Show that the partition function Z then factorizes into a product of three partition functions $Z^{(t)}, Z^{(r)}$ and $Z^{(v)}$.

1.2b Why is the pressure derived from $Z^{(t)}$ alone?

1.2c Show that the entropy is a sum of three terms $S^{(t)}, S^{(r)}$ and $S^{(v)}$. Explain the difference in form of $S^{(t)}$ from that of $S^{(r)}$ and $S^{(v)}$.

1.3a Determine the translational partition function $Z^{(t)}$.

1.3b Write the pressure P and entropy $S^{(t)}$ in terms of the variables T, V, N.

1.3c Hence derive an expression for $S^{(t)}$ in terms of T, P, N. Make use of the universal constant

$$i_0 = \ln \frac{(2\pi)^{3/2} k^{5/2}}{\mathcal{N}^{3/2} h^3} = 18.22 \text{ SI}$$

as well as the molar mass M of the gas.

1.3d Calculate the value of the molar entropy of translation $s^{(t)}$ for nitrogen at its boiling point at atmospheric pressure T_B, given that the molar mass of the gas is $M = 28.01$ g mol^{-1}.

1.4 The energy of the rotational level K for a nitrogen molecule can be written

$$\epsilon_K^{(r)} = K(K + 1)k\Theta_r ,$$

where $\Theta_r = 2.89$ K is the characteristic temperature of rotation and the associated degeneracy is $2K + 1$.

1.4a Determine the rotational partition function $Z^{(r)}$, using the symmetry factor σ of the molecule.

1.4b Hence deduce the entropy of rotation $S^{(r)}$.

1.4c What is the value of the molar entropy $s^{(r)}$ of nitrogen at 77.3 K?

1.5a The wave number of the vibration in nitrogen is $w_e = 2360$ cm^{-1}. Explain why the entropy of vibration can be taken to be zero at 77.3 K.

1.5b Hence deduce the value of the molar entropy of nitrogen at its normal boiling point.

1.6 To compare this so-called spectroscopic value of the entropy with its experimental, or calorimetric, value, the following data at atmospheric pressure are available.

i. Changes in molar entropy found by numerical integration of c_P/T for the two solid phases β and α and the liquid phase

$$
\begin{aligned}
s\,(35.61\text{ K}) - s\,(0\text{ K}) &= 27.16 \text{ J K}^{-1}\text{ mol}^{-1} \ , \\
s\,(63.14\text{ K}) - s\,(35.61\text{ K}) &= 23.38 \text{ J K}^{-1}\text{ mol}^{-1} \ , \\
s\,(77.3\text{ K}) - s\,(63.14\text{ K}) &= 11.41 \text{ J K}^{-1}\text{ mol}^{-1} \ .
\end{aligned}
$$

ii. Molar latent heats of transition

$$
\begin{aligned}
(\beta \to \alpha) \quad L\,(35.61\text{ K}) &= 229.0 \text{ J mol}^{-1} \ , \\
(\alpha \to \text{liq.}) \quad L\,(63.14\text{ K}) &= 721.1 \text{ J mol}^{-1} \ , \\
(\text{liq.} \to \text{vap.}) \quad L\,(77.3\text{ K}) &= 5577 \text{ J mol}^{-1} \ .
\end{aligned}
$$

1.6a Calculate the changes in molar entropy involved in the three phase changes.
1.6b Hence calculate the difference $s\,(77.3\text{ K}) - s\,(0\text{ K})$.
1.6c Compare and discuss the spectroscopic and experimental values of the entropy.

2 Ortho and Para-nitrogen

In this part we examine the effect on the entropy of the fact that the nitrogen molecule is composed of two identical atoms with nuclear spin $I = 1$.

2.1 Calculate the nuclear degeneracy of each of the two atoms, then that of the molecule. Hence derive the term for the molar entropy of the gas corresponding to this degeneracy. This term is usually omitted.

2.2a Knowing the value of the nuclear spin of nitrogen atoms, indicate the symmetry of the total wave function of the molecule with respect to exchange of the nuclei.

2.2b Given that the total wave function is the product of the wave functions of translation $\psi^{(t)}$, rotation $\psi^{(r)}$, vibration $\psi^{(v)}$, and of the electronic and nuclear wave functions $\psi^{(e)}$ and $\psi^{(n)}$, derive the symmetry of the product $\psi^{(r)}\psi^{(n)}$.

2.2c The wave function $\psi^{(r)}$ has parity $(-1)^K$, where K is the rotational quantum number. Show how the symmetry of the nuclear wave function depends on the value of K.

2.3a Using the notation $|m >$ ($m = 1, 0, -1$) for the nuclear wave functions of a nitrogen atom, define and list the independent symmetric and antisymmetric combinations of the nuclear wave functions $|m_1 m_2 >$ of the molecule (ortho-nitrogen and para-nitrogen respectively).

2.3b Hence derive the fraction of ortho and para molecules in the temperature range $T \gg \Theta_r$.

2.3c Make a diagram of the rotational energy levels for the ortho and para molecular species and indicate their degeneracy (rotational and nuclear). In particular, state the degeneracy of the lowest level of both species.

2.4 The state of the solid at low temperatures can be inferred from the agreement between the spectroscopic and the calorimetric entropy, taking into account the nuclear spin degeneracy.

2.4a Determine the thermodynamic probability W and then the entropy of the solid at absolute zero, assuming that the para $\rightarrow$ ortho transition is complete and that all the molecules are in their lowest rotational state.

2.4b What would be the entropy of the solid at absolute zero if the transition para $\rightarrow$ ortho did not occur, and all the ortho- and para-nitrogen molecules were respectively in their lowest rotational energy states? As the solid is then a mixture containing N_O ortho and N_P para molecules, the mixing entropy of the two species is $S = k \ln(N!/N_O!\, N_P!)$.

2.4c What would be the entropy at absolute zero if the para $\rightarrow$ ortho transition did not occur, but the rotation of the para molecules were frozen in the solid?

2.4d Given the experimental results, which of the two previous assumptions is correct?

Physical constants: Gas constant $R = 8.314$ J K^{-1} mol^{-1}, $hc/k = 1.44$ cm K, atmospheric pressure 1 atm $= 1.013 \times 10^5$ Pa.
Reference: W.F. Giauque and J.O. Clayton, *J. Am. Chem. Soc.* **55**, 875 (1933).

SOLUTION
1 Comparison between Spectroscopic and Calorimetric Entropy
1.1 See §5.1.2, equations (5.5,6).

1.2a The additivity of the energies $\epsilon^{(t)}, \epsilon^{(v)}$ and $\epsilon^{(r)}$ leads to factorization of the partition function (§5.2).

1.2b Of the three partition functions, only $Z^{(t)}$ depends on the volume. Hence

$$P = NkT \left(\frac{\partial \ln Z}{\partial V} \right)_T = NkT \left(\frac{\partial \ln Z^{(t)}}{\partial V} \right)_T .$$

1.2c Since the partition function appears as a logarithm in the equation for the entropy, we may write $S = S^{(t)} + S^{(r)} + S^{(v)}$, with

$$\frac{S^{(t)}}{Nk} = \ln \frac{Z^{(t)}}{N} + 1 + T \left(\frac{\partial \ln Z^{(t)}}{\partial T} \right)_V$$

$$\frac{S^{(r)}}{Nk} = \ln Z^{(r)} + T \frac{d \ln Z^{(r)}}{dT}$$

$$\frac{S^{(v)}}{Nk} = \ln Z^{(v)} + T \frac{d \ln Z^{(v)}}{dT} .$$

Here, the indistinguishability term $-k \ln N!$ has been placed together with the entropy of translation $S^{(t)}$.

1.3a The partition function for translation was calculated in (5.18).

1.3b Writing the expression for $Z^{(t)}$ yields equations (5.22,26), apart from the term in $\ln(2J + 1)$.

1.3c Transformation into the variables T, P, N, gives

$$\frac{S^{(t)}}{Nk} = \frac{5}{2} \ln T - \ln P + \frac{5}{2} + i_0 + \frac{3}{2} \ln M .$$

1.3d Setting $T = 77.3$ K, $P = 1.013 \times 10^5$ Pa and $M = 28.01 \times 10^{-3}$ kg mol^{-1}, we get $s^{(t)} = 122.23$ J K^{-1} mol^{-1}.

1.4a Since the boiling point of nitrogen is much higher than the rotation temperature Θ_r, K can be taken to be a continuous variable. If the molecule were asymmetric we would have $Z^{(r)} = T/\Theta_r$ (§5.3.4). But as the nitrogen molecule is symmetric, this expression must be divided by the symmetry factor $\sigma = 2$, i.e., $Z^{(r)} = T/2\Theta_r$ (§5.4.2).

1.4b The entropy of rotation is therefore

$$S^{(r)} = Nk \left[\ln \frac{T}{2\Theta_r} + 1 \right] .$$

1.4c The value obtained for the molar entropy of rotation is $s^{(r)} = 29.88$ J K^{-1} mol^{-1}.

1.5a The wave number for vibration is $w_e = 2360$ cm^{-1}, and the characteristic temperature of vibration $\Theta_v = hcw_e/k$ is therefore 3395 K. Nitrogen at 77.3 K occupies a state for which $T \ll \Theta_v$. Only the vibrational ground level is populated and the entropy of vibration is zero. Calculation shows that its value is of order 10^{-17} J K^{-1} mol^{-1}.

1.5b The molar entropy of nitrogen is therefore $s = s^{(t)} + s^{(r)} = 152.11$ J K^{-1} mol^{-1}.

1.6a At a phase transition $\Delta s = L/T$, which gives respectively for the three transitions $\Delta s = 6.43$, 11.42 and 72.15 J K^{-1} mol^{-1}.

1.6b Altogether, we have therefore s (77.3 K) $-$ s (0 K) $= 151.95$ J K^{-1} mol^{-1}.

1.6c Comparison of the two values shows that the statistical method used is valid, as is also the Third Law. A more accurate theoretical calculation, taking into account interactions and the deformation of the molecule during rotation, does not alter these conclusions. It will be observed that we have not explicitly taken into account the nuclear spin degeneracy. This aspect will be discussed in question 2.4.

2 Ortho and Para-nitrogen

2.1 As the nuclear degeneracy of the atom is $2I + 1 = 3$, that of the molecule is $3^2 = 9$. Hence, according to the Boltzmann relation, the molar entropy of the gas contains an extra term $R \ln 9 = 18.27$ J K^{-1} mol^{-1}. By convention, we omitted this term in question 1. It therefore follows that the same term must be considered in the solid phase at low temperatures. Therefore s (0 K)$= R \ln 9$. Note that this result is not in conflict with the Third Law, since the nuclear spins become ordered around 1μK. This phenomenon is taken into account neither in the model nor in the extrapolation of the calorimetric measurements below 1 K.

2.2a Nuclei with spin $I = 1$ are bosons and the total wave function must be symmetric under exchange of the nuclei.

2.2b The product $\psi^{(r)}\psi^{(n)}$ must therefore be symmetric (§5.4.1).

2.2c As shown in Table 5.2, $\psi^{(n)}$ is symmetric for even K and antisymmetric for odd K.

2.3a 9 nuclear wave functions exist of the type $|m_1, m_2 >$. Of these, three are symmetric, i.e. $|1, 1 >$, $|0, 0 >$, $|-1, -1 >$, while the remaining 6 have no symmetry. By taking the 6 independent linear combinations

$$\frac{1}{\sqrt{2}} [|1, 0 > \pm |0, 1 >], \frac{1}{\sqrt{2}} [|1, -1 > \pm |-1, 1 >], \frac{1}{\sqrt{2}} [|0, -1 > \pm |-1, 0 >]$$

of which 3 are symmetric (+ sign) and 3 are antisymmetric (− sign), a total of 6 symmetric wave functions (ortho molecules) and 3 antisymmetric wave functions (para molecules) is obtained.

2.3b Therefore 2/3 of the molecules are ortho, and 1/3 are para (§5.4.3).

2.3c The rotational level diagram for nitrogen is the same as for deuterium (Fig. 5.8). The degeneracies of the ortho and para ground states are 6 and 9 respectively.

2.4a If all the molecules were of ortho type, the thermodynamic probability of the solid at low temperatures would be $W = 6^N$, owing to the 6 possible spin states. The molar entropy of the solid would then be $s_a = R \ln 6 = 14.90$ J K^{-1} mol^{-1}.

2.4b In this case, in the solid there are $N_0 = 2N/3$ ortho molecules with degeneracy 6 and $N_p = N/3$ para molecules with degeneracy 9. If the solid is taken to be this mixture of ortho and para molecules, its molar entropy at absolute zero is

$$s_b = \frac{2}{3} R \ln 6 + \frac{1}{3} R \ln 9 + k \ln \frac{\mathcal{N}!}{\left(\frac{2}{3}\mathcal{N}\right)! \left(\frac{1}{3}\mathcal{N}\right)!}$$

or, using Stirling's reduced formula (1.57), $s_b = (7R/3) \ln 3 = 21.31$ J K^{-1} mol^{-1}.

2.4c If the rotation of the para molecules is frozen, the degeneracy of the para molecules coming from the spins alone is equal to 3 (instead of 9). The same calculation as above then gives $s_c = 2R \ln 3 = R \ln 9 = 18.27$ J K^{-1} mol^{-1}.

2.4d Clearly assumption c is in agreement with the result of 2.1. Note that assumptions a and b would give rise to a discrepancy with experiment of the order of 3 J K^{-1} mol^{-1} and must therefore be rejected. Crystalline nitrogen at low temperatures is thus a disordered mixture of 2/3 ortho molecules and 1/3 para molecules whose rotation is frozen (in contrast to solid hydrogen for which the rotation of the ortho molecules persists). Nevertheless, the molecules in the solid perform rotational oscillations (called librations).

PROBLEM 5.3 MOLECULAR ASSOCIATION IN SODIUM VAPOUR

Discrepancies in alkali metal vapours from the equation of state of ideal gases reveal a phenomenon of polymerization. We investigate the production of the sodium dimer Na$_2$ in the reaction

$$2\,\text{Na} \;\rightarrow\; \text{Na}_2 \;,$$

where we make the approximation that the vapour is an ideal gas mixture.

1.1 Show that the chemical potential has the form

$$\mu = -kT \ln \left[kT f(T)/P \right] + \epsilon^0 \;,$$

where ϵ^0 is the energy of the ground state. Recall that in corrected Maxwell-Boltzmann statistics $\mu = -kT \ln Z/N$.

1.2 What is the law of mass action for the above reaction? Use the subscripts 1 and 2 for the functions involving Na and Na$_2$.

1.3 Define the equilibrium constant K_p for the reaction investigated. What are its SI units?

1.4 Deduce the constant $K_p(T)$ from the law of mass action. The binding energy of the molecule Na_2 is denoted by $D_0 = 2\epsilon_1^0 - \epsilon_2^0$.

2.1 Derive $f_1(T)$ by calculating the partition function for the monatomic gas Na. Include the degeneracy $g_e = 2$ of the electronic ground state of the atom and neglect the excited electronic states.

2.2 Write down $f_2(T)$ for the symmetric diatomic molecule Na_2, whose electronic ground state is not degenerate. Use Θ_r and Θ_v for the characteristic temperatures of rotation and vibration in Na_2, and neglect the excited electronic states. Give a simplified expression for $f_2(T)$ in the approximation $T \gg \Theta_v$.

2.3 Hence derive $\ln K_p(T)$ for sodium as a function of the atomic mass M, Θ_r, Θ_v and D_0. In this expression use the universal constant

$$i_0 = \ln \frac{(2\pi)^{3/2} k^{5/2}}{N^{3/2} h^3} = 18.22 \text{ SI} .$$

2.4 Use the van't Hoff relation

$$\frac{d}{dT} (\ln K_p) = -\frac{Q}{RT^2} ,$$

to find the heat of reaction Q as a function of temperature. Deduce this result independently by noting that the heat of this reaction is equal to the difference between twice the molar enthalpy of sodium Na and the molar enthalpy of the dimer Na_2, and then applying the theorem of equipartition of energy.

3.1 The atomic mass of sodium is $M = 23.0$ g mol^{-1}. The molecular constants of Na_2, determined by spectroscopy, are $\Theta_r = 0.223$ K, $\Theta_v = 229$ K and $D_0/k = 8551$ K. Calculate the values of K_p and Q for $T = 1400$ and $T = 1700$ K.

3.2 Between 1400 K and 1700 K, experimental measurements give

$$\log_{10}[K_p /\text{atm}^{-1}] = -4.3249 + \frac{4002.3}{T} .$$

Compare the previously calculated theoretical values with those obtained from this experimental relation.

4.1 Express the partial pressures P_1 and P_2 as a function of the total pressure and of the molar fraction $X \equiv X_2$ of dimer Na_2.

4.2 Deduce the relation giving $X(T, P)$ at equilibrium as a function of the product $PK_P = K_X$.

4.3 Evaluate X in the three states defined by

$$(\alpha) \quad T = 1400 \text{ K and } P = 6.50 \text{ atm}$$
$$(\beta) \quad T = 1700 \text{ K and } P = 31.2 \text{ atm}$$
$$(\gamma) \quad T = 1700 \text{ K and } P = 6.50 \text{ atm} .$$

The points α and β lie on the vaporization curve of sodium.

4.4 Using the relations for the shift in equilibrium as a function of temperature and pressure, give a qualitative interpretation of the above results.

Values of the constants: gas constant $R = 8.3145$ J K^{-1} mol^{-1}; 1 atm $= 1.01325 \times 10^5$ Pa.

References: K.P. Huber and G. Herzberg, *op. cit.*; R.F. Barrow et al., *Chem. Phys. Lett.*, **104**, 179 (1984); C.T. Ewing et al. *J. Phys. Chem.* **71**, 473 (1967).

SOLUTION

1.1 For an ideal gas, $Z = Vf(T)$. Use of the equation of state, $PV = NkT$, yields the required expression.

1.2 In this case the law of mass action (5.125) is $2\mu_1 = \mu_2$.

1.3 The equilibrium constant is defined as $K_p = P_2/P_1^2$, where P_1 and P_2 are the partial pressures at equilibrium. It is measured in Pa^{-1}.

1.4 Expressing μ_1 and μ_2 as a function of P_1, P_2 and T in the law of mass action gives

$$K_p = \frac{P_2}{P_1^2} = \frac{f_2(T)}{[f_1(T)]^2}(kT)^{-1}e^{D_0/kT} \quad .$$

2.1 For an atom of degeneracy $g^{(e)} = 2$ we have (§5.2)

$$Z_1 = 2\frac{V}{h^3}(2\pi mkT)^{3/2} \quad ,$$

from which we obtain $f_1(T) = Z_1/V$.

2.2 The partition function Z_2 is a product of the partition functions of translation, rotation, and vibration. Taking into account the symmetry factor $\sigma = 2$ of the molecule, we get

$$Z_2 = \frac{V}{h^3}(4\pi mkT)^{3/2} \times \frac{T}{2\Theta_r} \times \frac{1}{1 - \exp(-\Theta_v/T)} \quad ,$$

where the mass of the molecule is $2m$. In the range $T \gg \Theta_v$ the vibrational partition function reduces to T/Θ_v (classical limit), whence

$$Z_2 = \frac{V}{h^3}(2\pi mkT)^{3/2}\frac{2^{1/2}}{\Theta_r\Theta_v}T^2$$

and $f_2(T) = Z_2/V$.

2.3 Substituting f_1 and f_2 into the equation for K_p, we get

$$K_p = \frac{h^3\mathcal{N}^{3/2}}{8\pi^{3/2}k^{5/2}} \times \frac{1}{M^{3/2}\Theta_r\Theta_v} \times \frac{\exp(D_0/kT)}{T^{1/2}}$$

where $M = \mathcal{N}m$. Hence

$$\ln K_p = -\left[i_0 + \frac{3}{2}\ln 2M + \ln\Theta_r\Theta_v\right] + \frac{D_0}{kT} - \frac{1}{2}\ln T \quad .$$

2.4 Differentiation of $\ln K_p$ gives

$$Q = -RT^2\left[-\frac{D_0}{kT^2} - \frac{1}{2T}\right] = R\left(\frac{D_0}{k} + \frac{T}{2}\right) \quad .$$

This result can be obtained otherwise by evaluating the molar enthalpies of the gases Na and Na_2. For Na gas, since the only form of energy is the translational kinetic energy $(3RT/2)$, we have

$$h_1 = u_1 + RT = \mathcal{N}\epsilon_1^0 + \frac{3}{2}RT + RT = \mathcal{N}\epsilon_1^0 + \frac{5}{2}RT .$$

For the molecular gas Na_2, however, the energy of rotation (RT) and vibration (RT) must be included. Thus

$$h_2 = u_2 + RT = \mathcal{N}\epsilon_2^0 + \frac{7}{2}RT + RT = \mathcal{N}\epsilon_2^0 + \frac{9}{2}RT .$$

Therefore

$$Q = 2h_1 - h_2 = \mathcal{N}\left(2\epsilon_1^0 - \epsilon_2^0\right) + \frac{1}{2}RT = \mathcal{N}D_0 + \frac{1}{2}RT .$$

3.1 Inserting the values of M, Θ_r, Θ_v and D_0, we get

$$\ln[K_p /Pa^{-1}] = -17.53 + \frac{8551}{T /K} - \frac{1}{2}\ln[T /K]$$

$$\text{and} \quad Q /J\ mol^{-1} = 71\ 100 + 4.157 \times (T /K) .$$

This yields respectively

$$K_p = 2.93 \times 10^{-7}\ Pa^{-1} \quad \text{and} \quad Q = 76.9\ kJ\ mol^{-1} \quad \text{at} \quad T = 1400\ K$$
$$\text{and} \quad K_p = 9.04 \times 10^{-8}\ Pa^{-1} \quad \text{and} \quad Q = 78.2\ kJ\ mol^{-1} \quad \text{at} \quad T = 1700\ K .$$

3.2 Since $K_p /atm^{-1} = K_p /Pa^{-1} \times 1.01325 \times 10^5$ and $\ln x = 2.3026 \log_{10} x$, the experimental relation can be written

$$\ln[K_p /Pa^{-1}] = -21.485 + \frac{9215.7}{T /K} .$$

For the two temperatures considered, we have

$$K_p\ (1400\ K) = 3.37 \times 10^{-7}\ Pa^{-1} , \quad K_p\ (1700\ K) = 10.6 \times 10^{-8}\ Pa^{-1} .$$

The theoretical values are about 15 % smaller than the experimental ones. The approximation $T \gg \Theta_v$ accounts for half of this error.
The experimental heat of reaction, obtained from van't Hoff's law, is equal to

$$Q = -RT^2 \frac{d\ln K_p}{dT} = 9215.7\ R = 76.6\ kJ\ mol^{-1} .$$

In the range under consideration this heat of reaction is constant and approximately equal to the calculated theoretical values. It can be seen that the theoretical model yields results that are in satisfactory agreement with experiment.
4.1 We have $P_2 = PX_2 = PX$ and $P_1 = PX_1 = P(1 - X)$.

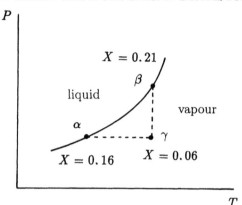

Figure 5.16: Vaporization curve of sodium

4.2 The molar fraction X at equilibrium is then determined by

$$K_p = \frac{P_2}{P_1^2} = \frac{X}{P(1-X)^2} \quad .$$

The solution of the above quadratic equation is

$$X = 1 + \frac{1}{2K_X} - \left(\frac{1}{K_X} + \frac{1}{4K_X^2} \right)^{1/2} \quad \text{where } K_X = PK_p \; .$$

4.3 Using the experimental values for K_p calculated in 3.2, we find respectively

$$X_\alpha = 0.16 \; ; \quad X_\beta = 0.21 \; ; \quad X_\gamma = 0.06 \; .$$

4.4 On moving from α to β on the vaporization curve (Fig. 5.16), the molar fraction varies weakly in spite of the large change in pressure. It can be seen that on going from α to γ, this fraction decreases, as expected from van't Hoff's law in exothermic reactions ($Q > 0$), but it increases on going from γ to β, since the dimer is favoured by the increase in pressure (Le Chatelier's principle). The two variations partially compensate. The existence of a tetramer Na_4 reduces the values of X calculated above.

Chapter 6

Kinetic Theory of Gases

6.1 INTRODUCTION

In the last chapter we investigated the properties of gases in thermodynamic equilibrium, as predicted from their molecular parameters. The distribution relation, in which only the energy of the molecules appears, formed the basis of our investigation. To provide an explanation of phenomena in non equilibrium states, however, such as gas discharges through an orifice, molecular beams or transport phenomena in gases (diffusion, thermal conduction and viscosity), it is essential to take into account the position of the molecule as well as the magnitude and direction of its velocity.

The ideal gas model will be used here in the framework of classical mechanics. This is a first approximation to the rigorously correct quantum mechanics, for which the essential arguments are much more complex. Note that classical theory yields the equation of state and the expression for the kinetic energy of translation in an ideal gas, as already found in chapter 5.

We therefore undertake a statistical study of the molecules, taking into account their particular motion. In the simplest kinetic theory, molecules are considered as being small spherically symmetrical corpuscles of mass m, having no internal degrees of freedom and obeying corrected Maxwell-Boltzmann statistics. In this chapter, we perform approximate calculations, but we also provide exact results, which in general differ only by a numerical factor close to 1. For the exact calculations, we refer the reader to the textbooks of E.H. Kennard and of D.A. McQuarrie. A deeper investigation into transport phenomena involves the use of Boltzmann's equation and lies outside the scope of this work.

6.2 DISTRIBUTION RELATIONS

6.2.1 Distribution in Phase Space

The basic equation of kinetic theory is found from relation (2.40) established for corrected Maxwell-Boltzmann statistics

$$N_i = \frac{N}{Z} g_i e^{-\beta \epsilon_i} , \quad \beta = \frac{1}{kT} , \tag{6.1}$$

in which the partition function Z is reduced to the translational partition function $Z^{(t)}$ (5.18) of an ideal gas

$$Z = Z^{(t)} = \frac{V}{h^3} \, (2\pi mkT)^{3/2} \, . \tag{6.2}$$

In the classical limit

$$N_i \rightarrow d^6 N_{\mathbf{r},\mathbf{p}} \, , \quad g_i \rightarrow \frac{d^3\mathbf{r} \, d^3\mathbf{p}}{h^3} \, , \quad \epsilon_i \rightarrow \epsilon = \frac{p^2}{2m} \, ,$$

the distribution in phase space $\mathbf{r}, \mathbf{p}$ is

$$d^6 N_{\mathbf{r},\mathbf{p}} = \frac{n}{(2\pi mkT)^{3/2}} \, e^{-\beta p^2/2m} \, d^3\mathbf{r} \, d^3\mathbf{p} \, , \tag{6.3}$$

where $n = N/V$ is the number of molecules per unit volume, $d^3\mathbf{r}$ the volume element in geometrical space, and $d^3\mathbf{p}$ the elementary volume in momentum space, variously expressed as

$$\begin{aligned} d^3\mathbf{p} \; &= \; dp_x \, dp_y \, dp_z && \text{in Cartesian co-ordinates,} \\ &= \; p^2 \, dp \, d^2\Omega = p^2 \, dp \sin\theta \, d\theta \, d\phi && \text{in spherical co-ordinates.} \end{aligned}$$

It can be seen that the function appearing in (6.3) depends neither on the molecular co-ordinates (uniform distribution in geometrical space), nor on the direction of $\mathbf{p}$ (isotropic distribution), but only on its modulus. Moreover, this function is proportional to the density of molecules and depends on the temperature. It is to be remarked that Planck's constant does not appear in this distribution, as is to be expected in a classical limit.

It can be seen that the normalization condition

$$\int d^6 N_{\mathbf{r},\mathbf{p}} = N \tag{6.4}$$

is verified. Integration yields

$$\begin{aligned} \int_0^\infty \exp\left(-\frac{\beta p^2}{2m}\right) d^3\mathbf{r} \, d^3\mathbf{p} \; &= \; \int_V d^3\mathbf{r} \int_{4\pi} d^2\Omega \int_0^\infty \exp\left(-\frac{\beta p^2}{2m}\right) p^2 \, dp \\ &= \; V \times 4\pi \times I_2 \left(\frac{\beta}{2m}\right) \, , \end{aligned}$$

where I_2 is one of the Gaussian integrals listed in Table 6.1.

The integrals I_n, which will be used below frequently, are obtained from I_0 and I_1 by a recurrence formula. Substitution for I_2 into the above expression yields the normalization condition (6.4). Conversely, the condition of normalization yields the pre-exponential factor in the distribution relation (6.3).

6.2.2 Distribution in Velocity Space

For the majority of applications the velocity variable $\mathbf{v}$ enters more naturally than that of momentum $\mathbf{p}$. Changing the variables $\mathbf{v} = \mathbf{p}/m$ in the distribution relation

Table 6.1: Integrals $I_n = \int_0^\infty x^n e^{-ax^2} \, dx = \frac{1}{2} a^{-(n+1)/2} \Gamma\left((n+1)/2\right)$

Recurrence Formulae $I_{n+2} = -dI_n/da = (n+1)/2a \times I_n$	
n even	n odd
$I_0(a) = \frac{1}{2} \sqrt{\pi/a}$	$I_1(a) = 1/2a$
$I_2(a) = \frac{1}{4} \sqrt{\pi/a^3}$	$I_3(a) = 1/2a^2$
$I_4(a) = \frac{3}{8} \sqrt{\pi/a^5}$	$I_5(a) = 1/a^3$
$- - - - -$	$- - - - -$
$\int_{-\infty}^{+\infty} x^n e^{-ax^2} \, dx = 2I_n$	$\int_{-\infty}^{+\infty} x^n e^{-ax^2} \, dx = 0$

(6.3) gives

$$d^6 N_{\mathbf{r},\mathbf{v}} = n \left(\frac{m}{2\pi kT}\right)^{3/2} e^{-\beta m v^2/2} \, d^3\mathbf{r} \, d^3\mathbf{v} \ . \tag{6.5a}$$

Integration over the space co-ordinates yields the number of particles per elementary volume having velocity $\mathbf{v}$ to within $d^3\mathbf{v}$

$$d^3 n_{\mathbf{v}} = \frac{d^3 N_{\mathbf{v}}}{V} = n \left(\frac{m}{2\pi kT}\right)^{3/2} e^{-\beta m v^2/2} \, d^3\mathbf{v} \ , \tag{6.5b}$$

$$\text{with} \quad d^3\mathbf{v} = dv_x \, dv_y \, dv_z = v^2 \, dv \, d^2\Omega \ .$$

J.C.Maxwell (1860) established this distribution for velocities before the development of statistical thermodynamics. His assumptions were, first, that the distribution of directions is isotropic, and second, that the distributions of the different components of velocity are independent (§6.2.5).

It is helpful to introduce the characteristic velocity

$$v_m = \left(\frac{2kT}{m}\right)^{1/2} = \left(\frac{2RT}{M}\right)^{1/2} \tag{6.6}$$

which allows dimensionless quantities to be used. In particular Maxwell's distribution becomes

$$d^3 n_{\mathbf{v}} = \frac{n}{\pi^{3/2}} e^{-(v/v_m)^2} \frac{d^3\mathbf{v}}{v_m^3} \ . \tag{6.7}$$

6.2.3 Distribution of Speeds

We first investigate the distribution relation for the modulus of the velocity. This allows us to introduce the following quantities: most probable velocity, average velocity and root mean square velocity, which play an important role in elementary kinetic theories. It also allows us to express them as functions of the temperature T and the molar mass M.

Integration of the Maxwell distribution (6.5) over the solid angle $d^2\Omega = \sin\theta\, d\theta\, d\phi$ ($0 \le \theta \le \pi$ and $0 \le \phi < 2\pi$) yields the number of particles with velocity between v and $v + dv$

$$dn_v = 4\pi n \left(\frac{m}{2\pi kT}\right)^{3/2} e^{-\beta mv^2/2}\, v^2\, dv = \frac{4n}{\pi^{1/2}} e^{-(v/v_m)^2} \left(\frac{v}{v_m}\right)^2 d\left(\frac{v}{v_m}\right). \quad (6.8)$$

This distribution gives the probability dP_v that the modulus of the velocity of a molecule lies between v and $v + dv$, i.e.

$$
\begin{aligned}
dP_v &= \frac{dn_v}{n} = 4\pi \left(\frac{m}{2\pi kT}\right)^{3/2} e^{-\beta mv^2/2}\, v^2\, dv \\
&= \frac{4}{\pi^{1/2}} e^{-(v/v_m)^2} \left(\frac{v}{v_m}\right)^2 d\left(\frac{v}{v_m}\right). \quad (6.9)
\end{aligned}
$$

The associated probability density $dP_v/d(v/v_m)$ is shown in Figure 6.1a as a function of the dimensionless variable (v/v_m). The curve exhibits a maximum at $v/v_m = 1$, of width at half height $\Delta(v/v_m) \simeq 1.15$. In terms of the variable v (Fig. 6.1b), the density dP_v/dv is maximum at

$$v = v_m = \left(\frac{2kT}{m}\right)^{1/2},$$

the value at the maximum being $0.83/v_m$. When the temperature increases, the position of the maximum increases as $T^{1/2}$, and its height decreases as $T^{-1/2}$. The curve thus becomes broader, its area remaining constant and equal to 1. The physical meaning of the parameter v_m becomes clear: it is the most probable velocity of the gas molecules. We are now in a position to determine the average velocity $\bar{v}$ of the gas molecules, as well as the root mean square velocity u, defined as the square root of the mean value of the square of the velocity, $u = (\overline{v^2})^{1/2}$. The average velocity is given by

$$\bar{v} = \int_0^\infty v\, dP_v = \frac{\int_0^\infty v\, dP_v}{\int_0^\infty dP_v} = \frac{\int_0^\infty \exp(-v^2/v_m^2) v^3\, dv}{\int_0^\infty \exp(-v^2/v_m^2) v^2\, dv} = \frac{I_3(v_m^{-2})}{I_2(v_m^{-2})}$$

where the integrals I_2 and I_3 are listed in Table 6.1. We thus find

$$\bar{v} = \frac{2}{\pi^{1/2}} v_m = \left(\frac{8kT}{\pi m}\right)^{1/2} = \left(\frac{8RT}{\pi M}\right)^{1/2}, \quad (6.10)$$

where M is the molar mass of the gas.

Similarly, the average value of the square of the velocity is

$$\overline{v^2} = \int_0^\infty v^2\, dP_v = \frac{\int_0^\infty v^2\, dP_v}{\int_0^\infty dP_v} = \frac{I_4(v_m^{-2})}{I_2(v_m^{-2})} = \frac{3}{2}v_m^2 = \frac{3kT}{m}. \quad (6.11)$$

The root mean square velocity is then

$$u = (\overline{v^2})^{1/2} = \left(\frac{3}{2}\right)^{1/2} v_m = \left(\frac{3kT}{m}\right)^{1/2} = \left(\frac{3RT}{M}\right)^{1/2}. \quad (6.12)$$

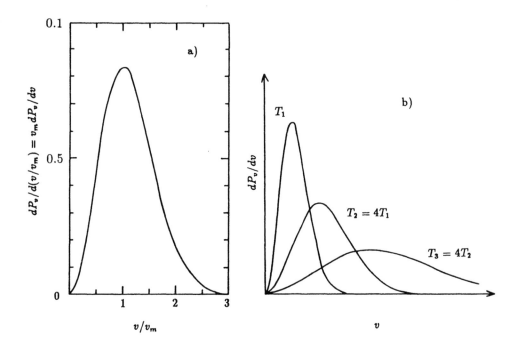

Figure 6.1: Distribution of the modulus of the velocity a) in dimensionless variables and b) in the variable v for three temperatures. The characteristic velocity v_m is given by $v_m = (2kT/m)^{1/2}$.

The velocities v_m, $\bar{v}$ and u lie in the ratio $1 : 1.128 : 1.225$. As they are of the same order of magnitude, they are characteristic of the velocity of the molecules in the gas. For example, for hydrogen H_2 ($M = 2.032$ g mol^{-1}) at 20°C, these velocities are of the order of $\bar{v} = 1\ 750$ m s^{-1} and, for argon ($M = 131.30$ g mol^{-1}) about $\bar{v} = 220$ m s^{-1}. Note that these velocities are of the same order of magnitude as that of sound c, given by the general formula

$$c = \left(\frac{\gamma RT}{M}\right)^{1/2} \quad , \quad \gamma = \frac{c_P}{c_V} \quad . \tag{6.13}$$

Finally, we observe that these velocities are independent of the pressure of the gas and proportional to the square root of the temperature.

In approximate models it is often assumed that the average velocity $\bar{v}$ or the root mean square velocity u is an attribute of all the particles, rather than considering the velocity distribution given by Maxwell. The velocity $\bar{v}$ is used when the displacement of the molecules is being considered, while u is used when their energy is involved.

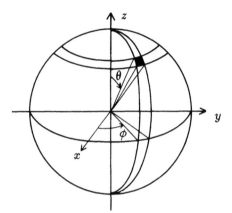

Figure 6.2: Polar coordinates

The mean kinetic energy of the molecules is

$$\bar{\epsilon} = \frac{1}{2}m\overline{v^2} = \frac{1}{2}mu^2 = \frac{3}{2}kT \quad ,$$

in agreement both with (5.24) and with the theorem of equipartition of energy (§3.5.1).

6.2.4   Distribution of the Direction of the Velocity

For some experiments, the variables θ and ϕ defining the orientation of the velocities must be included. Maxwell's distribution relation (6.5), expressed in spherical co-ordinates, is then

$$d^3 n_{v,\,\theta,\,\phi} = d^3 n_{v,\,\Omega} = n \left(\frac{m}{2\pi kT}\right)^{3/2} e^{-\beta mv^2/2} v^2 \; dv \; d^2\Omega = dn_v \frac{d^2\Omega}{4\pi} \quad , \quad (6.14)$$

with $d^2\Omega = \sin\theta \; d\theta \; d\phi$. This expression defines the number of molecules per unit volume with velocity modulus lying between v and $v + dv$ and for which the velocity vector points into the solid angle $d^2\Omega$ around the direction θ, ϕ (Fig. 6.2). Written in this way, the distribution relation explicitly states the isotropy of the molecular velocities through the factor $d^2\Omega/4\pi$, which is the fraction of molecules having their velocity directed into the solid angle $d^2\Omega$.

6.2.5   Distribution of Velocity Components

In some special problems, only one component of the velocity (e.g. perpendicular to a wall) is involved. In Cartesian co-ordinates, Maxwell's distribution law (6.5) is

$$d^3 n_{\mathbf{v}} = d^3 n_{v_x,\,v_y,\,v_z} = n \left(\frac{m}{2\pi kT}\right)^{3/2} e^{-\beta m(v_x^2 + v_y^2 + v_z^2)/2} \; dv_x \; dv_y \; dv_z \quad .$$

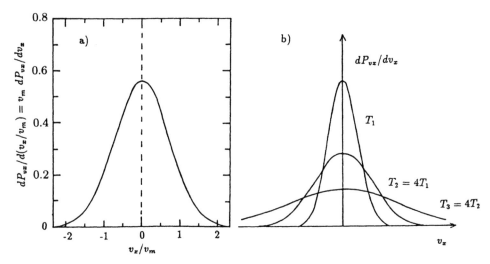

Figure 6.3: Distribution of a component of the velocity a) in dimensionless variables and b) in the variable v_x for three temperatures. The characteristic velocity v_m is given by $v_m = (2kT/m)^{1/2}$.

Integration over the components v_y and v_z yields the number of molecules per unit volume whose velocity component along the x axis lies between v_x and $v_x + dv_x$,

$$dn_{v_x} = n\left(\frac{m}{2\pi kT}\right)^{3/2} e^{-\beta m v_x^2/2}\, dv_x \int_{-\infty}^{+\infty} \int_{-\infty}^{+\infty} e^{-\beta m(v_y^2 + v_z^2)/2}\, dv_y\, dv_z$$

$$dn_{v_x} = n\left(\frac{m}{2\pi kT}\right)^{1/2} e^{-\beta m v_x^2/2}\, dv_x = \frac{n}{\pi^{1/2}} e^{-(v_x/v_m)^2}\, d\left(\frac{v_x}{v_m}\right) \quad . \tag{6.15}$$

The probability that a molecule has the x - component of its velocity lying between v_x and $v_x + dv_x$ is

$$
\begin{aligned}
dP_{v_x} &= \frac{dn_{v_x}}{n} = \left(\frac{m}{2\pi kT}\right)^{1/2} e^{-\beta m v_x^2/2}\, dv_x \\
&= \frac{1}{\pi^{1/2}} e^{-(v_x/v_m)^2}\, d\left(\frac{v_x}{v_m}\right) \quad .
\end{aligned}
\tag{6.16}
$$

The probability density in the variable v_x, shown in Figure 6.3, satisfies the Gaussian relation

$$p(x) = \frac{1}{(2\pi)^{1/2}\sigma} e^{-x^2/2\sigma^2} \quad , \tag{6.17}$$

with root mean square deviation $\sigma = v_m/\sqrt{2}$. This density, which is even in v_x, has a maximum equal to $1/(\sqrt{\pi}\, v_m)$ at $v_x = 0$. As the temperature increases, the curve becomes broader while keeping a constant area equal to 1.

We now determine the average values of a velocity component v_x and its square,

$\overline{v_x^2}$. First, for reasons of symmetry, we get

$$\overline{v_x} = \int_{-\infty}^{+\infty} v_x \, dP_{v_x} = 0 \ ,$$

$$\text{and} \quad \overline{v_x^2} = \int_{-\infty}^{+\infty} v_x^2 \, dP_{v_x} = \frac{\int_{-\infty}^{+\infty} v_x^2 \, dP_{v_x}}{\int_{-\infty}^{+\infty} dP_{v_x}} = \frac{2I_2(v_m^{-2})}{2I_0(v_m^{-2})} = \frac{v_m^2}{2} = \frac{kT}{m} \quad (6.18)$$

Since the velocity distribution is isotropic, we have

$$\overline{v_x} = \overline{v_y} = \overline{v_z} = 0 \quad \text{and} \quad \overline{v_x^2} = \overline{v_y^2} = \overline{v_z^2} = \frac{kT}{m} \ ,$$

giving the result already found in (6.11),

$$\overline{v^2} = \overline{v_x^2} + \overline{v_y^2} + \overline{v_z^2} = \frac{3kT}{m} \ .$$

In certain problems the average of the positive values of a velocity component must be evaluated, i.e.

$$\overline{v_{x+}} = \frac{\int_0^{+\infty} v_x \, dP_{v_x}}{\int_0^{+\infty} dP_{v_x}} = \frac{I_1(v_m^{-2})}{I_0(v_m^{-2})} \ .$$

Substitution of I_1 and I_0 (Table 6.1) yields

$$\overline{v_{x+}} = \frac{v_m}{\pi^{1/2}} = \frac{\overline{v}}{2} \ . \tag{6.19}$$

6.3   MOLECULAR BEAMS

Molecular beams are frequently used in modern experimental physics. An example is the famous experiment of Stern and Gerlach (1921), which revealed the spatial quantization of atomic magnetic moments, and that of Rabi (1939) which yielded measurements of nuclear magnetic moments. Molecular beams have also been used to verify the Maxwellian distribution of velocities. In this paragraph we investigate the properties of beams.

6.3.1   Distribution of Velocities in a Beam

We consider a gas inside a thin-walled vessel that is pierced by a small hole (the meaning of small will be defined later). Let Oz be the axis perpendicular to the wall and dS the elementary surface area of the orifice. In what follows we no longer use a superscript to define the order of certain differentials, i.e. we write $dS, d\Omega, d\tau$ instead of $d^2S, d^2\Omega, d^3\tau$. We calculate the number of molecules that, owing to their movement in the gas, pass through the orifice in a time interval dt, and count only those whose velocity is equal to $\mathbf{v}$ to within $d^3\mathbf{v}$. These molecules, which leave between t and $t + dt$, are contained in an oblique cylinder (Fig.6.4) of base dS, whose axis is parallel to $\mathbf{v}$, and the length of whose side is $v \, dt$ (maximum distance

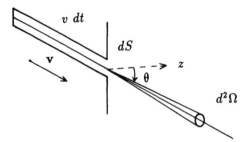

Figure 6.4: Diagram of a molecular beam

covered by a molecule inside the container in the time interval dt). Since the axis of the cylinder makes an angle θ with Oz, its volume is

$$d\tau = v \, dt \cos\theta \, dS = v_z \, dt \, dS . \tag{6.20a}$$

The number of molecules $d^6 N_v'$ leaving the orifice in dt is

$$d^6 N_v' = d^3 n_v \, d\tau \tag{6.20b}$$

where $d^3 n_v$, the number of molecules per unit volume having velocity $\mathbf{v}$ (to within $d^3\mathbf{v}$), is given by Maxwell's distribution (6.5). Substitution for $d^3 n_v$ yields

$$
\begin{aligned}
d^6 N_v' &= \frac{n}{\pi^{3/2}} \frac{1}{v_m^3} e^{-(v/v_m)^2} v_z \, d^3\mathbf{v} \, dt \, dS \\
&= \frac{n}{\pi^{3/2}} \frac{1}{v_m^3} e^{-(v/v_m)^2} v^3 \cos\theta \, dv \, d\Omega \, dt \, dS .
\end{aligned} \tag{6.21}
$$

The factor v^3, and not v^2 as in the Maxwell distribution, means that there are more molecules of higher velocity in the beam, since more of them pass through the hole in any given time. Therefore the root mean square velocity u_{beam} in a beam of given direction, defined by

$$u_{\text{beam}}^2 = \overline{v^2}_{\text{beam}} = \frac{\int_0^\infty v^2 \, d^6 N_v'}{\int_0^\infty d^6 N_v'}$$

is

$$u_{\text{beam}}^2 = \frac{\int_0^\infty v^5 \, e^{-(v/v_m)^2} \, dv}{\int_0^\infty v^3 \, e^{-(v/v_m)^2} \, dv} = \frac{I_5(v_m^{-2})}{I_3(v_m^{-2})} = 2v_m^2 = \frac{4kT}{m} .$$

Hence

$$u_{\text{beam}} = \sqrt{\frac{4kT}{m}} = \sqrt{\frac{4}{3}} \, u > u .$$

   This result, which applies to a molecular beam making an angle θ with the normal to the surface of the vessel, is independent of θ. The number of molecules leaving the orifice into the solid angle $d\Omega$ is, however, from (6.21), proportional to $\cos\theta$, in accordance with a well known principle of optics known as Lambert's law. To produce an intense molecular beam, therefore, the direction should be chosen perpendicular to the wall of the vessel.

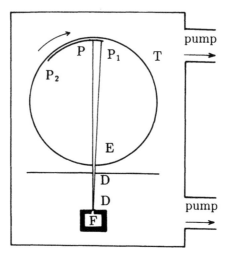

Figure 6.5: Diagram of a molecular beam apparatus for investigating velocity distributions. The metal atoms emitted by the oven F pass through the diaphragms D. After selection into bunches at slit E, they enter the rotating drum T. The atoms are deposited on plate P, the fastest towards P_1, and the slowest towards P_2. The vessel is evacuated by pumps.

6.3.2 Verification of Maxwell's Distribution

The first verifications of Maxwell's distribution were performed by Zartman (1931) and by Ko (1934) who investigated associations of molecules in metal vapours using the molecular beam technique developed by Stern from 1920 onwards. The experimental arrangement is outlined in Figure 6.5. A thermostated oven has a slit through which the vapour atoms above a molten metal can escape. Diaphragms define a molecular beam which strikes a rotating drum of radius R. A slit in the drum periodically intercepts the beam and allows a bunch of atoms to enter whose velocities are distributed according to (6.21). The atoms move along their linear trajectory inside the drum and are deposited on a plate in accordance with their velocity. Molecules with infinite velocity would be deposited on a line diametrically opposed to the slit E, while those with velocity v land on a line located on the circular arc at distance $l = 2R^2\omega/v$ from the starting line, where ω is the angular velocity of rotation of the drum. By measuring the thickness of the metallic layer as a function of l by photometric absorption, the distribution of velocities in the beam could be verified, and hence the Maxwellian distribution in the oven.

In more recent experiments, velocity selectors have been used, which, when coupled to highly sensitive detectors, yield a more direct and accurate measurement of the distribution (Fig. 6.6).

6.3.3 Molecular Discharges

Prior to investigations into molecular beams, Knudsen's experiments (1909) on molecular discharges provided an overall verification of the velocity distribution.

In these experiments, the number of molecules leaving a container via an orifice

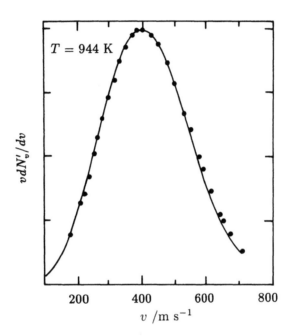

Figure 6.6: Experimental verification of the Maxwell distribution. The method uses a molecular beam of thallium ($M = 204.4$ g mol^{-1}), for which the velocity distribution is given by (6.21), and a velocity selector. The measured intensity, in arbitrary units on the vertical axis, is proportional to $v \, dN'_v/dv$. [R.C. Miller and P. Kush, *Phys. Rev.* **99**, 1314 (1955).]

is measured, whatever their velocity $\mathbf{v}$. This is done by measuring the flow per unit area, or current density, which is the ratio of this number to the surface area of the orifice and the time of observation. From equations (6.20), the total number of molecules leaving the surface dS in the interval dt is

$$d^3 N' = dt \; dS \int v_z \; d^3 n_{\mathbf{v}} \; ,$$

the integral being taken over all values of v_x and v_y, but only over positive values of v_z. The flow per unit area is then

$$j = \frac{d^3 N'}{dt \; dS} = \int_{v_z > 0} v_z \; d^3 n_{\mathbf{v}} = \overline{v_{z+}} \times \frac{n}{2} = \frac{1}{4} n\bar{v}$$

where $\overline{v_{z+}}$ is the average given in (6.19). Substituting for $\bar{v}$, and writing for the pressure inside the vessel $P = NkT/V$, we find

$$j = \frac{1}{4} n\bar{v} = \frac{1}{4} \frac{N}{V} \sqrt{\frac{8kT}{\pi m}} = \frac{P}{(2\pi mkT)^{1/2}} \; . \tag{6.22}$$

The flow per unit area is proportional to the pressure P and to $T^{-1/2}$. Moreover, it varies with the molar mass M of the gas as $M^{-1/2}$. For nitrogen at normal conditions of pressure and temperature, we have $j = 3 \times 10^{27} \; \text{m}^{-2} \; \text{s}^{-1}$. In other words, a square of side 1 Å receives 3×10^7 impacts per second.

For his experiments, Knudsen used platinum foil of thickness approximately 4 μm for the walls, in which holes of roughly $0.001 \; \text{mm}^2$ were pierced. The agreement to within 2 % between the kinetic theory and the experimental results provided at that time a good check of the Maxwell distribution.

These experiments also showed the limits within which the kinetic theory of molecular discharges applies. The dimensions of the hole must be much larger than the size of the molecules, which is always the case. They must also be much smaller than the mean free path of the molecules in the gas (§6.4.3). In this case there is no general movement of the gas (which would then be subject to hydrodynamic effects) near the hole, and the molecular motion is not perturbed by the surroundings.

Industrial applications of this phenomenon to isotope separation have been undertaken on a large scale. An example can be found in the nuclear industry, where the uranium isotopes ^{235}U and ^{238}U are separated in the form of uranium hexafluoride gas UF_6 by diffusion through porous walls. Given the small difference in molar mass, the number of transits made by the gas needs to be very large.

Exercise 6.1 *Kinetic Interpretation of Pressure in a Gas*

In kinetic theory it is supposed that the pressure exerted by a gas on the walls of the vessel owing to their translational motion is caused by collisions of the gas particles with the walls. Consider an element of the wall of area dS and the particles that strike it during the interval dt.

1 A particle with velocity $\mathbf{v}$ strikes the wall. Calculate the momentum $\mathbf{q}$ transferred to the wall, assuming that the collision is perfectly elastic. Choose the z axis perpendicular to the wall.

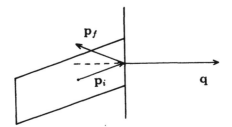

Figure 6.7: Kinetic pressure

2 Write in integral form the momentum transfer $d^3\,\mathcal{P}$ due to all the particles striking the area dS during dt.

3 Apply Newton's Second Law, and hence show that the pressure exerted on the wall by the gas is

$$P = \frac{1}{3}\,nmu^2\ . \tag{6.23}$$

4 Formulate this result for a gas of molecules.

Solution

1 In an elastic collision the component of the impulse $\mathbf{p} = m\mathbf{v}$ parallel to the wall remains unchanged, while the perpendicular component changes sign. The momentum transferred to the wall is thus equal to (Fig. 6.7)

$$(\mathbf{q})_z = (\mathbf{p_i})_z - (\mathbf{p_f})_z = 2mv_z\ .$$

2 Particles of velocity $\mathbf{v}$ having a spread $d^3\mathbf{v}$ and impinging on an area dS in the interval dt fit into a cylinder of volume $d\tau = v_z\,dt\,dS$. There are $d^6N'_v = d^3n_\mathbf{v}\,d\tau$ such particles. The total momentum transferred to the wall is therefore

$$(d^3\,\mathcal{P}\,)_z = \int q_z\,d^6N'_v = 2m\,dt\,dS\int v_z^2\,d^3n_\mathbf{v}\ .$$

The integral is taken over all v_x and v_y, but only over positive values of v_z.

3 The force exerted on dS by the pressure is found by applying Newton's law

$$(d^2\mathbf{F})_z = \frac{(d^3\,\dot{\mathcal{P}}\,)_z}{dt}.$$

This corresponds to a pressure

$$P = \frac{(d^2\mathbf{F})_z}{dS} = 2m\int v_z^2\,d^3n_\mathbf{v}\ .$$

Since the integrand is even in v_z, the integral is just one half of that taken over all values of v_z, and therefore

$$P = mn\overline{v_z^2} = \frac{1}{3}\,nmu^2\ .$$

Note that the elementary area remains at rest, since it is part of a rigid vessel for which the total momentum transfer is zero.

4 For a molecular gas $u^2 = 3kT/m$ (6.11). This yields

$$P = nkT = \frac{NkT}{V} \quad ,$$

giving once again the equation of state of ideal molecular gases.

6.4 MOLECULAR COLLISIONS

6.4.1 Molecular Interactions, Hard Sphere Model

In the ideal gas model we have considered so far, the size of the molecules and the interactions between them have been neglected, insofar as they were assumed to be point-like and independent. To explain transport phenomena such as viscosity (§6.5), interactions and the size of the molecules must be considered, as well as molecular collisions. After each collision, the individual velocities of molecules are modified, even though the statistical distribution of the velocities remains Maxwellian (§6.2.3). Indeed it is the collisions that " thermalize " an out-of-equilibrium gas and bring it back to the equilibrium state that corresponds to the constraints.

The interaction forces vary with the distance r between molecules. At large distances, they are electrostatic in origin (van der Waals forces) and attractive. At distances comparable to the size of the molecule, they are strongly repulsive and arise from quantum effects related to the Pauli exclusion principle. These forces derive from a potential $V(r)$, sketched in Figure 6.8a. In kinetic theory, an ideal model is used, that of hard spheres, in which the attractive contribution of the potential is zero and the repulsive contribution is infinite (Fig. 6.8b). Molecules are thus taken as being non interacting spherical particles of diameter d that collide perfectly elastically when they make contact ($r = d$). The collision between two molecules can equally well be considered as that of a molecule of radius d with a point-like molecule (Fig. 6.8b). The sphere of radius d is called the exclusion sphere of the molecule.

As will be seen below, molecular collisions are extremely frequent and, statistically speaking, the molecules are distributed uniformly in the volume. The mean distance a between two neighbouring molecules can be estimated by considering that each molecule occupies a cube of volume V/N of side $a = (V/N)^{1/3}$. For a gas at NTP, this mean distance is approximately 30 Å, i.e., roughly ten times the molecular diameter, which, for ordinary gases, is of the order of $d = 3$ Å. On average, the molecules are therefore fairly far from each other, which justifies the approximation in the hard sphere model that neglects attractions between molecules.

Exercise 6.2 *Distribution of distances between neighbouring molecules*
   Consider one molecule located at the origin of the axes and the $N - 1 \simeq N$ others distributed statistically uniformly in the volume V. Set $r_0 = (3V/4\pi N)^{1/3}$.
   1 Calculate the probability that

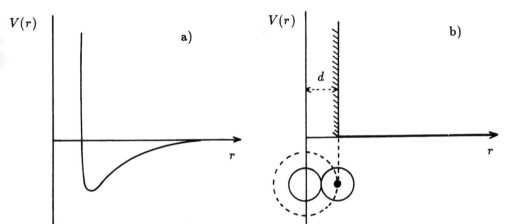

Figure 6.8: Interaction potentials of two molecules as a function of the separation r of their centres a) general shape; b) model of hard spheres with diameter d.

1.1 a given molecule lies at a distance less than r from the molecule at the origin;
1.2 a given molecule lies at a distance greater than r;
1.3 the N molecules lie at a distance greater than r;
1.4 the closest molecule lies between r and $r+dr$. Rewrite the last two expressions using the relation $(1 + \epsilon)^N \simeq e^{N\epsilon}$.
2 Calculate the mean distance $\bar{r}$ between a molecule and its closest neighbour. Make use of the relation

$$\Gamma\left(\frac{4}{3}\right) = \int_0^\infty e^{-t}\, t^{1/3}\, dt = 0.893 \ .$$

Compare with the approximate result $a = (V/N)^{1/3}$.

Solution
1 The probability of finding a molecule in the elementary volume $d\tau$ is $d\tau/V$. Hence
1.1

$$P_a \ = \ \frac{4\pi r^3}{3V} \ ;$$

1.2

$$P_b \ = \ 1 - P_a = 1 - \frac{4\pi r^3}{3V} \ ;$$

1.3

$$P_c \ = \ (P_b)^N = \left(1 - \frac{4\pi r^3}{3V}\right)^N \simeq \exp\left(-\frac{r^3}{r_0^3}\right) \ ;$$

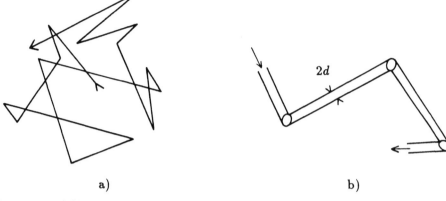

Figure 6.9: a) Sketch of a molecular trajectory. b) Successive cylinders of exclusion swept out by a molecule.

1.4 The required probability is such that all molecules are at a distance greater than r and that at least one is at a distance less than $r + dr$. This is equal to

$$dP = P_c(r) - P_c(r + dr) = -\frac{dP_c}{dr}\, dr = 3\frac{r^2}{r_0^3} \exp\left(-\frac{r^3}{r_0^3}\right)\, dr \ . \qquad (6.24)$$

2 The mean distance between a molecule and its nearest neighbour is

$$\bar{r} = \int_0^\infty r\, dP = 3 \int_0^\infty \frac{r^3}{r_0^3} \exp\left(-\frac{r^3}{r_0^3}\right)\, dr \ .$$

With the change of variable $t = r^3/r_0^3$, we get

$$\bar{r} = r_0 \Gamma\left(\frac{4}{3}\right) = \left(\frac{3}{4\pi}\right)^{1/3} \Gamma\left(\frac{4}{3}\right)\left(\frac{V}{N}\right)^{1/3} = 0.55 \left(\frac{V}{N}\right)^{1/3} \ .$$

This distance is one half of what would be found if the volume V were divided into N cubes.

6.4.2   Frequency of Collisions

During its motion, a molecule experiences many collisions. Between two successive collisions it travels in a straight line and at constant velocity. At each collision both the direction and the magnitude of the velocity change, and the path of a molecule is therefore a broken line (Fig. 6.9a). Over an interval dt during which the molecule has experienced many collisions, the length of the path travelled is $dl = \bar{v}\, dt$, since the distribution of the successive velocities of a molecule is the same as that of all the molecules at a given instant. The exclusion sphere of the molecule thus sweeps out a series of cylinders (Fig. 6.9b) of base

$$\sigma = \pi d^2 \qquad (6.25)$$

(collision cross-section) and total height $\bar{v}\, dt$, i.e. the total volume is $d\tau = \sigma\bar{v}\, dt$.

If we now make the simplifying assumption that the other molecules are immobile, then the volume swept out encloses on average $n\, d\tau$ centres of molecules, corresponding to $n\, d\tau$ collisions experienced by the moving molecule. In this model, the collision frequency ν, which is the number of collisions experienced by a molecule per second, is equal to $n\, d\tau/dt$, i.e.

$$\nu = n\sigma\bar{v} \quad \text{(approximately)} . \tag{6.26a}$$

If we now no longer assume that the molecules are immobile but instead move with a Maxwellian distribution of velocities, the correct expression for the frequency of collision is (Pb. 6.1)

$$\nu = \sqrt{2}n\sigma\bar{v} = \frac{4\mathcal{N}\sigma P}{(\pi M RT)^{1/2}} \; , \tag{6.26b}$$

where n and $\bar{v}$ have been expressed as functions of temperature and pressure. For nitrogen ($d = 3.75$ Å) under normal conditions $\nu = 7.6 \times 10^9$ s^{-1}, or several thousand million collisions per second for one molecule.

In any time interval dt the number of collisions fluctuates around the mean value $\nu\, dt$. For sufficiently long times, however, the fluctuations around the average are of the order of the square root of this average value (central limit theorem). A nitrogen molecule, for example, makes on average 7.6×10^6 collisions in a millisecond, with a fluctuation equal to 2.8×10^3, i.e., a relative amplitude of 3.6×10^{-4}. These fluctuations are thus negligible and no further distinction will be made between the frequency ν and its average value.

6.4.3 Mean Free Path

The free path of a molecule is the length of its trajectory between two successive collisions. Its mean value λ is called the mean free path. This concept, which is very important in transport phenomena, was introduced by Clausius around 1858 to explain the slow rate of diffusion of one gas into another, in spite of the high molecular velocities.

The mean free path λ is equal to the ratio of the distance $\bar{v}\, dt$ travelled by a molecule in the time interval dt to the number of collisions, $\nu\, dt$, i.e., from (6.26b)

$$\lambda = \frac{\bar{v}}{\nu} = \frac{1}{\sqrt{2}n\sigma} \; . \tag{6.27}$$

For nitrogen under standard conditions, the mean free path is $\lambda = 6.0 \times 10^{-8}$ m $=$ 600 Å. For these conditions, the diameter of the molecules (~ 3 Å), the intermolecular distances (~ 30 Å) and the mean free paths lie approximately in the ratios $1 : 10 : 100$.

If n is expressed as a function of temperature and pressure, it can be seen that λ is proportional to T/P, i.e. the mean free path varies inversely with the pressure. Thus, at the low pressures regularly attained by mechanical pumps ($P = 10^{-4}$ torr), the mean free path for nitrogen at $T = 273$ K is $\lambda = 45$ cm, even though the number of molecules per unit volume is still enormous ($n = 3.5 \times 10^{12}$ molecules par cm^3). The mean free path is therefore of the order of magnitude of the size of the container, and since the molecules collide more often with the walls than among each other, relation (6.27) no longer holds. The mean free path is then a characteristic of the geometry of the container.

We recall that the concepts introduced above and the results obtained are strictly valid only for the hard sphere model. If account is taken of the real shape of the interaction potential, the trajectory of the molecules is no longer a straight line and the notion of collisions must be replaced by that of diffusion. Nevertheless the above results are a good approximation, since interactions between molecules become significant only when they are close (short range interactions).

6.4.4   Distribution of Free Paths

In kinetic theory, it is important to know the distribution of molecular free paths around their mean value λ. To determine this relation, we consider a model in which all the molecules move at the same velocity $\bar{v}$, and we calculate the number of molecules $N(t)$ that have not collided between an instant $t = 0$ and a later time t. This number decreases in the course of time from its initial value $N(0) = N$ to its final value $N(\infty) = 0$. As the collision frequency of one molecule is ν, the number of collisions of $N(t)$ molecules in the interval between t and $t + dt$ is $N(t)\, \nu\, dt$, and hence

$$N(t) - N(t + dt) = -dN = N(t)\, \nu\, dt \ .$$

Integration gives the relation for the exponential decrease,

$$N(t) = N(0)\, e^{-\nu t} \ . \tag{6.28}$$

According to (6.27), the number of molecules that have travelled a distance $l = \bar{v}t$ without colliding is therefore

$$N(l) = N(0)\, e^{-\nu l/\bar{v}} = N(0)\, e^{-l/\lambda} \ . \tag{6.29}$$

The probability that the free path of a molecule lies between l and $l + dl$ is then

$$dP = \frac{N(l) - N(l + dl)}{N(0)} = e^{-l/\lambda}\, \frac{dl}{\lambda} \ . \tag{6.30}$$

This probability is the required distribution (Fig. 6.10). It can be confirmed that we indeed have

$$\int_0^\infty dP = 1 \ , \qquad \int_0^\infty l\, dP = \lambda \quad \text{and} \quad \int_0^\infty l^2\, dP = 2\lambda^2 \ , \tag{6.31}$$

and that the most probable free path length is zero.

In this demonstration it is not assumed that the molecules undergo a collision at $t = 0$.

The above arguments were framed for a model in which the molecules have a velocity $\bar{v}$. To be completely rigorous, the probability that a molecule with velocity

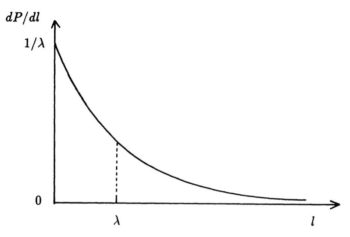

Figure 6.10: Distribution of free paths

v has a given free path depends on v. For example, it is certain that a particle with zero velocity will have a free path equal to zero. Relation (6.30) remains valid for molecules with velocity v if λ is replaced by some function $\lambda(v)$. For a hard sphere model having a Maxwellian distribution of velocities, the following expression for $\nu(v) = v/\lambda(v)$ is found

$$\nu(v) = \frac{\nu}{2\sqrt{2}} \left[e^{-x^2} + \left(2x + \frac{1}{x} \right) \int_0^x e^{-y^2}\, dy \right] , \tag{6.32}$$

where ν is the collision frequency (6.26b) and $x = v/v_m$.

It is interesting to examine the limits of $\nu(v)$ and $\lambda(v)$. For $v \to 0$, we have

$$\nu(v) = \frac{\nu}{\sqrt{2}} \quad \text{and} \quad \lambda(v) = \frac{\sqrt{2}v}{\nu} \to 0 ,$$

while for $v \to \infty$

$$\nu(v) = \frac{\nu}{\sqrt{2}} \frac{v}{v_m} \frac{\sqrt{\pi}}{2} = n\sigma v \quad \text{and} \quad \lambda(v) = \frac{1}{n\sigma} .$$

The last result is the same as was found in paragraph 6.4.2, since if one molecule is moving fast the velocities of the others are negligible in comparison.

Exercise 6.3 *Flow of molecules through an orifice subsequent to a collision*

We now rederive expression (6.22) for the flow per unit area j by considering molecules that experience a collision between t and $t + dt$ and then emerge from an orifice without further collision.

1 Calculate the number of molecules that, in the interval dt after a collision, leave an elementary volume $d\tau$ of gas, with their velocity directed into the solid angle $d\Omega$ subtended by the orifice of area dS (Fig. 6.11).

2 Hence deduce the number of these molecules reaching the orifice without experiencing another collision.

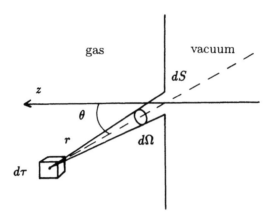

Figure 6.11: Solid angle for molecules leaving an orifice

3 Calculate the total flow per unit area j of molecules through the orifice.

4 Determine the mean value of the height $z = r \cos \theta$ at which the molecules undergo their last collision.

Solution

1 There are $n d\tau$ molecules in the volume element $d\tau$. As the collision frequency is ν, the number of molecules that experience a collision in the interval dt is $n \, d\tau \, \nu \, dt$. Of these, a fraction

$$\frac{d\Omega}{4\pi} = \frac{dS \cos \theta}{4\pi r^2}$$

have their velocity directed towards the orifice.

2 The number of molecules reaching the orifice is therefore

$$d^6 N' = n \, d\tau \, \nu \, dt \, \frac{dS \cos \theta}{4\pi r^2} \, e^{-r/\lambda} \, ,$$

where the exponential is the probability that the free path of a molecule exceeds r.

3 The total number of molecules that experience their last collision between t and $t + dt$ and then leave through the orifice is found by integrating $d^6 N'$ over $d\tau = r^2 \, dr \, \sin \theta \, d\theta \, d\phi$. The flow per unit area is therefore

$$\begin{aligned}
j &= \int \frac{d^6 N'}{dt \, dS} = \frac{n\nu}{4\pi} \int_0^\infty e^{-r/\lambda} \, dr \int_0^{\pi/2} \sin \theta \cos \theta \, d\theta \int_0^{2\pi} d\phi \\
&= \frac{n\nu}{4\pi} \times \lambda \times \frac{1}{2} \times 2\pi = \frac{1}{4} n\bar{v} \, .
\end{aligned}$$

This result is identical to that found in (6.22).

4 The average height $\bar{z}$ at which the molecules encounter their last collision is

$$\bar{z} = \frac{\int z \, d^6 N'}{\int d^6 N'} = \frac{\int_0^{\pi/2} \sin \theta \cos^2 \theta \, d\theta \int_0^\infty r \, e^{-r/\lambda} \, dr}{\int_0^{\pi/2} \sin \theta \cos \theta \, d\theta \int_0^\infty e^{-r/\lambda} \, dr} = \frac{2}{3} \lambda \, . \tag{6.33}$$

6.4.5   Total Displacement of a Labelled Molecule

We consider a labelled molecule located at the origin of the co-ordinate axes at time $t = 0$, and seek the probability $F(\mathbf{r},\ t)\ d^3\mathbf{r}$ of finding this molecule in the neighbourhood of the point $\mathbf{r}$ at a time t much greater than the mean period $1/\nu$ of a molecular free path. If we call $N = \nu t$ the number of collisions experienced by this molecule during the interval t, we then have

$$\mathbf{r} = \mathbf{l}_1 + \mathbf{l}_2 + \ldots + \mathbf{l}_N \ ,$$

where the $\mathbf{l}_i$ are the successive trajectories of the molecule. The modulus of these trajectories has a distribution given by expression (6.30), and their directions are distributed isotropically. Each component of the vector $\mathbf{r}$, for example

$$x = l_{1x} + l_{2x} + \ldots + l_{Nx} \ ,$$

is then the sum of N random variables l_x whose average value is zero (isotropy) and the mean square deviation, from (6.31), is

$$\overline{l_x^2} - (\overline{l_x})^2 = \overline{l_x^2} = \frac{1}{3}\overline{l^2} = \frac{2}{3}\lambda^2 \ .$$

According to the central limit theorem, the variable x is then a Gaussian variable of mean value $\bar{x} = N\overline{l_x} = 0$ and standard deviation equal to $\overline{x^2} = N\overline{l_x^2} = 2N\lambda^2/3$. Its density distribution is

$$(2\pi\overline{x^2})^{-1/2} \exp(-x^2/2\overline{x^2}) = \left(\frac{4\pi}{3}\nu\lambda^2 t\right)^{-1/2} \exp(-3x^2/4\nu\lambda^2 t) \ .$$

As the three components x, y and z are independent, the required probability density is then

$$F(\mathbf{r},\ t) = (4\pi Dt)^{-3/2} \exp(-r^2/4Dt) \quad \text{with} \quad D = \nu\lambda^2/3 = \lambda\bar{v}/3 \ . \qquad (6.34)$$

The coefficient D, called the diffusion coefficient, will be investigated in the next paragraph.

The distribution F, shown in figure 6.12, is similar to the Maxwell velocity distribution. The mean displacement is thus

$$\bar{r} = 4 \left(\frac{Dt}{\pi}\right)^{1/2} \ .$$

The mean displacement does not increase linearly with time. For nitrogen under normal conditions, where D is equal to 2.0×10^{-5} m^2 s^{-1}, a labelled molecule covers a distance $\bar{r} = 1$ cm in one second and $\bar{r} = 1$ m in approximately 3 hours. This example shows that the total displacement of a labelled molecule in a gas is very small in spite of its high instantaneous velocity. This is a consequence of the large number of collisions.

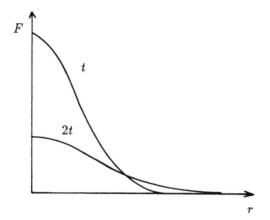

Figure 6.12: Distribution of the total displacement $F(\mathbf{r}, t)$ as a function of $r = |\mathbf{r}|$. This isotropic distribution is plotted for two values of time, t and $2t$. Note that $\int 4\pi r^2 \, F(r, t) \, dr = 1$ and $F(\mathbf{r}, 0) = \delta(\mathbf{r})$.

6.5   TRANSPORT PHENOMENA IN GASES

Until now we have been studying kinetic theory for gases that are in complete thermodynamic equilibrium, i.e., in particular having uniform temperature and density. Away from equilibrium (presence of a temperature gradient for example), irreversible phenomena occur in the gas, such as

- transport of molecular species, or diffusion,
- energy transport or thermal conduction,
- momentum transport or viscosity.

Kinetic theory provides an explanation for phenomenological relations using simple models in which local thermodynamic equilibrium is assumed to hold, i.e. each elementary volume of gas obeys the equations for a gas at equilibrium. It also relates the phenomenological coefficients to molecular quantities (mass, diameter) and thus allows relations to be derived among these coefficients.

6.5.1   Diffusion

When a chemical species in a given medium is distributed inhomogeneously, the species is transported from zones of high concentration towards those of lower concentration. In 1855 the physiologist Fick showed that the particle current density $\mathbf{j}_N$ is proportional to the gradient of the number of particles per unit volume $n(\mathbf{r})$

$$\mathbf{j}_N = -D \, \mathbf{grad} \, n = -D \, \boldsymbol{\nabla} \, n \qquad (6.35)$$

where D is called the diffusion coefficient of the species under consideration in the medium.

This relation and the expression for the coefficient D reappear in a model of diffusion in a gas in which the number of labelled molecules per unit volume $n^*(z)$ is a function of the co-ordinate z only (Fig. 6.13). We calculate the number of

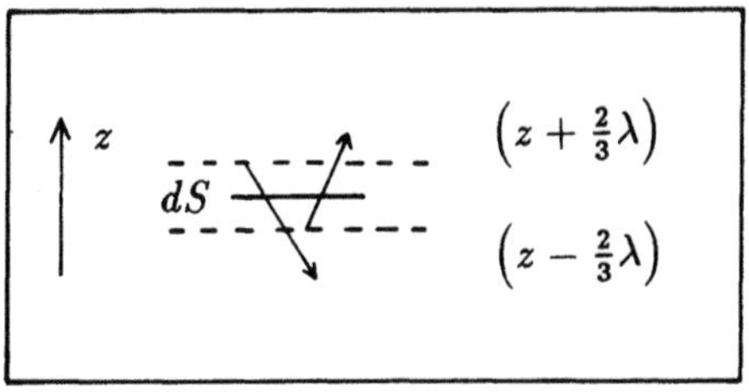

Figure 6.13: Transport of molecules through an area dS

labelled molecules dN^* crossing an imaginary surface dS perpendicular to Oz in a time interval dt, at a height z. We assume that all molecules passing through dS have undergone their last collision at a mean distance $2\lambda/3$ (6.33) from the surface. The number of molecules crossing from below during dt is then (§6.3.3)

$$dN^*(\uparrow) = \frac{1}{4}\, n^*\left(z - \frac{2\lambda}{3}\right)\, \bar{v}\, dS\, dt \quad,$$

while the number of those going in the other direction is

$$dN^*(\downarrow) = \frac{1}{4}n^*\left(z + \frac{2\lambda}{3}\right)\, \bar{v}\, dS\, dt \quad,$$

i.e., in total

$$dN^* = dN^*(\uparrow) - dN^*(\downarrow) = -\frac{1}{4}\frac{dn^*(z)}{dz}\frac{4\lambda}{3}\, \bar{v}\, dS\, dt \quad.$$

In this simplified model, the current density is thus

$$j_N = \frac{dN^*}{dt\, dS} = -\frac{1}{3}\lambda\bar{v}\,\frac{dn^*}{dz} \quad.$$

This yields Fick's law again, with an approximate expression for the diffusion coefficient

$$D = \frac{1}{3}\lambda\bar{v} \quad \text{(approximately)} \quad. \tag{6.36}$$

An exact calculation using the hard sphere model gives the modified result

$$D = \frac{3\pi}{16}\lambda\bar{v} = 0.589\lambda\bar{v} \quad. \tag{6.37}$$

On substituting the expressions for λ and $\bar{v}$ as a function of temperature and pressure, we get

$$D = \frac{3\sqrt{\pi}}{8}\,\frac{1}{\sigma\sqrt{m}}\,\frac{(kT)^{3/2}}{P} \quad. \tag{6.38}$$

For nitrogen at normal temperature and pressure, $D = 1.6 \times 10^{-5}$ m^2 s^{-1}, which is very close to the experimental value $D = 2.0 \times 10^{-5}$ m^2 s^{-1}. Experimentally, equation (6.38) gives a good description of the inverse pressure relationship, but the dependence on T indicates an exponent between 1.75 and 2.0 rather than 1.5. We note that experiments on molecular diffusion within a given chemical species (isotopic diffusion) are uncommon and that data for diffusion of chemical species in a different gas are more numerous. Under these conditions, for a given gas, each species should be considered as having its own diffusion coefficient.

To determine the way in which the concentration of a chemical species varies with time, the continuity equation is used

$$\frac{\partial n}{\partial t} + \text{div } \mathbf{j}_N = 0 \ . \tag{6.39}$$

This ensures conservation of the number of molecules. Eliminating $\mathbf{j}_N$ by means of Fick's equation, and assuming that D is constant, we obtain a relation called the diffusion equation

$$\frac{\partial n}{\partial t} = D \ \Delta n = D \ \left(\frac{\partial^2 n}{\partial x^2} + \frac{\partial^2 n}{\partial y^2} + \frac{\partial^2 n}{\partial z^2} \right) \ . \tag{6.40}$$

If the concentration at time $t = 0$ is known, i.e., $n(\mathbf{r}, 0) = n_0(\mathbf{r})$, concentrations at later times can be calculated from the diffusion equation, through

$$n(\mathbf{r}, \ t) = \int n_0(\mathbf{r}') \ F(\mathbf{r} - \mathbf{r}', \ t) \ d^3\mathbf{r}'$$

where the distribution $F(\mathbf{r}, \ t)$ is that found in (6.34).

In particular, if at zero time the labelled substance is concentrated at a point taken to be the origin, i.e., $n_0(\mathbf{r}) = N_0 \ \delta(\mathbf{r})$, then the concentration at a later time t will be

$$n(\mathbf{r}, \ t) = N_0 \ F(\mathbf{r}, \ t) \ .$$

The concentration is then a Gaussian with standard deviation $(2Dt)^{1/2}$ and height $(4\pi DT)^{-3/2}$ (Fig. 6.12). These results reveal the relationship between diffusion and the displacement of an individual molecule (§6.4.5). They also allow us to comprehend the slowness of the diffusion process and the experimental difficulties involved in observing the phenomenon, which is often obscured by convective movements.

Exercise 6.4 *Rate of evaporation of alcohol*

A test-tube contains ethyl alcohol. The surface of the liquid is at a distance $h = 10$ cm from the top of the tube, which is open to the outside air.

1 In the stationary regime, derive an expression for the number density $n(z)$, assuming that it is zero at the top of the tube. P_0 is the saturated vapour pressure of alcohol at room temperature T_0 ($P_0 = 44$ torrs at $T_0 = 293$ K).

2 Hence find the rate of descent of the surface of the liquid in the tube. The molar volume of liquid alcohol is $v = 60$ cm^3 mol^{-1} and the diffusion coefficient of alcohol in air is $D = 0.1$ cm^2 s^{-1}.

Solution

1 For this permanent regime problem in one dimension the diffusion equation (6.40) is

$$\frac{\partial n}{\partial t} = D\frac{\partial^2 n}{\partial z^2} = 0 \ .$$

It follows that n takes the form $n = n_0(1 - z/h)$, where the boundary conditions $n(h) = 0$, and $n(0) = n_0$ have been introduced, and where n_0 is equal to P_0/kT_0, the number of alcohol molecules per unit volume in the saturated vapour.

2 From Fick's equation, the current density of alcohol molecules is

$$j_N = -D\frac{dn}{dz} = \frac{Dn_0}{h} \ .$$

The number of molecules leaving through the top of the tube in time t is thus $j_N St = Dn_0 St/h$, which corresponds to a volume of liquid

$$V = \frac{Dn_0 St}{h} \times \frac{v}{\mathcal{N}} \ .$$

The rate of descent of the level of the liquid is

$$\frac{V}{St} = \frac{Dn_0 v}{h\mathcal{N}} = \frac{D}{h}\frac{P_0 v}{RT_0} = 1.4 \times 10^{-6} \ \text{cm s}^{-1} \ ,$$

or 1.2 mm per day. This result applies to pure diffusion of the molecules in the absence of any convection.

6.5.2 Heat Conduction

When the temperature in a material is not uniform a heat current develops from the hot towards the cold regions with a density $\mathbf{j}_U$ that is proportional to the gradient of the temperature $T(\mathbf{r})$

$$\mathbf{j}_U = -K \ \mathbf{grad} \ T = -K \ \boldsymbol{\nabla} T \ . \tag{6.41}$$

K is the thermal conductivity of the substance (Fourier law).

Using a method analogous to that used for diffusion, it is straightforward to obtain an approximate expression for the thermal conductivity of a gas. Here we are interested in the transport of kinetic energy of the molecules. We consider a gas whose temperature is a function of z only, and establish the kinetic energy balance crossing a surface area dS perpendicular to Oz in the time interval dt (Fig. 6.13). Once again we assume that all the molecules have experienced their last collision at an average distance $2\lambda/3$ from the surface. The energy transported in dt by molecules crossing dS from below is then

$$dU(\uparrow) = \frac{1}{4}n\bar{v} \ dS \ dt \times \bar{\epsilon}\left(z - \frac{2\lambda}{3}\right) ,$$

where $\bar{\epsilon}(z)$ is the mean kinetic energy of the molecules at height z. Similarly, the energy transported by molecules travelling in the other direction is

$$dU(\downarrow) = \frac{1}{4}n\bar{v} \, dS \, dt \times \bar{\epsilon}\left(z + \frac{2\lambda}{3}\right) \; ,$$

or, in total

$$dU = dU(\uparrow) - dU(\downarrow) = \frac{1}{4}n\bar{v} \, dS \, dt \times \left(\frac{-4\lambda}{3}\frac{d\bar{\epsilon}}{dz}\right) \; .$$

In this simplified model, the heat current density is therefore

$$j_U = \frac{dU}{dt \, dS} = -\frac{1}{3}n\lambda\bar{v}\frac{d\bar{\epsilon}}{dz} = -\frac{1}{3}n\lambda\bar{v}\frac{d\bar{\epsilon}}{dT}\frac{dT}{dz} \; .$$

The general expression (6.41) is thus obtained once again, in which the approximate expression for the thermal conductivity is

$$K = \frac{1}{3}n\lambda\bar{v}\frac{d\bar{\epsilon}}{dT} = \frac{1}{3}n\lambda\bar{v}\frac{c_V}{\mathcal{N}} \quad \text{(approximately)} \; , \tag{6.42}$$

c_V being the molar heat capacity of the gas. An exact calculation based on the hard sphere model, taking into account the position of the last collision, gives

$$K = \frac{25\pi}{64}n\lambda\bar{v}\frac{c_V}{\mathcal{N}} = 1.227n\lambda\bar{v}\frac{c_V}{\mathcal{N}} \; . \tag{6.43}$$

Expressing n, λ and $\bar{v}$ as functions of temperature and pressure, we find

$$K = \frac{25\sqrt{\pi}}{32}\frac{1}{\mathcal{N}\sigma m^{1/2}}(kT)^{1/2}c_V \; . \tag{6.44}$$

For argon ($M = 40.0 \, \text{g mol}^{-1}$) at 273 K, with an atomic diameter of 3.64 Å, the value calculated for the coefficient K is 1.64×10^{-2} J m^{-1} s^{-1} K^{-1}. The experimental value is 1.66×10^{-2} SI. The agreement is not as good for nitrogen, 3.08×10^{-2} SI instead of 2.40×10^{-2} SI. This discrepancy arises from the fact that the exact formula (6.43) applies to molecules that possess only translational kinetic energy (§6.5.4).

The hard sphere model explains why light gases are the most strongly conducting. This is the reason that helium is used as the exchange gas in cryostats. It also explains why the conductivity is independent of P at ordinary pressures. However, the increase of K as $T^{1/2}$ is slower than that observed experimentally. The difference is a result of the approximation made for the interaction potential in this model.

6.5.3   Viscosity

Consider a gas confined between two planes perpendicular to an axis Oz (Fig. 6.14), one of which is fixed and the other moves at constant velocity V_0 parallel to itself. Each layer of gas at height z is then dragged along at the mean velocity $V(z)$, while

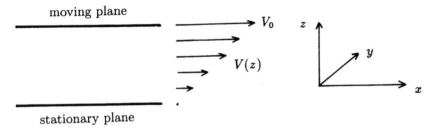

Figure 6.14: Diagram of Couette flow

both boundary layers of gas have the same velocity as the two planes with which they are in contact. In the permanent regime, the velocity $V(z)$ varies linearly with z between 0 and V_0 (planar Couette flow).

In kinetic theory the velocity distribution is a shifted Maxwell distribution, in which the average velocity is parallel to the plates and equal to $V(z)$. In their motion, the molecules carry from one level to another the mean momentum that they possessed at their last collision. For the interval dt, we now calculate the momentum balance $d^3\mathcal{P}$ through a surface of area dS at height z. As before, we assume that the molecules undergo their last collision before crossing the surface at the heights $z \pm 2\lambda/3$. The momentum carried by the molecules crossing the surface from below is then

$$d^3\mathcal{P}\,(\uparrow) = \frac{1}{4}n\bar{v}\,dS\,dt \times mV\left(z - \frac{2\lambda}{3}\right)$$

and that carried by those crossing dS from above is

$$d^3\mathcal{P}\,(\downarrow) = \frac{1}{4}n\bar{v}\,dS\,dt \times mV\left(z + \frac{2\lambda}{3}\right) \quad .$$

In total, this gives

$$d^3\mathcal{P} = d^3\mathcal{P}\,(\uparrow) - d^3\mathcal{P}\,(\downarrow) = \frac{1}{4}n\bar{v}\,dS\,dt \times \left(-\frac{4\lambda}{3}m\frac{dV}{dz}\right) \quad .$$

In this problem, where the flow is in only one dimensional, the momentum current density is

$$j_{\mathcal{P}} = \frac{d^3\mathcal{P}}{dt\,dS} = -\frac{1}{3}n\lambda\bar{v}m\frac{dV}{dz} = -\eta\frac{dV}{dz} \tag{6.45}$$

where η is the dynamic viscosity coefficient given by

$$\eta = \frac{1}{3}n\lambda m\bar{v} \quad \text{(approximately)} \quad . \tag{6.46}$$

An exact calculation using the hard sphere model yields

$$\eta = \frac{5\pi}{32}n\lambda m\bar{v} = 0.491n\lambda m\bar{v} \quad . \tag{6.47}$$

Table 6.2: Viscosity η of some gases in normal conditions. The molecular diameter d is calculated from (6.48) with $\sigma = \pi d^2$; d_{sol} is the nearest neighbour distance in the crystal. d_∞ and C are the constants defined in (6.51) with $\sigma_\infty = \pi d_\infty^2$.

gas	η	d	d_{sol}	d_∞	C
	$/10^{-6}$ SI	$/10^{-10}$ m	$/10^{-10}$ m	$/10^{-10}$ m	$/K$
He	18.6	2.18	-	1.83	116
Ne	29.7	2.58	3.20	2.35	56
A	21.0	3.64	3.83	2.95	142
Kr	23.3	4.16	4.0	3.20	188
H_2	8.35	2.74	-	2.20	149
N_2	16.6	3.75	-	3.19	104
CO_2	13.9	4.59	-	3.30	254

Inserting the expressions for n, λ and $\bar{v}$, we get

$$\eta = \frac{5\sqrt{\pi}}{16} \frac{m^{1/2}}{\sigma} (kT)^{1/2} . \tag{6.48}$$

For nitrogen at 273 K, the coefficient is equal to 1.66×10^{-5} kg m^{-1} s^{-1}, in agreement with experiment. In the international system, the unit of viscosity is the poiseuille. The poise, a cgs unit, is also commonly used, and is equal to 0.1 poiseuille.

We see that the viscosity is independent of the pressure (or the density of the gas). This result, obtained by Maxwell in 1860, surprised many physicists of that period who expected the coefficient of viscosity to increase with the density of the gas. The experimental verification of this result proved to be a great success for the kinetic theory of gases. However, the lack of dependence on P applies only in a pressure range between about 10^{-2} torr and 10 atmospheres. At higher pressures, interactions are no longer negligible, and viscosity increases with pressure. At lower pressures, when the free path λ becomes comparable with the size of the container, the viscosity (6.47) no longer depends upon λ , and η is proportional to the number of particles per unit volume n, and hence to the pressure P. As for its variation with temperature, the viscosity increases with T (contrary to liquids). The experimentally observed increase is, however, faster than the $T^{1/2}$ dependence predicted by the hard sphere model.

We remark that ideal gases, which are devoid of attractive forces, are viscous. The viscosity is not the result of "frictional forces" between the molecules, but of momentum transport. The viscosity of gases ($\sim 10^{-5}$ poiseuille) is smaller than that of liquids ($\sim 10^{-3}$ poiseuille).

In the problem considered above, the gas flows in one direction only, say x. Since the velocity depends only on z (Fig. 6.14), the current density j_P corresponds to a transfer along z of momentum parallel to the x axis. This shows that this quantity is neither a scalar nor a vector but that it is the $(j_P)_{zx}$ component of a second order

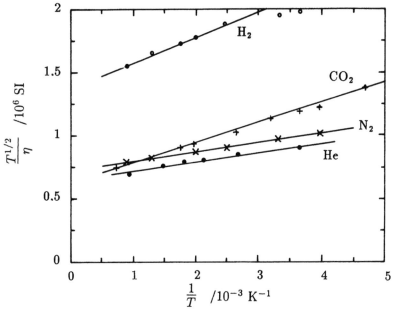

Figure 6.15: Variation of the viscosity η with temperature T. The straight lines are given by Sutherland's equation (6.51) with $A = 5(\pi m k)^{1/2}/16\sigma_\infty$.

tensor $(j_P)_{ij}$. For an arbitrary velocity field $\mathbf{V}(\mathbf{x})$, equation (6.45) generalizes to

$$(j_P)_{ij} = -\eta \left(\frac{\partial V_i}{\partial x_j} + \frac{\partial V_j}{\partial x_i} \right) \quad (i,\, j = x,\, y,\, z) \, . \tag{6.49}$$

In hydrodynamics, the stress tensor τ_{ij} is introduced in such a way that the force $d\mathbf{f}$ exerted by one part of the fluid on another over a surface dS with normal $\mathbf{n}$ is

$$df_i = n_j \tau_{ji} \, dS \, .$$

Using the equation of conservation of momentum and Newton's Second Law, it can be shown that $j_P = -\tau$, i.e.

$$\tau_{ij} = \eta \left(\frac{\partial V_i}{\partial x_j} + \frac{\partial V_j}{\partial x_i} \right) \, , \tag{6.50}$$

which is the equation used in textbooks on hydrodynamics to define viscosity.

6.5.4   Transport Coefficients in General

Experimentally, viscosity is the transport coefficient that has been the most extensively investigated. It is easier to measure than conduction and diffusion coefficients.

Viscosity

Methods for measuring the viscosity coefficient are exposed in paragraph 9.2.3, which discusses the remarkable property of superfluidity found in liquid helium.

Table 6.3: Thermal conductivity K of several gases under normal conditions. This table can be used to verify Eucken's relation (6.53).

gas	$K/$ 10^{-2} SI	$\eta/$ 10^{-6} SI	$M/$ g mol^{-1}	$c_V/$ J K^{-1} mol^{-1}	γ	$\dfrac{KM}{\eta c_V}$	$(9\gamma - 5)/4$
He	14.2	18.6	4.003	12.6	1.66	2.43	2.49
Ne	4.55	29.7	20.18	12.7	1.66	2.43	2.49
A	1.66	21.0	39.95	12.8	1.67	2.47	2.51
Kr	0.88	23.3	83.80	12.4	1.67	2.55	2.51
H$_2$	17.3	8.35	2.016	20.2	1.41	2.07	1.92
N$_2$	2.40	16.6	28.01	20.9	1.40	1.94	1.90
CO$_2$	1.46	13.9	44.01	27.8	1.30	1.66	1.68

Table 6.2 lists the viscosity of several gases at normal conditions and Figure 6.15 shows their temperature dependence. The kinetic theory interpretation of viscosity measurements enabled Loschmidt in 1865 to make the first determination of Avogadro's number and also of molecular diameters. With the presently accepted value of $\mathcal{N}$, molecular diameters obtained in this way agree well with those obtained from investigations of crystal structures (Table 6.2).

Figure 6.15 shows that equation (6.48), predicting a viscosity variation in $T^{1/2}$, is incorrect, because interactions were neglected in its derivation. By taking into account the attractive interaction forces, Sutherland (1893) showed that equation (6.48) remains valid if the cross-section σ has a temperature dependence of the form

$$\sigma = \sigma_\infty \left(1 + \frac{C}{T} \right) .$$

Sutherland's formula then becomes

$$\eta = \frac{AT^{1/2}}{1 + \frac{C}{T}} . \tag{6.51}$$

This predicts that $T^{1/2}/\eta$ varies linearly with $1/T$, which is in better agreement with experiment (Fig. 6.15). More complex theories, based on Boltzmann's equation and involving more realistic interaction potentials, yield more refined comparisons between experiment and theory.

Thermal conduction

The thermal conductivities K of several gases are given in Table 6.3 and plotted as a function of temperature in Figure 6.16. It can be seen that, like the viscosity coefficient, the thermal conductivity approximately obeys a relation of the same form as that of Sutherland (6.51).

It is interesting to compare the coefficients of thermal conduction and viscosity, which, according to kinetic theory (6.43 et 47), are related by

$$\frac{K}{\eta} \frac{M}{c_V} = \frac{5}{2} . \tag{6.52}$$

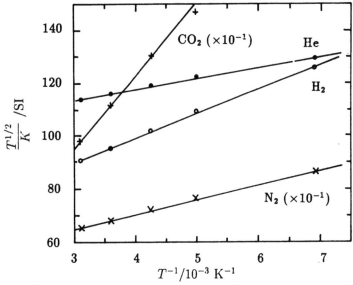

Figure 6.16: Variation of the thermal conductivity K with temperature T. The straight lines represent Sutherland's relation (6.51) with $A = 5(\pi m k)^{1/2}/16\sigma_\infty$.

Molecular quantities do not appear in this equation. Relation 6.52 holds well for monatomic gases. For argon, for example, the experimental ratio (6.52) varies between 2.42 and 2.61 for temperatures between 100 and 1000 K.

Equation (6.52) is not valid for polyatomic gases since internal kinetic energy (rotation, vibration, ...) is not transported in the same way as kinetic energy of translation. To take account of this, Eucken (1913) suggested that relation (6.52) should be written in the form

$$\frac{KM}{\eta} = \frac{5}{2} \times c_V^{tr} + 1 \times c_V^{int} ,$$

where $c_V^{tr} = 3R/2$ is the heat capacity of translation and $c_V^{int} = c_V - 3R/2$ is the internal heat capacity. The factor 1 in front of c_V^{int} is that which would be obtained for the ratio $KM/\eta c_V$ using the approximate formulae (6.42 and 46), for which it is assumed that the energy transported is not correlated with the displacement velocity. On setting the coefficient $\gamma = c_P/c_V = 1+R/c_V$, which gives $R/c_V = \gamma-1$, Eucken's formula becomes

$$\frac{KM}{\eta} = c_V + \frac{9}{4}R = c_V\frac{9\gamma - 5}{4} \quad \text{or} \quad \frac{K}{\eta}\frac{M}{c_V} = \frac{9\gamma - 5}{4} . \tag{6.53}$$

For monatomic gases ($\gamma = 5/3$), expression (6.52) is recovered. Table 6.3 shows that Eucken's formula is in good agreement with experiment.

Diffusion

Measurements of diffusion of labelled molecules in a gas are carried out using different isotopic species or different states of the same species (ortho and para-hydrogen).

Table 6.4: Diffusion coefficient D of some gases at normal conditions. The density is $\rho = M/v_0$, with $v_0 = 22.4$ L mol^{-1}.

gas	$D/$ 10^{-5} m^2 s^{-1}	$\eta/$ 10^{-6} SI	$M/$ g mol^{-1}	$\rho D/\eta$
Ne	4.7	29.7	20.18	1.43
A	1.58	21.0	39.95	1.34
Kr	0.9	23.3	83.80	1.45
H$_2$	12.9	8.35	2.016	1.39
N$_2$	2.0	16.6	28.01	1.51

Some experimental values of the diffusion coefficient D are listed in Table 6.4.

As for thermal conduction, a simple theoretical ratio holds between the diffusion coefficient and the viscosity of a gas. According to (6.37 and 47)

$$\frac{\rho D}{\eta} = \frac{6}{5} \ , \tag{6.54}$$

where $\rho = nm$ is the density of the gas. Comparisons between this relation and experiment can be made using Table 6.4.

BIBLIOGRAPHY

J.H. Jeans, *The Dynamical Theory of Gases*, Dover, New York (1954).
E.H. Kennard, *Kinetic Theory of Gases*, McGraw-Hill, New York (1938).
M. Knudsen, *The Kinetic Theory of Gases*, Wiley, New York (1950).
D.A. McQuarrie, *op. cit.*
R.D. Present, *Kinetic Theory of Gases*, McGraw-Hill, New York (1958).
K.F. Smith, *Molecular Beams*, Wiley, New York (1955).
Historical References
R. Clausius, *Ann. Physik* **105**, 239 (1858).
A. Eucken, *Phys. Z.* **14**, 324 (1913).
A.E. Fick, *Ann. Physik* **94**, 59 (1855).
M. Knudsen, *Ann. Physik* **28**, 75 (1909).
C.C. Ko, *J. Franklin Inst.* **217**, 173 (1934).
J. Loschmidt, *Wien. Ber.* **52**, 395 (1865).
J.C. Maxwell, *Phil. Mag.* **19**, 31 (1860); *Scientific Papers*, Dover, New York (1952).
R.C. Miller and P.Kush, *Phys. Rev.* **99**, 1314 (1955).
I.I. Rabi et al., *Phys. Rev.* **55**, 526 (1939).
O. Stern, *Z. Physik* **39**, 751 (1926).
O. Stern and W.Gerlach, *Z. Physik* **8**, 110 (1921); **9**, 349, 353 (1922); **41**, 563 (1927).
W. Sutherland, *Phil. Mag.* **36**, 507 (1893).
I.F. Zartman, *Phys. Rev.* **37**, 383 (1931).

COMPREHENSION EXERCISES

6.1 Calculate explicitly the integrals I_0 and I_1 of Table 6.1 and verify the results of the table.

6.2 Starting from the Boltzmann relation in the form

$$d^6 N_{\mathbf{r},\mathbf{p}} = A\, e^{-\beta\epsilon}\, d^3\mathbf{r}\, d^3\mathbf{p} \ ,$$

calculate the constant A by noting that the integral $\int d^6 N_{\mathbf{r},\mathbf{p}}$ is equal to N.

6.3 Calculate the most probable velocity v_m, the mean velocity $\bar{v}$ and the root mean square velocity u of nitrogen molecules at 25 °C. Compare with the velocity of sound c in this gas [Ans.: 421; 475; 515; 352 m s^{-1}].

6.4 Calculate the mean velocity of helium atoms in the gas at 4 K and at 300 K [Ans.: 146; 1260 m s^{-1}].

6.5 Inspect Figure 6.1b and confirm that the probability density for a given velocity displays a maximum as the temperature changes. Show that this maximum occurs at $T_m = mv^2/3k$.

6.6 Show that the width at half height of the curve dP_{v_x}/dv_x is equal to $2(\ln 2)^{1/2}v_m = 1.67\, v_m$.

6.7 Radiation of frequency ν_0 emitted by a moving atom undergoes a Doppler shift such that the frequency observed in one direction, Ox, is $\nu = \nu_0(1 + v_x/c)$. Using the distribution of velocities of the molecules in a gas, determine the spectrum of observed radiation and show that it is centred at ν_0 and that its width at half height is $\Delta\nu = \nu_0 \left(8\ln 2 kT/mc^2\right)^{1/2}$.

6.8 Write the distribution relation for molecules as a function of their kinetic energy ϵ. Hence show that the most probable energy is $\epsilon_m = kT/2$.

6.9 What is the explanation for the distribution relation

$$d^3 n_{\mathbf{v}} = n\left(\frac{m}{2\pi kT}\right)^{3/2} \exp\left\{-\frac{m}{2kT}\left[(v_x - a)^2 + (v_y - b)^2 + (v_z - c)^2\right]\right\} d^3\mathbf{v} \ ,$$

where a, b and c are constants? [Ans.: a, b, c are the components of the overall velocity of the gas].

6.10 Find the velocity distribution of a gas in two dimensions. What are the most probable velocity, the mean velocity and the root mean square velocity? [Ans.: $v_m = (kT/m)^{1/2}$, $\bar{v}/v_m = (\pi/2)^{1/2}$, $u/v_m = 2^{1/2}$].

6.11 Calculate the most probable velocity, the mean velocity and the mean kinetic energy in a beam. Compare with the corresponding quantities for the molecules inside the oven that generates the beam [Ans.: $v_{m\ beam}/v_m = (3/2)^{1/2}$, $\bar{v}_{beam}/\bar{v} = 3\pi/8$, $\bar{\epsilon}_{beam}/\bar{\epsilon} = 4/3$].

6.12 Show that in experiments of the type of Zartman and of Ko (§6.3.2), the thickness of the metal deposit has the form $A\exp(-B/l^2)/l^5$.

6.13 In Figure 6.6 verify that the position of the maximum corresponds to that of the Maxwell distribution.

6.14 Calculate the mass and energy flow per unit area in a molecular discharge [Ans.: $j_m \simeq nm\bar{v}/4$; $j_U = \pi nm\bar{v}^3/16$].

6.15 Show that the mean relative velocity of two molecules having the same speed (magnitude of velocity) v but arbitrary orientations is $v_r = 4v/3$.

6.16 Evaluate the fraction of free paths of a molecule that exceed 2λ and 4λ [Ans.: 0.135 and 0.018].

6.17 Compare the orders of magnitude of the molecular diameter d, the distance between two nearest neighbours and the mean free path in nitrogen at 273 K and at a pressure of 10^{-4} torr ($d = 3.75$ Å, $M = 28.0$ g mol^{-1}) [Ans.: $a \sim 7000$ Å; $\lambda = 45$ cm].

6.18 In 1865 Loschmidt made the first estimate of the order of magnitude of the diameter d of a molecule and of Avogadro's number $\mathcal{N}$, using kinetic theory and measurements of macroscopic properties of substances (viscosity of the gas phase η and density of the solid phase ρ). Estimate d and $\mathcal{N}$ using the data for carbon dioxide CO_2, η (0°C) $= 13.6 \times 10^{-6}$ SI, $\rho = 1.53$ g cm^{-3} and $M = 44.0$ g mol^{-1} [Ans.: The numerical values depend on which formula is chosen. Typically, $d \simeq 3$ Å; $\mathcal{N} = 10^{24}$ mol^{-1}].

PROBLEM 6.1 CHEMICAL REACTIONS IN THE GAS PHASE

1 Collision Frequency in a Gas

An ideal monatomic gas is placed in a container of volume V at temperature T. The number of atoms per unit volume is $n = N/V$ and their mass is m.

1.1 Find the expression in Maxwell-Boltzmann statistics for the number $d^3 n_{\mathbf{p}}$ of atoms per unit volume having momentum $\mathbf{p}$ to within $d^3\mathbf{p}$. Hence obtain the equation for the number of atoms $d^3 n_{\mathbf{v}}$ per unit volume having velocity $\mathbf{v}$ to within $d^3\mathbf{v}$ (Maxwell distribution).

1.2 In this question we investigate collisions of the atoms with an element of area dS located inside the gas.

1.2a What is the number $d^6 N'_{\mathbf{v}}$ of atoms with velocity $\mathbf{v}$ (to within $d^3\mathbf{v}$) that strike the given element during an interval dt, in terms of the variables v_x, v_y, v_z? Denote by v_z the projection of $\mathbf{v}$ upon the normal to the wall.

1.2b Hence derive an expression for the number of collisions with the surface during dt, for which the component of velocity perpendicular to the wall lies between v_z and $v_z + dv_z$.

1.2c Calculate the total number of collisions per unit of time and per unit surface area, j.

1.2d Show that j can be expressed simply as a function of the number of atoms per unit volume n and of the average velocity $\bar{v} = (8kT/\pi m)^{1/2}$.

1.3 To estimate the frequency of collisions between one atom and the others in the gas, two limiting cases are investigated. In these two cases, the hard sphere model is used. d denotes the diameter of the spheres and σ the product πd^2.

1.3a Consider one atom held at rest (approximation $v \ll \bar{v}$), while the other atoms obey Maxwell's distribution. Calculate the number of collisions experienced by the atom in unit time (collision frequency) $\nu(0)$.

1.3b Consider an atom moving at velocity $v \gg \bar{v}$. The other atoms can thus be considered to be fixed. Calculate the collision frequency $\nu(v)$ in this model.

1.3c The exact collision frequency in a hard sphere model is $\nu = \sqrt{2}n\sigma\bar{v}$. Compare this value with those found in the two limiting cases.

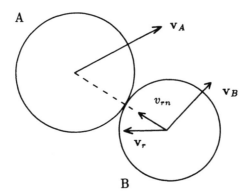

Figure 6.17: Collision between two hard spheres

2 Kinetic Theory of Chemical Reactions in the Gas Phase

Consider an ideal mixture of 2 ideal gases A and B composed of molecules of masses m_A and m_B, respectively, and held in a container of volume V at temperature T. Assume that the molecules are spherical and that their diameters are d_A and d_B respectively.

In a simple model, a chemical reaction is assumed to occur at the moment of impact between two colliding molecules A and B if the relative velocity $\mathbf{v}_r = \mathbf{v}_B - \mathbf{v}_A$ of the two molecules has a component v_{rn} along the line of centres (Fig. 6.17) that satisfies the condition (effective collision), $\frac{1}{2}\mu v_{rn}^2 \geq \epsilon_0$. $\mu = m_1 m_2/(m_1 + m_2)$ is the reduced mass of the molecules and ϵ_0 is an energy called the activation energy of the reaction.

We recall the formulae of classical mechanics

$$\mathbf{v}_G = \frac{m_A \mathbf{v}_A + m_B \mathbf{v}_B}{m_A + m_B} \quad \text{and} \quad \frac{1}{2}m_A v_A^2 + \frac{1}{2}m_B v_B^2 = \frac{1}{2}(m_A + m_B)v_G^2 + \frac{1}{2}\mu v_r^2$$

where $\mathbf{v}_G$ is the velocity of the centre of mass of the two molecules.

2.1 In the absence of all correlation, the number of pairs of A molecules having velocity $\mathbf{v}_A$ to within $d^3\mathbf{v}_A$ and of B molecules having velocity $\mathbf{v}_B$ to within $d^3\mathbf{v}_B$ is, for the whole volume, $d^3 N_{\mathbf{v}_A} \times d^3 N_{\mathbf{v}_B}$. Express this quantity as a function of $\mathbf{v}_A$ and $\mathbf{v}_B$. Assume that at all times the distribution of velocities for each gas is Maxwellian, and denote by N_A and N_B the instantaneous number of molecules A and B.

2.2 From the above result, derive an expression for the number $d^3 N_{\mathbf{v}_r}$ of pairs $A - B$ having relative velocity $\mathbf{v}_r$ to within $d^3\mathbf{v}_r$. Assume that $d^3\mathbf{v}_A \cdot d^3\mathbf{v}_B = d^3\mathbf{v}_G \cdot d^3\mathbf{v}_r$.

2.3a We now place ourselves in the reference frame of an A molecule. Use the results of 1.2b to find how many B molecules strike it during the interval dt with a relative normal velocity between v_{rn} and $v_{rn} + dv_{rn}$. Set $\sigma_{AB} = \pi(d_A + d_B)^2/4$.

2.3b Hence calculate the frequency of effective collisions ν_{eff} experienced by this molecule, then the total number of effective collisions per second ν_{AB} among all the molecules of gas A and gas B.

2.3c What happens to this result if the A and B molecules are identical?

2.4a If each effective collision gives rise to a chemical transformation of the A and B molecules and if all other chemical reactions are neglected, show that

$$\frac{dn_A}{dt} = \frac{dn_B}{dt} = -Kn_An_B \ ,$$

where n_A and n_B are the number of molecules of A and B per unit volume. How does K depend on the temperature, the molecular parameters and the activation energy?

2.4b We assume that at $t = 0$ the concentrations n_A and n_B are equal to n_0, and hence at all times $n_A = n_B$. Integrate the above equation and calculate the time τ required for n_A to reach the value $n_0/2$.

2.5 Measurements of the chemical kinetics of the decomposition reaction for hydroiodic acid $HI + HI \rightarrow H_2 + I_2$ show that the product $n_0\tau$ is independent of n_0 and, between 550 and 789 K, is given by $n_0\tau$ /s m$^{-3} = 3 \times 10^{17}T^{-1/2} \, e^{22\,000/T}$.

2.5a Are these results in qualitative agreement with the previous results?

2.5b Determine the molar activation energy (in J mol^{-1}) and compare this with the molar heat of the (endothermic) reaction 9.1 kJ mol^{-1}.

2.5c The diameter of the HI molecule obtained from viscosity measurements at 520 K is 4.65 Å. What is the probability that an effective collision, as defined above, does indeed produce a chemical reaction?

Algebraic data: $\int_0^\infty e^{-ax^2} \, dx = 1/2 \, \sqrt{\pi/a}$ and $\int_{x_0}^\infty xe^{-ax^2} \, dx = e^{-ax_0^2}/2a$.
Numerical data: Gas constant $R = 8.314$ J K^{-1} mol^{-1}, molar mass of hydroiodic acid $M = 127.9$ g mol^{-1}.

SOLUTION 1 Collision Frequencies in a Gas

1.1 See paragraphs 6.2.1,2.

1.2a The expression for $d^6N'_v$ is given in (6.21)

$$d^6N'_v = \frac{n}{\pi^{3/2}} \frac{1}{v_m^3} \exp\left(-\frac{v^2}{v_m^2}\right) \ v_z \, d^3\mathbf{v} \, dt \, dS \quad \text{with} \quad v_m = (2kT/m)^{1/2} \ .$$

1.2b Integration over v_x and v_y yields

$$d^4N'_{v_z} = \frac{n}{\pi^{1/2}} \frac{1}{v_m} \exp\left(-\frac{v_z^2}{v_m^2}\right) \ v_z \, dv_z \, dt \, dS \ . \tag{6.55}$$

1.2c The total number of collisions is found by integrating over positive v_z, thus

$$d^3N' = \frac{n}{2\pi^{1/2}} v_m \, dt \, dS \ .$$

Hence

$$j = \frac{d^3N'}{dt \, dS} = \frac{n}{2\pi^{1/2}} v_m = \frac{n}{2}\left(\frac{2kT}{\pi m}\right)^{1/2} \ .$$

1.2d Insertion of $\bar{v}$ yields equation (6.22)

$$j = \frac{1}{4}n\bar{v} \ .$$

1.3 Generally, the atom under consideration is assumed to be a sphere of radius d (sphere of exclusion), the other atoms being reduced to points.

1.3a In this case collisions occur on a stationary sphere of radius d and total area $4\pi d^2 = 4\sigma$. Hence

$$\nu(0) = j \times 4\sigma = n\sigma\bar{v} .$$

1.3b In the interval dt the atom sweeps out a cylinder of volume $\sigma v\, dt$ containing $n\sigma v\, dt$ atoms. The collision frequency is therefore

$$\nu(v) = n\sigma v .$$

1.3c The correct value is bounded by the values in these two limiting cases. It is obtained as an average of expression (6.32), whose limits $v \ll \bar{v}$ and $v \gg \bar{v}$ were found in 1.3a and 1.3b.

2 Kinetic Theory of Chemical Reactions in the Gas Phase

2.1 The number of pairs of A molecules with velocity $\mathbf{v}_A$ to within $d^3\mathbf{v}_A$ and of B molecules with velocity $\mathbf{v}_B$ to within $d^3\mathbf{v}_B$ is

$$d^3N_A\, d^3N_B \doteq N_A\, N_B\, \frac{(m_A m_B)^{3/2}}{(2\pi kT)^3} \exp\left(-\frac{m_A v_A^2 + m_B v_B^2}{2kT}\right) d^3\mathbf{v}_A\, d^3\mathbf{v}_B .$$

2.2 The change of variables $\mathbf{v}_A,\ \mathbf{v}_B\ \rightarrow\ \mathbf{v}_r,\ \mathbf{v}_G$ gives

$$d^3N_A\, d^3N_B = N_A\, N_B\, \frac{(m_A m_B)^{3/2}}{(2\pi kT)^3} \exp\left[-\frac{(m_A + m_B)v_G^2 + \mu v_r^2}{2kT}\right] d^3\mathbf{v}_r\, d^3\mathbf{v}_G .$$

Integration over $\mathbf{v}_G$ yields

$$d^3N_{\mathbf{v}_r} = N_A N_B \left(\frac{\mu}{2\pi kT}\right)^{3/2} \exp\left(-\frac{\mu v_r^2}{2kT}\right) d^3\mathbf{v}_r ,$$

which is comparable to Maxwell's distribution for a particle of mass μ.

2.3a As in part 1, a sphere of exclusion of radius $(d_A + d_B)/2$ is assigned to the A molecule and the B molecules are considered to be point-like. The number of collisions with B molecules experienced by the A molecule, for a relative velocity v_{rn} in the normal direction, is found from relation (6.55) by replacing dS with the area $4\sigma_{AB}$ of the sphere of exclusion. Thus

$$d^2N'_{v_{rn}} = \frac{4n_B\sigma_{AB}}{\pi^{1/2}}\, \frac{1}{v_m} \exp\left(-\frac{v_{rn}^2}{v_m^2}\right) v_{rn}\, dv_{rn}\, dt \quad \text{with} \quad v_m = (2kT/\mu)^{1/2} .$$

2.3b The total number of effective collisions is found by integrating the above number for $v_{rn} \geq (2\epsilon_0/\mu)^{1/2}$,

$$dN' = \frac{4n_B\sigma_{AB}}{\pi^{1/2}}\, \frac{1}{v_m} \exp\left(-\frac{\epsilon_0}{kT}\right) \frac{v_m^2}{2}\, dt .$$

The frequency of effective collisions for an A molecule is then

$$\nu_{\text{eff}} = \frac{dN'}{dt} = 2n_B\sigma_{AB}\left(\frac{2kT}{\pi\mu}\right)^{1/2}\exp\left(-\frac{\epsilon_0}{kT}\right) \ .$$

The total number of effective $A - B$ collisions per second occurring in the gas is then

$$\nu_{AB} = N_A\nu_{\text{eff}} = 2N_An_B\sigma_{AB}\left(\frac{2kT}{\pi\mu}\right)^{1/2}\exp\left(-\frac{\epsilon_0}{kT}\right) \ .$$

Note that this formula is, as it should be, symmetrical in A and B , since $N_An_B = n_AN_B = N_AN_B/V$. Furthermore, setting $\epsilon_0 = 0$ and A = B in the expression for ν_{eff} yields the exact formula (6.26b).

2.3c If the A and B molecules are identical the last result must be divided by 2, since the total number of pairs of molecules is then $N^2/2$ instead of $N_A\ N_B$.

2.4a If each effective collision produces a reaction, then

$$\frac{dN_A}{dt} = \frac{dN_B}{dt} = -\nu_{AB} \ .$$

Hence, dividing by V and substituting for ν_{AB}, we find

$$\frac{dn_A}{dt} = \frac{dn_B}{dt} = -Kn_An_B \quad \text{with} \quad K = 2\sigma_{AB}\left(\frac{2kT}{\pi\mu}\right)^{1/2}\exp\left(-\frac{\epsilon_0}{kT}\right) \ . \quad (6.56)$$

2.4b If we set $n_A = n_B = n$, the following differential equation is to be solved

$$\frac{dn}{dt} = -Kn^2 \quad \text{or} \quad -\frac{dn}{n^2} = K\,dt \ .$$

On integrating between 0 and t, this gives

$$\frac{1}{n} - \frac{1}{n_0} = Kt \ .$$

The time τ required to reach $n = n_0/2$ is then $\tau = 1/Kn_0$.

2.5a In the above model, $n_0\tau = 1/K$ is independent of n_0 and has the form $AT^{-1/2}\exp(\epsilon_0/kT)$. This is similar to the experimental relation, with

$$A = \frac{1}{\sigma}\left(\frac{\pi m}{4k}\right)^{1/2} = \frac{1}{\sigma}\left(\frac{\pi M}{4R}\right)^{1/2} \ ,$$

since, as the molecules are identical, the factor 2 vanishes from the equation for K, and we have $\mu = m/2$.

2.5b The molar activation energy is found by converting the factor 22 000 K into J mol^{-1}, namely 180 kJ mol^{-1}. This energy is greater than the heat of reaction to be supplied, since a potential barrier must be overcome for the reaction to take place.

2.5c Taking $d = (d_A+d_B)/2 = 4.65$ Å yields a value of 1.62×10^{17} SI for A, which is about one half of the experimental value. This can be explained by that fact that the probability of an effective collision yielding a chemical reaction is roughly 0.5.

Chapter 7

Bose-Einstein Statistics

7.1 INTRODUCTION

Bose-Einstein statistics introduced in Chapter 2 applies to systems of indistinguishable non-interacting particles of integer spin. When the population density of the levels is sufficiently low, however, the limit of corrected Maxwell-Boltzmann statistics provides a good approximation and is often preferred on account of its greater simplicity. This is the case for the majority of molecular gases (Chap. 5).

Boson systems fall into two categories, those composed of a definite number of molecular particles (helium for example), and systems containing an indeterminate number of particles. Photons, which provide a model for describing electromagnetic radiation, belong to the latter category. For such systems the constraint (2.1) does not exist, i.e. the multiplier $\bar{\mu} = \mu$ disappears from the formalism of chapter 2, thus simplifying the use of Bose-Einstein statistics. For this reason, we first investigate electromagnetic radiation in equilibrium with matter, so-called thermal radiation. We then consider the case of molecular boson gases, as well as the Einstein condensation phenomenon that these may exhibit. This condensation is a phase transition that occurs not in real space but in momentum space.

7.2 BOSE MODEL FOR THERMAL RADIATION

7.2.1 Introduction

In the Bose model, electromagnetic radiation inside a vessel can be quantized by assuming that it consists of a gas of massless particles, namely photons. The linearity of Maxwell's equations implies that these particles have no mutual interaction, and, owing to their translational motion, they are indistinguishable. Moreover, as they are associated with a vector field, they are bosons of spin 1 and Bose-Einstein statistics apply to them. Lastly, when the photons collide with the molecules of the enclosure walls, some are absorbed while others are emitted. Their number is thus not constant, and the constraint (2.1), $\sum N_i = \text{const.}$, no longer applies. The Lagrange multiplier $\bar{\mu}$ therefore vanishes from the formalism.

In particular, the equilibrium distribution relation (2.26) for photons becomes

$$N_i = \frac{g_i}{e^{\beta \epsilon_i} - 1} \qquad (\beta = 1/kT) .$$ (7.1)

The grand potential $\Omega = U - TS - N\mu$, given by (2.34), can be identified with the free energy $F = U - TS$ and is equal to

$$\Omega \equiv F = kT \sum_i g_i \ln \left(1 - e^{-\beta \epsilon_i}\right) .$$ (7.2)

7.2.2   Energy Levels

For photons in a rectangular parallelepiped enclosure, the energy levels and their degeneracy are determined using Maxwell's equations. In a manner analogous to paragraph 1.2, it is found that

$$\epsilon_{m_x, m_y, m_z} = \frac{hc}{2} \left(\frac{m_x^2}{a^2} + \frac{m_y^2}{b^2} + \frac{m_z^2}{c^2} \right)^{1/2} , \qquad m_x, m_y, m_z = 1, 2, 3, \ldots$$

For an enclosure of macroscopic size, for example $V = 1$ litre, the spacing between two energy levels is approximately

$$\epsilon_0 \simeq \frac{hc}{2V^{1/3}} = 9.93 \times 10^{-25} \text{ J} = 6.20 \times 10^{-6} \text{ eV} .$$

The levels are very closely spaced and ϵ can be taken to be a continuous variable, related to the momentum $\mathbf{p}$ of the photon through the *dispersion relation*

$$\epsilon = c|\mathbf{p}| = cp .$$ (7.3)

In order to count the states and find the degeneracies, we recall, as in paragraph 1.2, that each quantum state occupies a volume h^3 in phase space. The degeneracy of the translational levels then becomes

$$g_i \rightarrow \frac{d^3\mathbf{r}\, d^3\mathbf{p}}{h^3} .$$ (7.4)

In addition to this degeneracy, the spin degeneracy of the photon, $g_s = 2$, must also be included. Although, as a general rule, a particle of spin 1 possesses $2J + 1 = 3$ spin states, the situation is different for the photon, which has only two states, corresponding to the two circular polarization states of light. The third spin state corresponding to longitudinal polarization is excluded, owing to the transverse character of electromagnetic waves.

These properties allow us to calculate the free energy of a photon gas, with the correspondences

$$g_i \rightarrow 2 \frac{d^3\mathbf{r}\, d^3\mathbf{p}}{h^3} \quad \text{and} \quad \epsilon_i \rightarrow \epsilon = cp .$$ (7.5)

7.2.3   Thermodynamic Functions

Equation (7.2) for the free energy F becomes

$$F = \frac{2kT}{h^3} \int d^3\mathbf{r} \int d^3\mathbf{p} \ln\left(1 - e^{-\beta cp}\right) .$$

The integral in $\mathbf{r}$ is taken over the volume of the container and is therefore equal to V. Transforming to spherical co-ordinates for $\mathbf{p}$ and integrating over the angular variables, we find

$$\begin{aligned}
F &= \frac{2kTV}{h^3} \int_0^\infty 4\pi p^2 \ln\left(1 - e^{-\beta cp}\right) dp \\
&= \frac{8\pi(kT)^4 V}{h^3 c^3} \int_0^\infty x^2 \ln\left(1 - e^{-x}\right) dx ,
\end{aligned} \tag{7.6}$$

where we have set $x = \beta cp$. The numerical integral appearing in (7.6) is a member of a family of integrals. We have

$$-\int_0^\infty x^{n-1} \ln\left(1 - e^{-x}\right) dx = \frac{1}{n} \int_0^\infty \frac{x^n \, dx}{e^x - 1} = \Gamma(n)\, \zeta(n+1) . \tag{7.7a}$$

The gamma function $\Gamma(n)$ is equal to $(n-1)!$ when n is a positive integer, and has the following properties

$$\Gamma(n) = (n-1)\Gamma(n-1) , \quad \Gamma\left(\frac{1}{2}\right) = \sqrt{\pi} , \quad \Gamma(1) = 1 . \tag{7.7b}$$

The Riemann zeta function ζ is defined by

$$\zeta(n) = \sum_{p=1}^\infty p^{-n} , \quad n > 1 \tag{7.7c}$$

with $\zeta(3/2) = 2.612$, $\zeta(2) = \pi^2/6$, $\zeta(5/2) = 1.341$, $\zeta(3) = 1.202$, $\zeta(7/2) = 1.127$ and $\zeta(4) = \pi^4/90$.

The free energy is then

$$F = -\frac{a}{3}VT^4 \quad \text{with} \quad a = \frac{48\pi k^4}{h^3 c^3} \zeta(4) = \frac{8\pi^5 k^4}{15 h^3 c^3} = 7.57 \times 10^{-16} \text{ SI} . \tag{7.8}$$

The entropy and the pressure of the gas are obtained by differentiating F,

$$S = -\left(\frac{\partial F}{\partial T}\right)_V = \frac{4}{3} aVT^3 \quad \text{and} \quad P = -\left(\frac{\partial F}{\partial V}\right)_T = \frac{1}{3} aT^4 . \tag{7.9}$$

Finally, we find for the internal energy U and the heat capacity at constant volume C_V

$$U = F + TS = aVT^4 \tag{7.10}$$

$$\text{and} \qquad C_V = \left(\frac{\partial U}{\partial T}\right)_V = 4aVT^3 . \qquad (7.11)$$

Some remarks are in order about the above results.

- The expression for the free energy is an extensive quantity. This necessary result comes from the term $d^3\mathbf{r}$ in equation (7.5) for the degeneracy, which, on integration, gives rise to the volume factor V.

- The pressure exerted by the photon gas on the walls of the surrounding container (radiation pressure) does not depend on the volume V, but only on the temperature T. It follows that P and T cannot be considered as independent equilibrium variables.

- $PV = U/3$, while for ideal monatomic gases (5.22 and 23), $PV = 2U/3$. The result

$$PV = U/3 \qquad (7.12)$$

is general for systems of particles whose dispersion relation is $\epsilon = cp$.

- Radiation pressure is extremely weak at room temperature. It is equal to 2×10^{-11} atmosphere. Since, however, it varies as T^4, this pressure becomes very large at high temperatures, for example at the centre of stars $(T > 10^7 \text{ K})$, where it can exceed the kinetic pressure of the stellar plasma.

7.2.4   Number of Photons

Although the number of photons in the container is not fixed, we can calculate this number in the equilibrium state of the gas by eliminating the parameter μ in (2.26), i.e.,

$$N = \sum_i N_i = \sum_i \frac{g_i}{e^{\beta \epsilon_i} - 1} .$$

Transforming into continuous variables as in paragraph 7.2.3, we get

$$N = \int 2 \times \frac{d^3\mathbf{r}\, d^3\mathbf{p}}{h^3} \times \frac{1}{e^{\beta cp} - 1} = \frac{8\pi V}{h^3} \frac{(kT)^3}{c^3} \int_0^\infty \frac{x^2\, dx}{e^x - 1} .$$

According to (7.7a), the integral, which is purely numerical, is equal to $2\zeta(3) = 2.404$. We have therefore

$$N = 2.404 \times \frac{8\pi k^3}{h^3 c^3} VT^3 = 0.370 \, \frac{a}{k} VT^3 . \qquad (7.13)$$

This equilibrium photon number is an extensive function of the volume and of the temperature. Note that the entropy of the photon gas (7.9) can be expressed as a function of N alone in the form $S = 3.602 \, Nk$. These results show, for an isothermal expansion of the photon gas $(T = \text{const.})$, that the number of photons per unit volume N/V remains constant, but in an adiabatic expansion $(S = \text{const.})$, it is the total number of photons N that stays constant.

Exercise 7.1 *Thermal radiation of the universe*
1 The universe is a source of thermal radiation whose present temperature is close to $T_2 = 2.7$ K. Calculate the number of photons and the energy per unit volume, denoted n_2 and u_2. Hence deduce the mean energy of a photon. What wavelength of radiation does this correspond to?
2 In the Big Bang cosmological model, the universe is undergoing isentropic expansion and its volume varies as t^2, where t is the time since the beginning.
2a Express T, n and u as functions of t.
2b Calculate the age of the universe at which thermal radiation became "decoupled" from matter, i.e. when its temperature became too small to ionize atoms significantly ($T_1 \sim 3000$ K). The present age of the universe is estimated to be 15 thousand million years.
3 The average density of known matter in the universe is at present of the order of 10^{-30} g cm^{-3}. Calculate this density at the time of matter-radiation decoupling. Compare the energy density of matter with the energy density of radiation at that time.
Take $h = 6.63 \times 10^{-34}$ J s and $c = 3.00 \times 10^8$ m s^{-1}.

Solution
1 Equations (7.13) and (7.10) yield

$$n_2 = \left(\frac{N}{V}\right)_{T=T_2} = 3.99 \times 10^8 \text{ m}^{-3} = 399 \text{ cm}^{-3}$$

$$\text{and} \quad u_2 = \left(\frac{U}{V}\right)_{T=T_2} = 4.02 \times 10^{-14} \text{ J m}^{-3} = 2.51 \times 10^{-1} \text{ eV cm}^{-3} .$$

The average energy of a photon is thus $\bar{\epsilon} = u_2/n_2 = 6.3 \times 10^{-4}$ eV. The corresponding wavelength is $\lambda = hc/\bar{\epsilon} = 0.20$ cm.
2a Since the entropy remains constant during the expansion, the product VT^3 is constant and T is thus proportional to $V^{-1/3}$, i.e. to $t^{-2/3}$. Hence

$$T = T_2(t/t_2)^{-2/3} , \quad n = n_2(t/t_2)^{-2} , \quad u = u_2(t/t_2)^{-8/3} .$$

At the beginning of the universe, these numbers we´re enormous and decreased rapidly, which is why the expression Big Bang is used to describe the first instants.
2b $t_1 = t_2 \times (T_1/T_2)^{-3/2} = 400\ 000$ years.
3 The density of matter ρ varies as V^{-1} ($\rho V = M$ =const.), i.e. as t^{-2}. Therefore $\rho = \rho_2(t/t_2)^{-2}$, and hence $\rho_1 = 1.4 \times 10^{-21}$ g cm^{-3}. The energy density of the corresponding mass is equal to $\rho_1 c^2 = 0.13$ J m$^{-3} = 7.9 \times 10^{11}$ eV cm^{-3} while the energy density of the radiation at the same moment is $u_1 = 0.063$ J m$^{-3} = 4.0 \times 10^{11}$ eV cm^{-3}. At the time of decoupling, these two values were comparable. Before that time ($t \ll 10^5$ years) the energy density of matter, which varies as t^{-2}, was negligible in comparison with the radiation energy density ($u \propto t^{-8/3}$), while the reverse situation prevails at present.

7.3   PLANCK'S LAW. BLACK BODY RADIATION

7.3.1   Derivation of Planck's law

In the previous paragraph, we considered thermal radiation in its totality without enquiring into the contribution of the various energy levels. We now go into the details of this contribution by calculating the internal energy per frequency interval. This quantity has the advantage of being experimentally verifiable, since radiation can be resolved in frequency by spectroscopic methods.

Returning to equations (7.1 and 5), we see that the number of photons $d^6 N_{\mathbf{r},\,\mathbf{p}}$ present at $\mathbf{r}$ in the volume $d^3\mathbf{r}$ having momentum $\mathbf{p}$ in the element $d^3\mathbf{p}$ is equal to

$$d^6 N_{\mathbf{r},\,\mathbf{p}} = \frac{2}{h^3} \frac{d^3\mathbf{r}\, d^3\mathbf{p}}{e^{\beta\epsilon} - 1} \quad \text{where} \quad \epsilon = cp \ . \tag{7.14}$$

The number dN_p of photons in the whole volume having momentum between p and $p + dp$ is found by integrating over the volume $(d^3\mathbf{r} \to V)$ and over the angles θ and ϕ that define the direction of $\mathbf{p}$ $(d^3\mathbf{p} \equiv p^2\, dp\, \sin\theta\, d\theta\, d\phi \to 4\pi p^2\, dp)$. This number is

$$dN_p = \frac{8\pi V}{h^3} \frac{p^2\, dp}{e^{\beta cp} - 1} \ .$$

Introducing the frequency variable ν through the Planck relation $\epsilon = h\nu = cp$, we find the number of photons with frequency between ν and $\nu + d\nu$

$$dN_\nu = \frac{8\pi V}{c^3} \frac{\nu^2\, d\nu}{e^{\beta h\nu} - 1} \ . \tag{7.15}$$

The elementary energy corresponding to this frequency interval is thus

$$dU_\nu = h\nu\, dN_\nu = \frac{8\pi V h}{c^3} \frac{\nu^3\, d\nu}{e^{\beta h\nu} - 1} \ ;$$

and the spectral density of the energy per unit volume, defined by

$$u_\nu = \frac{1}{V} \frac{dU_\nu}{d\nu} \ ,$$

is equal to

$$u_\nu = \frac{8\pi h}{c^3} \frac{\nu^3}{e^{h\nu/kT} - 1} \ . \tag{7.16}$$

This equation is *Planck's law*, by means of which quantization of energy and the constant h were introduced for the first time (M. Planck, 1900).

The equation for the internal energy density U/V can be recovered by integrating the spectral density of the energy per unit volume u_ν over the whole range of frequencies. With the change of variable $x = \beta h\nu = h\nu/kT$, we have

$$\frac{U}{V} = \int u_\nu\, d\nu = \frac{8\pi h}{c^3} \int_0^\infty \frac{\nu^3\, d\nu}{e^{\beta h\nu} - 1} = \frac{8\pi}{h^3 c^3} k^4 T^4 \int_o^\infty \frac{x^3\, dx}{e^x - 1} \ .$$

The above numerical integral is equal to $6\,\zeta(4) = \pi^4/15$, and hence

$$\frac{U}{V} = \frac{8\pi^5 k^4}{15 h^3 c^3}\, T^4 = a T^4$$

in agreement with the result already found in (7.10).

7.3.2 Discussion of Planck's Law

Planck's Law unified two previous laws. First, the Rayleigh-Jeans relation

$$u_\nu = \frac{8\pi k T}{c^3}\, \nu^2 \tag{7.17}$$

is valid at low frequencies and is the limit of Planck's law in the approximation $h\nu \ll kT$. Planck's constant vanishes from this equation, which is the classical limit of Planck's law as h tends to zero.

Second, Wien showed that the spectral density of the energy obeys a scaling relation

$$u_\nu = T^3\, f\left(\frac{\nu}{T}\right) = \nu^3\, g\left(\frac{\nu}{T}\right)$$

which is satisfied by Planck's law (7.16), the empirical expression for which he had found at high frequencies

$$u_\nu = A\, \nu^3\, e^{-B\nu/T}\ . \tag{7.18}$$

This expression is the limiting case of Planck's law at $h\nu \gg kT$, with $A = 8\pi h/c^3$ and $B = h/k$. This result comes from neglecting the term 1 in the denominator of (7.16), and corresponds to the approximation of corrected Maxwell-Boltzmann statistics.

Planck's law can be expressed as

$$y = \frac{x^3}{e^x - 1}, \tag{7.19a}$$

where the following dimensionless (reduced) variables are used

$$x = \frac{h\nu}{kT} \quad \text{and} \quad y = \frac{h^2 c^3}{8\pi k^3}\, \frac{u_\nu}{T^3}. \tag{7.19b}$$

Numerically, this gives

$$x = 4.80 \times 10^{-11}\, \frac{\nu\ /\mathrm{s}^{-1}}{T\ /\mathrm{K}} \quad \text{and} \quad y = 1.79 \times 10^{26}\, \frac{u_\nu\ /(\mathrm{J\ m}^{-3}\ \mathrm{s})}{T^3\ /\mathrm{K}^3}\ .$$

Planck's law can thus be described by a universal curve that is independent of any parameter (Fig 7.1). This curve has a maximum $y_m = 1.42$ at $x_m = 2.82$.

If the function u_ν is known for a temperature T_1, then the property of scale invariance of Planck's law predicts its form at any other temperature T_2 through

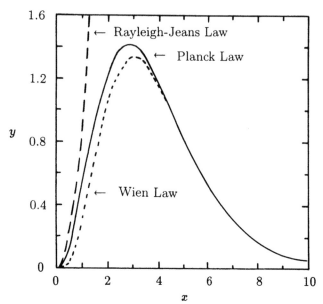

Figure 7.1: Planck's law in reduced variables (7.19). The curve displays a maximum at $x_m = 2.82$ and $y_m = 1.42$. The limiting cases of the Rayleigh-Jeans law $y = x^2$ and Wien's law $y = x^3 e^{-x}$ are shown as dashed lines.

the scaling transformations $\nu \rightarrow \nu\, T_2/T_1$ and $u_\nu \rightarrow u_\nu (T_2/T_1)^3$. In particular, the x and y co-ordinates of the maximum of u_ν vary respectively as T and T^3

$$\nu_m /\mathrm{s}^{-1} = 5.88 \times 10^{10}\, T\,/\mathrm{K} \quad \text{and} \quad u_{\nu_m} /(\mathrm{J\ m}^{-3}\ \mathrm{s}) = 7.94 \times 10^{-27}\, [T\,/\mathrm{K}]^3 \quad (7.20)$$

Note that the area under the curve u_ν varies as $T \times T^3 = T^4$, in accordance with equation (7.10).

7.3.3 Black Body Radiation

Consider a vessel at temperature T containing thermal radiation, in the wall of which there is a small hole by which radiation escapes (Fig 7.2). This orifice constitutes a "black body". As in the kinetic theory of gases (§6.3.1), we determine the properties of the emitted radiation by defining the radiative emittance $\mathcal{E}$ (or radiance) as the amount of energy radiated per unit time and per unit area (this is an energy current density). The spectral emittance e_ν is the emittance per unit of frequency.

First, we consider photons with frequencies between ν and $\nu + d\nu$ leaving the container through the hole during the interval dt. The area of the orifice is dS, and the photons enter a solid angle $d\Omega$ in a direction that makes an angle θ to the normal of the wall (Fig 7.2). These photons can be enclosed in an oblique cylinder of base dS and generator $c\, dt$ in the corresponding direction. As the photon distribution in the container is uniform and isotropic, the energy increment removed by these

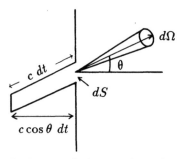

Figure 7.2: Radiation of photons through an orifice

photons is equal to

$$u_\nu \, d\nu \times dV \times \frac{d\Omega}{4\pi}$$

where $dV = dS \times c \cos\theta \, dt$ is the volume of the cylinder and $d\Omega/4\pi$ the fraction of photons that are directed into the solid angle $d\Omega$. The elementary energy radiated by the hole in the time interval dt and the frequency interval $d\nu$ is found by integrating the above expression over all directions that allow the photons to escape. This yields the integral

$$\int \cos\theta \, d\Omega = \int \cos\theta \sin\theta \, d\theta \, d\phi = \pi$$

where the integration over ϕ is from 0 to 2π and θ varies from 0 to $\pi/2$ (the values of θ between $\pi/2$ and π are directions for which the photons move away from the hole). Hence the energy emitted by the orifice of area dS, in the time interval dt and frequency interval $d\nu$, is

$$e_\nu \, dS \, dt \, d\nu = u_\nu \, d\nu \times \frac{dS \, c \, dt}{4} \quad .$$

On substituting (7.16) for u_ν, the spectral emittance e_ν becomes

$$e_\nu = \frac{c}{4} u_\nu = \frac{2\pi h}{c^2} \frac{\nu^3}{e^{h\nu/kT} - 1} \quad . \tag{7.21}$$

The total emittance is found by integrating over the complete frequency range,

$$\mathcal{E} = \int e_\nu \, d\nu = \frac{c}{4} \int u_\nu \, d\nu = \frac{c}{4} \frac{U}{V} \quad .$$

Using expression (7.10) for the internal energy of a photon gas, we find

$$\mathcal{E} = \frac{ac}{4} T^4 = \sigma T^4 \quad \text{where} \quad \sigma = \frac{2\pi^5 k^4}{15 h^3 c^2} = 5.67 \times 10^{-8} \text{ SI} \quad . \tag{7.22}$$

This result (the Stefan-Boltzmann law) shows that the emittance of a black body depends on its temperature only. The universal constant $\sigma = ac/4$ is called the

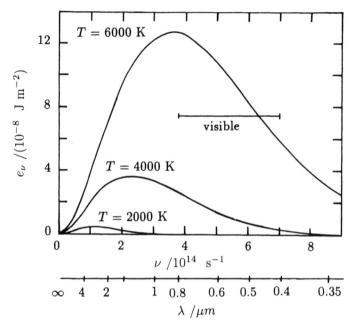

Figure 7.3: Black body spectral emittance e_ν at three temperatures. The total emittance, determined by the area under the curve and given by the Stefan-Boltzmann relation (7.22) is equal respectively to 9.07×10^5, 1.45×10^7 and 7.35×10^7 W m^{-2}. The scale of wavelengths and the visible region are indicated.

Stefan-Boltzmann constant. This law is in remarkable agreement with observations. In particular, the measured value of σ coincides with its theoretical value (7.22). Similarly, the spectral emittance of a black body conforms perfectly with the distribution relation (7.21). In fact Planck's law was confirmed by measuring the spectral emittance e_ν of a black body rather than the spectral density of the energy per unit volume u_ν of the photon gas. Figure 7.3 shows the spectral emittance of a black body at several temperatures. Figure 7.4 indicates the nomenclature of the various regions of radiation that require different emission and detection techniques.

Exercise 7.2 *Kinetic theory of radiation pressure*
    As in exercise 6.1, determine the kinetic pressure of a photon gas.

Solution
    Each photon has a momentum of modulus $|\mathbf{p}| = h\nu/c$. After reflection at the wall, the parallel component of this momentum remains unchanged while the normal component changes sign. The momentum transferred to the wall is therefore (Fig. 6.7)

$$(\Delta \mathbf{p})_n = (\mathbf{p}_i)_n - (\mathbf{p}_f)_n = 2p\cos\theta \ .$$

Photons with momentum $\mathbf{p}$ in the element $d^3\mathbf{p}$ that strike the surface dS in the interval dt fit into a cylinder of volume $dV = c\,dt\,dS\,\cos\theta$. Their number, found

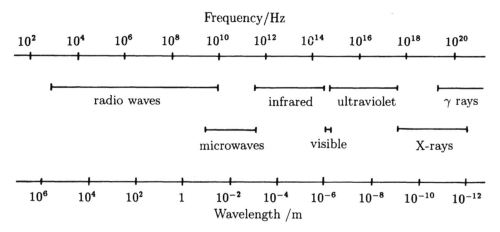

Figure 7.4: Different domains of electromagnetic radiation.

from $d^6 N_{\mathbf{r},\,\mathbf{p}}$ (7.1 and 5), is equal to

$$\frac{2}{h^3}\, c\, dt\, dS\, \frac{\cos\theta\, d^3\mathbf{p}}{e^{\beta\epsilon}-1}\quad.$$

The total momentum transferred to the wall is therefore

$$(d^3\mathcal{P})_n = \frac{4c}{h^3}\, dt\, dS \int \cos^2\theta\, \sin\theta\, d\theta\, d\phi \int \frac{p^3\, dp}{e^{\beta\epsilon}-1}\quad.$$

The integral over the angles in the range $0 \le \phi < 2\pi$ and $0 \le \theta \le \pi/2$, corresponding to the half space in which the photon gas is located, is equal to $2\pi/3$. Introducing the variable ν yields, with (7.16),

$$(d^3\mathcal{P})_n = \frac{8\pi h}{3c^3}\, dt\, dS \int_0^\infty \frac{\nu^3\, d\nu}{e^{\beta h\nu}-1} = \frac{1}{3}\, dt\, dS \int_0^\infty u_\nu\, d\nu = \frac{1}{3}\frac{U}{V}\, dt\, dS\ .$$

The radiation pressure is thus $P = U/3V$, which is identical to the result (7.12) obtained from the thermodynamic functions.

7.4 EXPERIMENTAL ASPECTS

The spectrum of the emitted radiation, given by (7.21) and shown in Figure 7.3, varies with temperature. Up to about 500 °C energy is radiated almost entirely in the infra-red and very little in the visible range. To the eye, a black body looks black in the usual sense of the word. Above 500 °C, radiation in the visible range becomes perceptible to the eye and the black body turns dark red (700 °C), then cherry red (900 °C), then bright orange (1200 °C). These colours are those observed through a hole bored into the oven. We see that the frequency ν_m at which the spectral emittance e_ν is a maximum (the same as that for which u_ν is maximum (7.20)), remains in the infra-red region up to about 6000 °C (Fig. 7.3).

By definition, a black body is one that can absorb all radiation energy that reaches it. This is the case for a hole drilled in the wall of an oven, since photons entering by this orifice are reflected, scattered and eventually absorbed by the walls. Other approximately black bodies can be produced by depositing soot or platinum black on a plate. Other bodies absorb only a fraction a_ν of the radiation with frequency ν that reaches them. a_ν is called the absorption factor. If this factor is constant in the visible range, the body appears grey to the eye in natural light ($a \simeq 0.8$ for graphite). If it varies strongly in the visible range, the substance appears coloured. Thus the absorption factor of copper is lower at the red end of the visible spectrum ($a_\nu = 0.28$ at $\lambda = 0.6$ μm and $a_\nu = 0.58$ at $\lambda = 0.5$ μm). This variation accounts for the red colour of copper.

Stellar radiation is composed of a continuous spectrum that is close to that of a black body. It is characteristic of the temperature of the external layer of the star (photosphere), which determines its "colour", e.g., red stars ($\sim 3\,000$ K), or blue stars ($\sim 35\,000$ K). Moreover, their spectra exhibit absorption lines whose position and intensity reveal the chemical elements present in the stellar atmosphere, and also their abundance.

Exercise 7.3 *Solar constant*

The temperature of the photosphere of the Sun is about 5950 K. Calculate the radiation power striking a surface of unit area placed outside the earth's atmosphere and perpendicular to the rays of the Sun (solar constant). The angle subtended by the Sun from the earth is $\alpha = 32'$.

Solution

The total power emitted by the Sun, expressed as a function of its emittance $\mathcal{E}$ (7.22), is

$$\mathcal{P} = 4\pi R^2 \mathcal{E} = 4\pi R^2 \times \sigma T^4 \, ,$$

where R is the radius of the Sun. The power received by an area S at a distance d from the Sun is

$$p = \mathcal{P} \times \frac{S}{4\pi d^2} = S \left(\frac{R}{d}\right)^2 \sigma T^4 = S \frac{\alpha^2}{4} \sigma T^4 \, .$$

The solar constant is therefore equal to

$$\frac{p}{S} = \frac{1}{4}\alpha^2 \sigma T^4 = 1540 \ \mathrm{W \ m^{-2}} = 2.21 \ \mathrm{cal \ min^{-1} \ cm^{-2}} \, .$$

The measured value of this constant (2.0 cal min^{-1} cm^{-2}) is slightly lower than that calculated, as the Sun is not a perfectly black body.

We see that an area of 1 m^2 on the ground receives about 1000 kWh per year, which corresponds to the combustion energy of 85 kg of petroleum or 0.085 equivalent tonnes of petroleum (1 etp $= 42 \times 10^9$ J). Atmospheric absorption has to be taken into account (approximately 50 %), as well as the inclination of the Sun (average factor 0.5) and an insolation of about 2500 hours per year.

The above formula allows the apparent angle α of the stars to be determined by measuring their apparent luminosity (p/S) and their spectral distribution, the position of whose maximum gives T. Their radius can be deduced if their distance is known, or *vice versa*.

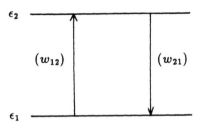

Figure 7.5: Absorption and emission between two states

7.5 INTERACTION BETWEEN MATTER AND RADIATION

7.5.1 Equation of Detailed Balance

We now consider the interaction of electromagnetic radiation with the atoms of the walls of the containing vessel. We restrict ourselves to a simple model in which each atom can be in only one of two different quantum states 1 and 2. The radiation excites the atoms from state 1 to 2 by absorption of a photon of frequency $\nu = (\epsilon_2 - \epsilon_1)/h$ and, conversely, they are de-excited from state 2 to 1 by emission of a photon of the same frequency (Fig. 7.5).

In the detailed balance of the exchange occurring in a time interval dt, we can write for the change in the numbers N_1 and N_2 of atoms in each state

$$dN_2 = -dN_1 = w_{12}N_1\,dt - w_{21}N_2\,dt \;\; ; \tag{7.23}$$

here the term $w_{12}N_1$ describes transitions from level 1 to 2, and the term $-w_{21}N_2$ transitions in the other direction.

The detailed balance equation is very general and can be applied to more complex systems than that considered here. It has the property of linearity in the interval t and is not invariant under time reversal $(t \rightarrow -t)$, unlike Newton's equation or that of Schrödinger. It therefore describes irreversible processes.

7.5.2 Einstein Model of Thermal Radiation

In his 1917 theory of detailed balance, A. Einstein assumed that the probability per unit time w_{12} of a photon of frequency ν being absorbed by an atom is proportional to the spectral density of the radiation u_ν

$$w_{12} = B_{12}u_\nu \;\; . \tag{7.24}$$

He also assumed that atoms emit photons by two processes, namely
- spontaneous emission, whose probability per unit time A_{21} is constant;
- emission stimulated by the radiation, whose probability per unit time is proportional to u_ν, i.e., $B_{21}u_\nu$.

The total emission probability for an atom in unit time is then

$$w_{21} = A_{21} + B_{21}u_\nu \;\; . \tag{7.25}$$

The coefficients B_{12}, A_{21} and B_{21} are called the Einstein coefficients.

Einstein was able to derive Planck's law from the thermal equilibrium between matter and radiation, using the condition of detailed balance. At equilibrium, N_1 and N_2 are constant and equal to N_1^0 and N_2^0, and the equation of balance (7.23) becomes

$$w_{12} N_1^0 - w_{21} N_2^0 = 0 \ ,$$

which expresses the fact that the number of transitions from $1 \to 2$ is equal to that from $2 \to 1$. Substitution of equations (7.24,25) for the probabilities w yields

$$u_\nu = \frac{A_{21}/B_{12}}{N_1^0/N_2^0 - B_{21}/B_{12}} \ .$$

The equilibrium ratio N_1^0/N_2^0 of the number of atoms in each of the levels is found on applying Maxwell-Boltzmann statistics. It is equal to

$$\frac{N_1^0}{N_2^0} = e^{\beta(\epsilon_2 - \epsilon_1)} = e^{\beta h\nu} \ .$$

By identifying the above expression for u_ν with Planck's law (7.16), Einstein found

$$B_{12} = B_{21} \quad \text{and} \quad \frac{A_{21}}{B_{12}} = \frac{8\pi h}{c^3} \ \nu^3 \ . \tag{7.26}$$

These results were subsequently confirmed by quantum theory, according to which
- the elementary probabilities of two inverse processes are equal ($B_{12} = B_{21}$),
- the coefficients are given by

$$A_{21} = \frac{64 \ \pi^4 \nu^3}{3hc^3} \ |\mathbf{p}_{21}|^2 \quad \text{and} \quad B_{12} = B_{21} = \frac{8\pi^3}{3h^2} |\mathbf{p}_{21}|^2, \tag{7.27}$$

where $|\mathbf{p}_{21}|^2$ is the square of the matrix element between states 1 and 2 of the electric dipolar moment of the atom.

Einstein introduced the notion of stimulated emission in order to recover Planck's law. In the absence of stimulated emission ($B_{21} = 0$), the spectral density of the energy would have the form

$$u_\nu = \frac{A_{21}}{B_{12}} e^{-\beta h\nu}$$

corresponding to Wien's law (7.18), which does not account for all the experimental facts.

The relative strength of the two kinds of emission is given by the ratio

$$\frac{B_{21} \ u_\nu}{A_{21}} = \frac{1}{e^{\beta h\nu} - 1} \ .$$

If $h\nu/kT$ is very large, stimulated emission is negligible. If, on the contrary, $h\nu/kT$ is very small, then stimulated emission is preponderant. Stimulated emission is therefore encountered primarily in the radio frequency region ($\lambda \gtrsim 30$mm). We recall here that stimulated emission is the underlying physical process involved in lasers.

It is also to be noted that, in the theory of quantization of the electromagnetic field, spontaneous and stimulated emission are part of one single emission process. In particular, it can be shown that

$$\frac{w_{21}}{w_{12}} = \left| \frac{<n+1 \, |a^+| \, n>}{<n-1 \, |a| \, n>} \right|^2 = \frac{n+1}{n}$$

where $| \, n >$ is the quantum state of the electromagnetic field containing n photons allowing the transition $1 \to 2$ and a, and a^+ are the annihilation and creation operators of the harmonic oscillator. At equilibrium, we have

$$\frac{w_{21}}{w_{12}} = \frac{N_1^0}{N_2^0} = e^{\beta h \nu} \; ,$$

from which we find

$$n = \frac{1}{e^{\beta h \nu} - 1}$$

in accordance with the Bose-Einstein distribution (7.1).

7.6   IDEAL MOLECULAR BOSON GASES

7.6.1   Distribution Relation

For a molecular gas of non-interacting bosons, the Bose-Einstein distribution (2.26) is

$$n_i = \frac{N_i}{g_i} = \frac{1}{e^{\beta(\epsilon_i - \mu)} - 1} \; . \tag{7.28}$$

The physical condition $n_i \geq 0$ implies $\epsilon_i - \mu \geq 0$ for all i. In particular, if the origin for the energy is chosen to be that of the ground state ($\epsilon_0 = 0$), a choice that we adopt from now on, the chemical potential must be negative, i.e. $\mu \leq 0$. We shall discuss later how the value of μ is determined. In general, this cannot be done explicitly. It is recalled that if $n_i \ll 1$ for all i, the limit of Bose-Einstein statistics is corrected Maxwell-Boltzmann statistics (§2.5.5). In particular, this happens when

$$e^{-\beta\mu} \gg 1 \quad \text{or} \quad \mu \ll -kT \; . \tag{7.29}$$

7.6.2   Thermodynamic Functions in the Variables T, V, μ

According to the general method, we must determine the grand potential Ω of (2.34)

$$\Omega(T, \, V, \, \mu) = kT \sum_i g_i \ln \left[1 - e^{\beta(\mu - \epsilon_i)} \right] \tag{7.30a}$$

which yields the thermodynamic properties of the gas. To evaluate Ω when the molecules possess only translational movement ($\epsilon = p^2/2m$), we transform to continuous variables (§5.2). Thus

$$\Omega = g_s \frac{kT}{h^3} \int d^3\mathbf{r}\, d^3\mathbf{p} \ln\left[1 - e^{\beta(\mu-\epsilon)}\right] , \qquad (7.30\text{b})$$

where $g_s = 2J + 1$ is the degeneracy due to the spin J of the molecules.

Integration over $\mathbf{r}$ ($d^3\mathbf{r} \to V$) and over the directions of $\mathbf{p}$ ($d^3\mathbf{p} \to 4\pi p^2\, dp$), together with the change of variable $x = \beta\epsilon = \beta p^2/2m$, leads to

$$\Omega = kT\frac{2\pi V}{h^3}g_s(2mkT)^{3/2} \int_0^\infty \sqrt{x}\ln\left(1 - e^{\beta\mu-x}\right)\, dx .$$

Introducing the notations

$$Z(T,\, V) = g_s\frac{V}{h^3}(2\pi mkT)^{3/2} \text{ and } f(\nu) = -\frac{2}{\sqrt{\pi}} \int_0^\infty \sqrt{x}\ln\left(1 - e^{\nu-x}\right)\, dx, \qquad (7.31)$$

we can write for the grand potential

$$\Omega(T,\, V,\, \mu) = -kTZ(T,\, V)\, f(\nu) \quad \text{with} \quad \nu = \mu/kT . \qquad (7.32)$$

We can now write down the expressions for the entropy S, pressure P and number of particles N, these quantities being related to Ω by

$$d\Omega = -SdT - PdV - Nd\mu .$$

Hence, using equations (7.31,32) for Z, ν and for $f(\nu)$, we find

$$S \;=\; -\left(\frac{\partial\Omega}{\partial T}\right)_{\mu,\, V} = kZ\left(\frac{5}{2}f(\nu) - \nu f'(\nu)\right) , \qquad (7.33)$$

$$P \;=\; -\left(\frac{\partial\Omega}{\partial V}\right)_{T,\, \mu} = kT\frac{Z}{V}f(\nu) = -\frac{\Omega}{V} , \qquad (7.34)$$

$$N \;=\; -\left(\frac{\partial\Omega}{\partial\mu}\right)_{T,\, V} = Zf'(\nu) . \qquad (7.35)$$

Also, from the above expressions, we obtain for the internal energy

$$U = \Omega + TS + N\mu = \frac{3}{2}kTZf(\nu) = \frac{3}{2}PV , \qquad (7.36)$$

which can be rewritten as

$$PV = \frac{2}{3}U . \qquad (7.37)$$

This equation, previously obtained for corrected Maxwell-Boltzmann statistics (5.22) and (5.23), is general for non-interacting gases in which the energy takes the form $\epsilon = p^2/2m$. This is true whatever kind of statistics applies. In the various types of statistics, the expression for Ω always has the form (7.32), and only the equation for the function $f(\nu)$ changes.

Finally, we may calculate the heat capacity at constant volume

$$C_V = \left(\frac{\partial U}{\partial T}\right)_{N, V} = \frac{3}{2}k\left[\frac{5}{2}Zf(\nu) + TZf'(\nu)\left(\frac{\partial \nu}{\partial T}\right)_{N, V}\right] .$$

To calculate $\partial\nu/\partial T$, expression (7.35) is differentiated with respect to T while N and V are kept fixed,

$$0 = \frac{3}{2}\frac{Z}{T}f'(\nu) + Zf''(\nu)\left(\frac{\partial \nu}{\partial T}\right)_{N, V} .$$

Substitution of $\partial\nu/\partial T$ into the formula for C_V gives

$$C_V = \frac{3}{2}kZ\left[\frac{5}{2}f(\nu) - \frac{3}{2}\frac{f'(\nu)^2}{f''(\nu)}\right] = \frac{3}{2}Nk\left[\frac{5}{2}\frac{f(\nu)}{f'(\nu)} - \frac{3}{2}\frac{f'(\nu)}{f''(\nu)}\right] . \tag{7.38}$$

7.6.3   Description of the Function $f(\nu)$

Analysis shows that for $\nu < 0$, the function $f(\nu)$ introduced in (7.31) takes the following equivalent forms

$$f(\nu) = -\frac{2}{\sqrt{\pi}}\int_0^\infty \sqrt{x}\ln\left(1 - e^{\nu-x}\right)\,dx = \frac{4}{3\sqrt{\pi}}\int_0^\infty \frac{x^{3/2}}{e^{x-\nu} - 1}\,dx \tag{7.39a}$$

$$f(\nu) = \sum_{n=1}^\infty \frac{e^{n\nu}}{n^{5/2}} = e^\nu + \frac{e^{2\nu}}{2^{5/2}} + \frac{e^{3\nu}}{3^{5/2}} + \dots \tag{7.39b}$$

Figure 7.6 shows $f(\nu)$ and its first two derivatives. From these, the thermodynamic functions of the last paragraph can be evaluated numerically.

In the limit $\nu \to -\infty$, corresponding to corrected Maxwell-Boltzmann statistics, $f(\nu) \simeq e^\nu$. Figure 7.6 shows that this limit is reached in practice when $\nu \sim -2$.

In the limit $\nu \to 0$, $f(\nu)$ is approximately

$$f(\nu) = 2.363(-\nu)^{3/2} + 1.341 + 2.612\nu - 0.730\nu^2 - 0.0347\nu^3 + \mathcal{O}(\nu^4) \tag{7.40a}$$

which shows that

$$f(0) = \zeta(5/2) \quad \text{and} \quad f'(0) = \zeta(3/2) \tag{7.40b}$$

in agreement with equations (7.7 and 7.31).

7.6.4   Thermodynamic Functions in the Variables T, V, N

To substitute N for μ (or ν) in the expressions for the thermodynamic functions, the following equation must be solved for ν (7.35)

$$f'(\nu) = \alpha \equiv \frac{N}{Z(T, V)} = \frac{Nh^3}{g_s V(2\pi mkT)^{3/2}} . \tag{7.41}$$

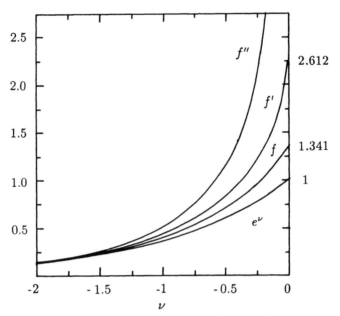

Figure 7.6: Graph of $f(\nu)$ (7.35) and its first two derivatives. For $\nu < -2$, the three functions coincide with e^{ν}, the value of $f(\nu)$ in the region where corrected Maxwell-Boltzmann statistics applies. $f(0) = 1.341$, $f'(0) = 2.612$ and $f''(0) = \infty$.

The solution can be found graphically from Figure 7.6. $f'(\nu)$ can also be found by differentiating equation (7.39b) for $f(\nu)$, i.e.,

$$\alpha = f'(\nu) = e^{\nu} + 2^{-3/2}e^{2\nu} + 3^{-3/2}e^{3\nu} + \dots \, ,$$

which, on inversion, yields e^{ν} as a function of α, i.e.,

$$e^{\nu} = \alpha - 2^{-3/2}\alpha^2 + 0.058\alpha^3 + \dots$$

The chemical potential thus becomes

$$\mu = kT \ln e^{\nu} = kT \left[\ln \alpha - 2^{-3/2}\alpha + \dots \right] \quad \text{with} \ \ \alpha = N/Z \, , \tag{7.42}$$

and the expression for $f(\nu)$ is

$$f(\nu) = e^{\nu} + 2^{-5/2}e^{2\nu} + \dots = \alpha - 2^{-5/2}\alpha^2 + \dots$$

These results allow all the thermodynamic functions to be written in terms of the variables T, V and N. For example, the free energy, which is a characteristic energy function, is, in terms of the variables T, V, N

$$
\begin{aligned}
F &= \Omega + N\mu = -kTZf(\nu) + N\mu \\
&= -NkT \left[\left(1 + \ln \frac{Z}{N} \right) + \frac{N}{2^{5/2}Z} + \dots \right] \, .
\end{aligned}
\tag{7.43}
$$

Similarly, either from equations (7.33-38) or by differentiating F, we get

$$S = -\left(\frac{\partial F}{\partial T}\right)_{N,V} = Nk\left[\left(\frac{5}{2} + \ln\frac{Z}{N}\right) - \frac{N}{2^{7/2}Z} + \cdots\right] \qquad (7.44)$$

$$P = -\left(\frac{\partial F}{\partial V}\right)_{T,N} = \frac{NkT}{V}\left[1 - \frac{N}{2^{5/2}Z} + \cdots\right] \qquad (7.45)$$

$$U = F + TS = \frac{3}{2}NkT\left[1 - \frac{N}{2^{5/2}Z} + \cdots\right] \qquad (7.46)$$

$$C_V = \left(\frac{\partial U}{\partial T}\right)_{N,V} = \frac{3}{2}Nk\left[1 + \frac{N}{2^{7/2}Z} + \cdots\right] . \qquad (7.47)$$

In the limiting case $\alpha = N/Z \ll 1$, these results coincide with those found for corrected Maxwell-Boltzmann statistics (cf. Chap. 5), in which the quantum effects connected with the nature of the particles are neglected. In expression (7.41) for α it can be seen that this limit is reached when

$$\left(\frac{V}{N}\right)^{1/3} \gg \lambda_T \equiv \frac{h}{(2\pi mkT)^{1/2}} ,$$

that is, when the mean distance between particles is very much greater than the *thermal wavelength* λ_T of the gas. This wavelength $\lambda = h/p$ is that associated with a particle of mass m and kinetic energy (5.24) $3kT/2$.

Exercise 7.4 *Pressure and heat capacity of a boson gas*
    An ideal boson gas with zero spin, molar mass $M = 4$ g mol^{-1} and molar volume $v = 150$ cm^3 mol^{-1} is maintained at a temperature $T = 2.1$ K. Determine its pressure and heat capacity at constant volume, firstly by a graphical method using Figure 7.6, and then by an approximate calculation using the finite series expansions (7.45 and 47).

Solution
    The value of α must first be determined from (7.41)

$$\alpha = \frac{N}{Z} = \frac{\mathcal{N}}{v}\left(\frac{\mathcal{N}^2 h^2}{2\pi MRT}\right)^{3/2} = 0.877 .$$

From Figure 7.6, for the graphic solution of equation (7.41) $f'(\nu) = \alpha$, we read

$$\nu = -0.46 \quad ; \quad f(\nu) = 0.74 \quad ; \quad f''(\nu) = 1.27 .$$

Substitution of these values into equations (7.34 and 38) yields

$$P = \frac{RT}{v}\frac{f(\nu)}{\alpha} = 9.82 \times 10^4 \text{ Pa} = 0.97 \text{ atm} ;$$

$$c_V = \frac{3}{2}R\left[\frac{5}{2}\frac{f(\nu)}{\alpha} - \frac{3}{2}\frac{\alpha}{f''(\nu)}\right] = 13.4 \text{ J K}^{-1} \text{ mol}^{-1} .$$

Restricting ourselves to the linear terms in $\alpha = N/Z$ in the series expansions (7.45 and 47), we find

$$P = 1.15 \text{ atm} \times (1 - 0.16) = 0.97 \text{ atm} \quad ;$$

$$c_V = \frac{3}{2}R \times (1 + 0.08) = 13.4 \text{ J K}^{-1} \text{ mol}^{-1} \quad .$$

The graphic solution and calculation by a first order limited expansion in α lead to identical results. It can be seen that the pressure of a boson gas in these conditions is lower than that given by the equation of state of an ideal gas in corrected Maxwell-Boltzmann statistics, i.e., $P = RT/v = 1.15$ atm. The quantum character of the particles (bosons) causes a reduction in pressure, in the same way as would an attractive interaction.

7.7 EINSTEIN CONDENSATION

7.7.1 Chemical Potential for $\alpha > \alpha_c$

We saw above how to change from T, V, μ to the variables T, V, N by using the quantity $\alpha = N/Z(T,V)$, which is determined by the state of the gas. However, when α becomes greater than

$$\alpha_c \equiv f'(0) = 2.612 \quad , \tag{7.48}$$

equation (7.41) determining ν (or $\mu = kT\nu$)

$$f'(\nu) = \alpha \tag{7.49}$$

no longer has a solution. This result is unphysical since, first, there is no reason to place a bound on α and, second, the chemical potential of the gas must have a definite value.

This apparent contradiction arises from the mathematical transformation whereby the discrete sum over energy levels (7.30a) is replaced by the continuous integral (7.30b). This transformation is valid as long as the integrand is "well-behaved". When μ tends to zero, however, the first term of the sum (7.30a), which is equal to $kT \ln(1 - e^{\nu})$, diverges. (From now on we take $g_s = 1$ corresponding to a spin $J = 0$, and we take into account the non-degeneracy of the translational ground state, i.e., $g_0 = 1$). This corresponds to the fact that the number of particles in the ground state

$$N_0 = \frac{1}{e^{-\nu} - 1} \qquad (\nu = \mu/kT) \tag{7.50}$$

varies as $-1/\nu$ as $\nu \rightarrow 0^-$, i.e., a large fraction of particles occupy the ground state $\epsilon_0 = 0$.

The contribution of this level must then be calculated explicitly, the change to the continuous limit being allowed only for the other levels. We thus find

$$\Omega = kT \ln(1 - e^{\nu}) + \frac{kT}{h^3} \int_{\epsilon_1}^{\infty} d^3\mathbf{r} \, d^3\mathbf{p} \ln\left((1 - e^{\nu - \beta\epsilon}\right) \quad ,$$

where ϵ_1 is the energy of the first excited state. The integral can now be transformed as in paragraph 7.2. Moreover, the lower limit $x_1 = \beta\epsilon_1 = \epsilon_1/kT$ of this new integral is extremely small (§1.2.1) and can be taken equal to zero. The expression for Ω that replaces (7.32) as $\nu \to 0^-$ is then

$$\Omega = kT \ln(1 - e^\nu) - kTZf(\nu) \ , \tag{7.51}$$

and the equation defining μ becomes

$$N = -\left(\frac{\partial\Omega}{\partial\mu}\right)_{T,\,V} = \frac{1}{e^{-\nu} - 1} + Zf'(\nu) \tag{7.52a}$$

or alternatively

$$f'(\nu) + \frac{1}{Z}\frac{1}{e^{-\nu} - 1} = \alpha \ . \tag{7.52b}$$

If $\alpha < \alpha_c$, this equation reduces to equation (7.49), employed in the previous paragraph. In the thermodynamic limit ($N \to \infty$, $V \to \infty$, where V/N is fixed), the function Z (7.31) indeed tends to infinity, and as ν is different from zero, the second term on the left hand side vanishes.

In contrast, when $\alpha > \alpha_c = f'(0)$, equation (7.52b) has a solution only when ν tends to zero in the thermodynamic limit. Thus

$$f'(0) + \frac{1}{Z \times (-\nu)} = \alpha \quad \text{and therefore} \quad -\nu = \frac{1}{N}\frac{1}{1 - \alpha_c/\alpha} \quad (\alpha > \alpha_c) \ \ (7.53)$$

In this form, it can be seen that ν disappears in the thermodynamic limit. This justifies our replacement of $f(\nu)$ and its derivatives by $f(0)$, $f'(0)$ and $f''(0)$ when $\alpha > \alpha_c$.

7.7.2   Thermodynamic Functions for $\alpha > \alpha_c$

We now determine the thermodynamic functions in the region $\alpha > \alpha_c$ from the free energy $F(T, V, N)$. According to (7.51), this function is

$$F = \Omega + N\mu = -kTZf(\nu) + kT\left[\ln(1 - e^\nu) + N\nu\right] \ .$$

In the thermodynamic limit this becomes $N \times \lim_{N\to\infty} F/N$, or

$$F = -kTZf(0) = -C(kT)^{5/2}V \quad \text{with} \quad C = f(0)\frac{(2\pi m)^{3/2}}{h^3} \ . \tag{7.54}$$

From this expression we find

$$S = -\left(\frac{\partial F}{\partial T}\right)_{V,\,N} = \frac{5}{2}kC(kT)^{3/2}V \ , \tag{7.55}$$

$$P = -\left(\frac{\partial F}{\partial V}\right)_{T,\,N} = C(kT)^{5/2} \ , \tag{7.56}$$

$$U = F + TS = \frac{3}{2}C(kT)^{5/2}V = \frac{3}{2}PV \ , \tag{7.57}$$

$$C_V = \left(\frac{\partial U}{\partial T}\right)_{V,\,N} = \frac{15}{4}kC(kT)^{3/2}V \ . \tag{7.58}$$

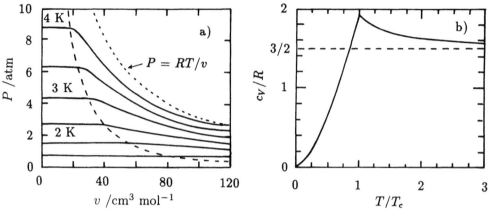

Figure 7.7: a) $(v,\ P)$ isotherms for an ideal boson gas $(M = 4\ \mathrm{g\ mol^{-1}})$. The $T = 4$ K isotherm for an ideal gas $(Pv = RT)$ is shown as a dotted line. The dashed line is the phase transition curve (7.61). b) Molar heat capacity at constant volume for the same gas. The horizontal line c_V/R is the Maxwell-Boltzmann molar heat capacity.

When $\alpha > \alpha_c$, these equations replace (7.33, 34, 36 and 38), which are valid for $\alpha < \alpha_c$. Note that these quantities, in particular the entropy and the heat capacity, vanish at absolute zero in accordance with the Third Law.

7.7.3 Phase Transition

We have just seen that the thermodynamic quantities have two different forms depending on the relative values of $\alpha(T,\ V,\ N)$ and α_c. This situation corresponds to the existence of two different phases called I $(\alpha < \alpha_c)$ and II $(\alpha > \alpha_c)$, whose respective domains are separated by the curve

$$\alpha \equiv \frac{Nh^3}{V(2\pi mkT)^{3/2}} = \alpha_c = 2.612 \ . \tag{7.59}$$

In each domain, the thermodynamic functions take different analytical forms. This is the case for the equation of state and for the molar heat capacity c_V shown in Figure 7.7.

For each value of molar volume v, the phase transition I $\rightleftharpoons$ II occurs at a temperature

$$T_c(v) = \frac{h^2}{2\pi mk}\left(\frac{\mathcal{N}}{\alpha_c v}\right)^{2/3} \tag{7.60}$$

and at a pressure, given by (7.56),

$$P_c(v) = C(kT_c)^{5/2} = \frac{1.341\ h^2}{2\pi m}\left(\frac{\mathcal{N}}{\alpha_c v}\right)^{5/3} \ . \tag{7.61}$$

In phase II, the pressure of the gas varies as $T^{5/2}$ and is independent of volume. This property is analogous to that observed during condensation of a gas into a liquid. For this reason the name Einstein condensation (1925) is given to the transformation I $\rightarrow$ II. In Einstein condensation, however, there is no coexistence between two phases that are separated in space and progressively transform from one into the other, but a change of the properties of the substance in the transition I $\rightleftharpoons$ II. In the terminology of Ehrenfest, this is a phase transition of order greater than 1. In fact, it is of order 3. The third derivatives of the free enthalpy $G \equiv N\mu$ display a discontinuity at the transition (cusp in the curve $c_V(T)$, while those of first order (entropy, volume) and second order (heat capacity, ...) remain continuous (Fig. 7.7).

In modern terminology (L.D. Landau, 1937), phase transitions of order greater than 1 are characterized by an order parameter, whose value is zero in phase I and stable at higher temperatures, and which increases progressively from 0 to 1 in phase II as the temperature decreases from $T = T_c$ down to absolute zero. In an Einstein condensation, the fraction of particles in the ground state, N_0/N, plays the role of the order parameter. For $T > T_c$ (or $\alpha < \alpha_c$), the chemical potential μ is non-zero, and the ratio

$$\frac{N_0}{N} = \frac{1}{N} \times \frac{1}{e^{-\beta\mu} - 1}$$

is zero in the thermodynamic limit. For $T < T_c$ (or $\alpha > \alpha_c$), however, the chemical potential $\mu = kT\nu$ tends to zero (7.53) and the ratio

$$\frac{N_0}{N} = \frac{1}{N \times (-\beta\mu)} = 1 - \frac{\alpha_c}{\alpha}$$

adopts a finite non-zero value that can be written in the form

$$\frac{N_0}{N} = 1 - \left(\frac{T}{T_c}\right)^{3/2}. \tag{7.62}$$

It can thus be seen that below T_c, the ground state is populated by a finite and significant fraction of particles (Fig. 7.8). In this respect, Einstein condensation is a condensation into the ground state energy level, or alternatively, a condensation in momentum space. In contrast, condensation of a gas into a liquid occurs in real space.

The fact that the ground state is the only one to be significantly populated can be verified by calculating the fraction N_1/N of particles in the first excited state ($\epsilon = \epsilon_1 \sim 10^{-20}$ eV according to (1.3)). For $T < T_c$, this fraction

$$\frac{N_1}{N} = \frac{g_1}{N(e^{\beta\epsilon_1} - 1)} \simeq \frac{g_1}{N\beta\epsilon_1}$$

is zero in the thermodynamic limit and is approximately equal to 10^{-8} for $N = 6 \times 10^{23}$ and $T = 1$ K. The same is true for the other excited levels. However, the total number of particles in these levels, $N'(T, V, N)$

$$N' = N - N_0 = N\left(\frac{T}{T_c}\right)^{3/2}$$

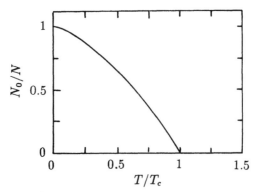

Figure 7.8: Fraction N_0/N of particles in the ground state as a function of the ratio of the temperature to the critical temperature T_c given by (7.60).

is a finite and significant fraction of the gas particles.

It is of interest to note that the thermodynamic functions (7.54-58) can be rewritten in the form

$$F = -aN'kT \qquad S = \frac{5}{2}aN'k \qquad P = a\frac{N'kT}{V}$$
$$U = \frac{3}{2}aN'kT \qquad C_V = \frac{15}{4}aN'k \tag{7.63}$$

where $a = f(0)/f'(0) = 0.513$. It is remarkable that all these functions depend only on the number N' of "uncondensed" particles. This result can be explained by observing that these particles are the only ones that contribute to the internal energy and the pressure, and that this contribution is approximately one half of that of a Maxwell-Boltzmann gas.

Experimentally, no boson system exhibits the Einstein condensation phenomenon that we have just described, since at the temperatures and densities at which the condensation could occur, interactions between particles are no longer negligible and most of them have crystallized. Nonetheless, the phase transition of order higher than 1 that occurs in liquid helium around 2 K (Chapter 9) can be considered as an Einstein condensation process that is perturbed by the presence of interaction forces.

Exercise 7.5 *Application of Einstein condensation to helium ^{4}He*

At what temperature would Einstein condensation occur in liquid helium if interactions between atoms were negligible? $M = 4$ g mol^{-1} and the density is $\rho = 0.14$ g cm^{-3}.

Solution

The critical temperature of the phase transition, given by (7.60), is $T_c = 3.07$ K for helium. The value is in fairly good agreement with the experimentally measured value of the phase change observed in liquid helium at 2.17 K. The difference is caused by the fact that interactions are neglected.

BIBLIOGRAPHY

Historical References
L. Boltzmann, *Ann. Physik* **22**, 31 (1884); **22**, 291 (1884); **22**, 616 (1884).
S. N. Bose, *Z. Physik* **26**, 178 (1924).
A. Einstein, *Physik. Z.* **18**, 121 (1917) (detailed balance).
A. Einstein, *Berliner Ber.* 3 (1925) (Einstein condensation).
A. Einstein, *Berliner Ber.* 261 (1924); 3 and 18 (1925) (Bose-Einstein distribution).
J.H. Jeans, *Phil. Mag.* **49**, 539 (1900).
L.D. Landau, *Phys. Z. Sowjun.* **11**, 26, 545 (1937); *Zh. Eksp. Teor. Fiz.* **7**, 19, 627 (1937).
M. Planck, *Verh. Dtsch. Phys. Ges.* **2**, 202 (1900); **2**, 237 (1900).
Lord Rayleigh, *Phil. Mag.* **10**, 91 (1905).
J. Stefan, *Wien. Ber.* **79**, 391 (1879).
W. Wien, *Ann. Physik* **58**, 662 (1896).

COMPREHENSION EXERCISES

7.1 By integrating by parts, verify that the two sides of equation (7.7a) are equal.

7.2 Consider a 2 dimensional "photon" gas enclosed in a region of area A. Show that its free energy F and the number of photons at equilibrium are given by

$$F = -\frac{4\pi\,\zeta(3)A}{h^2 c^2}(kT)^3 \quad \text{and} \quad N = \frac{4\pi\,\zeta(2)A}{h^2 c^2}(kT)^2 \ .$$

7.3 What is the value of the radiation pressure P at $T = 300$ K and $T = 10^6$ K? At what temperature is $P = 1$ atm? [Ans.: 2.0×10^{-11} atm; 2500 atm; 1.4×10^5 K].

7.4 Show that Planck's equation expressed in the variable λ is

$$u_\lambda = \frac{1}{V}\frac{dU_\lambda}{d\lambda} = \frac{8\pi hc}{\lambda^5}\frac{1}{e^{\beta hc/\lambda} - 1} \ .$$

Show that the wavelength λ_m at which u_λ is a maximum is not the same as the frequency ν_m for which u_ν is maximum. [Ans.: $\lambda_m \nu_m = 0.568\ c$ and not c].

7.5 Show that in an isentropic transformation, $PV^{4/3}$ = constant for thermal radiation and $PV^{5/3}$ = constant for a molecular gas of bosons.

7.6 Show that the equilibrium curve (7.61) separating phase I and II of a boson gas is isentropic.

7.7 Calculate the molar heat capacity c_V of a boson gas for $T = T_c$. Check that the same value is found for the two phases. [Ans.: $c_V/R = 15f(0)/4f'(0) \simeq 1.92$].

7.8 For a free ^{4}He gas in two dimensions, determine the chemical potential at an arbitrary temperature as a function of S/N, where S is the surface area of the gas. Compute the number N_0 of particles on the fundamental level and discuss the possibility of Einstein condensation. [Ans.: $\mu = kT\ln(1 - e^{-u})$ with $u = h^2/(2\pi mkT) \times N/S$; $N_0 = e^u - 1$; for $N/S \sim 1/(4\text{Å})^2$, one has $u \simeq 6K/T$ and Einstein condensation occurs below 0,1K.

Chapter 8

Crystals : Structure, Phonons, Thermal Properties

8.1 INTERATOMIC BONDS AND CRYSTAL STRUCTURES

Interaction forces between atoms are mainly electrostatic in origin and can be considered as the sum of an attractive and a repulsive term. The latter, which arises when the electron clouds of the atoms start to overlap, is a manifestation of the Pauli exclusion principle. It depends strongly on the distance r between atoms and is often expressed as a power law r^{-n} where $n \simeq 12$. The attractive term, which depends on the electronic structure, determines the mode of bonding.

In molecules, atoms are bound in two ways:

• *covalent bonding* (H_2, CH_4, ...) is a result of valence electron sharing such that the electronic layers of each atom are full. It involves the exchange interaction, and the bond is directional.

• *ionic bonding* (NaCl, LiF, ...) arises from the Coulomb attraction between ions of opposite charge that are formed when certain atoms (Na, Li) transfer their electrons to others (Cl, F).

In these two modes and in intermediate modes, the bond energy is several electron-volts.

In solids, both of these bonding modes are encountered, in addition to others. The following can be distinguished:

• *covalent crystals*: each atom is surrounded by a small number of other atoms, the angles between the bonds being determined by the electronic structure. In diamond, for example, each carbon atom is at the centre of a regular tetrahedron, surrounded by four neighbours located at the vertices (Fig. 8.1a). These crystalline edifices are particularly stable and have low densities, great hardness and high melting points. In addition, they are electrical insulators.

• *ionic crystals*: these are composed of ions stacked in the most compact possible way. For example, for sodium chloride NaCl, the Cl^- ions are located at the corners of cubes and at the centre of their faces ("face-centred cubic" lattice), while the Na^+ ions, which have a smaller radius, are housed between them (Fig. 8.1b). Ionic crystals are fairly dense, are averagely hard and do not have very high melting points. They conduct electricity only at high temperatures.

• *metallic crystals*: these are composed of positive ions located at the vertices of a cubic or hexagonal lattice (Fig. 8.1c). The ions are kept apart by Coulomb

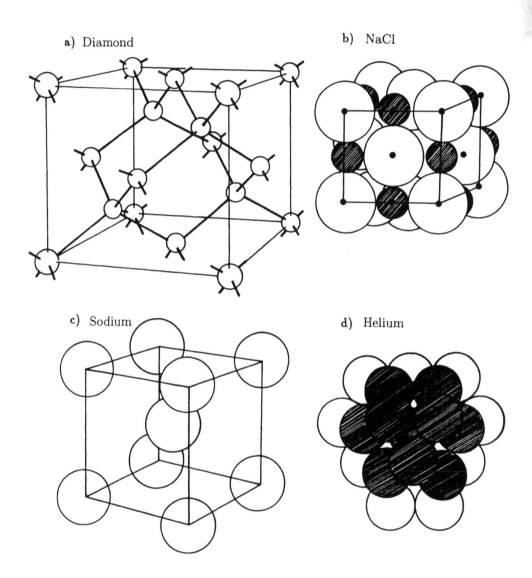

Figure 8.1: Types of crystalline structures. a) Covalent crystal of diamond (edge $a = 3.56$ Å); the radius of the carbon atoms ($r = 0.70$ Å) is reduced in this sketch. b) Ionic crystal of NaCl ($a = 5.63$ Å); the radius of the Cl$^-$ ions is $r = 1.81$ Å and that of the Na$^+$ ions (hatched) $r = 0.98$ Å. c) Crystal of metallic sodium ($a = 4.28$Å); the radius of the Na$^+$ ions is $r = 0.98$ Å. d) Molecular crystal of helium; the side of the equilateral triangles is $a = 3.57$ Å and the vertical distance between two layers is $c/2 = 2.91$ Å.

Table 8.1: Physical properties of different types of crystals .

Type of crystal	Substance	Density /g cm^{-3}	Bond energy /eV	Melting temperature /K
Covalent	Diamond	3.51	7.3	> 3500
Ionic	NaCl	2.17	7.81	1074
Metallic	Na	0.97	1.13	1165
Molecular	He	∼ 0.2	∼ 0.03	∼ 1
Hydrogen Bond	H$_2$O	1	0.52	273

repulsion and occupy only a small fraction of the volume (10 % in the case of sodium). Binding is ensured by the electrons liberated by the ions. The density of metallic crystals increases with increasing atomic number. They are very malleable and are good conductors.

• *molecular crystals*: forces between molecules or rare gas atoms are of van der Waals type. These electrostatic forces between instantaneous dipoles derive from a potential in $1/r^6$ and give rise to compact stacking of the molecules. For example, helium crystallizes into a hexagonal compact structure (Fig. 8.1d) and the other rare gases into face-centred cubic structures. In these crystals, each atom has twelve nearest neighbours in a highly symmetrical arrangement. In other crystals, the stacking of the molecules depends upon their shape. Such crystals are very compressible and have fairly low melting points. They are insulators.

• *hydrogen bonded crystals*: in a molecular crystal, "hydrogen bonds" can exist between hydrogen atoms and highly electronegative atoms (F, O, N) on neighbouring molecules. Such bonds greatly increase the binding strength of molecular crystals. The best known example is that of water (Fig. 1.9). Hydrogen bonds are of great structural importance for crystals of organic compounds.

Table 8.1 lists a few properties of the various types of crystal. Note that a large number of cases arise in which different types of bonding coexist. In graphite, covalent and metallic bonds exist simultaneously.

8.2 ELEMENTS OF CRYSTALLOGRAPHY

Here we outline the elements of crystallography that are essential for an understanding of the physical properties of crystals. A more detailed account can be found in specialist textbooks, such as that of F. C. Philipps.

8.2.1 Lattice and Primitive Cell

Crystals display regular atomic arrangements that admit of symmetry operations involving translations on the atomic scale. Such translations transform any point in the crystal into another point having the same surroundings with the same orientation (translational invariance). All such translations constitute an infinite group

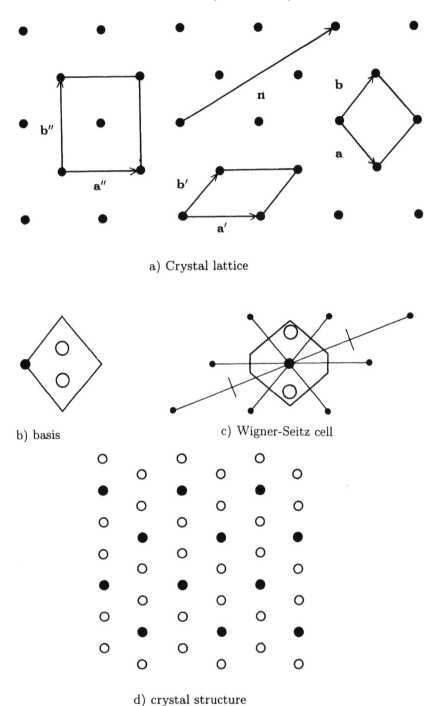

a) Crystal lattice

b) basis   c) Wigner-Seitz cell

d) crystal structure

Figure 8.2: Two dimensional model of a crystalline structure. Three sets of vectors (**a**, **b**), (**a′**, **b′**) and (**a″**, **b″**) and the translation vector $\mathbf{n} = \mathbf{a} + 3\mathbf{b} = \mathbf{a′} + 2\mathbf{b′}$ are shown. The cells constructed on the first and second sets of vectors are primitive, whereas the unit cell constructed on the third set is double and reveals the rectangular structure of the crystal.

whose elements can be represented by

$$\mathbf{n} = n_1\mathbf{a} + n_2\mathbf{b} + n_3\mathbf{c} \tag{8.1}$$

where n_1, n_2 and n_3 are integers and $\mathbf{a}$, $\mathbf{b}$ and $\mathbf{c}$ are three independent translations called primitive vectors or fundamental translation vectors. Figure 8.2 illustrates these considerations for a two-dimensional lattice (n_3 and $\mathbf{c}$ do not exist). Several triplets ($\mathbf{a}$, $\mathbf{b}$, $\mathbf{c}$) can be used as primitive vectors, but the volume of the parallelepiped ("primitive cell") constructed on these primitive vectors is always the same.

To describe a crystal, an origin in space is chosen and a lattice is built whose points, characterized by three integers (n_1, n_2, n_3), can be deduced from the origin by the translations (8.1). The whole of the crystal is defined if, in addition to the lattice, the contents of a cell are known, called the basis (Fig. 8.2b).

The Wigner-Seitz cell is also used; this includes all points in space that lie closer to one given site than to any other site. It is constructed by drawing the perpendicular bisecting planes of the translation vectors from the point to all its neighbours. In certain cases a multiple cell may be chosen that represents the symmetry of the crystal more clearly (Fig. 8.2). Primitive and multiple cells are called unit cells.

When the primitive vectors and the associated basis have been selected, the centre of any atom in the crystal is then defined by the vector

$$\mathbf{r} = n_1\mathbf{a} + n_2\mathbf{b} + n_3\mathbf{c} + \mathbf{r}_\alpha \tag{8.2}$$

where n_1, n_2 and n_3 are three integers that locate the position of the cell containing the atom and $\mathbf{r}_\alpha$ is a vector defining the position of the atom in the cell.

8.2.2 Crystal Symmetry

In addition to translations, crystals can have other operations that leave them invariant. In particular,

• rotations about an axis ("direct" rotations). The only rotations that are compatible with the periodic structure of a lattice are through an angle $\alpha = 2\pi/n$ (and its multiples), where $n = 2, 3, 4$ or 6. It is established that a surface can be paved with equilateral triangles ($n = 3$), squares ($n = 4$), regular hexagons ($n = 6$), but not with regular pentagons ($n = 5$) or octagons ($n = 8$);

• inverse rotations: direct rotations ($n = 1, 2, 3, 4, 6$) followed by a symmetry operation with respect to a point on the axis ("inversion"). Particular cases are pure inversion ($n = 1$), reflection or symmetry with respect to a plane perpendicular to the axis ($n = 2$).

• operations containing a translation: direct rotation followed by a translation along the rotation axis ("helicoidal" rotations); reflection followed by a translation parallel to the plane.

The symmetry operations of a crystal form a group. E.S. Fedorov (1891) and A.M. Schoenflies (1891) enumerated a total of 230 possible crystalline symmetry groups, called space groups (see *International Tables for X-ray Crystallography*).

In the symmetry operations considered here, translations refer to the atomic scale (~ 1 Å$= 10^{-1}$ nm). However, for the majority of macroscopic properties investigated, such translations are undetectable, whereas rotations and other symmetry operations (orientation of cleavage planes, optical birefringence, ...) are easily detected. This means that as far as their macroscopic properties are concerned, crystals can be thought of as homogeneous anisotropic substances. If therefore we eliminate translation operations from the 230 space groups, the number of groups reduces to 32, called point groups. Among these, we note

- n ($n = 1$, 2, 3, 4, 6) groups generated by rotation through an angle $\alpha = 2\pi/n$, containing n rotations through angles $m\alpha$ ($m = 1, \ldots n$) (symmetry axis of order n);

- $\bar{n}$ groups generated by an inverse rotation through $2\pi/n$; in particular the $\bar{1}$ group that comprises inversion and the identity operation (centre of symmetry), and the $\bar{2}$ group, denoted m, which comprises a mirror reflection and the identity operation (plane of symmetry).

The remaining groups are generated by several symmetry operations.

The 32 crystal point groups are classified into seven families, called crystal systems, depending on the nature and number of symmetry axes. All crystals whose point group belongs to a given crystal system have the same lattice symmetry. The following systems can be distinguished:

- triclinic: no symmetry axis;
- monoclinic: one diad axis (2 or $\bar{2}$);
- orthorhombic: three orthogonal diad axes (2 or $\bar{2}$);
- trigonal: one triad axis (3 or $\bar{3}$);
- tetragonal: one tetrad axis (4 or $\bar{4}$);
- hexagonal: one hexad axis (6 or $\bar{6}$);
- cubic: four triad axes (3 or $\bar{3}$).

Bravais showed that these seven crystal systems correspond to 14 types of lattice (Fig. 8.3 and Table 8.2). For some of these, multiple cells are chosen in order to display the symmetry properties of the lattice.

8.2.3   Tensor Quantities

The macroscopic properties of a homogeneous isotropic system are independent of the orientation of its co-ordinate axes. This is the case for the density, which is a scalar property, as well as for the magnetic susceptibility, electrical conductivity, refractive index, ... The same is not true for the properties of a crystal, which can have tensor character, that is, the characteristic values may depend on the orientation of the crystal.

For example, the electrical conductivity of a crystal is described by a symmetric tensor of order 2, denoted $[\sigma]$ or σ_{ij}, and defined by 3×3 numbers, only 6 of which are independent ($\sigma_{ji} = \sigma_{ij}$). It relates the current density $\mathbf{j}$ to the electric field $\mathbf{E}$ through Ohm's law

$$\mathbf{j} = [\sigma]\mathbf{E} \qquad \text{or} \qquad j_i = \sum_j \sigma_{ij} E_j \tag{8.3}$$

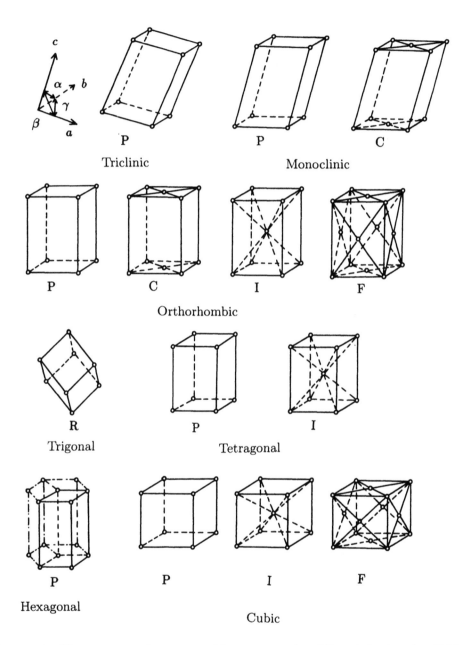

Figure 8.3: The seven crystal systems and fourteen associated Bravais lattices (see Table 8.2).

Table 8.2: The fourteen Bravais lattices and the conventional cells. P or R = primitive cell, I = body-centred, C = (**a**, **b**) base-centred, F = face-centred.

Crystal system	Symbol	Multiplicity	Properties	
Triclinic	P	1	$a \neq b \neq c$	$\alpha \neq \beta \neq \gamma$
Monoclinic	P	1	$a \neq b \neq c$	$\alpha = \gamma = 90° \neq \beta$
	C	2		
Orthorhombic	P	1	$a \neq b \neq c$	$\alpha = \beta = \gamma = 90°$
	C	2		
	I	2		
	F	4		
Trigonal	R	1	$a = b = c$	$\alpha = \beta = \gamma \neq 90°$
Tetragonal	P	1	$a = b \neq c$	$\alpha = \beta = \gamma = 90°$
	I	2		
Hexagonal	P	1	$a = b \neq c$	$\alpha = \beta = 90° \ \gamma = 120°$
Cubic	P	1	$a = b = c$	$\alpha = \beta = \gamma = 90°$
	I	2		
	F	4		

and can be diagonalized. This second order tensor is thus specified by its three eigenvalues σ_{11}, σ_{22} and σ_{33} (principal conductivities) and by the three Euler angles defining the eigenvectors of the tensor (principal axes of the ellipsoid associated with $[\sigma]$) with respect to the primitive vectors of the lattice.

If the crystal possesses symmetry axes, the number of independent parameters is reduced. Depending on the crystalline system, there remain

- 6 for the triclinic system, which has no symmetry axis.
- 4 for the monoclinic system: σ_{11}, σ_{22}, σ_{33} and the angle between the first eigenvector and the **a** axis in the (**a**, **c**) plane, where the second eigenvector is directed along **b**.
- 3 for the orthorhombic system: σ_{11}, σ_{22}, σ_{33}. The three eigenvectors coincide with **a**, **b** and **c**.
- 2 for trigonal, tetragonal and hexagonal systems: $\sigma_{11} = \sigma_{22}$ and σ_{33}. The third axis lies along the symmetry axis (of order 3, 4 or 6 depending on the system), while the other two axes in the perpendicular plane are arbitrary.
- 1 for the cubic system: $\sigma_{11} = \sigma_{22} = \sigma_{33} = \sigma$. In this case, and for tensor quantities of rank 2 only, the crystal appears isotropic; the corresponding properties are therefore independent of direction, just as for a scalar property. Note that if the sample is not a single crystal but an aggregation of microcrystals, or a powder, its properties are isotropic and a measurement of the second order tensor yields the average of the diagonal elements. For the conductivity, this gives

$$\sigma = \frac{\sigma_{11} + \sigma_{22} + \sigma_{33}}{3} \quad ; \tag{8.4}$$

this result is valid for any crystal system.

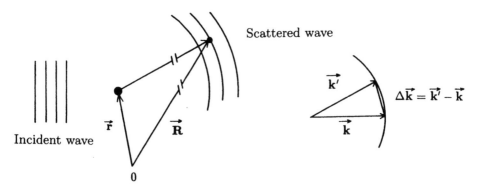

Figure 8.4: Laue scattering

Another second order tensor property is the refractive index of the medium. It has the same symmetry properties as the electrical conductivity. Crystals are called
- biaxial: crystals with 3 different indices (triclinic, monoclinic and orthorhombic).
- uniaxial: crystals with 2 different indices (trigonal, tetragonal and hexagonal).
- isotropic: crystals with 3 equal indices (cubic). All other crystalline properties are defined by tensors of order between 0 and 4. For example,
- 0 (scalar property): heat capacity, density;
- 1 (vector property): electric or magnetic dipole moment;
- 2: thermal and electrical conductivity, permittivity, magnetic susceptibility, refractive index, thermal expansion, sound velocity;
- 3: piezoelectricity tensor;
- 4: elasticity tensor.

Depending on the point group, the number of independent parameters of a given tensor can be reduced. In particular, tensor properties of order 1 and 3 may vanish for certain groups. For instance, a crystal possessing a centre of symmetry cannot be piezoelectric.

8.3   DIFFRACTION OF WAVES BY CRYSTALS

The structure of crystalline solids can be deduced from observations of the diffraction patterns of X-rays, or of beams of electrons or neutrons whose wavelength ($\lambda = h/p$ for material particles) is of the order of interatomic distances, $\lambda \sim 1$ Å.

8.3.1   von Laue Theory (1912)

When a plane wave of frequency ω, wave vector $\mathbf{k}$ ($|\mathbf{k}| = 2\pi/\lambda$) and amplitude

$$E(\mathbf{r}) = E_0 \exp\left[i(\omega t - \mathbf{k}.\mathbf{r})\right] \tag{8.5}$$

impinges on an isolated atom at $\mathbf{r}$, the atom scatters a wave of the same frequency. At large distances, this wave is spherical, and its amplitude at $\mathbf{R}$ (Fig. 8.4) is

$$e(\mathbf{R}) \;=\; E(\mathbf{r})\, f(\mathbf{k}')\, \frac{e^{ik\,|\mathbf{R}-\mathbf{r}|}}{|\mathbf{R}-\mathbf{r}|}$$

$$=\; \frac{E_0}{R} e^{i(\omega t - kR)} \times f(\mathbf{k}') \times e^{i(\mathbf{k}'-\mathbf{k})\cdot\mathbf{r}} \;\; ; \tag{8.6}$$

in this expression, $\mathbf{k}'$ is a vector lying in the direction of observation, with the same magnitude as $\mathbf{k}$, while $f(\mathbf{k}')$ is the *scattering factor*, which depends on the direction of observation and on the nature of the atom.

If the plane wave falls on a crystal, the total amplitude diffracted by all the atoms in the crystal is the sum of the amplitudes scattered by each atom. It is given by

$$e(\mathbf{R}) = \frac{E_0}{R} e^{i(\omega t - kR)} \times F(\mathbf{k},\ \mathbf{k}') \tag{8.7}$$

where

$$F(\mathbf{k},\ \mathbf{k}') = \sum_{\mathbf{r}} f_{\mathbf{r}}(\mathbf{k}') e^{i\,\Delta\mathbf{k}\cdot\mathbf{r}} \qquad (\Delta\mathbf{k} = \mathbf{k}' - \mathbf{k}) \ . \tag{8.8}$$

The factor F, found by summing over all positions $\mathbf{r}$ defined by (8.2), is called the *structure factor* of the crystal and determines the intensity diffracted by the crystal in a given direction. On substituting (8.2), it can be separated into two factors

$$F = \sum_{n_1,\, n_2,\, n_3} e^{i\,\Delta\mathbf{k}\cdot(n_1\mathbf{a}+n_2\mathbf{b}+n_3\mathbf{c})} \times \sum_{\alpha} f_\alpha(\mathbf{k}') e^{i\,\Delta\mathbf{k}\cdot\mathbf{r}_\alpha} = F_r \times F_m \ . \tag{8.9}$$

the first of which, F_r, depends only on the lattice, while the second, F_m, involves the cell. We examine these two terms separately.

8.3.2   Reciprocal Lattice

The term F_r describing the lattice can be separated into three factors of the form

$$\sum_{n_1} \exp(in_1\,\Delta\mathbf{k}.\mathbf{a}),$$

each of which contains a large number of terms. If now $\Delta\mathbf{k}.\mathbf{a}$ is equal to $2\pi h_1$, where h_1 is an integer, then this factor is non zero. In all other cases, the phase of each term in the sum is different, and F_r vanishes.

It follows that the crystal diffracts only along certain directions $\mathbf{k}' = \mathbf{k} + \Delta\mathbf{k}$ such that

$$\Delta\mathbf{k}.\mathbf{a} = 2\pi h_1 \ ; \quad \Delta\mathbf{k}.\mathbf{b} = 2\pi h_2 \ ; \quad \Delta\mathbf{k}.\mathbf{c} = 2\pi h_3 \ , \tag{8.10}$$

where h_1, h_2 and h_3 are integers. These three conditions can be solved for $\Delta\mathbf{k}$ to yield

$$\Delta\mathbf{k} = h_1\mathbf{A} + h_2\mathbf{B} + h_3\mathbf{C} \tag{8.11}$$

in which

$$\mathbf{A} = \frac{2\pi}{\tau}\,\mathbf{b}\times\mathbf{c} \ ; \quad \mathbf{B} = \frac{2\pi}{\tau}\,\mathbf{c}\times\mathbf{a} \ ; \quad \mathbf{C} = \frac{2\pi}{\tau}\,\mathbf{a}\times\mathbf{b} \tag{8.12}$$

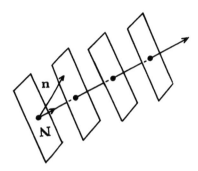

Figure 8.5: A family of lattice planes

where $\tau = (\mathbf{a} \times \mathbf{b}).\mathbf{c}$ is the volume of the lattice cell under consideration. Diffraction therefore occurs only when the vector $\Delta\mathbf{k}$ connects two sites of a lattice that is constructed from the vectors $\mathbf{A}$, $\mathbf{B}$, $\mathbf{C}$ and is called the reciprocal lattice of the crystal.

8.3.3 Cell Structure Factor

All the allowed diffraction directions can be found from the factor F_r discussed above. The intensity of the wave diffracted in those directions, however, is defined by the structure factor F_m, since the $f_\alpha(\mathbf{k}')$ involve the atoms of the cell. The existence of symmetry elements in the cell can give rise to vanishing intensities in certain directions that are allowed by the lattice.

8.3.4 Properties of the Reciprocal Lattice

Consider a real lattice and its reciprocal lattice. $\mathbf{N}$ is a vector defining a point in the reciprocal lattice

$$\mathbf{N} = h_1\mathbf{A} + h_2\mathbf{B} + h_3\mathbf{C}$$

such that the integers h_1, h_2 and h_3 have no common factor. We form the scalar product of this vector with the set of vectors $\mathbf{n}$ that define all the sites in the direct lattice (8.1), i.e.

$$
\begin{aligned}
\mathbf{N}.\mathbf{n} &= (h_1\mathbf{A} + h_2\mathbf{B} + h_3\mathbf{C})(n_1\mathbf{a} + n_2\mathbf{b} + n_3\mathbf{c}) \\
&= 2\pi(h_1n_1 + h_2n_2 + h_3n_3) .
\end{aligned}
$$

Since h_1, h_2 and h_3 have no common factor, a theorem in arithmetic states that as n_1, n_2 and n_3 vary, the expression in brackets adopts all integer values. It follows that all the sites in the real lattice are projected into $\mathbf{N}$ as equally spaced points (Fig. 8.5); these lie in planes perpendicular to $\mathbf{N}$, separated by

$$d = 2\pi/|\mathbf{N}| . \tag{8.13}$$

These planes make up the family of lattice planes denoted $(h_1\ h_2\ h_3)$, with Miller indices h_1, h_2 and h_3, and whose equation is

$$h_1x_1 + h_2x_2 + h_3x_3 = n \qquad n = 0, \pm1, \pm2, \ldots$$

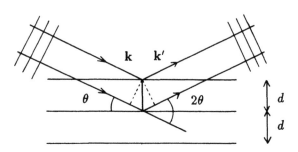

Figure 8.6: Bragg scattering

in the co-ordinate system **a**, **b**, **c**. They intersect the crystal axes at points whose abscissas are n/h_1, n/h_2 and n/h_3 respectively and they are perpendicular to the $[h_1\ h_2\ h_3]$ direction of the reciprocal lattice.

The notion of reciprocal lattice is very important in solid state physics. As the vectors **A**, **B** and **C** have the dimensions of a wave vector, they enter into the description of wave propagation in crystals. The Wigner-Seitz cell of the reciprocal lattice, called the first Brillouin zone, plays an essential role, e.g. in electron band theory (Ch. 11).

8.3.5 Bragg Condition

W.L. Bragg (1912) gave a simple explanation for the diffraction condition (8.11). Consider a family of lattice planes $(h_1\ h_2\ h_3)$ and suppose that an incident wave is reflected at each plane as in a mirror (Fig. 8.6). The reflected wave will have non zero intensity only if the contributions from two consecutive planes are in phase, that is if the Bragg condition holds

$$2d\sin\theta = n\lambda \ . \tag{8.14}$$

This condition, which can also be written

$$|\Delta\mathbf{k}| = 2k\sin\theta = \frac{4\pi}{\lambda}\sin\theta = \frac{2\pi}{d}n \ ,$$

shows that $\Delta\mathbf{k}$, which is orthogonal to the lattice plane, is equal to $n\mathbf{N}$, where $\mathbf{N} = h_1\mathbf{A} + h_2\mathbf{B} + h_3\mathbf{C}$. $\Delta\mathbf{k}$ is therefore a vector of the reciprocal lattice.

It can be seen from the Bragg formula that for a given family of lattice planes diffraction occurs only in a small number of crystal orientations. It follows that when the crystal is oriented at random, there is no diffraction. This explains why methods such as crystal rotation or powder diffraction are necessary to obtain diffraction patterns from crystals.

8.4 VIBRATIONS IN CRYSTALS. PHONONS

8.4.1 Monatomic Crystals in One Dimension

We start by investigating a one dimensional model of a crystal that highlights the basic features of vibrational movements in real crystals.

Line of atoms

Consider a linear crystal (line of atoms) composed of N identical interacting atoms of mass m. Their binary interaction potential is $\epsilon(r)$. At equilibrium, the N atoms are equally spaced with a separation $a = L/(N-1) \simeq L/N$, where L is the length of the row. Their abscissas are thus $x_0 + na$ (Fig. 8.7). When they vibrate, the position of each atom is

$$x_n = x_0 + na + u_n \qquad (8.15)$$

where u_n is the displacement of the atom from its equilibrium position and $\dot{x}_n = \dot{u}_n$ is its velocity. The kinetic energy and potential energy of the line are equal to

$$T = \frac{1}{2} m \sum_n \dot{u}_n^2 \quad \text{and} \quad E = \frac{1}{2} \sum_{n,n' \neq n} \epsilon(|x_{n'} - x_n|) \qquad (8.16)$$

where n and n' are two integers whose values lie between 1 and N. The factor $1/2$ is inserted to avoid counting the interactions twice. The potential energy can be written in the form

$$E = \sum_{n,n'>n} \epsilon(x_{n'} - x_n)$$

$$\text{or} \qquad E = \sum_{n,p>o} \epsilon(x_{n+p} - x_n) = \sum_{n,p>0} \epsilon(pa + u_{n+p} - u_n) \ . \qquad (8.17)$$

In this expression, p is a positive integer that will be varied from 1 to infinity, given that $\epsilon(r)$ quickly tends to zero with increasing r. This approximation is valid in the thermodynamic limit $N \to \infty$ where edge effects are negligible.

Harmonic approximation

If the atoms do not move far from their equilibrium positions, i.e., if $|u_n| \ll a$, the function ϵ in (8.17) can be expressed as a Taylor expansion around $x = pa$. The potential energy is then

$$E = E^{(0)} + E^{(1)} + E^{(2)} + \dots$$

with

$$E^{(0)} = \sum_{n,p} \epsilon(pa) = N \sum_p \epsilon(pa) \qquad (8.18)$$

$$E^{(1)} = \sum_{n,p} \epsilon'(pa)(u_{n+p} - u_n)$$

$$E^{(2)} = \frac{1}{2} \sum_{n,p} \epsilon''(pa)(u_{n+p} - u_n)^2 \ . \qquad (8.19)$$

The term $E^{(0)}$ is the energy of the line of atoms at equilibrium. The term $E^{(1)}$ is zero, since for constant p, each displacement u appears twice in the sum over n, with opposite signs. This corresponds to the fact that the energy is minimum at equilibrium ($u_n = 0$). The term $E^{(2)}$ is quadratic in u_n and hence to this order (harmonic approximation), the atoms vibrate like coupled harmonic oscillators.

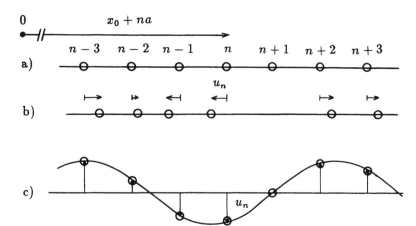

Figure 8.7: Line of atoms. a) Equilibrium. b) Vibration. c) Longitudinal displacements represented on a perpendicular axis (magnification 2).

Normal co-ordinates

To find the independent harmonic oscillators, the quadratic expression (8.19) for $E^{(2)}$ must be diagonalized. The following quantities are therefore introduced

$$Q_k = \frac{1}{\sqrt{N}} \sum_n u_n e^{inka} \qquad (8.20)$$

where k is a real number having the dimensions of a wave number. These quantities obey the relations

$$Q_k^* = Q_{-k} \qquad \text{and} \qquad Q_{k+2\pi/a} = Q_k \quad . \qquad (8.21)$$

We apply the change of variables (8.20) to transform the N displacements u_n into N quantities Q_k such that

$$k = \frac{2\pi}{Na}p \quad \text{with} \quad -\frac{N}{2} < p \le \frac{N}{2} \quad . \qquad (8.22)$$

This corresponds to N regularly spaced values of k between $-\pi/a$ and π/a (Born-von Kármán boundary conditions). These N quantities Q_k, called normal co-ordinates, are complex. However, in view of (8.21), they yield N real quantities that are independent. Using the mathematical relations for $N \to \infty$

$$\sum_n e^{in(k-k')a} = N\delta_{kk'} \qquad \text{and} \qquad \sum_k e^{i(n-n')ka} = N\delta_{nn'} \quad , \qquad (8.23)$$

we find

$$u_n = \frac{1}{\sqrt{N}} \sum_k Q_k e^{-inka} \quad . \qquad (8.24)$$

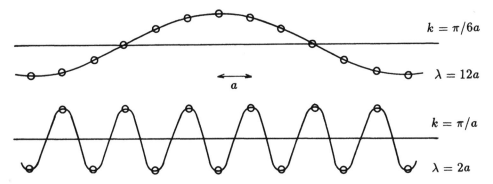

Figure 8.8: Elastic waves for two values of k. The displacement of the atoms is depicted along a transverse axis.

The kinetic and potential energies can then be written

$$T = \frac{m}{2} \sum_n \dot{u}_n^2 = \frac{m}{2} \sum_k \dot{Q}_k \dot{Q}_k^*$$

and

$$E - E^{(0)} = \frac{1}{2} \sum_{np} \epsilon''(pa)(u_{n+p} - u_n)^2 = \frac{m}{2} \sum_k \omega_k^2 Q_k Q_k^*$$

where we have set

$$m\omega_k^2 = \sum_p \epsilon''(pa)|1 - e^{ipka}|^2 = 2 \sum_p \epsilon''(pa)(1 - \cos pka) \quad . \tag{8.25}$$

The total energy of the vibrating crystal is then

$$E_t = T + E = E^{(0)} + \sum_k \left(\frac{1}{2} m \dot{Q}_k \dot{Q}_k^* + \frac{1}{2} m \omega_k^2 Q_k Q_k^* \right) \quad . \tag{8.26}$$

This expression describes the energy of a set of N independent harmonic oscillators each having angular frequency equal to ω_k. It follows that each Q_k varies sinusoidally in time, i.e.

$$Q_k(t) = A_k e^{i\omega_k t} + B_k e^{-i\omega_k t} \tag{8.27}$$

where A_k and B_k are complex constants.

Elastic waves

Equations (8.24 and 27) show that the displacement of atom n takes the form

$$u_n = \frac{1}{\sqrt{N}} \sum_k \left[A_k e^{i(\omega_k t - nka)} + B_k e^{-i(\omega_k t + nka)} \right] \quad . \tag{8.28}$$

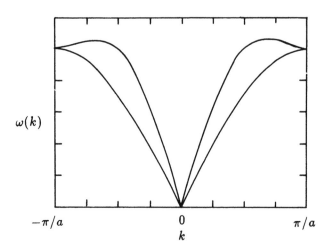

Figure 8.9: Dispersion curve for a line of atoms. The lower curve corresponds to inter-actions between nearest neighbours only, while the upper curve includes second nearest neighbours with $\epsilon''(2a)/\epsilon''(a) = 0.5$.

The displacements of all the atoms are therefore a superposition of progressive longitudinal waves called elastic waves (Fig. 8.8), such that

$$u_n = \frac{A_k}{\sqrt{N}} \exp\left[i(\omega_k t - nka)\right] \qquad (8.29)$$

where k is the wave vector and ω_k the angular frequency. The function $\omega_k = \omega(k)$ (8.25), called the *dispersion relation*, is sketched in Figure 8.9. When $k \to 0$, i.e. $ka \ll 1$, this becomes

$$\omega(k) = ck \qquad \text{with} \qquad c^2 = \frac{a^2}{m} \sum_p \epsilon''(pa)p^2 \; . \qquad (8.30)$$

This low frequency region lies in the domain of sound waves. As the phase velocity c_ϕ and the group velocity c_g of a wave are defined by

$$c_\phi = \frac{\omega}{k} \qquad \text{and} \qquad c_g = \frac{d\omega}{dk} \; , \qquad (8.31)$$

it can be seen that the sound waves propagate with phase and group velocities that are independent of frequency and equal to c, the velocity of sound (8.30). Their wavelength, $\lambda = 2\pi/k$, is then much larger than a. Observe that in the limit $k = 0$, the atoms move in phase, which corresponds to a translation of the crystal as a whole.

When $k \to \pm\pi/a$, ω tends to a non zero limiting value, and neighbouring atoms vibrate in opposite phase (Fig. 8.8) at a frequency that is comparable to optical frequencies ($\omega \sim 10^{14}$ Hz). For this value of k, the group velocity is zero and propagation no longer occurs.

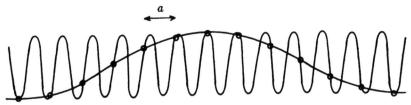

Figure 8.10: Atomic displacements generated by two sinusoidal waves, $k = \pi/6a$ ($\lambda = 12a$) and $k = \pi/6a + 2\pi/a$ ($\lambda = 12a/13$). The wave selected satisfies $|k| < \pi/a$.

The N values of k make up a discrete sequence (8.22) between $-\pi/a$ and $+\pi/a$. However, since N is very large, k can be considered as a continuous variable. The sums over k transform into an integral, where

$$\sum_k \to \int_{-\pi/a}^{+\pi/a} g(k)\, dk \qquad \text{with} \qquad g(k) = \frac{Na}{2\pi} = \frac{L}{2\pi}\ . \tag{8.32}$$

The function $g(k)$, which is a constant here, is the density of k values between $-\pi/a$ and π/a. The values of k that contribute to the integrals correspond to wavelengths in the range between $\lambda = 2a$ ($|k| = \pi/a$) and $\lambda = \infty$ ($k = 0$). Waves of length $\lambda < 2a$ have no real existence, since they can always be interpreted as having a wavelength greater than $2a$ (Fig. 8.10); replacement of k by $k + 2\pi/a$ in (8.29) produces the same physical displacements u_n.

Phonons

At the atomic scale, systems must be treated quantum mechanically, which means that the energy is quantized. For a harmonic oscillator of eigenfrequency ω, the energy is

$$\epsilon = \left(n + \frac{1}{2}\right)\hbar\omega\ , \qquad n = 0,\ 1,\ 2,\ \dots$$

where $\hbar$ is the reduced Planck constant $h/2\pi$. Hence, the energy of a crystal (8.26) must be

$$E_t = E^{(0)} + \sum_k \left(n_k + \frac{1}{2}\right)\hbar\omega_k. \tag{8.33}$$

The N quantum numbers n_k are zero or positive integers that define the vibrational state of the one-dimensional crystal. This energy, which varies through creation or annihilation of quanta of energy $\epsilon_k = \hbar\omega_k$, can also be written

$$E_t = U_0 + \sum_k n_k \epsilon_k \qquad \text{with} \qquad U_0 = E^{(0)} + \frac{1}{2}\sum_k \hbar\omega_k\ . \tag{8.34}$$

In this form, the energy can be understood as being the sum
  • of a term U_0 called the cohesive energy, which is equal to the energy of the unexcited crystal ($n_k = 0$) (crystal at absolute zero),

• plus a term $\sum n_k \epsilon_k$, which can be described as the energy of a gas of n_k quanta of energy ϵ_k.

By analogy with photons, which are electromagnetic wave quanta, the energy quanta associated with elastic waves are called *phonons*. The physical existence of phonons has been demonstrated by observations of inelastic neutron scattering in crystals, where energy changes are detected corresponding to the creation or annihilation of phonons.

The general properties of phonons are as follows:

• in the harmonic approximation (§8.4.1), there are no interactions. These make their appearance only at higher order (§8.7).

• their total number

$$N_{ph} = \sum_k n_k$$

is indefinite because they can be created and annihilated just like photons.

• they obey Bose-Einstein statistics, since for each value of k in the quantization, there are two mutually adjoint operators, a_k (annihilation operator) and a_k^+ (creation operator). These obey the commutation relation for bosons

$$[a_k, \ a_{k'}^+] = \delta_{kk'} \ . \tag{8.35}$$

These operators possess the property of transforming a crystal state with n_k phonons into states with $n_k - 1$ and $n_k + 1$ phonons respectively, and are such that the eigenvalues of the operator $N_k = a_k^+ a_k$ are the numbers n_k. The quantum mechanical Hamiltonian can then be written

$$H = U_0 + \sum_k N_k = U_0 + \sum_k a_k^+ a_k \ . \tag{8.36}$$

• their density of states in phase space $(x, \ p = \hbar k)$ is

$$g(x, \ p) \ dx \ dp = \frac{dx \ dp}{h} \ . \tag{8.37}$$

On integrating over x $(dx \rightarrow L)$ and setting $p = \hbar k$, we find

$$g(k) \ dk = \frac{L}{2\pi} \ dk$$

in agreement with (8.32). It should be pointed out that p is not a momentum in the usual sense. In particular, it is not a conserved quantity (§8.7).

The concept of phonons has been very fruitful. By replacing a discussion in terms of waves by one in terms of particles, it shows up more clearly the quantization of energy. Other particles, quasi-particles, or even excitations have also been introduced into solid state physics, such as magnons (§13.4.5), excitons, plasmons, rotons (§7.4).

8.4.2 Monatomic Crystals in Three Dimensions

The discussion of the above paragraph can be generalized to real monatomic crystals. The atoms in the crystal now undergo three-dimensional vibrations such that

$$u_n \rightarrow u_{n_1 n_2 n_3} \equiv \mathbf{u_n} \ .$$

The expression for the total energy in the harmonic approximation contains N three-dimensional coupled oscillators. As in paragraph 8.4.1, we introduce N vectors $\mathbf{Q_k}$. These are linear combinations of the $\mathbf{u_n}$ associated with the N regularly spaced vectors $\mathbf{k}$ in the Wigner-Seitz cell (§8.2.1) of the reciprocal lattice (first Brillouin zone). The N vectors $\mathbf{Q_k}$ represent N independent three-dimensional oscillators. For each value of $\mathbf{k}$, however, the three components of $\mathbf{Q_k}$ are coupled. Nevertheless, three linear combinations of the components exist, $Q_{\mathbf{k}s}$ ($s = 1$, 2, 3), giving three independent one-dimensional harmonic oscillators with angular frequency $\omega_s(\mathbf{k})$. The directions of vibration of the atoms in these modes are almost parallel to $\mathbf{k}$ (longitudinal vibrations L) or almost perpendicular to $\mathbf{k}$ (transverse vibrations T_1 and T_2).

Figure 8.11a shows the three functions $\omega_L(\mathbf{k})$, $\omega_{T1}(\mathbf{k})$ and $\omega_{T2}(\mathbf{k})$ in copper for 3 directions of $\mathbf{k}$. In certain directions, the transverse branches T_1 and T_2 are degenerate (T) for reasons of symmetry. At small values of k (long wavelengths), the angular frequency $\omega_s(\mathbf{k})$ is proportional to $\mathbf{k}$

$$\omega_s(\mathbf{k}) = c_s(\mathbf{k})\mathbf{k} \ , \tag{8.38}$$

where, in good agreement with experiment, the proportionality constant $c_s(\mathbf{k})$ can be identified with the velocity of sound in the direction $\mathbf{k}$ for the branch $s = L$, T_1, T_2.

Quantization of the normal modes $\mathbf{Q_k}$ gives rise to phonons of momentum $\hbar\mathbf{k}$ and energy $\hbar\omega_s(\mathbf{k})$. These phonons obey Bose-Einstein statistics and their number is indeterminate. Their density of states is

$$g(\mathbf{r}, \ \mathbf{p}) \, d^3\mathbf{r} \, d^3\mathbf{p} = 3\frac{d^3\mathbf{r} \, d^3\mathbf{p}}{h^3} = \frac{3}{(2\pi)^3} \, d^3\mathbf{r} \, d^3\mathbf{k} \tag{8.39}$$

where the factor 3 corresponds to the three directions of polarization. The values of $\mathbf{k}$ to be used lie inside the Brillouin zone (Fig. 8.11b) and the total number of states is

$$\int g(\mathbf{r}, \ \mathbf{p}) \, d^3\mathbf{r} \, d^3\mathbf{p} = 3N \ . \tag{8.40}$$

In terms of the variable ω, it can be shown that the density of states is

$$g(\omega) \ = \ \frac{V}{(2\pi)^3} \sum_s \int d^3\mathbf{k} \, \delta \left(\omega - \omega_s(\mathbf{k}) \right) \tag{8.41}$$

with
$$\int g(\omega) \, d\omega = 3N \ . \tag{8.42}$$

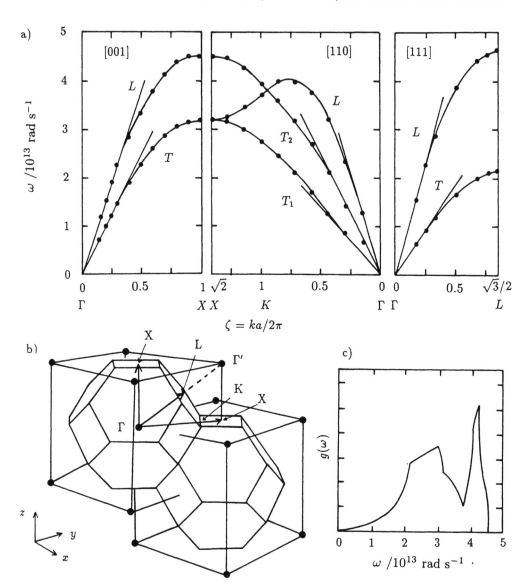

Figure 8.11: a) Dispersion curves $\omega(k)$ for phonons in copper in the three directions [001], [110] and [111]. The edge of the face-centred cubic cell in copper is $a = 3.61$ Å. The straight lines $\omega = ck = 2\pi c\zeta/a$ are shown, the values of the velocity of sound $c \equiv c_s(k)$ in each direction and each branch L, T (or T_1, T_2) being calculated from the elastic constants of copper. b) Sketch of the Brillouin zones in copper found from the cubic centred reciprocal lattice. The directions [001] (Γ X), [110] (ΓKX) and [111] (ΓLΓ') are shown. c) Densities of states $g(\omega)$ in copper (arbitrary units) determined from the experimental data of a) by measurements of scattering from a monochromatic neutron beam [E.C. Svensson et al., *Phys. Rev.* **155**, 619 (1967)].

Figure 8.11c shows a typical density of states $g(\omega)$, measured experimentally for copper. The distribution, which is continuous up to a maximum frequency, displays van Hove discontinuities, and at small ω varies as ω^2. From (8.41) it can be shown that

$$g(\omega) = \frac{V}{2\pi^2} \frac{3\omega^2}{\bar{c}^3} \quad \text{with} \quad \frac{3}{\bar{c}^3} = \sum_s \frac{1}{4\pi} \int \frac{d\Omega}{[c_s(\mathbf{k})]^3} \ . \tag{8.43}$$

The mean velocity $\bar{c}$ defined in this way has been determined for copper (Fig. 8.11c). It is equal to $\bar{c} = 2535$ m s^{-1}. Since it is classical in origin, equation (8.41) for $g(\omega)$ does not contain $\hbar$.

Exercise 8.1 *Density of states in the Debye model*
  P. Debye assumed that the atoms in isotropic solids oscillate as waves with a constant phase velocity $c_\phi = \omega/k$ that is equal to the velocity of sound waves, i.e., c_L for longitudinal waves and c_T for transverse waves. Also, since the number of waves must be limited to $3N$, Debye assumed that the two types of wave have the same cut-off frequency ω_D such that

$$\int_0^{\omega_D} g(\omega) \, d\omega = 3N \ . \tag{8.44}$$

1 From the density of states (8.39) derive the function $g(\omega)$.
2 Calculate the cut-off frequency ω_D.
3 Measurements in copper yield $c_L = 4760$ m s^{-1} and $c_T = 2325$ m s^{-1}. Calculate the mean velocity $\bar{c}$ (8.43) and the cut-off frequency ω_D. The molar mass of copper is $M = 63.54$ g mol^{-1} and its density at 0 K is $\rho = 9.018$ g cm^{-3}.

Solution
  1 To find $g(\omega)$, we must first integrate (8.39) over $\mathbf{r}$ and over the directions of $\mathbf{k}$, and then apply the change of variable $k \rightarrow \omega$. This change depends on the type of wave. For longitudinal and transverse waves respectively,

$$\omega = c_L k \qquad \text{and} \qquad \omega = c_T k \ .$$

This gives

$$g(\mathbf{r}, \mathbf{p}) \, d^3\mathbf{r} \, d^3\mathbf{p} \rightarrow \frac{3}{(2\pi)^3} V \times 4\pi k^2 \, dk = \frac{V}{2\pi^2} \left[\frac{\omega^2 \, d\omega}{c_L^3} + 2 \frac{\omega^2 \, d\omega}{c_T^3} \right] \ ,$$

and hence the density of modes is

$$g(\omega) = \frac{V}{2\pi^2} \frac{3}{\bar{c}^3} \omega^2 \quad \text{with} \quad \frac{3}{\bar{c}^3} = \frac{1}{c_L^3} + \frac{2}{c_T^3} \ . \tag{8.45}$$

This density (cf. Fig. 8.15) can be found directly from (8.41) by setting $\omega_s(\mathbf{k}) = c_s|\mathbf{k}|$. It is an extrapolation of the limiting analytical form (8.43), which is valid at low frequencies, over the whole frequency range.

2 On integrating $g(\omega)$ from 0 to ω_D, the normalization condition (8.42) becomes

$$3N = \frac{V}{2\pi^2} \frac{3}{\bar{c}^3} \int_0^{\omega_D} \omega^2 \, d\omega = \frac{V}{2\pi^2} \frac{\omega_D^3}{\bar{c}^3} \quad ,$$

which gives for the cut-off frequency

$$\omega_D = \bar{c} \left(6\pi^2 \frac{N}{V} \right)^{1/3} \quad . \tag{8.46}$$

3 The numerical value of $\bar{c}$ is $\bar{c} = 2612$ m s^{-1}. This value is fairly close to that found from neutron scattering, $\bar{c} = 2535$ m s^{-1}. Since $N/V = \mathcal{N}\rho/M$, it follows that $\omega_D = 4.49 \times 10^{13}$ rad s^{-1}. The cut-off frequency is similar to the maximum frequency in the spectrum of Figure 8.11. It can be seen that the wavelengths corresponding to this frequency ω_D, namely $\lambda = 2\pi c_s/\omega_D$, are respectively equal to 3.24 Å (L) and 6.60 Å (T) and are close to $\lambda = 2d$, where $d = a/\sqrt{2} = 2.55$ Å is the nearest neighbour distance.

8.4.3 Polyatomic Crystals in Three Dimensions

Let p be the number of atoms in the primitive cell of a polyatomic crystal. With N primitive cells, such a crystal has $3pN$ degrees of freedom and therefore $3pN$ vibration modes distributed in $3p$ branches (Fig. 8.12 and Problem 8.1). Three of these branches are acoustic, that is, they have the same properties as the three acoustic branches in the monatomic crystal (§8.4.2). In particular, when $k \to 0$, we have $\omega = ck$ and the different atoms of the elementary cell vibrate in phase as if the medium were continuous.

For the $3(p-1)$ other branches, as $k \to 0$, then $\omega(k) \to$ constant, the atoms belonging to the same cell moving relative to each other. If the atoms are electrically charged, such vibration modes create an oscillating electric dipole moment and can be excited by electromagnetic waves, which is the reason they are called optical branches.

The density of normal modes $g(\omega)$ is still given by (8.41), where the sum over s now contains $3p$ terms. As $\omega \to 0$, only the three acoustic branches contribute, and the limiting form (8.43) remains valid. Note that for the optical branches, $\omega_s(\mathbf{k}) \simeq \omega_s$ varies relatively little, and equation (8.41) indicates that the contribution to $g(\omega)$ from these branches is approximately

$$g_s(\omega) = N\delta(\omega - \omega_s) \quad . \tag{8.47}$$

The Einstein model described earlier (§3.6) is then applied to the optical branches, each containing N oscillators of frequency ω_s.

8.5   EXPERIMENTAL THERMODYNAMIC PROPERTIES OF SOLIDS

Experimental measurements of the thermodynamic properties of solids encompass
  • the adiabatic compressibility $\chi_S(T, P) = -(\partial V/\partial P)_S/V$,

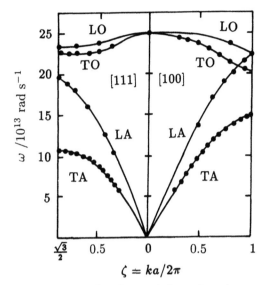

Figure 8.12: Dispersion curves for diamond ($p = 2$ carbon atoms per primitive cell, $a = 3.57$ Å) for the two directions [100] and [111]. In these directions, the transverse branches T are pair-wise degenerate [J.L. Warren, R.G. Wenzel and J.L. Yarnell, *Inelastic Scattering of Neutrons* (Vienna, IAEA, 1965)].

- the thermal expansion coefficient $\alpha(T, P_0) = (\partial V/\partial T)_{P_0}/V = 3(\partial L/\partial T)_{P_0}/L$,
- the molar heat capacity at constant pressure $c_P(T, P_0)$, where P_0 is generally taken as being atmospheric pressure. From these three quantities, all the thermodynamic functions can be found. Note that the molar heat capacity at constant volume c_V is given by

$$c_P - c_V = Tv\frac{\alpha^2}{\chi_T},\tag{8.48}$$

and the isothermal compressibility, $\chi_T = -(\partial V/\partial P)_T/V = \chi_S c_V/c_P$, is given by

$$\chi_T - \chi_S = Tv\frac{\alpha^2}{c_P}$$

(v being the molar volume).   For the different solids, the curves of $\chi \equiv \chi_T$, α and c_V are similar (Fig. 8.13):
- the isothermal compressibility varies weakly with pressure $((\partial\chi/\partial P)/\chi \sim 10^{-3}$ atm^{-1}) and temperature;
- the coefficient of thermal expansion α and the heat capacity c_V increase rapidly from zero, but their variation at room temperature is small. The shapes of $\alpha(T)$ and $c_V(T)$ are similar.
Experimental observations establish the following general behaviour:
- the Grüneisen ratio

$$\gamma = \frac{v\alpha}{\chi_T c_V} = \frac{v\alpha}{\chi_S c_P}\tag{8.49}$$

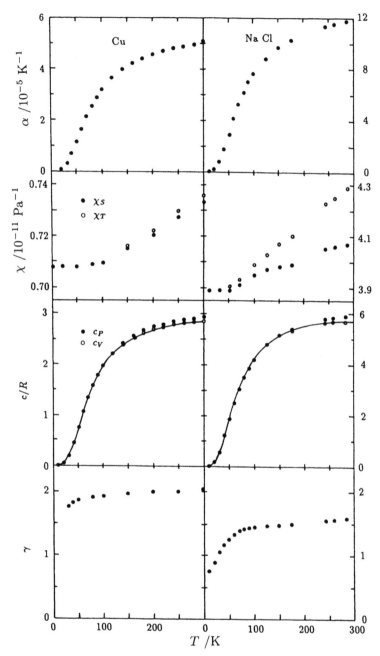

Figure 8.13: Thermal expansion coefficient α, compressibility χ, heat capacity c and Grüneisen constant γ in copper and sodium chloride [V.J. Johnson (Natl. Bur. Stand.), *Properties of Materials at Low Temperatures*, Pergamon Press (1961); J.A. Rayne, *Phys. Rev.* **115**, 63 (1959); R.O. Simmons and R.N. Balluffi, *Phys. Rev.* **108**, 278 (1957); P.P.M. Meincke and G.M. Graham, *Proc.* 8th *Int. Conf. Low Temp. Phys.*, London, 1962, 401-412]. The curves show the heat capacity c_V in the Debye model for Θ_D (Cu) = 315 K and Θ_D (NaCl) = 280 K.

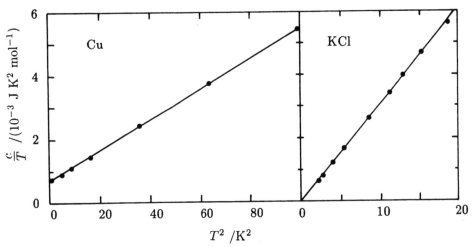

Figure 8.14: Molar heat capacity at low temperature in copper and in potassium chloride [V.J. Johnson (Natl. Bur. Stand.), *Properties of Materials at Low Temperatures*, Pergamon Press (1961); P.H. Keesom and N. Pearlman, *Phys. Rev.* **91**, 1354 (1953)].

varies weakly with temperature and its value is close to 2 for various substances (Grüneisen law, 1908);

 • at a sufficiently high temperature the molar heat capacity c_V is approximately $3R \times n$, where n is the number of atoms per molecule (Dulong and Petit law, §3.5.2);

 • at low temperatures, the heat capacity c_V varies as T^3 for non metallic solids (or dielectrics). An additional linear term in T arises in the case of metals (Fig. 8.14). As will be seen, this is due to the presence of free electrons in metals (§10.3.4).

8.6 STATISTICAL INTERPRETATION OF PROPERTIES OF SOLIDS

8.6.1 Thermodynamic Functions

In paragraph 8.4.1 we saw that the energy of a crystal is the sum of its energy at absolute zero U_0 (cohesive energy) and the energy of a gas of phonons, that is, particles that are indeterminate in number and that obey Bose-Einstein statistics. The free energy is also the sum of two terms

$$F = F_0(V) + F_{\text{ph}}(T, V) \tag{8.50}$$

where the cohesive energy $F_0 \equiv U_0$ depends only on the volume, and the free energy of the phonons is, from (7.2)

$$F_{\text{ph}} \equiv \Omega = kT \int g(\omega) \, d\omega \ln \left[1 - \exp \left(-\frac{\hbar\omega}{kT} \right) \right] . \tag{8.51}$$

Note that the dependence of F_{ph} on V enters through the density of states $g(\omega)$. $g(\omega) \, d\omega$ is found from (8.39) or (8.41) and the dispersion relations $\omega(\mathbf{k})$ depend implicitly on the interatomic distance (cf. 8.25).

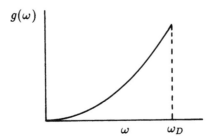

Figure 8.15: Density of states in the Debye model

The other thermodynamic functions are found from F by differentiating with respect to T and V. In particular, the internal energy is equal to

$$
U = F + TS = F - T \left(\frac{\partial F}{\partial T} \right)_V = -T^2 \frac{\partial}{\partial T} \left(\frac{F}{T} \right)_V
$$

$$
= U_0(V) + \int \hbar \omega \frac{g(\omega) \, d\omega}{\exp\left(\frac{\hbar \omega}{kT} \right) - 1} \ . \tag{8.52a}
$$

The phonon contribution is identified with $\sum N_i \epsilon_i$ for a boson gas, and the heat capacity at constant volume is equal to

$$
C_V = k \int \left(\frac{\hbar \omega}{kT} \right)^2 \frac{e^{\beta \hbar \omega}}{(e^{\beta \hbar \omega} - 1)^2} \, g(\omega) \, d\omega \ . \tag{8.52b}
$$

8.6.2 Debye Model

The Debye model was introduced in 1912 as an improvement on the Einstein model (§3.6), long before the densities of states $g(\omega)$ were measured experimentally (Fig. 8.11c). By considering isotropic solids (e.g. a polycrystalline metal), P. Debye extended the dispersion relation $\omega = ck$, which is valid for acoustic waves (small k), to all values of k. He also introduced a cut-off frequency ω_D in order to restrict the number of modes to $3N$. In this model, (Ex. 8.1), the density of states is

$$
g(\omega) = \begin{cases} 3V\omega^2 \, / \, 2\pi^2 \bar{c}^3 = 3N \times 3\omega^2 \, / \, \omega_D^3 & (\omega < \omega_D) \\ 0 & (\omega > \omega_D) \end{cases} \tag{8.53}
$$

with

$$
\omega_D = \bar{c} \left(6\pi^2 \frac{N}{V} \right)^{1/3} \quad \text{and} \quad \frac{3}{\bar{c}^3} = \frac{1}{c_L^3} + \frac{2}{c_T^3} \ ; \tag{8.54}
$$

where c_L and c_T are the velocities of the longitudinal and transverse sound waves, respectively. This density, which satisfies (8.44), is shown in Figure 8.15.

In this model the free energy of the phonons (8.51) is then

$$
F_{\text{ph}} = \frac{3NkT}{\omega_D^3} \int_0^{\omega_D} 3\omega^2 \ln \left[1 - \exp\left(-\frac{\hbar \omega}{kT} \right) \right] \, d\omega \tag{8.55a}
$$

Table 8.3: Debye Function (8.59)

y	$D(y)$	y	$D(y)$	y	$D(y)$
0	0	0.15	0.213	0.5	0.825
0.025	0.00122	0.175	0.293	0.6	0.874
0.05	0.00974	0.2	0.369	0.8	0.926
0.075	0.0328	0.25	0.503	1	0.952
0.1	0.0758	0.3	0.610	1.5	0.980
0.125	0.138	0.4	0.746	2	0.988

$$= 3NkT \times 3 \left(\frac{T}{\Theta_D}\right)^3 \int_0^{x_D} x^2 \ln\left(1 - e^{-x}\right) \, dx \tag{8.55b}$$

where we have set $x = \hbar\omega/kT$ and introduced the Debye temperature

$$\Theta_D = \frac{\hbar\omega_D}{k} = \frac{\hbar}{k}\bar{c}\left(6\pi^2\frac{N}{V}\right)^{1/3} \tag{8.56}$$

where $x_D = \Theta_D/T$. The other thermodynamic functions are found from F. Differentiation of (8.55b) with respect to T yields

$$S = -\left(\frac{\partial F}{\partial T}\right)_V = \frac{9Nk}{x_D^3}\left[\int_0^{x_D}\frac{x^3 \, dx}{e^x - 1} - \int_0^{x_D} x^2 \ln(1 - e^{-x}) \, dx\right] \tag{8.57a}$$

$$U = -T^2\frac{\partial}{\partial T}\left(\frac{F}{T}\right)_V = U_0(V) + \frac{9NkT}{x_D^3}\int_0^{x_D}\frac{x^3 \, dx}{e^x - 1} \tag{8.57b}$$

$$C_V = \left(\frac{\partial U}{\partial T}\right)_V = \frac{9Nk}{x_D^3}\int_0^{x_D}\frac{x^4 e^x \, dx}{(e^x - 1)^2} \ . \tag{8.57c}$$

The molar heat capacity c_V is then

$$c_V = 3R\, D\left(\frac{T}{\Theta_D}\right) \tag{8.58}$$

where the Debye function

$$D(y) = 3y^3 \int_0^{1/y}\frac{x^4 e^x \, dx}{(e^x - 1)^2} \tag{8.59}$$

is listed in Table 8.3 and plotted in Figure 8.16.
    The limiting expressions for $D(y)$ are

$$D(y) = \frac{4\pi^4}{5}y^3 \quad (y \to 0) \quad \text{and} \quad D(y) = 1 - \frac{1}{20y^2} \quad (y \to \infty) \ .$$

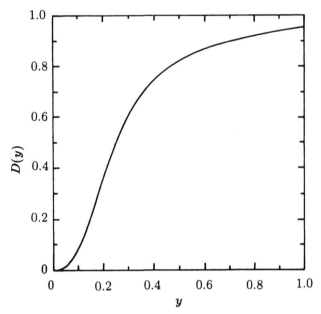

Figure 8.16: Debye function (8.59). For $y > 1$, $D(y) = 1 - 1/20y^2$ to within 10^{-3}.

The Debye model, with its single parameter Θ_D, gives quite a good description of the main characteristics of heat capacities of solids (Fig. 8.13). In particular, it yields the Dulong and Petit law at high temperatures ($T \gtrsim \Theta_D$) and the T^3 behaviour (Fig. 8.14),

$$c_V = \frac{12\pi^4}{5} R \frac{T^3}{\Theta_D^3}, \tag{8.60}$$

at low temperatures ($T \ll \Theta_D$). Table 8.4 lists values of Θ_D found from the T^3 law for several non metallic substances. Tables 8.5 and 10.1 contain values of Θ_D for other substances.

Exercise 8.2 *Heat capacity of copper*
The velocities of sound in copper are respectively $c_L = 4760$ m s^{-1} and $c_T = 2325$ m s^{-1}. Calculate Θ_D and determine the heat capacity of copper at low temperature. Compare with the experimental result c_V /(J K^{-1} mol^{-1})= $6.86 \times 10^{-4}(T$ /K) + 4.76 \times 10^{-5}(T$ /K)^3. Take $M = 63.54$ g mol^{-1} for copper and $\rho = 9.018$ g cm^{-3}.

Solution
Equations (8.56) and (8.54) yield (cf. Ex. 8.1) $\Theta_D = 343$ K. Equation (8.60) then gives $c_V = 4.82 \times 10^{-5}T^3$. The agreement with experiment is satisfactory. The linear term in T comes from the free electrons in copper.

8.6.3   Discussion of the Debye Model

The model accounts for the experimental data even though the density of states used is very crude. The reason for the agreement at low temperatures is understandable, since only the lowest frequencies are excited. In that case the Debye density of states (8.53) adopts the limiting form (8.43) of the real density of states. Similarly, at sufficiently high temperatures, the internal energy (8.52) has the limiting form

$$U = U_0(V) + kT \int g(\omega)\, d\omega = U_0(V) + 3NkT \ ,$$

which reduces to the Dulong and Petit law, $c_V = 3R$, independently of the density of states. The Debye model provides a simple interpolation between these two regions. This is why it gives a satisfactory description of heat capacities using a single parameter Θ_D that is characteristic of the substance.

The Debye model is frequently used in the following way to display experimental heat capacity data (Fig. 8.17). By measuring c_V at each temperature T and solving (8.58) numerically, a temperature Θ_D can be calculated. The curve $\Theta_D(T)$ thus obtained contains the same information as $c_V(T)$. If the Debye model were exact we would expect that $\Theta_D(T)$ =constant. This, however, is never the case, since the Debye density of states (8.53) is too crude. On substituting the true density of states found experimentally (Fig. 8.11c), the phonon model faithfully reproduces the variations of $\Theta_D(T)$, i.e., of $c_V(T)$.

When a crystal cell contains several atoms, in addition to the acoustic branches, there exist optical branches for which the frequency varies relatively little with the wave vector k (Fig. 8.12). From (8.41,42), the density of states for each of these optical branches can then be written

$$g_s(\omega) = N\delta(\omega - \omega_s) \ , \tag{8.61}$$

where ω_s is the mean frequency of the branch. The contribution of an optical branch to the free energy of the phonons is (cf. 8.51)

$$F_s = NkT \ln \left[1 - \exp\left(-\frac{\hbar\omega_s}{kT} \right) \right] \ .$$

In this expression can be seen the free energy (3.32) of the Einstein model for N oscillators of frequency ω_s. The heat capacity of such a crystal can be expressed as the sum of a Debye term for the acoustic branches and of as many Einstein terms (3.34) as there are optical branches.

8.6.4   Equation of State

The equation of state is obtained from the free energy (8.50). We have

$$P(T,\ V) = - \left(\frac{\partial F}{\partial V} \right)_T = P_0(V) + P_{ph}(T,\ V) \tag{8.62}$$

where $P_0 = -dF_0/dV$ is the pressure at absolute zero caused by interactions between atoms, while $P_{ph} = -(\partial F_{ph}/\partial V)_T$ is a term related to thermal agitation, which in general is much smaller than P_0.

Table 8.4: Debye temperatures /K found from the T^3 law (8.60) for dielectrics.

He	26	Te	153	MgO	946
Ne	67	As	282	SiO_2	470
A	93	Diamond	2230	TiO_2	760
Kr	72	Graphite	420	Fe_2O_3	660
Xe	55	Si	640	CaF_2	510
Se	90	Ge	370	ZnS	315

The isothermal compressibility $\chi \equiv \chi_T$ of the solid therefore contains two terms

$$\chi(T,V) = -\frac{1}{V}\left(\frac{\partial V}{\partial P}\right) = \chi_0(V) + \chi_{ph}(T,V) \ ,$$

the second being smaller than the first. This explains why χ depends only weakly on temperature (Fig. 8.13). Moreover, as χ is small, an increase in pressure produces little change in V and hence in χ. In conclusion, χ depends only weakly on T and P.

To find the thermal expansion coefficient α from the equation of state (8.62), we use the cyclic relation

$$\left(\frac{\partial V}{\partial T}\right)_P = -\left(\frac{\partial V}{\partial P}\right)_T \left(\frac{\partial P}{\partial T}\right)_V$$

which yields

$$\alpha = \frac{1}{V}\left(\frac{\partial V}{\partial T}\right)_P = \chi_T \left(\frac{\partial P}{\partial T}\right)_V = \chi_T \left(\frac{\partial P_{ph}}{\partial T}\right)_V \ . \tag{8.63}$$

We state this relation for the case of the Debye model (§8.6.2), where the free energy F_{ph} (8.55) depends on the volume only through Θ_D. Differentiation of F with respect to V gives

$$P_{ph} = -\left(\frac{\partial F_{ph}}{\partial V}\right)_T = -\frac{9Nk}{x_D^4}\frac{d\Theta_D}{dV}\int_0^{x_D}\frac{x^3\,dx}{e^x - 1} \ .$$

On comparing this equation with the internal energy (8.57b), we may write

$$P_{ph} = \gamma \frac{(U - U_0)}{V} \tag{8.64}$$

with

$$\gamma = -\frac{d\ln\Theta_D}{d\ln V} \ . \tag{8.65}$$

Substituting into equation (8.63), we get for α

$$\alpha = \chi_T \frac{\gamma}{V}\left(\frac{\partial U}{\partial T}\right)_V = \gamma \frac{c_V \chi_T}{v} \ , \tag{8.66}$$

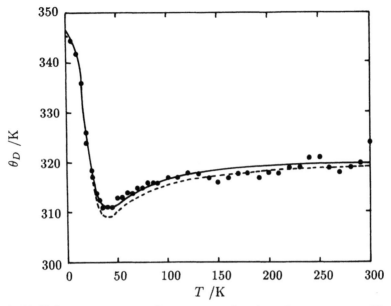

Figure 8.17: Debye temperature of copper as a function of temperature. The curve is obtained from the experimentally measured variation of $g(\omega)$ [G. Nilsson and S. Rolandson, *Phys. Rev. B* **7**, 2393 (1973)].

which, when compared with (8.49), shows that γ is the Grüneisen ratio. From equation (8.65) it is easy to understand that γ hardly varies with T or with the type of crystal. In particular, it can be shown that if the interaction energy between atoms varies as r^{-n}, the Grüneisen constant is equal to

$$\gamma = \frac{n+2}{6} \ . \tag{8.67}$$

For $\gamma \sim 2$, this corresponds to a value of the exponent $n \sim 10$ in the repulsive part of the potential, the attractive part being negligible.

In conclusion, equation (8.66) explains the fact that, since γ, χ_T and v have weak variations, the temperature dependence of α and c_V are similar. In particular

$$c_V \text{ or } \alpha \propto T^3 \qquad \text{for} \qquad T \ll \Theta_D$$
$$c_V \text{ or } \alpha \sim \text{const.} \qquad \text{for} \qquad T \gtrsim \Theta_D \ . \tag{8.68}$$

8.7 ANHARMONICITY

The phonon model in the harmonic approximation explains a large number of properties of solids. However, it does not account for

- the nearly linear increase in c_V at high temperatures (Problem 3.2);
- the fact that the thermal conductivity is not infinite;
- establishment of thermal equilibrium of the phonons.

To explain these effects, terms of higher order than the harmonic approximation (§8.4.1) must be included in the series expansion of the interatomic interaction

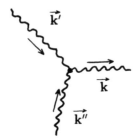

Figure 8.18: Phonon-phonon scattering

energy. After quantization, the anharmonic terms correspond to interactions between phonons, accompanied by creation and annihilation. The simplest coupling (Fig. 8.18) corresponds to annihilation of two incident phonons with wave vectors $\mathbf{k'}$ and $\mathbf{k''}$ and creation of one phonon $\mathbf{k}$ such that

$$\epsilon(\mathbf{k}) = \epsilon(\mathbf{k'}) + \epsilon(\mathbf{k''})$$
$$\mathbf{k} = \mathbf{k'} + \mathbf{k''} + \mathbf{K} \qquad (8.69)$$

where $\mathbf{K}$ is a vector of the reciprocal lattice of the crystal. These interactions can be shown to account for the three properties stated above.

BIBLIOGRAPHY

M.J. Buerger, *Elementary Crystallography*, John Wiley, New York (1956).
International Tables for X-ray Crystallography, Kynoch Press, Birmingham (1952).
C. Kittel, *Introduction to Solid State Physics*, Wiley (1983)
F.C. Philipps, *An Introduction to Crystallography*, Longmans (1963).
A.F. Wells, *Structural Inorganic Chemistry*, Clarendon Press, Oxford (1962).
Historical References
W.L. Bragg, *Nature* **90**, 402 (1912).
A. Bravais, *Etudes cristallographiques* (1848), Gauthier-Villars, Paris (1866).
P. Debye, *Ann. Physik* **39**, 789 (1912).
E.S. Fedorov, *Bull. Soc. Imp. russe* (2), **28**, 1 (1891).
E. Grüneisen, *Ann. Physik* **26**, 211 (1908).
M. von Laue, *Sitzungsber. math. phys. Kl. Akad. Wiss.* München 363 (1912).
A.M. Schoenflies, *Krystallsystem und Krystallstruktur*, Leipzig (1891).

COMPREHENSION EXERCISES

8.1 Show that the diamond lattice is face-centred cubic, that the primitive cell contains two atoms, and the cubic cell 8 atoms (Fig. 8.1a).

8.2 Compare the distance between two neighbouring sodium atoms and the diameter of the Na^+ ion in metallic sodium (Fig. 8.1c).

8.3 Verify that the lattice of helium (Fig. 8.1d) is hexagonal. Explain the value of the ratio $c/a = \sqrt{8/3}$.

8.4 For each of the three cubic Bravais lattices (Fig. 8.3), determine the number of points per cubic cell, the number of nearest neighbour points to a given point, and their distance in terms of a, the edge of the cell [Ans.: 1, 2 and 4; 6 at a; 8 at $a\sqrt{3}/2$; 12 at $a/\sqrt{2}$].

8.5 Show that the maximum filling ratios for hard spheres in face-centred cubic, body-centred cubic, simple cubic and diamond structures are respectively $\pi/3\sqrt{2}$, $\sqrt{3}\pi/8$, $\pi/6$ and $\sqrt{3}\pi/16$.

8.6 Show that the reciprocal lattice of a simple cubic lattice is simple cubic and that of a face-centred cubic lattice is body-centred cubic (and *vice versa*).

8.7 Calculate the volume of a cell in the reciprocal lattice as a function of that of the direct lattice, τ [Ans.: $8\pi^3/\tau$].

8.8 In a cubic lattice, what do the directions [100], [110], [111] correspond to? [Ans.: edge, diagonal of a face, diagonal of the cube].

8.9 The density of silicon Si is $\rho = 2.34$ g cm^{-3}. X-ray scattering indicates a cubic lattice with edge $a = 5.43$ Å. How many Si atoms are there per cubic cell ($M = 28.09$ g mol^{-1})? What might the structure be? [Ans.: 8; diamond]

8.10 Show that the total momentum of the atoms in a crystal is zero for displacements of the type (8.28). The momentum of a phonon is therefore zero.

8.11 In a cubic crystal of side $a \sim 3$ Å sound propagates at ~ 3000 m s^{-1}. What is the range of variation of the wave vector k and of the energy ϵ of the phonons in the acoustic branch? [Ans.: $k < 10^{10}$ m^{-1}; $\epsilon < 0.02$eV]

8.12 In the same crystal, the frequency of sound waves is $\nu < 20\,000$ s^{-1}. What are the corresponding ranges of k and ϵ [Ans.: $k < 40$ m^{-1} and $\epsilon < 10^{-10}$ eV].

8.13 Calculate the number of phonons in the Debye model for $T \ll \Theta_D$ and $T \gg \Theta_D$.

8.14 In the Debye model show that the most probable energy of the phonons at temperature T is $\epsilon \simeq 1.6\,kT$.

8.15 Calculate the energy of vibration of the crystal at absolute zero in the Debye model and show that it is equal to U_{ph} at $T = 0.67\,\Theta_D$.

8.16 Investigate the properties of a one-dimensional monatomic crystal in the Debye approximation.

PROBLEM 8.1 HEAT CAPACITY OF LINES OF ATOMS

1 Monatomic line

A one-dimensional lattice consists of N identical atoms of mass m whose equilibrium positions have an equal spacing of a. The atoms are assumed to interact only with their nearest neighbours, the binary interaction energy being $\phi(r)$.

1.1a Show that the potential energy of the line is

$$E = \sum_{p=1}^{N-1} \phi(a + u_{p+1} - u_p)$$

where u_p is the displacement of the pth atom from its equilibrium position.

1.1b Perform a series expansion on ϕ around a and show that E has the form

$$E = E_0(a) + \frac{m}{2}\omega_0^2(a) \sum_p (u_{p+1} - u_p)^2 .$$

Express E_0 and ω_0 as functions of N, ϕ and m.

1.1c Hence show, to the same precision, that the equation of motion of the pth atom is

$$\ddot{u}_p = \omega_0^2 (u_{p+1} + u_{p-1} - 2u_p) .$$

1.2 We seek solutions to these equations having the oscillatory form

$$u_p = u_0 \cos(kpa - \omega t) .$$

1.2a What is the dispersion relation $\omega = \omega(k)$ that these solutions obey? It is recalled that $\cos 2\alpha + \cos 2\beta = 2\cos(\alpha + \beta)\cos(\alpha - \beta)$.

1.2b What is the physical meaning of ω and k? Why is it sufficient to restrict the range of solutions to $|k| < \pi/a$?

1.2c Plot $\omega(k)$ and specify its behaviour in the neighbourhood of $k = 0$ and $k = \pm\pi/a$. Derive expressions for the velocity of sound c and the cut-off frequency ω_m in this model.

1.3 The length of the chain $L = (N - 1)a$ is fixed.

1.3a What values of k satisfy this condition? Calculate $g(k)$, the density of allowed values of k.

1.3b Hence derive the density of frequencies, $g(\omega)$. Plot this function.

1.4 It is known from quantum mechanics that the energy of a wave of angular frequency ω is

$$\epsilon_n = \left(n + \frac{1}{2}\right)\hbar\omega , \qquad n \geq 0 .$$

By applying Maxwell-Boltzmann statistics to the energy states of a wave, find the probability P_n that its energy is ϵ_n at temperature T. Hence derive an expression for the average energy of a wave, $< \epsilon(\omega, T) >$, at this temperature. k_B is the Boltzmann constant.

1.5a Show that the total internal energy U of the crystal can be written in the form

$$U = U_0(a) + U_{\text{ph}}(a, T)$$

and explain U_{ph} in terms of phonons, i.e., particles of indeterminate number obeying Bose-Einstein statistics.

1.5b Calculate U_{ph} in the high temperature limit. Express the characteristic temperature Θ_m in terms of ω_m, then as a function of the experimental parameters c and a.

1.5c Determine the dependence of U_{ph} upon T at low temperature, as well as that of the molar heat capacity c_V of the line.

2 Diatomic line

Consider now a one-dimensional lattice composed of N atoms of type A alternating with $N-1$ atoms of type B ($ABAB \ldots BA$). The AB interaction energy is $\phi(r)$ and only nearest neighbour interactions are retained. The mass of the $A(B)$ atoms is $m_1 (m_2)$ and $u_{2p-1} (v_{2p})$ is their displacement from their equilibrium positions, which are separated by $a/2$. The index p runs from 1 to $N(N-1)$ for the $A(B)$ atoms.

2.1 Generalize the results of 1.1 to show that the equations of motion of atoms $2p-1$ and $2p$ are respectively

$$\ddot{u}_{2p-1} = \omega_1^2 (v_{2p-2} + v_{2p} - 2u_{2p-1})$$
$$\ddot{v}_{2p} = \omega_2^2 (u_{2p-1} + u_{2p+1} - 2v_{2p}) \ .$$

Find expressions for ω_1 and ω_2.

2.2 Look for solutions to the above equations having the form

$$u_{2p-1} = u_0 \cos \left[k \left(p - \frac{1}{2} \right) a - \omega t \right] \quad \text{and} \quad v_{2p} = v_0 \cos(kpa - \omega t) \ .$$

2.2a Show that for each value of k the allowed values of ω^2 cause a 2×2 determinant to vanish. Write down an expression for the ratio v_0/u_0.

2.2b Find explicitly the values of $\omega^2(k)$, denoted $\omega_A^2(k)$ and $\omega_O^2(k)$ ($\omega_A < \omega_O$).

2.3a Investigate the limits of $\omega_A(k)$ for $k \to 0$ and $k = \pm\pi/a$ (take $m_1 < m_2$). Plot the branch $\omega = \omega_A(k)$ and comment. In particular, give the velocity of sound c.

2.3b Calculate the ratio v_0/u_0 for this branch in the limit $k \to 0$. Why is it called the acoustic branch?

2.4 Redo question 2.3 for the $\omega = \omega_O(k)$ branch. Why is this called the optical branch?

2.5 Investigate the limiting cases $m_1 = m_2$ and $m_1 \ll m_2$.

2.6 In the case of real crystals such as the alkali halides (A =Li$^+$, Na$^+$, K$^+$, Rb$^+$ and B =F$^-$, Cl$^-$, Br$^-$, I$^-$), the characteristic Debye temperature Θ_D, which generalizes to Θ_m, is given by

$$\Theta_D = 2(3\pi^2)^{1/3} \frac{\hbar \bar{c}}{k_B a}$$

where a is the edge of the cubic cell. For the velocity of sound $\bar{c}$, use the expression for c found in 2.3a for the one-dimensional model, and show that the product $\Theta_D \sqrt{m_1 + m_2}$ is a function of a only. Plot this function using the experimental data for the alkali halide family (Table 8.5).

SOLUTION 1 Monatomic line

1.1a The distance between atoms p and $p+1$ is $a + u_{p+1} - u_p$ and the total potential energy is the sum of the $(N-1)$ interaction energies between neighbours. Hence the expression given for E.

1.1b Expanding ϕ yields

$$\phi(a + u_{p+1} - u_p) = \phi(a) + \phi'(u)(u_{p+1} - u_p) + \frac{\phi''(a)}{2}(u_{p+1} - u_p)^2 + \ldots$$

Table 8.5: Debye temperatures in Kelvin, found from the T^3 dependence (upper numbers), and lattice parameters in Ångström units (lower numbers) of alkali halides. The atomic masses (in g mol^{-1}) are given in parentheses [J.T. Lewis et al., *Phys. Rev.* **161**, 877 (1967)].

	Li (6.9)	Na (23.0)	K (39.1)	Rb (85.5)
F	730	492	336	
(19.0)	4.03	4.63	5.34	5.63
Cl	422	321	231	165
(35.5)	5.14	5.64	6.28	6.55
Br		224	173	131
(79.9)		5.97	6.58	6.86
I		164	131	103
(126.9)		6.46	7.07	7.34

The linear terms in u_p cancel in the sum over p (only the difference $u_n - u_1$ remains, and this also vanishes when the length of the chain is fixed). Finally, we have

$$E = (N - 1)\phi(a) + \frac{\phi''(a)}{2} \sum_p (u_{p+1} - u_p)^2 \ ,$$

and hence

$$E_0(a) = (N - 1)\phi(a) \simeq N\phi(a) \qquad \text{and} \qquad \omega_0^2(a) = \frac{\phi''(a)}{m} \ . \qquad (8.70)$$

1.1c The force acting on the pth atom is

$$F_p = -\frac{\partial E}{\partial u_p} = -m\omega_0^2 \left[-(u_{p+1} - u_p) + (u_p - u_{p-1}) \right] \ ,$$

so that its equation of motion, $m\ddot{u}_p = F_p$, is the one that is given.

1.2a Inserting the expression for the displacements into the equation of motion, and noting that

$$\begin{aligned} u_{p+1} + u_{p-1} &= u_0 \left[\cos(k(p+1)a - \omega t) + \cos(k(p-1)a - \omega t) \right] \\ &= 2u_0 \cos(kpa - \omega t) \cos ka \ , \end{aligned}$$

we find

$$-u_0\omega^2 \cos(kpa - \omega t) = 2\omega_0^2 u_0 \cos(kpa - \omega t) (\cos ka - 1) \ ,$$

which yields the dispersion relation

$$\omega(k) = \omega_0 \sqrt{2(1 - \cos ka)} = 2\omega_0 \left| \sin \frac{ka}{2} \right| \ . \qquad (8.71)$$

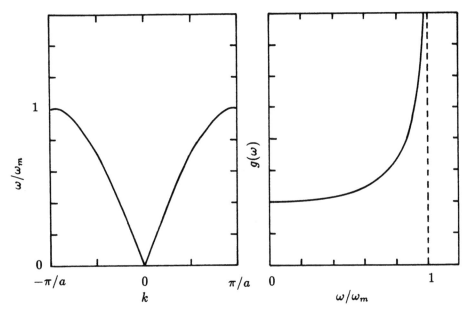

Figure 8.19: Dispersion relation and density of states in the model of a line of identical atoms with only nearest neighbour interactions.

1.2b During the movement, each atom oscillates about its equilibrium position with the same angular frequency ω and the same amplitude u_0, the phase difference between two neighbours being ka. The corresponding motion is a wave (Fig. 8.8) in which k is the wave vector. Since the same motion of the atoms is described by values of k differing by $2\pi/a$, we may restrict ourselves to the interval $]-\pi/a,\ \pi/a]$ (Fig. 8.10).

1.2c The dispersion curve is shown in Figure 8.19. Near the origin, $\omega \simeq \omega_0 a|k|$ and at the boundaries of the interval

$$\omega = \omega_m \equiv 2\omega_0 \ , \tag{8.72}$$

the slope is zero. The velocity of sound, which is equal to the phase velocity and the group velocity for $k \to 0$ (long wavelengths), is

$$c = \omega_0 a \ . \tag{8.73}$$

1.3a Since L is fixed, it follows that $u_N = u_1$, or, in the general expression for u_p, $\cos(kNa - \omega t) = \cos(ka - \omega t)$. We thus have $kNa = ka + 2n\pi$ (n integer) and hence

$$k = \frac{2n\pi}{(N-1)a} = \frac{2n\pi}{L} \ . \tag{8.74}$$

The values of k are spaced equally by $2\pi/L$ between $-\pi/a$ and $+\pi/a$. The density $g(k)$ is thus

$$g(k) = \frac{1}{2\pi/L} = \frac{L}{2\pi} \ . \tag{8.75}$$

1.3b On changing the variable, we have

$$g(\omega) = g(k) \left| \frac{dk}{d\omega} \right| \quad .$$

Since $k = \pm(2/a) \ \arcsin(\omega/\omega_m)$, it follows that

$$g(\omega) = \frac{L}{2\pi} \times 2 \times \frac{2}{a\sqrt{\omega_m^2 - \omega^2}} \quad ,$$

where the factor 2 takes account of the positive and negative values of k . As $L = (N-1)a \simeq Na$, the density is therefore (Fig. 8.19)

$$g(\omega) = \begin{cases} 2N/\pi \sqrt{\omega_m^2 - \omega^2} & \omega < \omega_m \\ 0 & \omega > \omega_m \end{cases} \quad . \tag{8.76}$$

1.4 In Maxwell-Boltzmann statistics, the occupation probability of the energy level ϵ_n is

$$P_n = \frac{e^{-\beta\epsilon_n}}{Z} \qquad \text{with} \quad Z = \sum_n e^{-\beta\epsilon_n} \quad .$$

With the expression given for ϵ_n, the partition function Z is equal to (cf. Ex. 3.1)

$$Z = \frac{e^{-\beta\hbar\omega/2}}{1 - e^{-\beta\hbar\omega}} \quad , \quad \text{and hence} \qquad P_n = e^{-n\beta\hbar\omega} \left(1 - e^{-\beta\hbar\omega}\right) \quad .$$

It can be shown that $\sum P_n = 1$. The average energy of a wave is therefore

$$< \epsilon(\omega, T) > = \sum \left(n + \frac{1}{2}\right) \hbar\omega P_n = \hbar\omega \sum n P_n + \frac{\hbar\omega}{2} \quad .$$

Since

$$\sum_{n=0}^{\infty} n e^{-n\beta\hbar\omega} = -\frac{d}{d(\beta\hbar\omega)} \sum e^{-n\beta\hbar\omega} \quad ,$$

we have

$$\begin{aligned} < \epsilon(\omega, T) > &= -\hbar\omega \frac{d}{d(\beta\hbar\omega)} \ln \left(1 - e^{-\beta\hbar\omega}\right) + \frac{\hbar\omega}{2} \\ &= \hbar\omega \left[\frac{1}{e^{\beta\hbar\omega} - 1} + \frac{1}{2}\right] \quad . \end{aligned} \tag{8.77}$$

1.5a The total internal energy thus becomes

$$U = E_0(a) + \int_0^{\omega_m} g(\omega) \, d\omega \times < \epsilon(\omega, T) > \quad .$$

On substituting expression (8.77) for $< \epsilon(\omega, T) >$, we find $U = U_0(a) + U_{\text{ph}}(a, T)$, with

$$U_0(a) = E_0(a) + \int_0^{\omega_m} \frac{\hbar\omega}{2} g(\omega) \, d\omega$$

$$\text{and} \qquad U_{\text{ph}}(a, T) = \int_0^{\omega_m} \frac{\hbar\omega}{e^{\beta\hbar\omega} - 1} g(\omega) \, d\omega \quad . \tag{8.78}$$

This expression for U_{ph} can be identified with the internal energy $U = \sum N_i \epsilon_i$ found in Bose-Einstein statistics for an indeterminate number of particles ($\mu = 0$), where

$$N_i = \frac{g_i}{e^{\beta \epsilon_i} - 1}$$

with $\epsilon_i \rightarrow \epsilon = \hbar\omega$. These particles are called phonons.

1.5b When $\beta\hbar\omega \ll 1$ with $\omega < \omega_m$, i.e., for

$$T \gg \Theta_m = \hbar\omega_m/k_B \quad , \tag{8.79a}$$

then

$$U_{\text{ph}} \simeq \int \frac{\hbar\omega}{\beta\hbar\omega} g(\omega) \, d\omega = k_B T \int_0^{\omega_m} g(\omega) \, d\omega \quad .$$

The integral is equal to the total number of modes N, and thus $U_{\text{ph}} \simeq N k_B T$. In the range under consideration, the molar heat capacity c_V is then equal to $\mathcal{N} k_B = R$. This is the Dulong and Petit law for a one-dimensional solid. From (8.72,73), the characteristic temperature Θ_m can also be written

$$\Theta_m = \frac{2\hbar}{k_B} \frac{c}{a} \quad . \tag{8.79b}$$

1.5c When $T \rightarrow 0$, the factor $1/\left(e^{\beta\hbar\omega} - 1\right)$ is very small if $\omega \gg k_B T/\hbar$. Therefore in (8.78) the upper limit of the integral can be set equal to infinity, and $g(\omega)$ replaced by $g(0) = 2N/\pi\omega_m$. Thus

$$U_{\text{ph}} = \int_0^\infty \frac{\hbar\omega}{e^{\beta\hbar\omega} - 1} \times \frac{2N}{\pi\omega_m} \, d\omega \quad .$$

Substitution of the variable $x = \beta\hbar\omega$ gives

$$U_{\text{ph}} = \frac{2N}{\pi\hbar\omega_m} (k_B T)^2 \int_0^\infty \frac{x}{e^x - 1} \, dx,$$

where the purely numerical integral is equal to $\pi^2/6$. The internal energy varies as T^2, and therefore c_V varies as T,

$$c_V = \frac{2\pi}{3} R \frac{T}{\Theta_m} \qquad (T \ll \Theta_m) \quad .$$

This corresponds to the T^3 law for 3-dimensional solids.

2 Diatomic line

2.1 The potential energy of the line is now

$$U = \sum_{p=1}^{N-1} \phi\left(\frac{a}{2} + v_{2p} - u_{2p-1}\right) + \sum_{p=1}^{N-1} \phi\left(\frac{a}{2} + u_{2p+1} - v_{2p}\right) \quad .$$

In the harmonic approximation, this energy becomes

$$U = 2(N-1)\phi\left(\frac{a}{2}\right) + \frac{1}{2}\phi''\left(\frac{a}{2}\right) \sum_{p=1}^{N-1} (v_{2p} - u_{2p-1})^2 + (u_{2p+1} - v_{2p})^2 \quad .$$

Writing the equation of motion for u_{2p-1} and v_{2p}, the required equations are obtained with

$$\omega_1^2 = \frac{\phi''(a/2)}{m_1} \quad \text{and} \quad \omega_2^2 = \frac{\phi''(a/2)}{m_2} \quad . \tag{8.80}$$

2.2a On inserting the given displacements into the equations of motion, we find, as in question 1.2,

$$-\omega^2 u_0 = 2\omega_1^2\left(v_0 \cos\frac{ka}{2} - u_0\right) \quad \text{and} \quad -\omega^2 v_0 = 2\omega_2^2\left(u_0 \cos\frac{ka}{2} - v_0\right) \quad .$$

This set of two linear homogeneous equations has non-zero solutions u_0 and v_0 only if its determinant vanishes, i.e., if

$$\begin{vmatrix} \omega^2 - 2\omega_1^2 & 2\omega_1^2 \cos\frac{ka}{2} \\ 2\omega_2^2 \cos\frac{ka}{2} & \omega^2 - 2\omega_2^2 \end{vmatrix} = 0 \quad .$$

For each solution in ω, we then have

$$\frac{v_0}{u_0} = \frac{2\omega_1^2 - \omega^2}{2\omega_1^2 \cos(ka/2)} \quad . \tag{8.81}$$

2.2b The values of ω^2 for which the determinant vanishes are solutions of

$$\omega^4 - 2\omega^2(\omega_1^2 + \omega_2^2) + 4\omega_1^2\omega_2^2 \sin^2\frac{ka}{2} = 0 \quad ,$$

and therefore

$$\omega^2 = (\omega_1^2 + \omega_2^2) \pm \left[(\omega_1^2 + \omega_2^2)^2 - 4\omega_1^2\omega_2^2 \sin^2\frac{ka}{2}\right]^{1/2} \quad . \tag{8.82}$$

The solution ω_A (ω_O) corresponds to the sign $-$ (+).

2.3a When k tends to zero, the square root in the expression for ω^2 is equivalent to

$$(\omega_1^2 + \omega_2^2) - \frac{2\omega_1^2\omega_2^2}{\omega_1^2 + \omega_2^2} \sin^2\frac{ka}{2} \quad ,$$

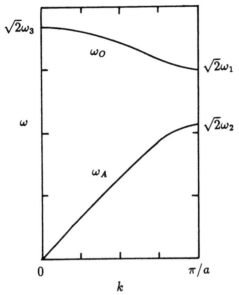

Figure 8.20: Acoustic and optical branches of a diatomic line of atoms for $\omega_1 = \sqrt{2}\omega_2$ ($m_1 = m_2/2$). Here $\omega_3 = (\omega_1^2 + \omega_2^2)^{1/2}$.

and hence

$$\omega_A^2 \simeq \frac{2\omega_1^2\omega_2^2}{\omega_1^2 + \omega_2^2}\left(\frac{ka}{2}\right)^2 .$$

As $k \to 0$, we therefore have

$$\omega_A = c|k| \qquad \text{with} \qquad c^2 = \frac{\omega_1^2\omega_2^2}{2(\omega_1^2 + \omega_2^2)}a^2 \qquad (8.83)$$

and the frequency varies linearly with k, as for sound waves. When $k \to \pm\pi/a$, the square root is equal to $(\omega_1^2 - \omega_2^2)$, since m_1 is smaller than m_2 and hence ω_1 is greater than ω_2. Thus $\omega \to \sqrt{2}\omega_2$. Note that $d\omega/dk$ vanishes at this point. The curve $\omega_A(k)$ is shown in Figure 8.20.

2.3b The ratio v_0/u_0, defined by (8.81), tends to 1 as k tends to zero. The A and B atoms therefore vibrate with the same amplitude and the nature of the atoms does not affect the wave propagation (Fig. 8.21a). This result and the asymptotic relation (8.83) show the analogy with sound waves in continuous media, which accounts for the name acoustic branch.

2.4 When k tends to zero, then, as in question 2.3a,

$$\omega_O^2 \simeq 2(\omega_1^2 + \omega_2^2) - c^2k^2,$$

and when k tends to $\pm\pi/a$, $\omega_O \simeq \sqrt{2}\omega_1$, with $d\omega/dk = 0$ at this point. The curve $\omega_O(k)$ is shown in Figure 8.20.

Note that when $k \to 0$,

$$\frac{v_0}{u_0} \to -\frac{\omega_2^2}{\omega_1^2} = -\frac{m_1}{m_2}$$

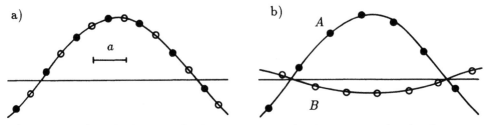

a)   b)

A

a

B

Figure 8.21: Sketch of atomic displacements in an AB diatomic line for the a) acoustic and b) optical branches. Here $\lambda = 10a$ and $m_1 = m_2/4$.

and neighbouring A and B atoms vibrate with opposite phase (Fig. 8.21b) in such a way that their centre of mass remains fixed. If the A and B atoms are ions (of opposite sign), their displacements can be excited by the electric field of an optical wave of frequency $\omega = \omega_O$. This accounts for the name optical branch. This property can be investigated by infrared absorption spectroscopy.

2.5 When $m_1 = m_2 \equiv m$, then $\omega_1 = \omega_2 \equiv \omega_0$ and equation (8.82) for the frequencies becomes

$$\omega^2 = 2\omega_0^2 \left(1 \pm \cos \frac{ka}{2}\right) .$$

Therefore, for the acoustic and optical branches,

$$\omega_A = 2\omega_0 \left|\sin \frac{ka}{4}\right| \quad \text{and} \quad \omega_O = 2\omega_0 \cos \frac{ka}{4} .$$

These branches, which are shown in Figure 8.22, can be derived from the single branch of a monatomic line with spacing $a/2$.

When $m_1 \ll m_2$, the square root in (8.82) is approximately equal to

$$\omega_1^2 + \omega_2^2 \left(1 - 2\sin^2 \frac{ka}{2}\right) = \omega_1^2 + \omega_2^2 \cos ka ,$$

and therefore

$$\omega_A^2 = \omega_2^2(1 - \cos ka) \quad \text{and} \quad \omega_O^2 = 2\omega_1^2 + \omega_2^2(1 + \cos ka) \simeq 2\omega_1^2 .$$

The acoustic branch takes the form that is has in the monatomic case, and the frequency of the optical branch, $\sqrt{2}\omega_1$, is constant, corresponding to the A atoms oscillating almost alone, the amplitude of the B atoms being much smaller.

2.6 On setting $\bar{c}$ equal to expression (8.83) for c, the Debye temperature can be written

$$\Theta_D = \sqrt{2}(3\pi^2)^{1/3} \frac{\hbar}{k_B} \frac{\omega_1\omega_2}{(\omega_1^2 + \omega_2^2)^{1/2}} .$$

On insertion of equations (8.80) for ω_1 and ω_2, this becomes

$$\Theta_D = \sqrt{2}(3\pi^2)^{1/3} \frac{\hbar}{k_B} \sqrt{\phi'' \left(\frac{a}{2}\right)} \times (m_1 + m_2)^{-1/2} , \tag{8.84}$$

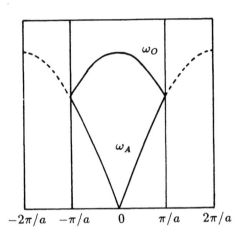

Figure 8.22: Acoustic and optical branches of an AB diatomic line for the case $\omega_1 = \omega_2$ ($m_1 = m_2$).

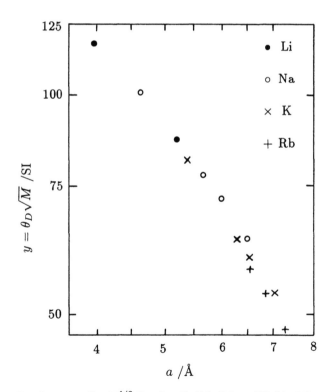

Figure 8.23: Product $y = \Theta_D M^{1/2}$ for the alkali halides of Table 8.5 as a function of the length of the edge of the cubic cell. Θ_D and M are respectively the Debye temperature and the molar mass ($M = M_1 + M_2$) of the substance.

which shows that the product $y = \Theta_D \sqrt{m_1 + m_2}$ depends only on a via ϕ''. For a family of substances, such as the alkali halides, we expect the values of y to vary in a regular way with a. This is indeed what is observed (Fig. 8.23).

PROBLEM 8.2 IMPROVEMENTS TO THE DEBYE MODEL

In the phonon model, the thermodynamic properties of solids are determined from the properties of a gas of phonons, which are particles of indeterminate number obeying Bose-Einstein statistics. In this problem, we investigate a particular model of monatomic solids due to P. Debye, as well as some improvements to the model.

1 In Bose-Einstein statistics, the grand potential $\Omega(T, V, \mu) = U - TS - N\mu$ is given by

$$\Omega = k_B T \sum_i g_i \ln \left[1 - e^{\beta(\mu - \epsilon_i)} \right] , \qquad \beta = 1/k_B T$$

where ϵ_i are the energy levels and g_i their degeneracy (k_B is the Boltzmann constant).

1a Derive the general expression for the free energy F of a phonon gas, then that for the internal energy U and the heat capacity at constant volume C_V. Comment on the expression for U.

1b Knowing that the momentum and the energy of a phonon are given by $\mathbf{p} = \hbar \mathbf{k}$ and $\epsilon = \hbar \omega$, with $\omega = \omega(\mathbf{k})$, determine the density of states $g(\mathbf{k})$ and express C_V as an integral over $\mathbf{k}$.

1c It is recalled that the N allowed values of $\mathbf{k}$ are restricted to the first Brillouin zone in reciprocal space, where N is the number of atoms in the crystal. Show that the volume of this zone is $\tau = (2\pi)^3 N/V$, where V is the volume of the crystal.

2a Determine the limiting expression for the molar heat capacity c_V of a monatomic solid as $T \to \infty$. Take into account the three kinds of phonons (one longitudinal L and two transverse, T_1 and T_2). The limiting value of c_V is denoted $c_V(\infty)$. Comment on its value.

2b When $k = |\mathbf{k}| \to 0$, the dispersion relation for each polarization takes the form $\omega = c_s(\Omega)k$, where c_s is the velocity of propagation of sound in the direction $\Omega = (\theta, \phi)$ for polarization s ($s = L, T_1, T_2$). Using the mean velocity $\bar{c}$ defined by

$$\frac{3}{\bar{c}^3} = \sum_s \int \frac{d\Omega}{4\pi} \times \frac{1}{[c_s(\Omega)]^3} ,$$

find the limiting value of C_V as $T \to 0$. What is its dependence on T? The integral $\int_0^\infty x^4 e^x \, dx/(e^x - 1)^2$ is equal to $4\pi^4/15$.

3 Debye model

This model is equivalent to a phonon model together with the three following assumptions.

(i) The three phonon polarizations are governed by only one dispersion relation.
(ii) The dispersion relation has the form $\omega = ck$.
(iii) The first Brillouin zone is replaced by a sphere of radius k_m and volume τ.

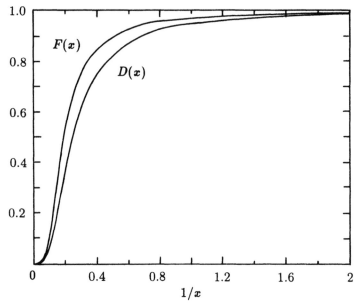

Figure 8.24: Debye function $D(x)$ and the function $F(x)$ defined in question 5b. Note that the variable of the abscissa is $1/x$.

3a Calculate the radius k_m of the sphere as a function of V and N. Hence deduce the existence of a cut-off frequency ω_D.

3b Determine the angular frequency density $g(\omega)$ from the density $g(\mathbf{k})$ of question 1b. Sketch this density in a graph.

3c In this model, show that c_V is equal to $c_V(\infty) \times D(x)$ where the function

$$D(x) = \frac{3}{x^3} \int_0^x \frac{u^4 e^u}{(e^u - 1)^2} \, du = 3x^2 \int_0^1 \frac{u^4 e^{xu}}{(e^{xu} - 1)^2} \, du$$

is shown in Figure 8.24. Express x as a function of $T_m = \hbar c k_m / k_B$.

3d In question 2b, write $\bar{c}$ as a function of c and express the molar heat capacity c_V at low temperatures as a function of T/T_m.

3e We wish to determine the parameter c in this model for copper from the following experimental data: molar mass, $M = 63.54$ g mol^{-1}; density at 0 K, $\rho = 9.02$ g cm^{-3}; molar heat capacity for $T \lesssim 10$ K, $c_V = 4.76 \times 10^{-5} T^3$ J K^{-1} mol^{-1}. Calculate the values of k_m and T_m and hence deduce that of c.

4 Model I

Consider a model in which assumptions (ii) and (iii) of the Debye model are retained, but where the longitudinal dispersion relation $\omega = c_L k$ is taken to be different from the two transverse relations $\omega = c_T k$ ($c_L > c_T$).

4a Show that there are two cut-off frequencies ω_L and ω_T depending on the type of wave. Find an expression for and plot the total density $g(\omega)$. Compare this with the Debye model. What is $\bar{c}$ in this model?

4b For the three types of phonon, express each contribution to the heat capacity c_V in terms of the Debye function D. Use two characteristic temperatures T_L and

T_T and find equations for these. Then write the formula for c_V.

4c What are the limiting expressions for c_V as $T \to \infty$ and $T \to 0$? Compare with the Debye model.

4d In copper $c_L = 4760$ m s^{-1} and $c_T = 2325$ m s^{-1}. Calculate $\bar{c}$, T_L and T_T. Hence derive the equation for c_V as $T \to 0$ and compare with the experimental variation of part 3.

4e Using Figure 8.24, calculate the ratio $c_V/3R$ in this model for $T = 100$, 200 and 300 K. Hence, for each of these temperatures, deduce the value of $1/x$ that yields the same ratio $c_V/3R$ in the Debye model. Calculate the corresponding temperatures $\Theta_D = xT$ and plot the shape of the curve $\Theta_D(T)$.

5 Model II

We now consider a model in which assumptions (i) and (iii) of the Debye model are maintained, but for the dispersion relation we take the more realistic form $\omega = \omega_m \sin(\pi k/2k_m)$ with $\omega_m = 2ck_m/\pi$.

5a Plot the above dispersion relation and compare it with that of the Debye model. Find an expression for the frequency density $g(\omega)$ and make a graph.

5b In this model, show that c_V is equal to $c_V(\infty)F(x)$, where x is the variable defined in 3c and $F(x)$ is an integral which is to be determined. The function $F(x)$ is shown in Figure 8.24.

5c Same question as 4e for this model and for the temperatures 25 K, 100 K and 300 K. Try to explain the experimental curve of $\Theta_D(T)$ (Fig. 8.17) for copper in terms of both models I and II.

Numerical data: Avogadro's number $\mathcal{N} = 6.02 \times 10^{23}$ mol^{-1}; Boltzmann constant $k_B = 1.38 \times 10^{-23}$ J K^{-1}; Planck's constant $\hbar = 1.05 \times 10^{-34}$ J s.

SOLUTION

1a As the number of phonons is indeterminate, their chemical potential μ is zero. Consequently

$$F \equiv \Omega = k_B T \sum_i g_i \ln \left[1 - e^{-\beta \epsilon_i}\right] .$$

From the Helmholtz relation, the internal energy is equal to

$$U = -T^2 \frac{\partial}{\partial T}\left(\frac{F}{T}\right) = \sum_i \frac{g_i \epsilon_i}{e^{\beta \epsilon_i} - 1}$$

and the heat capacity C_V is given by

$$C_V = \frac{\partial U}{\partial T} = k_B \sum_i g_i \left(\frac{\epsilon_i}{k_B T}\right)^2 \frac{e^{\beta \epsilon_i}}{(e^{\beta \epsilon_i} - 1)^2} . \tag{8.85}$$

The internal energy thus has the form $U = \sum N_i \epsilon_i$ of a Bose-Einstein distribution.

1b As the density of states in phase space is $d^3r \, d^3p/h^3$, on integrating over $\mathbf{r}$ ($d^3r \to V$) and changing variables $\mathbf{p} \to \mathbf{k}$ we get

$$g(\mathbf{k}) \, d^3k = \frac{V}{(2\pi)^3} d^3k . \tag{8.86}$$

Changing now to continuous variables in equation (8.85) for C_V

$$\sum_i \rightarrow \int \ , \qquad g_i \rightarrow g(\mathbf{k}) \, d^3\mathbf{k} \ , \qquad \epsilon_i \rightarrow \epsilon = \hbar\omega(\mathbf{k}) \ ,$$

we find

$$C_V = k_B \frac{V}{(2\pi)^3} \int d^3\mathbf{k} (\beta\hbar\omega)^2 \frac{e^{\beta\hbar\omega}}{(e^{\beta\hbar\omega} - 1)^2} \ . \tag{8.87}$$

1c We must write $\sum g_i = N$, i.e.,

$$\int g(\mathbf{k}) \, d^3\mathbf{k} = \frac{V}{(2\pi)^3} \int d^3\mathbf{k} = N \ .$$

The volume of the first Brillouin zone is therefore

$$\tau = \int d^3\mathbf{k} = (2\pi)^3 \frac{N}{V} \ . \tag{8.88}$$

2a As T tends to infinity, $\beta\hbar\omega$ tends to zero and expression (8.87) for C_V adopts the limiting form

$$C_V = k_B \frac{V}{(2\pi)^3} \int d^3\mathbf{k} = k_B \frac{V}{(2\pi)^3} \tau = N k_B \ .$$

Taking into account the three phonon polarizations, the molar heat capacity $c_V(\infty)$ is equal to $3R$, in conformity with the Dulong and Petit law.

2b As T tends to zero, only the lowest levels are populated, and we can use the relation $\omega = c_s(\Omega)k$ in C_V. On setting $x = \beta\hbar c_s k$, this expression becomes

$$\begin{aligned}
C_V &= \sum_s \frac{k_B V}{(2\pi)^3} \int \frac{x^2 \, dx \, d\Omega}{(\beta\hbar c_s)^3} x^2 \frac{e^x}{(e^x - 1)^2} \\
&= \frac{k_B V}{h^3} (k_B T)^3 \sum_s \int \left(\frac{d\Omega}{c_s(\Omega)^3} \times \int x^4 \frac{e^x}{(e^x - 1)^2} \, dx \right) \ .
\end{aligned}$$

On extending the upper limit of the second integral to infinity and introducing the mean velocity $\bar{c}$, we find

$$C_V = k_B \frac{V}{h^3} (k_B T)^3 \times \frac{12\pi}{\bar{c}^3} \times \frac{4\pi^4}{15} = \frac{16}{5} \pi^5 k_B V \left(\frac{k_B T}{h\bar{c}} \right)^3 \ . \tag{8.89}$$

3a We have

$$\frac{4}{3} \pi k_m^3 = \tau = (2\pi)^3 \frac{N}{V} \ .$$

The radius k_m is therefore equal to

$$k_m = \left(6\pi^2 \frac{N}{V} \right)^{1/3} \tag{8.90}$$

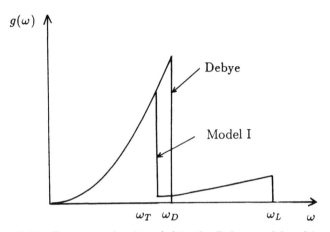

Figure 8.25: Frequency density $g(\omega)$ in the Debye model and in model I.

and the corresponding maximum angular frequency is

$$\omega_D = ck_m = c\left(6\pi^2 \frac{N}{V}\right)^{1/3} . \tag{8.91}$$

3b From (8.86), the density $g(\omega)$ is given by

$$g(\omega)\,d\omega = \frac{V}{(2\pi)^3}\,4\pi k^2\,dk = \frac{V}{(2\pi)^3}4\pi\frac{\omega^2\,d\omega}{c^3} ,$$

and hence, taking account of the three polarizations,

$$g(\omega) = \frac{3V}{2\pi^2}\frac{\omega^2}{c^3} \qquad (\omega < \omega_D) . \tag{8.92}$$

This density is shown in Figure 8.25.

3c On setting $\omega = ck$ in equation (8.87) and taking into account the three polarizations, we find for the heat capacity C_V

$$C_V = 3k_B\frac{V}{(2\pi)^3}\int_0^{k_m} 4\pi k^2\,dk\,(\beta\hbar ck)^2\,\frac{e^{\beta\hbar ck}}{(e^{\beta\hbar ck}-1)^2} .$$

With the substitution $u = \beta\hbar ck$ and $x = \beta\hbar ck_m$, this expression becomes

$$C_V = k_B\frac{Vk_m^3}{2\pi^2}\,D(x) .$$

If now we replace k_m by (8.90), the molar heat capacity c_V takes the form

$$c_V = 3R \times D\left(\frac{\hbar ck_m}{kT}\right) = 3R\,D\left(\frac{T_m}{T}\right) . \tag{8.93}$$

3d In this model, the velocities $c_s(\Omega)$ defined in 2b are equal to c, so that $\bar{c} = c$. The heat capacity C_V at low temperature, which is given by (8.89), becomes accordingly

$$C_V = \frac{16\pi^5}{5} k_B V \left(\frac{k_m}{2\pi}\right)^3 \left(\frac{T}{T_m}\right)^3 .$$

On substitution of expression (8.90) for k_m, it follows that

$$c_V = \frac{12\pi^4}{5} R \left(\frac{T}{T_m}\right)^3 . \tag{8.94}$$

3e The value of k_m calculated from (8.90) is 1.7×10^{10} m^{-1}, and the associated wavelength, $\lambda_m = 2\pi/k_m$, is of the order of magnitude of the interatomic distances. T_m is found by comparing the theoretical expression for c_V with the experimental measurement at low temperature, which gives $T_m = 344$ K. The relation $T_m = \hbar c k_m/k_B$ then yields $c = 2650$ m s^{-1}.

4a Since the maximum value of k is k_m, the maximum values of ω are respectively $\omega_L = c_L k_m$ and $\omega_T = c_T k_m$, where $\omega_T < \omega_L$. As in 3b, the density for each polarization s is given by

$$g_s(\omega) = \frac{V}{2\pi^2} \frac{\omega^2}{c_s^3} \qquad (s = L, T_1, T_2) .$$

For $\omega < \omega_T$, the total density is $g_L(\omega) + 2g_T(\omega)$ and for $\omega_T < \omega < \omega_L$, it is equal to $g_L(\omega)$. Thus

$$g(\omega) = \frac{V}{2\pi^2} \times \begin{cases} \omega^2 \times (c_L^{-3} + 2\,c_T^{-3}) & \omega < \omega_T \\ \omega^2/c_L^3 & \omega_T < \omega < \omega_L \\ 0 & \omega_L < \omega . \end{cases}$$

This density, shown in Figure 8.25, exhibits two maxima, while the Debye density has only a single maximum.

In this improved model, the velocity $\bar{c}$ defined in 2b is such that

$$\frac{3}{\bar{c}^3} = \frac{1}{c_L^3} + \frac{2}{c_T^3} . \tag{8.95}$$

If the same value is chosen for $\bar{c}$ as in the Debye model, the two densities coincide for $\omega < \omega_T$.

4b For each polarization, the density of states has the same form as in the Debye model. The corresponding contribution to c_V is then $RD(T_s/T)$, where $T_s = \hbar c_s k_m/k_B$. We thus find

$$c_V = R \left[D \left(\frac{T_L}{T}\right) + 2D \left(\frac{T_T}{T}\right) \right] \tag{8.96}$$

with

$$T_L = \frac{\hbar c_L k_m}{k_B} \qquad \text{and} \qquad T_T = \frac{\hbar c_T k_m}{k_B} . \tag{8.97}$$

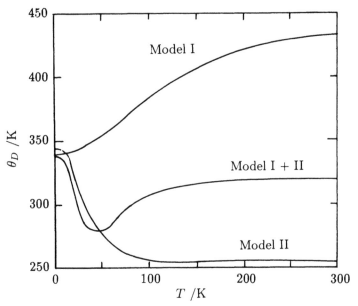

Figure 8.26: $\Theta_D(T)$ for the three models defined in problem 8.2, in 4 (model I) and in 5 (model II and model I + II).

4c When $T \to \infty$, the Debye function tends to 1 and c_V takes the limiting value $3R$ of Dulong and Petit, as in the Debye model. When $T \to 0$, each contribution has a T^3 behaviour of the form (8.94), and therefore

$$c_V = \frac{4\pi^4}{5} R \left[\left(\frac{T}{T_L} \right)^3 + 2 \left(\frac{T}{T_T} \right)^3 \right] = \frac{4\pi^4}{5} RT^3 \left[\frac{1}{T_L^3} + \frac{2}{T_T^3} \right]. \qquad (8.98)$$

Given (8.95 and 97), this T^3 law is identical with equation (8.94) of the Debye model.

4d From the experimental values of c_L and c_T we find $\bar{c} = 2612$ m s^{-1}, $T_L = 620$ K and $T_T = 302$ K. The T^3 law then becomes $c_V = 4.97 \times 10^{-5}\ T^3$, in satisfactory agreement with the experimental relationship.

4e c_V can be calculated from equation (8.96). At $T = 100$ K, we have $T/T_L = 0.161$ and $T/T_T = 0.331$. The corresponding values of the Debye function, read from Figure 8.24, are 0.23 and 0.67. Thus we have

$$c_V = R \times [0.23 + 2 \times 0.67] = 1.56\ R$$

and hence $c_V/3R = 0.52$. Similarly, this ratio is equal to 0.81 at $T = 200$ K and 0.90 at $T = 300$ K. In the simple Debye model, these ratios are equal to $D(x)$. The numerical values of $1/x$, read from Figure 8.24, are respectively equal to 0.26, 0.48 and 0.69. Thus $\Theta_D(100) = 385$ K, $\Theta_D(200) = 420$ K, and $\Theta_D(300) = 435$ K.

The curve of $\Theta_D(T)$ is shown in Figure 8.26. It can be seen that, as $T \to 0$, Θ_D is found by equating expressions (8.98 and 94) ($T_m \to \Theta_D$). Thus

$$\frac{3}{[\Theta_D(0)]^3} = \frac{1}{T_L^3} + \frac{2}{T_T^3},$$

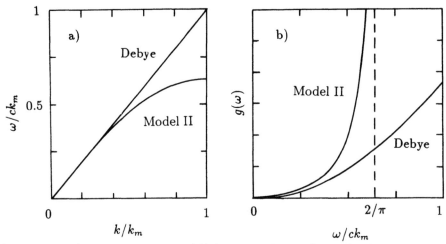

Figure 8.27: a) Dispersion curve and b) frequency density for model II and for the Debye model.

which gives $\Theta_D(0) = 339$ K. Similarly, with the limiting form $D(x) = 1 - x^2/20$ as $x \to 0$ (§8.6.2), we get

$$3[\Theta_D(\infty)]^2 = T_L^2 + 2T_T^2 \ ,$$

which yields $\Theta_D(\infty) = 434$ K.

5a The dispersion curve shown in Figure 8.27a is isotropic, and when $k \to 0$, is the same as the Debye dispersion relation $\omega = ck$. However, the slope vanishes for $k = k_m$, which is in better agreement with experiment.

According to (8.86), the frequency density is given by

$$g(\omega) \ d\omega = 3 \times \frac{V}{(2\pi)^3} \times 4\pi k^2 \frac{dk}{d\omega} \ d\omega \ .$$

Inverting the dispersion relation, we find $k = (2k_m/\pi)\arcsin(\omega/\omega_m)$, and hence

$$g(\omega) = \frac{12}{\pi^5} V k_m^3 \frac{[\arcsin(\omega/\omega_m)]^2}{(\omega_m^2 - \omega^2)^{1/2}} \ .$$

This density, which is shown in Figure 8.27b, is the same as the Debye form (8.92) in the limit $\omega \to 0$, but it diverges at $\omega = \omega_m$.

5b The general expression (8.87) for C_V for a single polarization becomes

$$C_V = 3k_B \frac{4\pi V}{(2\pi)^3} \int_0^{k_m} k^2 \ dk (\beta\hbar\omega)^2 \frac{e^{\beta\hbar\omega}}{(e^{\beta\hbar\omega} - 1)^2}$$

where $\omega = \omega(k)$ is the dispersion relation in model II. On applying the change of variable $k \to u = k/k_m$ and using equation (8.90) for k_m, we get

$$C_V = 3R \ 3 \int_0^1 u^2 \ du \ (\beta\hbar\omega)^2 \frac{e^{\beta\hbar\omega}}{(e^{\beta\hbar\omega} - 1)^2} \ .$$

Since $\beta\hbar\omega$ can also be written as $2/\pi\; x\sin(\pi u/2)$, where $x = T_m/T$, we obtain finally $c_V = 3RF(x)$ with

$$F(x) = \frac{12}{\pi^2}x^2 \int_0^1 u^2\; du\; \sin^2\frac{\pi u}{2}\; \frac{\exp\left(\frac{2}{\pi}u\sin\frac{\pi}{2}x\right)}{\left[\exp\left(\frac{2}{\pi}u\sin\frac{\pi}{2}x\right) - 1\right]^2}\;.$$

It can be shown that this function takes the limiting forms $F(x) \simeq D(x)$ for $x \to \infty$ ($T \to 0$) and $F(x) \simeq 1 - 2.72 \times 10^{-2}x^2$ for $x \to 0$ ($T \to \infty$) while $D(x) \simeq 1 - 5 \times 10^{-2}x^2$.

5c Taking $T_m = 344$ K as found in question 3e, the values of $1/x = T/T_m$ are 0.073, 0.29 and 0.87. At these three points $c_V/3R = F(x)$ is equal to 0.04, 0.74 and 0.97 (Fig. 8.24). In the Debye model, these ratios $c_V/3R$ would occur at $1/x = 0.08$, 0.39 and 1.2. The corresponding values of Θ_D are therefore $\Theta_D(25) = 310$ K, $\Theta_D(100) = 255$ K and $\Theta_D(300) = 250$ K. Observe that as T tends to zero, $F(x) \simeq D(x)$ and thus $\Theta_D(0) = T_m = 344$ K. Furthermore, as $T \to \infty$, comparison of the limiting expressions for $F(x)$ and $D(x)$ yields $\Theta_D(\infty) = T_m \times (2.72/5)^{1/2} = 254$ K. $\Theta_D(T)$ is shown in Figure 8.26.

We might also consider a model "I + II" in which longitudinal and transverse dispersion relations (model I) are allowed with the same form as in model II. In such a model, the molar heat capacity is given by

$$c_V = R\left[F\left(\frac{T_L}{T}\right) + 2F\left(\frac{T_T}{T}\right)\right]\;.$$

This expression yields the curve of Debye temperatures $\Theta_D(T)$ shown in Figure 8.26. The shape of this curve is similar to the experimental curve $\Theta_D(T)$ for copper (Fig. 8.17). The decrease in Θ_D at low temperatures arises from a deviation from linearity in the dispersion curves, and the increase at higher temperatures is caused by the presence of several dispersion relations.

Chapter 9

Helium ^{4}He

9.1 INTRODUCTION

Since it was first liquefied by H. Kamerlingh-Onnes in 1908, helium has opened up a new range of temperature for physical investigations. This field has proved rich in unexpected phenomena that display quantum effects on a macroscopic scale (superconductivity, superfluidity, the existence of a liquid at absolute zero).

Helium possesses two stable isotopes, ^{4}He ($\simeq 100$ %) and ^{3}He ($\simeq 1.3 \times 10^{-4}$ %). The former is composed of boson-type and the latter of fermion-type atoms. The peculiar properties of superfluidity in helium ^{4}He are due to its quantum mechanical character. Consequently we shall mainly discuss the properties of this isotope, which are intimately connected with the Einstein condensation (§7.7). Owing to interactions between atoms, however, the Einstein condensation theory must be reformulated. The introduction of the notion of elementary excitations by L.D. Landau (1941) provides an interpretation of the properties of helium below 2 K, including its transport properties.

For a more detailed discussion of the physics of helium, we recommend the book by J. Wilks (1967). Numerical data in this chapter for which the reference is not given are taken from this work.

9.2 PROPERTIES OF HELIUM ^{4}He

9.2.1 Phase Diagram

There are six known phases of helium ^{4}He:
- three solid phases: a hexagonal compact phase that is stable at low temperatures, a face-centred cubic phase that exists only at very high pressures ($P > 1000$ atm) and a body-centred cubic phase that occupies a very small area of the (T, P) plane in the neighbourhood of $T = 1.6$ K and $P = 28$ atm;
- two liquid phases, denoted He I and He II (Fig. 9.1a);
- one gas phase.

Helium is the only substance that exhibits a liquid phase at absolute zero, the solid phase existing only at pressures in excess of 25 atm. This property is related to the fact that helium is the smallest and lightest of the monatomic molecules. In

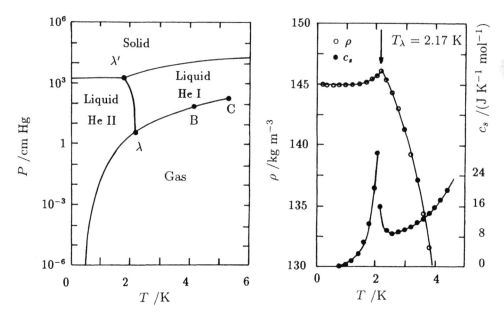

Figure 9.1: a) Phase diagram of helium ^{4}He. B and C are the normal boiling point (P = 1 atm) and the critical point; λ and λ' are both triple points. b) Density ρ and molar heat capacity c_s of liquid ^{4}He measured at saturated vapour pressure. The phase transition occurs at T_λ = 2.17 K [J. Wilks, *op. cit.* p. 666].

the first place, induced electric dipole moments and the resulting van der Waals forces (§8.1) are small, and in the second place, the vibration energy of the solid at absolute zero (§8.4) is large, so that the substance can exist in the solid state only under strong pressure. These reasons also explain the particularly low normal boiling point T_B and critical temperature T_C (Table 9.1). As these properties do not rely on the quantum mechanical character of helium atoms, they are common to both isotopes ^{4}He (bosons) and ^{3}He (fermions).

Helium also has the remarkable property of possessing several liquid phases. The isotope ^{4}He exists as a "normal" liquid phase (He I) and at approximately 2 K (λ line) transforms into a phase (He II) that exhibits "superfluid" properties. This phase change I $\rightarrow$ II is related to the quantum mechanical nature of the ^{4}He isotope (boson). In the same temperature range, the ^{3}He isotope (fermion) does not become superfluid. In zero magnetic field at a much lower temperature ($T \sim 3$ mK), however, helium ^{3}He exhibits two superfluid phases denoted A and B, in which associations of ^{3}He atoms are in effect equivalent to bosons. In what follows, we shall be concerned only with the properties of the boson fluid ^{4}He.

9.2.2   He I $\rightarrow$ He II Phase Transition. Lambda Point

Liquid helium ^{4}He has been and still is used as a thermostat in the low temperature domain. It is produced by liquefaction at 4.2 K at atmospheric pressure. Reducing its saturated vapour pressure decreases the temperature to roughly 1 K (pumped

Table 9.1: Particular points for helium. B and C are the normal boiling point and the critical point, and λ and λ' the triple points of ^{4}He.

		T /K	P /atm	ρ /(kg m^{-3})
^{4}He	B	4.22	1	125
	C	5.20	2.3	67.5
	λ	2.17	4.9×10^{-2}	146
	λ'	1.76	30	180
^{3}He	B	3.19	1	59
	C	3.32	1.15	41

helium). For this reason, a great many of the properties of helium have been measured along its vaporization curve. This is the case for the density ρ and the molar heat capacity c_s under saturated vapour pressure shown in Figure 9.1b. From the molar entropy s we have

$$c(\text{sat}) \equiv c_s \;=\; T \left(\frac{ds}{dT} \right)_{\text{sat}} = T \left(\frac{\partial s}{\partial T} \right)_P + T \left(\frac{\partial s}{\partial P} \right)_T \left(\frac{dP}{dT} \right)_{\text{sat}}$$

$$= \; c_P - T \left(\frac{\partial v}{\partial T} \right)_P \left(\frac{dP}{dT} \right)_{\text{sat}} = c_P - \alpha T v \left(\frac{dP}{dT} \right)_{\text{sat}} \tag{9.1}$$

and, in the region $T < 2$ K, we have $c_s \simeq c_P \simeq c_V$ to within a few parts per thousand.

Figure 9.1b shows the existence of an abrupt change in the physical properties of liquid helium at the "lambda point", $T_\lambda = 2.17$ K, which derives its name from the resemblance between the c_s curve and the Greek letter λ. This change involves a transition of order higher than 1 between the two liquid phases, the first derivatives of the free enthalpy G (volume and entropy) being continuous, i.e. continuity of ρ and absence of latent heat in the transformation He I $\to$ He II. No coexistence is observed between the two liquid phases, but rather a transition in the properties of the liquid. This critical phenomenon is illustrated by the fact that the second derivatives of G (expansion coefficient α_s, adiabatic compressibility and heat capacity c_s on the saturation curve) diverge logarithmically at the λ point (Fig. 9.2). For example

$$\alpha_s /(10^{-2} \text{ K}^{-1}) \;=\; 0.247 + 1.684 \log_{10} |T - T_\lambda| \quad (T < T_\lambda)$$
$$\alpha_s /(10^{-2} \text{ K}^{-1}) \;=\; 3.792 + 1.688 \log_{10} |T - T_\lambda| \quad (T > T_\lambda)$$
$$c_s /(\text{J K}^{-1} \text{ mol}^{-1}) \;=\; 18.2 - 12.00 \log_{10} |T - T_\lambda| \quad (T < T_\lambda)$$
$$c_s /(\text{J K}^{-1} \text{ mol}^{-1}) \;=\; -2.60 - 12.00 \log_{10} |T - T_\lambda| \quad (T > T_\lambda) \,.$$

The two liquid phases He I and He II are separated in the (T, P) plane by a critical line $\lambda\lambda'$ (Fig. 9.1a), at every point of which a similar transition occurs. This line is bounded by the two triple points λ and λ' (Table 9.1).

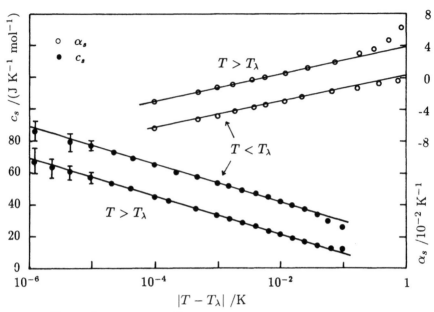

Figure 9.2: Thermal expansion coefficient α_s and molar heat capacity c_s under saturated vapour pressure near the lambda point in ^{4}He [E.C. Kerr and R.D. Taylor, *Ann. Phys.* (New York) **26**, 292 (1964) ; M.J. Buckingham and W.M. Fairbank, *Prog. Low Temp. Phys.* **3**, 80 (1961)].

9.2.3   Transport Properties of Superfluid He II

Thermal conduction

When the pressure of liquid helium at 4.2 K is reduced by pumping, it cools by boiling vigorously throughout its volume. As it goes through the lambda point, the boiling process ceases instantly and the liquid becomes calm, with evaporation occurring only at the surface. This fact can be explained by the huge increase in thermal conductivity K, whose value in phase I is approximately 2×10^{-4} W cm^{-1} K^{-1}. In phase II this value increases by a factor greater than 10^3 and in certain conditions close to 10^6, thus making He II the best heat conductor known.

The coefficient of thermal conductivity K of a substance is defined as the proportionality coefficient in the Fourier equation (§6.5.2)

$$\mathbf{J} = -K \ \boldsymbol{\nabla} T \tag{9.2}$$

which relates the heat current density $\mathbf{J}$ (amount of heat per unit time per unit area crossed) to the temperature gradient prevailing in the substance. In phase I of helium, Fourier's equation is valid. In phase II, however, it fails, and the ratio $J/|\ \boldsymbol{\nabla}\ T|$ depends on numerous factors (J, $|\ \boldsymbol{\nabla}\ T|$, the effective surface area traversed in a capillary tube, ...). These heat transport effects are explained by the existence of convective eddy movements that arise even inside capillaries. In other words, heat transport is accompanied by mass transport and it is therefore intimately related to the phenomenon of superfluidity which we shall now discuss.

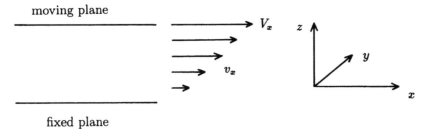

Figure 9.3: Couette flow

Superfluidity

The behaviour of the coefficient of viscosity η in liquid helium is remarkable. In phase I, η is already very small (about 3.5×10^{-5} poise $= 3.5 \times 10^{-6}$ kg m^{-1} s^{-1}) compared with that of water (1.3×10^{-2} poise). Moreover, in contrast to other liquids, this viscosity decreases as the temperature decreases. Phase II is even more remarkable in this regard, since fluid flows in capillary tubes without viscosity. This effect is called superfluidity (Kapitza, 1941; Allen and Misener, 1939).

To clarify the ensuing discussion, we recall the definition of the viscosity coefficient η (§6.5.3) and the methods used to measure it. Consider a fluid contained between two parallel planes that are perpendicular to an axis Oz (Fig. 9.3), one of which is kept fixed and the other moves parallel to itself at constant velocity V_x. The layers of fluid in immediate contact move at the same velocity as the planes, while, under stationary conditions, the velocity v_x of any layer is a linear function of z only ($\partial v_x / \partial z =$constant). A tangential force must be applied parallel to Ox in order to keep the mobile plane moving and the other fixed, since they are subject to the forces of viscosity. The stress (force per unit area) τ_{zx} that must be applied to the moving plane is related to the velocity gradient by the phenomenological equation

$$\tau_{zx} = \eta \, \frac{\partial v_x}{\partial z} \ . \tag{9.3}$$

The proportionality constant η is the dynamic viscosity coefficient.

Several methods are used for its measurement. The most convenient consists in measuring the volume flow Q of the fluid through a cylindrical tube of radius R and length l under a pressure difference ΔP. From Poiseuille's law

$$Q = \frac{\pi \, \Delta P}{8 \eta l} R^4, \tag{9.4}$$

η can be found. Another approach employs the Couette viscometer, which consists of two concentric cylinders of radii R_1 and R_2 and height h, the first of which rotates at a velocity ω_1, the second being fixed. The torque that must be applied to maintain the movement is given by

$$\Gamma = 4\pi \eta \omega_1 h \frac{R_1^2 R_2^2}{R_2^2 - R_1^2} \ , \tag{9.5}$$

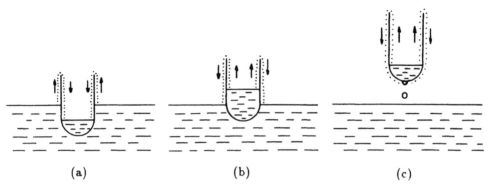

Figure 9.4: Daunt and Mendelssohn experiment.

which yields a measurement of η.

In all fluids, different methods yield the same results for the coefficient η, but in He II the situation is different. While the coefficient obtained from Poiseuille's law appears to be smaller than 10^{-11} poise, that found by the Couette method is of the order of 2×10^{-5} poise, a value close to that of helium I. The two-fluid model (§9.3) elucidates this discrepancy of a factor of 10^6 between the experimental values and explains the phenomenon of superfluidity.

The behaviour of helium in the experiment of Daunt and Mendelssohn (1939) has to do with its superfluidity. In this experiment, a beaker is first partially immersed in a bath of helium II (Fig. 9.4a). The beaker, initially empty, fills with liquid until, after a few minutes, the levels become equal. If now the beaker is raised (Fig. 9.4b), the flow occurs in the opposite direction. Finally, if the beaker is removed from the helium bath, it starts to drip until it empties entirely (Fig. 9.4c). An explanation for this effect is provided by the existence of a film of helium on the walls of the beaker. The film makes a continuous connection between the two levels and acts as a siphon. Even though the thickness of the film is of the order of 200 Å, it flows without resistance because the helium is superfluid.

Fountain effect

When a chemical substance is in equilibrium, its chemical potential (molar free enthalpy) has the same value everywhere. If the conditions of temperature and pressure are varied locally over a short period of time, the molar free enthalpy at a given point varies according to

$$dg = -sdT + vdP \ . \tag{9.6}$$

In normal liquids, the system quickly returns to mechanical equilibrium (equality of pressure), but the return to thermal equilibrium (equality of temperature) is slower because it is governed by thermal conduction. For this reason, the time taken for g to return to its equilibrium value is long.

In helium II, however, the very high conductivity due to superfluidity produces instantaneous equalization of the chemical potentials. For this reason any local

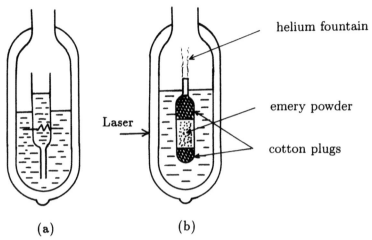

(a)   (b)

Figure 9.5: Experimental arrangements of Allen and Jones illustrating the thermomechanical effect in helium II.

increase in temperature is immediately compensated by an increase of pressure in such a way that $dg = 0$, i.e.

$$\frac{dP}{dT} = \frac{s}{v} .$$

(9.7)

This is the basis of the thermomechanical effect first observed by Allen and Jones (1938), who used the arrangement sketched in Figure 9.5a. An electric current passing through a resistance generates a local increase in temperature, and the resulting pressure increase shows up as an increase in the level of the helium in the tube. At 1.5 K, the value of dP/dT calculated from s and v corresponds to a rise in the helium level of 2.0 cm per millikelvin. This large value is the basis of a spectacular experiment called the fountain effect (Fig. 9.5b). The temperature of the helium increases when light energy is absorbed by emery powder. The resulting over-pressure causes a jet of helium to spout out of the tube that is placed at the top of the bulb. When the jet is strong, the helium does not fall back down, but evaporates in the warmer upper part of the cryostat.

9.3 LONDON MODEL AND TISZA MODEL

In 1924, Einstein discovered that at low temperatures an ideal boson gas experiences a phase change (§7.7), called the Einstein condensation. It was believed for a long time that this condensation had no application. In 1938, F. London applied the concept of Einstein condensation to liquid helium and suggested that the peculiar transport properties of this fluid are related to its quantum mechanical character. He determined the transition temperature ($T_C = 3.1$ K) and showed that the heat capacity at this temperature exhibits a cusp-like maximum (Fig. 7.7b). The agreement with the experimental results lent support to this model, in spite of its neglect of interactions.

Using the results of London, L. Tisza (1938) made the assumption that helium II is composed of a mixture of two fluids, a "normal" fluid, whose properties are the extrapolation of those of helium I, and a "superfluid", which he identified with all the atoms that are condensed in the ground state. In this two-fluid model, the fraction of superfluid is $1 - (T/T_C)^{3/2}$ (7.62), which varies from 0 at the transition temperature to 1 at absolute zero.

This model provides a simple explanation of superfluidity. Since the energy of the superfluid component is zero, it cannot be dissipated in friction, as its name indicates. This explains the discrepancy between measurements of the viscosity coefficient η obtained from Poiseuille flow and from rotating cylinder viscometers. In the former, the superfluid component flows freely in the tube, while in the latter the cylinders rub against the normal component. In other words, in the first case, what is measured is (cf. 9.4)

$$\frac{1}{\eta} = \frac{1}{\eta_{\text{super}}} + \frac{1}{\eta_{\text{normal}}} \quad,$$

and hence $\eta = 0$ if $\eta_{\text{super}} = 0$. In the second case, (cf. 9.5)

$$\eta = \eta_{\text{super}} + \eta_{\text{normal}} = \eta_{\text{normal}} \quad.$$

An interesting experiment performed by Daunt and Mendelssohn (1939) is directly explained by the two-fluid model. A container, whose base is composed of a large number of capillary tubes, is filled with helium II. As the fluid flowing through the capillaries contains mainly the superfluid component, it is, by virtue of (7.62), colder than the remaining liquid. The temperature of the latter, in contrast, increases. This experiment illustrates the mechano-thermal effect, which is the reverse of the thermomechanical effect investigated in paragraph 9.2.3. It might be thought that this phenomenon could be used for generating low temperatures. The heat capacity of helium at these temperatures is, however, too small for it to be useful as a thermostat.

The models of London and Tisza have nowadays been abandoned in favour of the Landau theory. They have nevertheless been successful in predicting and qualitatively explaining a large number of properties of helium II.

9.4 LANDAU THEORY

L. Landau (1941) developed a theory to explain the properties of helium in its superfluid phase He II. In this theory the system of moving helium atoms is replaced by a fluid having the properties of helium at absolute zero, upon which is superimposed a system of elementary excitations. This method, which is similar to that of phonons in solids (§8.4), is employed in many other fields of physics.

Landau proposed a special form for the dispersion relation $\epsilon(p)$ of these excitations. As predicted by Landau, this relation, which has been determined by neutron diffraction experiments on helium II (§8.3) (Fig. 9.6a), is composed of two regions. In the first region, for $p/\hbar < 0.6$ Å^{-1}, the relation is linear

$$\epsilon = cp \tag{9.8}$$

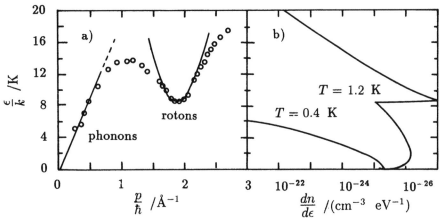

Figure 9.6: a) Dispersion relation of the elementary excitations in liquid helium at $T = 1.12$ K, obtained by neutron diffraction [D.G. Henshaw and A.D.B. Woods, *Phys. Rev.* **121**, 1266 (1961)]. b) Population densities $dn/d\epsilon$ of the energy levels of the excitations as a function of ϵ/k at $T = 0.4$ and 1.2 K (note that the x and y axes are interchanged).

where $c = 2.37$ m s^{-1} is the same as the velocity of sound in helium II. By analogy with the situation in solids, this region is called the phonon zone. In the second region, for $p/\hbar > 1$ Å^{-1}, the dispersion relation displays a minimum around which it has a parabolic form

$$\epsilon = \Delta + \frac{(p - p_0)^2}{2m_0} \tag{9.9}$$

where $\Delta/k = 8.65$ K, $p_0/\hbar = 1.91$ Å^{-1} and $m_0 \simeq 0.16\, m_{He}$. This region is called the roton zone, on the basis of Landau's idea that these are vortex excitations in the liquid. This explanation has since been abandoned.

Landau's theory can, in particular, give the form of the heat capacity for helium II below 2 K. For this, the free energy F of helium is evaluated by adding to the free energy at absolute zero ($F_0 \equiv U_0$) that of the excitations. As the latter are indeterminate in number and have no spin, Bose-Einstein statistics with zero chemical potential (§7.2) may be applied to them. Thus

$$F(T,\, V,\, N) = U_0(N,\, V) + \frac{kT}{h^3} \int d^3\mathbf{r}\, d^3\mathbf{p} \ln\left[1 - e^{-\beta\epsilon(p)}\right] . \tag{9.10}$$

In our search for an analytical expression for F, it is instructive to investigate the population density of the energy levels of the excitations. Application of the Bose-Einstein distribution with $\mu = 0$ (7.1) yields

$$d^6 N_{\mathbf{r},\, \mathbf{p}} = \frac{d^3\mathbf{r}\, d^3\mathbf{p}}{h^3}\, f(\epsilon) = \frac{d^3\mathbf{r}\, d^3\mathbf{p}}{h^3}\, \frac{1}{e^{\beta\epsilon(p)} - 1} . \tag{9.11}$$

Integration over $\mathbf{r}$ and over the directions of $\mathbf{p}$ then gives

$$dn = \frac{dN_p}{V} = \frac{4\pi p^2\, dp}{h^3}\, f(\epsilon) \tag{9.12a}$$

and the population density of the levels $dn/d\epsilon$ is obtained by changing the variables $p \rightarrow \epsilon = \epsilon(p)$. Bearing in mind that the inverse function $p = p(\epsilon)$ is multi-valued for $\epsilon \geq \Delta$, we obtain finally

$$\frac{dn}{d\epsilon} = \frac{4\pi}{h^3} f(\epsilon) \sum_i \frac{p_i^2}{|\epsilon'(p_i)|} \tag{9.12b}$$

where $\epsilon'(p)$ is the derivative of $\epsilon(p)$ and p_i is one of the solutions of $\epsilon = \epsilon(p)$. The population density of the excitation energy levels, calculated numerically with this formula from the experimentally measured variation $\epsilon(p)$ of Figure 9.6a, is shown in Figure 9.6b for $T = 0.4$ K and $T = 1.2$ K. At $T = 0.4$ K, it is found that only the phonon levels are populated. We may therefore expect to be able to apply the results of the Debye model (§8.6) to helium for $T \lesssim 0.6$ K. In contrast, at $T = 1.2$ K, rotons with $p \simeq p_0$ dominate, and we expect that a new regime will arise for $T \gtrsim 1$ K. Note that the density $dn/d\epsilon$ even diverges when $\epsilon'(p)$ vanishes. At such points, however, $dn/d\epsilon$ remains integrable and produces no divergence in any physical quantity. Thus, at $T = 0.4$ K, the divergence in $dn/d\epsilon$ for $\epsilon = \Delta$ occurs in such a narrow energy band that it contributes negligibly, owing to the smallness of $f(\Delta)$ at this temperature ($\sim 10^{-50}$). Similarly, the infinite value at $\epsilon/k \simeq 14$ K produces no perceptible effect.

This discussion shows that the physical quantities can be obtained with good precision by using for the dispersion relation the piece-wise analytical expression

$$\epsilon = \begin{cases} cp & p < p_1 \\ \Delta + (p - p_0)^2/2m_0 & p > p_1 \end{cases},$$

where $p_1/\hbar$ is of the order of 1 Å^{-1}. This expression gives a good fit for $\epsilon(p)$ in the respective regions where the phonon and the roton contributions dominate. Distinguishing the phonon contribution F_{ph} from that of the rotons F_r, we can therefore write for the free energy (9.10)

$$F = U_0 + F_{ph} + F_r$$

$$\text{with} \quad F_{ph} = \frac{kT}{h^3} V \int_0^{p_1} 4\pi p^2 \, dp \ln\left[1 - e^{-\beta cp}\right] \tag{9.14a}$$

$$\text{and} \quad F_r = \frac{kT}{h^3} V \int_{p_1}^{\infty} 4\pi p^2 \, dp \ln\left[1 - e^{-\beta\Delta - \beta(p-p_0)^2/2m_0}\right]. \tag{9.14b}$$

Borrowing from the Debye model (§8.6), we get for the phonons,

$$F_{ph} = \frac{4\pi V}{h^3 c^3} (kT)^4 \int_0^{\beta cp_1} x^2 \ln(1 - e^{-x}) \, dx .$$

Noting that $T < 2$ K and $p_1/\hbar \sim 1$ Å^{-1} gives $\beta cp_1 > 10 \gg 1$, we can therefore set the upper limit of the integral equal to infinity. Hence

$$F_{ph} = -\frac{4\pi^5 V}{45h^3 c^3} (kT)^4 . \tag{9.15}$$

For the rotons, we always have $\beta\epsilon(p) > \Delta/kT \gg 1$. The logarithm in (9.14b) can therefore be expanded as a series and the lower limit of the integral extended to $-\infty$. This gives

$$F_{\mathrm{r}} = -\frac{4\pi V}{h^3}kTe^{-\beta\Delta}\int_{-\infty}^{+\infty}p^2\exp\left[-\beta\frac{(p-p_0)^2}{2m_0}\right]\,dp \ .$$

The integral can be calculated in a straightforward way, but, as a first approximation, as the integrand is non-zero essentially in the vicinity of p_0, p^2 can be replaced by p_0^2 and, as the integral of the Gaussian exponential is equal to $(2\pi m_0 kT)^{1/2}$, we obtain finally

$$F_{\mathrm{r}} = -\frac{4\pi}{h^3}(2\pi m_0)^{1/2}p_0^2V(kT)^{3/2}e^{-\Delta/kT} \ . \tag{9.16}$$

The molar heat capacity of helium can now be calculated by differentiating the molar free energy ($V \to v$), giving

$$c_V \ = \ c_{V\,\mathrm{ph}} + c_{V\,\mathrm{r}}$$

$$\text{with} \ \ c_{V\,\mathrm{ph}} \ = \ \frac{16\pi^5 k^4}{15h^3c^3}vT^3 \tag{9.17}$$

$$\text{and} \ \ c_{V\,\mathrm{r}} \ = \ \frac{4\pi p_0^2 kv}{h^3}(2\pi m_0\Delta)^{1/2}\left(\frac{\Delta}{kT}\right)^{3/2}e^{-\Delta/kT}\left[1+\frac{kT}{\Delta}+\frac{3}{4}\left(\frac{kT}{\Delta}\right)^2\right] \ .$$

Figure 9.7 shows the curve of c_V in the Landau model. It is clear that the curve fits the experimental results very satisfactorily, confirming that phonons dominate up to about 0.6 K, while rotons are preponderant above about 1 K. Note that the T^3 law, characteristic of phonon systems, is obeyed below 0.6 K.

As we shall see later, the Landau theory provides an explanation for a large number of properties of superfluid helium. The fact that an interacting boson system can be described as a set of independent excitations (or quasiparticles) was proved by N.N. Bogoliubov [*J. Phys. Moscow* **11**, 23 (1947)] and the shape of the dispersion curve was obtained satisfactorily by R.P. Feynman [R.P. Feynman and M. Cohen, *Phys. Rev.* **102**, 1189 (1956)] by considering a system of interacting bosons in the framework of quantum mechanics.

9.5   SUPERFLUIDITY IN THE LANDAU THEORY

9.5.1   Absolute Zero

When an ordinary liquid flows in a pipe, the work done by the viscosity forces in the fluid produces an increase in internal energy which causes heating. In the stationary regime, a pressure gradient must be applied to maintain the movement, and the work supplied by the pressure forces generates the increase in internal energy. If the fluid flows freely (zero pressure gradient), the corresponding work is zero and this increase occurs at the expense of the kinetic energy.

In the Landau theory, viscosity is interpreted as the creation of elementary excitations (phonons and rotons) which increase the internal energy of the helium and

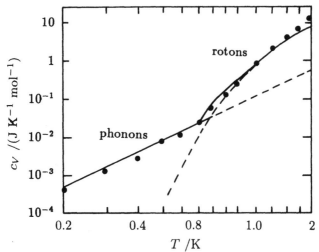

Figure 9.7: Heat capacity per unit mass at constant volume ($c_V \simeq c_s$) of helium II [H.C. Kramers, J.D. Wasscher and C.J. Gorter, *Physica* **18**, 329 (1952); J. Wiebes, C.G. Niels-Hakkenberg and H.C. Kramers, *Physica* **23**, 625 (1957)]. The curves show the predictions (9.17) of the Landau theory.

reduce its momentum if it is flowing freely. Superfluidity arises from the impossibility of creating excitations under certain conditions (L. Landau, 1941).

Now consider helium at absolute zero flowing at velocity $\mathbf{v}$ in a fixed capillary tube. In the reference frame moving at velocity $\mathbf{v}$ with respect to the tube, the helium is at rest and its energy and momentum are given by

$$E_0 = U_0 \qquad \text{and} \qquad \mathbf{P}_0 = 0 \ .$$

If an elementary excitation of momentum $\mathbf{p}$ is created, the above quantities become

$$E_1 = U_0 + \epsilon(p) \qquad \text{and} \qquad \mathbf{P}_1 = \mathbf{p} \ .$$

In the reference frame where the tube is at rest, the energy E' and momentum $\mathbf{P}'$ of the helium are found from the above values of E and $\mathbf{P}$ through the Galilean transformations

$$E' = E + \frac{1}{2}M\mathbf{v}^2 + \mathbf{P}.\mathbf{v} \qquad \text{and} \qquad \mathbf{P}' = \mathbf{P} + M\mathbf{v}$$

where M is the total mass of helium. Thus the energy and momentum of the excitation-free fluid are

$$E_0' = U_0 + \frac{1}{2}M\mathbf{v}^2 \qquad \text{and} \qquad \mathbf{P}_0' = M\mathbf{v} \qquad (9.18a)$$

and, when an excitation is present,

$$E_1' = U_0 + \epsilon(p) + \frac{1}{2}M\mathbf{v}^2 + \mathbf{p}.\mathbf{v} \qquad \text{and} \qquad \mathbf{P}_1' = \mathbf{p} + M\mathbf{v} \ . \qquad (9.18b)$$

We may therefore conclude that in the reference frame of the tube, the energy and momentum of the elementary excitation are

$$\epsilon'(p) = \epsilon(p) + \mathbf{p}.\mathbf{v} \qquad \text{and} \qquad \mathbf{p}' = \mathbf{p} \ . \tag{9.19}$$

These equations show that the momentum associated with an excitation is independent of the framework considered, but that its energy depends on the velocity $\mathbf{v}$ of the helium at absolute zero in the reference frame of the observation.

In the Landau theory, when helium flows freely in a tube, creation of an excitation does not affect the velocity $\mathbf{v}$ of the constituent possessing the properties of helium at absolute zero. The total momentum of the helium is nevertheless reduced, that is, $P_1' < P_0'$. Moreover, since the tube is fixed, the total energy of the helium remains constant when the excitation is created, i.e., $E_1' = E_0'$. From expressions (9.18) for E_0' and E_1', it follows that the excitation is subject to

$$\epsilon(p) + \mathbf{p}.\mathbf{v} = 0 \ .$$

Since we always have $\mathbf{p}.\mathbf{v} > -pv$, an excitation of momentum $\mathbf{p}$ occurs only if the helium flows at a velocity v such that

$$v > \frac{\epsilon(p)}{p} \ .$$

Geometrically, the ratio ϵ/p corresponds to the slope of a straight line in the (p, ϵ) plane (Fig. 9.6a) connecting the origin to a point on the dispersion curve. Since this slope has a minimum value v_c at $p/\hbar \simeq 1.9$ Å^{-1}, flows having a velocity $v < v_c$ cannot generate excitations. It follows that helium at absolute zero flows without viscosity provided its velocity is lower than a critical value v_c.

The value predicted by the Landau theory for the critical velocity for superfluid flow in a capillary tube is (cf. Fig. 9.6a)

$$v_c \simeq \frac{8.7 \ k}{1.9 \times 10^{10}\hbar} = 60 \text{ m s}^{-1} \ .$$

The experimental value of the critical velocity at absolute zero is of the order of 1 cm s^{-1}. This discrepancy comes from the assumption in the Landau theory that the flow is laminar and from neglecting the possibility of internal motions (vortices).

9.5.2 $T \neq 0$ K

For temperatures greater than zero, the Landau theory considers helium II to be equivalent to helium at absolute zero upon which a gas of elementary excitations has been superimposed. The above arguments demonstrating the existence of superfluidity in helium still remain valid above absolute zero, i.e. no new excitation can be created if $v < v_c$. Excitations that already exist, however, interact with the walls of the tube and produce the phenomenon of viscosity.

Let us now consider helium in stationary flow in a capillary tube at a velocity below the critical velocity. The "superfluid" component, which has the same properties as helium at zero Kelvin, flows without viscosity at a velocity that is denoted by $\mathbf{v}_s$. The "normal" component, which is composed of thermal excitations, is viscous and therefore flows at a velocity $\mathbf{v}_n$ that is smaller than $\mathbf{v}_s$. In the rest frame of the normal component, the distribution of elementary excitations is governed by the Bose-Einstein relation (9.11) in which the energy ϵ of an excitation is given as a function of its momentum $\mathbf{p}$ by equation (9.19) $\epsilon = \epsilon(p) + \mathbf{p} \cdot \mathbf{v}$, where

$$\mathbf{v} = \mathbf{v}_s - \mathbf{v}_n \qquad (9.20)$$

is the velocity of the superfluid component in this reference frame. On calculating the momentum associated with the normal component

$$\mathbf{P} = \int \mathbf{p} \, d^6 N_{\mathbf{r}, \, \mathbf{p}} = \int \mathbf{p} \, \frac{d^3 \mathbf{r} \, d^3 \mathbf{p}}{h^3} f \left(\epsilon(p) + \mathbf{p} \cdot \mathbf{v} \right)$$

for small $\mathbf{v}$, a series expansion of $f(\epsilon)$ can be made, giving

$$\mathbf{P} = \int \mathbf{p} \left[f(\epsilon(p)) + \mathbf{p} \cdot \mathbf{v} \frac{df}{d\epsilon} \right] \frac{d^3 \mathbf{r} \, d^3 \mathbf{p}}{h^3} \quad .$$

For reasons of symmetry, the first term in the integral gives a zero contribution, and the second yields a vector proportional to $\mathbf{v}$. Performing the integration over $d^3 \mathbf{r}$ and expressing $\mathbf{p}$ in spherical co-ordinates $(p, \, \theta, \, \phi)$ with the polar axis parallel to $\mathbf{v}$, we find

$$\mathbf{P} = \frac{V}{h^3} \int p \cos \theta \, \frac{df}{d\epsilon} \, p \cos \theta \, \mathbf{v} \, p^2 \, dp \, \sin \theta \, d\theta \, d\phi = \frac{4\pi V}{3h^3} \int p^4 \frac{df}{d\epsilon} dp \, \mathbf{v} \quad .$$

It is useful to introduce the notation

$$\rho_n = -\frac{4\pi}{3h^3} \int p^4 \frac{df}{d\epsilon} dp \qquad (9.21)$$

$(\rho_n > 0$ since $df/d\epsilon < 0)$, so that we can write the momentum associated with the normal component in the form

$$\mathbf{P} = \rho_n V(-\mathbf{v}) \qquad (9.22)$$

where $-\mathbf{v} = \mathbf{v}_n - \mathbf{v}_s$ is the velocity of this normal component with respect to the superfluid component. This shows that mass transport can be associated with the movement of the gas of excitations. The total momentum of the helium II is then equal to

$$\mathbf{P}_{\text{tot}} = V \left(\rho \mathbf{v}_s - \rho_n \mathbf{v} \right) \quad ,$$

which, with the notation

$$\rho_s = \rho - \rho_n \quad , \qquad (9.23)$$

and using (9.20), yields

$$\mathbf{P}_{\text{tot}} = V \left(\rho_s \mathbf{v}_s + \rho_n \mathbf{v}_n \right) \quad . \qquad (9.24)$$

Equations (9.21, 23 and 24) define the two-component model of Landau. The normal component, associated with the elementary excitations (phonons and rotons), behaves like an ordinary viscous fluid having a density ρ_n (9.21). The remaining density ρ_s is attributed to the superfluid component whose properties are those of helium at absolute zero. This separation corresponds neither to a division of the particles of the fluid into superfluid and normal particles nor, a *fortiori*, to separation into two phases. The two-component model has proved to be very fruitful, and has been used to predict and explain many of the properties of helium II.

To calculate ρ_n, we consider the contributions of the phonons and rotons separately. The density associated with the excitations of phonon type is calculated by integrating (9.21) by parts, i.e., using the relation $\epsilon = cp$,

$$\rho_{n\ ph} = -\frac{4\pi}{3h^3c^5}\left[\epsilon^4 f\right]_0^\infty + \frac{16\pi}{3h^3c^5}\int_0^\infty \frac{\epsilon^3\ d\epsilon}{e^{\beta\epsilon}-1} \ .$$

The fully integrated term is zero, and changing the variable $x = \beta\epsilon$ gives rise to a numerical integral equal to $\pi^4/15$ (§7.2.3). Finally, this gives for the phonons

$$\rho_{n\ ph} = \frac{16\pi^5}{45h^3c^5}(kT)^4 \ . \tag{9.25a}$$

The density associated with the rotons is found from (9.9). Observing that $f(\epsilon) \simeq e^{-\beta\epsilon}$ for $\epsilon > \Delta \gg kT$, we have

$$\rho_{n\ r} \simeq \frac{4\pi\beta}{3h^3}e^{-\beta\Delta}\int_{-\infty}^{+\infty} p^4 e^{-\beta(p-p_0)^2/2m_0} \ .$$

As in paragraph 9.4, p^4 can be replaced by p_0^4 and we find

$$\rho_{n\ r} = \frac{4\pi}{3h^3}p_0^4 e^{-\Delta/kT}\left(\frac{2\pi m_0}{kT}\right)^{1/2} \ . \tag{9.25b}$$

The density of the normal component

$$\rho_n = \rho_{n\ ph} + \rho_{n\ r} \tag{9.25c}$$

obtained from (9.25) is given by the curve in Figure 9.8. It varies as T^4 for $T \lesssim 0.5$ K where phonons are preponderant, and above this temperature it is dominated by rotons.

The two-fluid model has been confirmed by determining ρ_n experimentally using two techniques. Here we describe the more direct method, due to E.L. Andronikashvili [*Zh. Eksp. Teor. Fiz.* **16**, 780 (1946), **18**, 424 (1948)]. The experiment consists in placing a set of disks in liquid helium and making them oscillate around their common axis. Only the normal component is dragged along by the movement of the disks. Measuring the oscillation period yields the moment of inertia of the normal fluid present between the disks and hence its density. This method, which works for $\rho_n/\rho \gtrsim 10^{-2}$ ($T > 1.3$ K), gives values of ρ_n that are in satisfactory agreement with the two-fluid model (Fig. 9.8). Note that the agreement improves

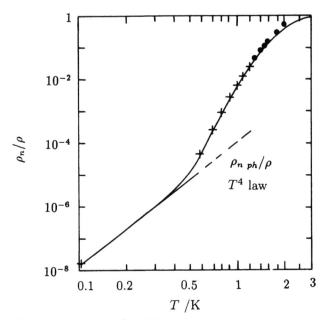

Figure 9.8: Relative density ρ_n/ρ of the normal component of He II. The full line is calculated from the Landau theory (9.25). The points are the experimental measurements of Andronikashvili obtained by an oscillation method and the crosses by Peshkov from measurements of the velocity of second sound.

around 2 K if the exact experimental value of Δ is used for each temperature [J.L. Yarnell et al., *Phys. Rev.* **113**, 1379 (1959)]. The second method, which extends to a lower temperature region, makes use of the phenomenon of second sound, described in the next paragraph. The experimental points [V.P. Peshkov, *Soviet Phys. JETP* **11**, 580, 1960] lie exactly on the curve in Figure 9.8.

Attention is drawn to the fact that the viscosity coefficient measured in helium with the Couette viscometer (§9.2.3) is that due to the normal component η_n (§9.3). This is of the same order of magnitude as the viscosity of helium I. For temperatures below about 1 K, however, the fraction of normal component is very small and the effects of viscosity are no longer felt.

9.5.3   Second Sound

Sound wave propagation in an ordinary fluid is a well known phenomenon. It consists of longitudinal waves in which the pressure P and density ρ vary at constant entropy S, and which transport mechanical energy. The phase velocity (velocity of sound) of these waves is given by

$$c^2 = \left(\frac{\partial P}{\partial \rho}\right)_S = \frac{1}{\rho \chi_S} = \frac{\gamma}{\rho \chi_T} \tag{9.26}$$

where $\gamma = c_P/c_V$, and where χ_S and χ_T are the adiabatic and isothermal compressibility.

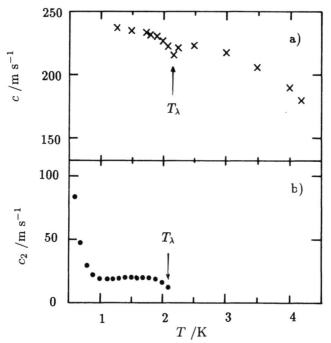

Figure 9.9: a) Velocity of sound c in liquid helium under saturated vapour pressure [Wilks, p. 670]. b) Velocity of second sound c_2 in helium II under saturated vapour pressure [Wilks, p. 669].

Sound waves also propagate in both He I and He II phases of liquid helium with a velocity whose temperature dependence is shown in Figure 9.9a. Note that at the λ point, the velocity tends to zero, concomitantly with the fact that χ_S tends logarithmically to infinity (§9.2.2) and that the absorption coefficient for sound waves also becomes infinite.

Another wave propagation effect in helium II was predicted by L. Tisza (1938) and L. Landau (1941) on the basis of the two-fluid models. It is known as second sound. Ordinary sound in helium II ("first sound") is composed of waves in which the two fluids of density ρ_n and ρ_s vibrate in phase, thus generating variations in the total density at constant entropy. Other waves also exist that propagate a "second sound" in which the two fluids, which can move without friction relative to each other, vibrate in opposite phase, in such a way that ρ is approximately constant. In this kind of wave, the variations in ρ_n (and in ρ_s) generate strong oscillations in the temperature (cf. (9.25)). These are not sound waves, therefore, since the energy is transported as heat (E.M. Lifschitz, 1944). Second sound is therefore generated by a resistance heater driven by an alternating current, and the oscillations in temperature are detected by a resistance thermometer (V.P. Peshkov, 1944). Measurements of the velocity of second sound are shown in Figure 9.9b. The equation

$$c_2^2 = \frac{\rho_s}{\rho_n} \frac{s^2 T}{M c_V} \tag{9.27}$$

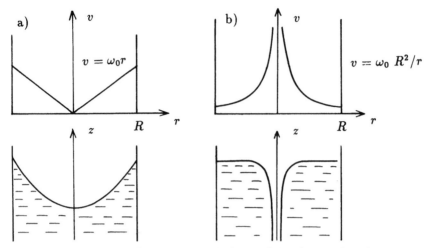

Figure 9.10: Distribution of velocities and shape of the free surface for two types of rotational movement in a viscous fluid: a) stable configuration; b) eddy or vortex.

for the velocity of second sound c_2 is obtained from hydrodynamic considerations [Wilks, p.47]. By means of this equation, ρ_s/ρ_n can be determined from the experimental values of the entropy and the molar heat capacity s and c_V. The resulting values of ρ_n/ρ, plotted in Figure 9.8, are in good agreement with the Landau theory. Note that in the region where only phonons are present ($T < 0.6$ K), second sound is no longer propagated, since the phonons have practically no mutual interaction.

9.6   ROTATION OF HELIUM. QUANTUM VORTICES

9.6.1   Rotation of an Ordinary Liquid

When a cylindrical vessel containing an ordinary liquid is rotated about its axis at angular velocity ω_0, the liquid is set in motion by viscosity, and, in the permanent regime, it turns at a uniform angular velocity equal to ω_0. The velocity $\mathbf{v}$ of a fluid element located at $\mathbf{r}$ is then

$$\mathbf{v} = \boldsymbol{\omega}_0 \times \mathbf{r} \ , \tag{9.28}$$

that is, the liquid rotates like a solid. Its free surface adopts a parabolic shape owing to the combined effect of gravity and the centrifugal force (Fig. 9.10a). This motion is subject to the condition

$$\mathrm{rot}\ \mathbf{v} = 2\omega_0 \ , \tag{9.29}$$

that is, using Stokes's theorem, the circulation of the vector $\mathbf{v}$ (vorticity) around any contour having a projected area s is equal to

$$\kappa = \oint \mathbf{v}.d\mathbf{l} = \int \mathrm{rot}\mathbf{v}.d\mathbf{s} = 2\ \omega_0 s \ . \tag{9.30}$$

Another solution exists, however, for the hydrodynamic equations describing a permanent movement. It has a velocity field in $1/r$ of the form

$$\mathbf{v} = \frac{R^2}{r^2} \, \omega_0 \times \mathbf{r} \, , \tag{9.31}$$

where R is the radius of the vessel. This motion corresponds to an eddy that creates a cylindrical hole of radius a in the liquid around the rotation axis (Fig. 9.10b), similar to that observed when a container is emptied. It obeys the condition

$$\mathrm{rot}\ \mathbf{v} = 0 \tag{9.32}$$

for $r > a$ (irrotational motion). This means that the vorticity of a circuit that excludes the axis is zero, while for any circuit that encompasses the axis once, it is equal to

$$\kappa = \oint \mathbf{v}.d\mathbf{l} = 2\omega_0 \pi R^2 \, . \tag{9.33}$$

This eddy motion, however, also designated by the name vortex, has a higher energy than the previous one, and only the "solid" movement (9.28) is observed experimentally.

9.6.2 Rotation Experiments in He II

When he published his theory, L. Landau predicted that in a rotating cylindrical vessel only the normal component of helium II would turn in the permanent regime, and the superfluid component would not be dragged along. In the temperature region below 1 K, since the normal component has a negligible density, it can therefore be expected that the free surface will be flat. D.V. Osborne (1950), however, observed a parabolic free surface, which corresponds to rotation of all the helium. Other measurements similarly contradicted Landau's prediction. In particular, H.E. Hall (1957) showed that the value of the angular momentum of the helium corresponds to rotation of all the liquid. It has nevertheless been confirmed that other properties are in agreement with the Landau theory for helium II in rotation, e.g. the excitation spectrum, velocity of second sound, fountain effect. This discrepancy could perhaps be explained by assuming the existence of weak frictional forces between the normal and superfluid components, but all the above facts can in fact be accounted for by the theory of quantum vortices.

9.6.3 Quantum Vortices

The explanation that is now generally accepted for the properties of rotating helium II was suggested by L. Onsager (1954) and R.P. Feynman (1955), who used general quantum mechanical arguments for the wave function of a set of helium atoms. They showed that the superfluid component of helium can exhibit eddies whose vorticity κ takes a value that is a multiple of an elementary quantum

$$\kappa_0 = \frac{h}{m} \, , \tag{9.34}$$

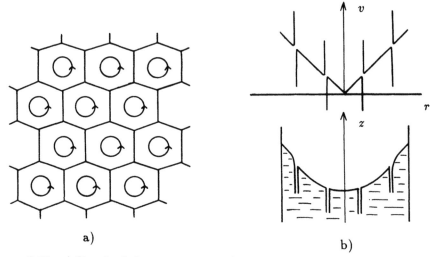

Figure 9.11: a) Sketch of the arrangement of quantum vortices seen in cross-section in the reference frame of the cylinder. b) Velocity distribution in the superfluid component of helium under rotation and shape of the free surface.

m being the mass of a helium atom.

The concept of quantum vortices provides an explanation for the properties of helium II under rotation. When the cylinder is set in motion, it drags the normal component with it. Quantum vortices of vorticity κ_0 with vertical axis then progressively appear in the superfluid component in contact with the walls. Their interaction with the elementary excitations of the normal phase sets these vortices in rotational motion and they move closer to the cylinder axis. This explains why at sufficiently low temperatures the outer zone of the free surface adopts a parabolic shape while the central region, which is flat, shrinks and vanishes as the final regime is reached. In this regime, the superfluid component contains N_v uniformly distributed vortices with their axis vertical (Fig. 9.11). On the basis of energy considerations, H.E. Hall (1960) showed that these vortices rotate about the cylinder axis with uniform angular velocity ω_0 and that their surface density is $n_v = 2\omega_0/\kappa_0$. This defines their outer radius

$$b \simeq \frac{1}{\sqrt{\pi n_v}} = \sqrt{\frac{\kappa_0}{2\pi\omega_0}} \ . \tag{9.35}$$

In such a movement, the velocity field of the superfluid component then becomes comparable with that of the normal component. The total vorticity of the superfluid component is equal to

$$\kappa = n_v \times \pi R^2 \times \kappa_0 = 2\omega_0\pi R^2$$

and is the same as that of a normal liquid given by (9.30). Furthermore, the angular velocity ω at the edge of a vortex, given by

$$\kappa_0 = \omega b \times 2\pi b \ ,$$

is, from (9.35), equal to ω_0. It follows that the velocities at the outer edge of each vortex are the same as those that would be found in a normal liquid. Given the distribution of velocities in a vortex, deviations from the velocity field (9.28) are appreciable only in the immediate neighbourhood of the axis. The resulting velocity field and shape of the free surface are shown schematically in Figure 9.11. For $\omega_0 = 1 \text{ s}^{-1}$, there are approximately $n_v = 2000$ vortices per square centimetre, where the radius of a vortex is $b = 0.1$ mm.

Quantum vortices, which were first detected by indirect methods, have been observed visually [E.J. Yarmchuk, M.J.V. Gordon and R.E. Packard, *Phys. Rev. Lett.* **43**, 213 (1979)], in confirmation of the above theory. In particular, it was found that the first vortex appears above a threshold rotation velocity corresponding to the condition of quantization for the vorticity.

9.7 DILUTION REFRIGERATOR

9.7.1 Evaporation Cooling

Temperatures down to 4 K are generally obtained by the use of liquefied gases (§4.9). Below this temperature, an order of magnitude can be achieved through cooling the liquefied gas (helium ^{4}He or ^{3}He) by evaporation.

In this method, the pressure P above the liquid helium is reduced by pumping the vapour above it. The liquid, which evaporates continuously, cools in accordance with the relation $P = P_e(T)$, where $P_e(T)$ is the saturated vapour pressure of the liquid at temperature T. Since $P_e(T)$ varies approximately as $\exp(-A/T)$, less and less gas is pumped as the temperature decreases, and the temperature attained depends on the power of the pump and the size of the heat leaks. Pressures normally used are of the order of one torr (1 torr = 1 mm of mercury = 133 Pa) and the lowest temperatures reached are 0.9 K for ^{4}He ($P_e = 4 \times 10^{-2}$ torr) and 0.3 K for ^{3}He ($P_e = 2 \times 10^{-3}$ torr). It is possible to reach 2 mK by making use of the properties of ^{4}He-^{3}He mixtures, which we describe below.

9.7.2 Properties of ^{4}He-^{3}He Mixtures

The properties of ^{4}He-^{3}He mixtures change progressively from those of helium ^{4}He as the molar fraction of ^{3}He

$$x = \frac{n(^3\text{He})}{n(^3\text{He}) + n(^4\text{He})}$$

increases. The temperature at which superfluidity occurs under saturated vapour pressure decreases with increasing x (Fig. 9.12).

Furthermore, for temperatures below about 0.8 K, depending on the value of x, the mixture can separate into two phases (Fig. 9.12), i.e. *phase separation* occurs. The first phase (line AB) is more impoverished in ^{3}He and its molar fraction at the lowest temperature reached is $x_B = 6.4\%$. The second phase (line AC'C) is practically pure ^{3}He below about 0.2 K. The molar enthalpies of each phase in the

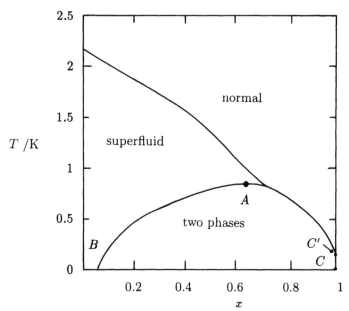

Figure 9.12: Phase diagram of ^{4}He-^{3}He mixtures under saturated vapour pressure. Below the curve BAC'C, the liquid separates into two phases whose molar fractions at a given temperature can be read off the curves AB and AC'C respectively.

neighbourhood of B and C are respectively

$$h_B /(\text{J mol}^{-1}) = 95 \, [T /K]^2 \quad \text{and} \quad h_C /(\text{J mol}^{-1}) = 13 \, [T /K]^2 \ . \tag{9.36}$$

Hence, when dn moles of ^{3}He leave the rich phase and enter the poor phase at constant temperature, there is a change in enthalpy

$$dH = dn(h_B - h_C) \ . \tag{9.37}$$

Since h_C is smaller than h_B, it can be seen that the transfer of helium ^{3}He from the rich phase to the poor phase is accompanied by absorption of heat, in the same way as a substance that transforms from the liquid to the vapour phase.

9.7.3   Description of the Dilution Refrigerator

The idea of transferring helium ^{3}He from the rich phase to the poor phase in order to produce low temperatures is due to H. London et al. (1960). The construction of the first dilution refrigerators dates back to 1965.

Figure 9.13 shows the operating principle of the corresponding heat machine. It works in a closed cycle, and the circulating fluid is almost pure ^{3}He. The cycle starts with ^{3}He gas extracted from the evaporator S at a pressure of 10^{-2} torr and at a temperature close to 1.3 K (a). This helium leaves the cryostat, is compressed at room temperature to 20 torrs (b) by the pump P and returns to the cryostat, where it is liquefied and cooled (c) by a ^{4}He bath at 1.3 K. This bath B is the warm source for the cycle. The ^{3}He then circulates in the counter-current exchanger E which

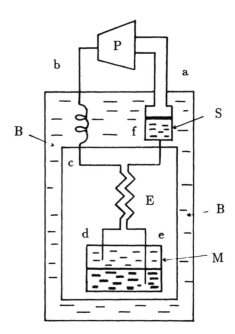

S - evaporator,
P - pump ,
B - ^{4}He bath at 1.3 K,
E - counter-current exchanger,
M - mixing chamber.

Figure 9.13: Outline diagram of a dilution refrigerator.

lowers its temperature, and enters the mixing chamber M inside the ^{3}He-rich phase (d) lying above the other phase. It then passes into the poor phase (e) and absorbs heat, the mixing chamber being the cold source of the cycle. The ^{3}He returns by diffusion through ^{4}He in the second branch of the exchanger back to the evaporator S (f). This diffusion, which works against gravity, is driven by a concentration gradient of ^{3}He in the return tube, thus producing an osmotic pressure. The cycle then finishes in the evaporator S where the almost pure ^{3}He evaporates. For a refrigerator operating reversibly, the power absorbed at the cold source is, from (9.36, 37),

$$\dot{Q} \ /\mathrm{W} = 82 \ [\dot{n} \ /(\mathrm{mol \ s}^{-1})] \ [T \ /\mathrm{K}]^2 \ . \tag{9.38}$$

A typical refrigerator uses $\dot{n} = 10^{-3}$ mol s^{-1} of ^{3}He. The refrigerative power is thus of the order of 2×10^{-3} W at 200 mK and 2×10^{-5} W at 20 mK.

Dilution refrigerators are at present in common use in low temperature laboratories. They span the same temperature range as refrigerators based on electronic adiabatic demagnetization (§4.6).

BIBLIOGRAPHY

F. London, *Superfluids*, Wiley (1954).

J. Wilks, *The Properties of Liquid and Solid Helium*, Clarendon Press, Oxford (1967).

Historical References

J.F. Allen and H. Jones, *Nature* **141**, 243 (1938).

J.F. Allen and A.D. Misener, *Proc. Roy. Soc. A* (London) **172**, 467 (1939).

J.G. Daunt and K. Mendelssohn, *Proc. Roy. Soc. A* (London) **170**, 423 (1939).

J.G. Daunt and K. Mendelssohn, *Nature* **143**, 719 (1939).

A. Einstein, *Berliner Ber.*, 261 (1924); 3, (1925).

R.P. Feynman, *Progress in Low Temperature Physics*, Gorter, North Holland **1**, 17 (1955).

H.E. Hall, *Phil. Trans. R. Soc. A* **250**, 359 (1957).

H.E. Hall, *Phil Mag. Supp.* **9**, 89 (1960).

H. Kamerlingh-Onnes, *Proc. Sect. Sci. K. ned. Akad. Wet.* **11**, 168 (1908).

P.L. Kapitza, *J. Phys. Moscow* **5**, 59 (1941).

L.D. Landau, *J. Phys. Moscow* **5**, 71 (1941).

E.M. Lifshitz, *J. Phys. Moscow* **8**, 110 (1944).

F. London, *Phys. Rev* **54**, 947 (1938).

H. London, G.R. Clarke and E. Mendoza, *Nature*, **185**, 349 (1960).

L. Onsager, note in *Superfluids*, F. London, Wiley (1954), p. 151.

D.V. Osborne, *Proc. Phys. Soc. A* **63**, 909 (1950).

V.P. Peshkov, *J. Phys. Moscow* **8**, 381 (1944).

L. Tisza, *C.R. Hebd. Séances Acad. Sci. Paris* **207**, 1035, 1186 (1938).

Chapter 10

Fermi-Dirac Statistics

10.1 INTRODUCTION

Fermi-Dirac statistics, introduced in chapter 2, applies to systems of indistinguishable non-interacting particles with half-integer spin (J = 1/2, 3/2,
...). In molecular gases, the corrected Maxwell-Boltzmann limit is used as it
gives a reasonable approximation. For helium ^{3}He ($J = 1/2$), however, which exists
in the gaseous state down to very low temperatures (normal boiling point $T_B = 3.19$
K), Fermi-Dirac statistics must be applied.

One very important application of this type of statistics is in an electron gas.
Indeed, a large number of physical properties of metals can be explained by assuming
that the electrons liberated by the atoms form a gas of non-interacting particles.
This model is not unrealistic, since, for example, it is demonstrated experimentally
that the free electrons move without colliding over distances much larger than the
interatomic spacing. As a first approximation (cf. Ch.11), the only effect of the
interactions is to create a potential well and hence confer on each electron a potential
energy that is uniform in space. By suitably choosing the origin of the energy, we
can take this potential energy to be zero.

10.2 DISTRIBUTION LAW. FERMI FUNCTION

We start by investigating the Fermi-Dirac distribution (2.27)

$$N_i = \frac{g_i}{e^{\beta(\epsilon_i - \mu)} + 1} \quad \text{or} \quad n_i = \frac{N_i}{g_i} = \frac{1}{e^{\beta(\epsilon_i - \mu)} + 1} . \tag{10.1}$$

For this, consider the Fermi function in the variable ϵ, defined by

$$n(\epsilon) = \frac{1}{e^{\beta(\epsilon - \mu)} + 1} \quad (\beta = 1/kT) \tag{10.2}$$

where T and μ are two parameters.

We first examine this function at the particular value $T = 0$ (for this value, μ is
denoted μ_0). In this case, the argument of the exponential is infinite with a sign
that depends on that of ϵ μ_0. It follows that if ϵ is smaller than μ_0 ($\epsilon < \mu_0$),

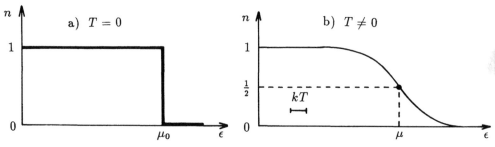

Figure 10.1: Fermi function for a) $T = 0$ and b) $T \neq 0$.

the exponential is zero and $n(\epsilon)$ is equal to 1. If, however, ϵ is greater than μ_0 ($\epsilon > \mu_0$), then the exponential is infinite and $n(\epsilon)$ is zero. Figure 10.1a shows the Fermi function when $T = 0$. It follows that in quantum states for which $\epsilon_i < \mu_0$, there are as many particles as there are quantum states ($N_i = g_i$) while states for which $\epsilon_i > \mu_0$ are not populated ($N_i = 0$). The Pauli exclusion principle provides an explanation for this distribution. At absolute zero, the system is in the lowest energy state that can be obtained by "filling" each quantum state with a single particle, starting from the lowest energy level until there are no particles left. All quantum states with energy less than μ_0 are therefore occupied by one particle, while higher levels are empty.

At other temperatures, the shape of the Fermi function $n(\epsilon)$ changes as shown in Figure 10.1b. Note that the point with co-ordinates $(\mu, \ 1/2)$ is the centre of symmetry of the curve, since

$$1 - n(\epsilon) = \frac{e^{\beta(\epsilon - \mu)}}{e^{\beta(\epsilon - \mu)} + 1} = \frac{1}{1 + e^{-\beta(\epsilon - \mu)}} \ .$$

This point, where the slope of the tangent is $-1/4kT$, is therefore also a point of inflection. The curve deviates significantly from its asymptotic values (1 and 0) only in a region whose width is of the order of $6kT$, as can be seen in the example

$$n(\mu - 3kT) - n(\mu + 3kT) = 0.9 \ .$$

It follows that when $kT \ll \mu$, the shape of the curve is very close to that for $T = 0$.

Finally, when ϵ is much larger than μ ($\epsilon \gg \mu$), the exponential in (10.1,2) becomes very large, and the following limit then applies,

$$n(\epsilon) = e^{\beta(\mu - \epsilon)} \qquad (\epsilon \gg \mu) \ , \tag{10.3}$$

which is similar to the corrected Maxwell-Boltzmann distribution.

10.3 IDEAL FERMION GAS

10.3.1 General Considerations

The thermodynamic functions of an ideal fermion gas in terms of the variables T, V, N are found from expression (2.34) for the grand potential

$$\Omega(T, \ V, \ \mu) = -kT \sum_i g_i \ln \left[1 + e^{\beta(\mu - \epsilon_i)} \right] \ ,$$

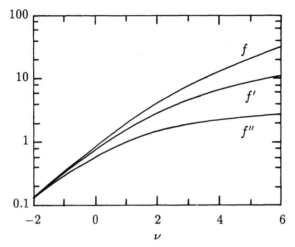

Figure 10.2: The function $f(\nu)$ (10.6) and its first two derivatives. For $\nu < -2$, these three functions are indistinguishable from e^ν and, for $\nu > 6$, $f(\nu)$ is given by (10.28) with a precision better than 10^{-3}.

using the same reasoning as for a boson gas (§7.5.2), i.e. keeping μ as an intermediate variable.

First we change to continuous variables (§5.2.2) with the transformations

$$\epsilon_i \rightarrow \frac{p^2}{2m} \ , \qquad g_i \rightarrow g_s \frac{d^3\mathbf{r}\, d^3\mathbf{p}}{h^3} \qquad \text{and} \qquad \sum_i \rightarrow \int \tag{10.4}$$

where $g_s = 2J + 1$ is the degeneracy due to the spin J of the particle. Integration over $\mathbf{r}$ ($d^3\mathbf{r} \rightarrow V$) and over the directions of $\mathbf{p}$ ($d^3\mathbf{p} \rightarrow 4\pi p^2 dp$) gives

$$\Omega = -kTg_s \frac{4\pi V}{h^3} \int_0^\infty p^2\, dp \ln\left[1 + \exp\beta\left(\mu - \frac{p^2}{2m}\right)\right] \ .$$

On applying the change of variable $x = \beta p^2/2m$ and setting

$$Z(T,\ V) = g_s \frac{V}{h^3}(2\pi mkT)^{3/2} \ , \tag{10.5}$$

we find

$$\Omega = -kT Z(T,\ V) \frac{2}{\sqrt{\pi}} \int_0^\infty \sqrt{x}\ln[1 + \exp(\beta\mu - x)]\ dx \ .$$

On introducing the function

$$f(\nu) = \frac{2}{\sqrt{\pi}} \int_0^\infty \sqrt{x}\ln\left(1 + e^{\nu - x}\right)\ dx = \frac{4}{3\sqrt{\pi}} \int_0^\infty \frac{x^{3/2}}{e^{x - \nu} + 1}\ dx \tag{10.6}$$

shown in Figure 10.2, the grand potential takes the form

$$\Omega(T,\ V,\ \mu) = -kT\ Z(T,\ V)\ f(\nu) \qquad (\nu = \mu/kT) \ . \tag{10.7}$$

These expressions are analogous to those for a boson gas (7.33,34).

As in Bose-Einstein statistics, (§7.6.2) we then obtain the thermodynamic functions in the variables T, V, μ

$$S = -\left(\frac{\partial \Omega}{\partial T}\right)_{\mu, V} = kZ\left[\frac{5}{2}f(\nu) - \nu f'(\nu)\right] \tag{10.8}$$

$$P = -\left(\frac{\partial \Omega}{\partial V}\right)_{T, \mu} = kT\frac{Z}{V}f(\nu) = -\frac{\Omega}{V} \tag{10.9}$$

$$N = -\left(\frac{\partial \Omega}{\partial \mu}\right)_{T, V} = Zf'(\nu). \tag{10.10}$$

For the internal energy, we also have

$$U = \Omega + TS + N\mu = \frac{3}{2}kTZf(\nu) = \frac{3}{2}PV \ , \tag{10.11a}$$

which is consistent with the general relation

$$PV = \frac{2}{3}U \tag{10.11b}$$

that applies to non-interacting gases in which the energy takes the form $\epsilon = p^2/2m$. Lastly, as in (7.38), the molar heat capacity c_V can be expressed as

$$c_V = \frac{3}{2}R\left[\frac{5}{2}\frac{f(\nu)}{f'(\nu)} - \frac{3}{2}\frac{f'(\nu)}{f''(\nu)}\right] \ . \tag{10.12}$$

To find the expressions for the thermodynamic functions in the variables T, V, N, equation (10.10) must be solved for $\nu = \mu/kT$,

$$f'(\nu) = \alpha \equiv \frac{N}{Z(T, V)} = \frac{Nh^3}{g_s V(2\pi mkT)^{3/2}} \ . \tag{10.13}$$

This solution, which can be found graphically from Figure 10.2, then allows the thermodynamic functions to be evaluated. Later in this section we shall encounter approximate methods for solving (10.13) that yield analytical expressions for these functions.

10.3.2   Molecular Gases

The chemical potential μ of molecular gases is generally negative. It therefore follows that ν is negative for these gases and we can use the series expansion

$$f(\nu) = \sum_{n=1}^{\infty}(-1)^{n-1}\frac{e^{n\nu}}{n^{5/2}} = e^{\nu} - 2^{-5/2}e^{2\nu} + 3^{-5/2}e^{3\nu} + \ldots \quad (\nu < 0) \ . \tag{10.14}$$

Solving equation (10.13), $f'(\nu) = \alpha$, then gives (cf. §7.6.4)

$$e^{\nu} = \alpha + 2^{-3/2}\alpha^2 + \ldots \ ,$$

from which we find

$$\mu = kT \left[\ln \alpha + 2^{-3/2} \alpha + \dots \right] \quad \text{with} \quad \alpha = N/Z \ . \tag{10.15}$$

We now eliminate μ to obtain the free energy in terms of N

$$F(T, V, N) = \Omega + N\mu = -NkT \left[\left(1 + \ln \frac{Z}{N} \right) - \frac{N}{2^{5/2} Z} + \dots \right] \tag{10.16}$$

and derive the various functions by differentiation. Thus

$$S = Nk \left[\left(\frac{5}{2} + \ln \frac{Z}{N} \right) + \frac{N}{2^{7/2} Z} + \dots \right] \tag{10.17}$$

$$P = \frac{NkT}{V} \left[1 + \frac{N}{2^{5/2} Z} + \dots \right] \tag{10.18}$$

$$U = \frac{3}{2} PV = \frac{3}{2} NkT \left[1 + \frac{N}{2^{5/2} Z} + \dots \right] \tag{10.19}$$

$$c_V = \frac{3}{2} R \left[1 - \frac{N}{2^{7/2} Z} + \dots \right] \ . \tag{10.20}$$

When N/Z is very much smaller than 1 ($\alpha = N/Z \ll 1$), the results of corrected Maxwell-Boltzmann statistics are recovered (Ch. 5). Note that the correction terms in the latter case are equal and of opposite sign in Bose-Einstein statistics (§7.6.4). Note also that the pressure of a molecular gas of fermions is higher than that in the equation of state of an ideal gas ($P = RT/v$), which is derived from corrected Maxwell-Boltzmann statistics. The fact that the particles are fermions increases the pressure in the same way as a repulsive interaction.

Exercise 10.1 *Pressure and heat capacity of a fermion gas*
    Using a graphical method and Figure 10.2, determine the pressure P and the molar heat capacity c_V of a perfect fermion gas of spin $J = 1/2$, molar mass $M = 3$ g mol^{-1} and molar volume $v = 113$ cm^3 mol^{-1}, at $T = 3.19$ K. Compare with the values found from the series expansions (10.18,20).

Solution
    The numerical value of α is

$$\alpha = \frac{N}{Z} = \frac{\mathcal{N}}{2v} \left(\frac{\mathcal{N}^2 h^2}{2\pi MRT} \right)^{3/2} = 0.478 \ .$$

Solving equation (10.13) $f'(v) = \alpha = 0.478$ graphically gives $v = -0.57$, and the values $f(v) = 0.52$ and $f''(v) = 0.41$ can be found by inspection. Substitution into the exact expressions (10.9 and 12) yields

$$P = 1.1 \times 10^5 \text{ Pa} = 1.09 \text{ atm} \qquad \text{and} \qquad c_V = \frac{3}{2} R \times 0.97 = 12.1 \text{ J K}^{-1} \text{ mol}^{-1} \ .$$

Restricting ourselves to the terms given in approximations (10.18 and 20), we have

$$P = 1 \text{ atm} \times (1 + 0.08) = 1.08 \text{ atm and } c_V = \frac{3}{2} R \times (1 - 0.04) = 11.9 \text{ J K}^{-1} \text{mol}^{-1}.$$

10.3.3   Absolute Zero

In this paragraph we inquire into the properties of an ideal fermion gas at absolute zero, firstly because of the simplification that arises in the calculations, and secondly because many electronic properties of metals display little variation between 0 and 500 K. The function $\Omega(T = 0, V, \mu)$ can be determined and hence expressions for the number of electrons N, the pressure P, ... can be derived. In this paragraph, however, we shall not pursue this method (cf. §10.3.4), but instead use a more direct approach.

Chemical potential. Fermi energy

The chemical potential at absolute zero μ_0 can be determined directly from the constraint $N = \sum N_i$. Given the form of the Fermi function at absolute zero, this condition can be written

$$N = \sum_{\epsilon_i < \mu_0} N_i + \sum_{\epsilon_i > \mu_0} N_i = \sum_{\epsilon_i < \mu_0} g_i \ . \tag{10.21a}$$

On changing to continuous variables (10.4) and integrating over $\mathbf{r}$ and the directions of $\mathbf{p}$, this equation becomes

$$N = g_s \frac{4\pi V}{h^3} \int_0^{p_F^0} p^2 \, dp = g_s \frac{4\pi V}{3h^3} p_F^{0\,3} \tag{10.21b}$$

where the Fermi momentum p_F^0 is related to μ_0 through

$$\mu_0 = \frac{p_F^{0\,2}}{2m} \ . \tag{10.22}$$

On substituting the Fermi momentum, defined by (10.21), into equation (10.22), we get

$$p_F^0 = h \left(\frac{3}{4\pi g_s} \frac{N}{V} \right)^{1/3} \quad \text{and} \quad \mu_0 = \frac{h^2}{2m} \left(\frac{3}{4\pi g_s} \frac{N}{V} \right)^{2/3} \ . \tag{10.23}$$

It can be seen that the chemical potential μ_0, which depends only on N/V, is indeed an intensive quantity.

For an ideal gas of free electrons in a metal ($J = 1/2$) the degeneracy is $g_s = 2$ and the chemical potential, denoted ϵ_F^0 and called the Fermi energy at absolute zero, is

$$\epsilon_F^0 = \frac{h^2}{8m} \left(\frac{3N}{\pi V} \right)^{2/3} = \frac{h^2}{8m} \left(\frac{3n}{\pi} \right)^{2/3} \ . \tag{10.24}$$

The numerical value of the Fermi energy in metals can then be calculated from the electron density $n = N/V$. If z is the number of free electrons per atom (e.g., $z = 1$ in copper), the ratio $n = N/V$ is equal to $z\mathcal{N}/v$ where v is the molar volume of the metal. Table 10.1 lists the values of n and of ϵ_F^0 for several metals.

Table 10.1: Thermal properties of monovalent metals. Free electron density $n = N/V$, Fermi energy at absolute zero (10.24) ($\epsilon^0_{F\ theor}$), electronic heat capacity constant γ (theoretical (10.37) and experimental values), ratio of the effective thermal mass m^* to the electron mass m, and experimental Debye temperature at absolute zero Θ_D [K.A. Gschneider, *Solid State Physics* **16**, 275 (1964)].

Metal	n $/10^{22}$ cm^{-3}	$\epsilon^0_{F\ theor}$ /eV	γ_{theor} $/10^{-3}$ SI	γ_{exp} $/10^{-3}$ SI	m^*/m	Θ_D /K
Li	4.60	4.7	0.76	1.69	2.22	352
Na	2.54	3.2	1.12	1.38	1.23	157
K	1.32	2.1	1.74	2.11	1.21	89.4
Rb	1.08	1.8	1.98	2.52	1.27	54
Cs	0.86	1.5	2.31	3.55	1.54	40
Cu	8.46	7.0	0.50	0.693	1.39	342
Ag	5.86	5.5	0.64	0.659	1.03	228
Au	5.90	5.5	0.64	0.748	1.17	165

It is of importance to remember that ϵ^0_F is of the order of several electron-volts. This energy, which is the maximum value of the kinetic energy of the electrons, corresponds to a velocity

$$v^0_F = \left(\frac{2\epsilon^0_F}{m}\right)^{1/2} = \frac{h}{2m}\left(\frac{3n}{\pi}\right)^{1/3} . \tag{10.25}$$

This velocity is of the order of a thousand kilometres per second ($v^0_F = 1.57 \times 10^6$ m s^{-1} in copper).

It can therefore be seen that even at absolute zero the electrons move at high velocities, some 10^4 faster than the average velocity of molecules in a gas such as nitrogen at room temperature. This is characteristic of fermions obeying the Pauli exclusion principle.

It may be observed that all the occupied quantum states at absolute zero lie inside a sphere of radius p^0_F in momentum space, called the Fermi sphere. The notion of the Fermi surface, which describes this sphere in the general case, plays an important role in solid state physics (Ch. 11).

Other thermodynamic functions

The internal energy of a free electron gas at absolute zero is calculated using the relation

$$U_0 = \sum_i N_i\epsilon_i = \sum_{\epsilon_i < \epsilon^0_F} g_i\epsilon_i .$$

On transformation to a continuous integral (10.4), this equation becomes

$$U_0 = 2 \times \frac{4\pi V}{h^3} \int_0^{p^0_F} p^2\,dp \times \frac{p^2}{2m} = \frac{4\pi}{5m}p^0_F{}^5\frac{V}{h^3} .$$

From equations (10.21,22), we can write N explicitly,

$$U_0 = \frac{3}{5}N\epsilon_F^0 = \frac{3h^2}{40m}N\left(\frac{3}{\pi}\frac{N}{V}\right)^{2/3} . \tag{10.26}$$

We now calculate the value of the entropy at absolute zero using the Boltzmann relation $S = k \ln W$, where $W = W_{FD}$ is the thermodynamic probability (2.21) in Fermi-Dirac statistics. In this probability, each factor

$$W_i = \frac{g_i!}{N_i!(g_i - N_i)!}$$

is equal to 1, whether N_i is zero ($\epsilon_i > \epsilon_F^0$) or equal to g_i ($\epsilon_i < \epsilon_F^0$). It follows that W_{FD} is equal to 1 and that the entropy is zero,

$$S_0 = 0 . \tag{10.27}$$

Finally, the equation of state at absolute zero can be recovered by noting that

$$P_0 V = -\Omega_0 = -U_0 + N\mu_0 = \frac{2}{5}N\epsilon_F^0 ,$$

in accordance with the general relation (10.11) $PV = 2U/3$.

Exercise 10.2 *Isothermal compressibility of a free electron gas*
Calculate $\chi_T = -(1/V) \times (\partial V/\partial P)_T$ for a free electron gas at absolute zero. The electron density of potassium is $N/V = 1.32 \times 10^{28}$ m^{-3} and its isothermal compressibility at absolute zero is $\chi_T^0 = 2.73 \times 10^{-10}$ Pa^{-1}. Compare this value with that given by the model, and comment.

Solution
The pressure of an electron gas at absolute zero is

$$P_0 = \frac{2}{5}\frac{N}{V}\epsilon_F^0 = \frac{h^2}{20m}\left(\frac{3}{\pi}\right)^{2/3}\left(\frac{N}{V}\right)^{5/3} ,$$

and hence

$$-V\frac{dP_0}{dV} = \frac{h^2}{12m}\left(\frac{3}{\pi}\right)^{2/3}\left(\frac{N}{V}\right)^{5/3} .$$

The isothermal compressibility of a free electron gas is therefore equal to

$$\chi_T^0 = -\frac{1}{V}\frac{dV}{dP_0} = \frac{12m}{h^2}\left(\frac{\pi}{3}\right)^{2/3}\left(\frac{V}{N}\right)^{5/3} .$$

It can be seen that as the density of a free electron gas increases, it becomes less compressible. Applying the above equation to potassium, we find $\chi_T^0 = 3.5 \times 10^{-10}$ Pa^{-1}. This compressibility is slightly larger than the experimental value, owing to the fact that we have neglected the binding energy of the crystal lattice.

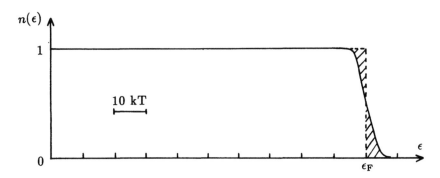

Figure 10.3: Fermi function for $\epsilon_F \equiv \mu = 100$ kT, e.g. $\epsilon_F = 3$ eV, $T = 300$ K.

10.3.4 Temperatures Above Absolute Zero

Thermodynamic functions

We have seen (§10.1) that the Fermi function $n(\epsilon)$ for $T \neq 0$ is different from that at $T = 0$ only in a region of a few kT. When ϵ_F (or μ) is of the order of a few electron-volts ($T_F = \epsilon_F/k \sim 5 \times 10^4$ K), the graph of $n(\epsilon)$ for T less than 300 K hardly differs from that at $T = 0$ (Fig. 10.3). This is the reason that the electronic properties of metals are practically constant in the temperature range 0 to 1000 K.

We now determine the differences in the thermodynamic functions from their values at absolute zero, as calculated in the preceding paragraph. To do this, we return to the general case (§10.3.1) and use the limiting form of the function $f(\nu)$ in (10.6) for $\nu = \mu/kT \gg 1$,

$$f(\nu) = \frac{8}{15\sqrt{\pi}} \nu^{5/2} \left(1 + \frac{5\pi^2}{8\nu^2} + \ldots \right) \qquad (\nu \gg 1) \qquad (10.28)$$

which comes from the asymptotic expansion of the Sommerfeld integrals

$$\int_0^\infty \frac{\phi(\epsilon)}{e^{\beta(\epsilon-\mu)} + 1} \, d\epsilon = \int_0^\mu \phi(\epsilon) \, d\epsilon + \frac{\pi^2}{6} (kT)^2 \phi'(\mu) + \ldots \qquad (10.29)$$

On substituting the limiting form (10.28) into the general expressions (10.8-11), we find

$$S = \frac{2}{3}\pi^{3/2} kZ\nu^{1/2} + \ldots \qquad (10.30)$$

$$U = \frac{3}{2}PV = \frac{4}{5\sqrt{\pi}} kTZ\nu^{5/2} \left(1 + \frac{5\pi^2}{8\nu^2} + \ldots \right) \qquad (10.31)$$

$$\frac{N}{Z} = f'(\nu) = \frac{4}{3\sqrt{\pi}} \nu^{3/2} \left(1 + \frac{\pi^2}{8\nu^2} + \ldots \right) . \qquad (10.32)$$

The last equation connecting $\nu = \mu/kT$ and $\alpha = N/Z(V, T)$ (cf. (10.13)) can be used in order to eliminate ν and find N. Solving (10.32) by successive approximations, we get for order zero

$$\nu^{(0)} = \left(\frac{3\sqrt{\pi}}{4}\frac{N}{Z}\right)^{2/3} = \frac{h^2}{8mkT}\left(\frac{3}{\pi}\frac{N}{V}\right)^{2/3} = \frac{\epsilon_F^0}{kT} \quad , \tag{10.33a}$$

where ϵ_F^0 is the Fermi energy at absolute zero (10.24), found directly in the previous paragraph. For the next order, we get

$$\nu = \nu^{(0)}\left(1 - \frac{\pi^2}{12\nu^{(0)2}}\right) \quad , \tag{10.33b}$$

from which we derive the expression for the Fermi energy $\epsilon_F \equiv \mu = kT\nu$

$$\epsilon_F = \epsilon_F^0\left[1 - \frac{\pi^2}{12}\left(\frac{kT}{\epsilon_F^0}\right)^2 + \ldots\right] \quad \text{with} \quad \epsilon_F^0 = \frac{h^2}{8m}\left(\frac{3N}{\pi V}\right)^{2/3} \quad . \tag{10.34}$$

From equation (10.33) for ν, the thermodynamic functions (10.30,31) can be written in terms of the variables T, V, N as

$$U = \frac{3}{2}PV = \frac{3}{5}N\epsilon_F^0\left[1 + \frac{5\pi^2}{12}\left(\frac{kT}{\epsilon_F^0}\right)^2 + \ldots\right] \tag{10.35}$$

$$S = Nk\frac{\pi^2}{2}\frac{kT}{\epsilon_F^0} + \ldots \tag{10.36}$$

and the equation for the molar heat capacity at constant volume becomes

$$c_V = R\frac{\pi^2}{2}\frac{kT}{\epsilon_F^0} = \gamma T \quad . \tag{10.37}$$

It can be verified from equations (10.34-36) that when $kT \ll \epsilon_F^0$ (which is the case for metallic solids), the thermodynamic functions such as ϵ_F, U and S differ only very slightly from their values (10.24, 26 and 27) at absolute zero.

We also note that the molar heat capacity c_V, which vanishes at absolute zero, varies linearly with temperature.

Application to metals

For metals, the Fermi energy is of the order of 5 eV, and the heat capacity of the free electrons c_V is approximately equal to 0.05 R at room temperature ($T = 300$ K). This contribution is therefore masked by the lattice heat capacity, which is about 3 R (§8.6.2). At low temperatures ($T \lesssim 5$ K), however, its value becomes greater than that of the lattice, which varies as T^3 in that region (Fig. 10.4a). The total heat capacity of the metal then takes the form

$$c_V = \gamma T + aT^3 \quad . \tag{10.38}$$

Experimentally, this equation is found to be remarkably well obeyed (Fig. 8.14 and 10.4b). Table 10.1 lists the values of the theoretical and experimental coefficients

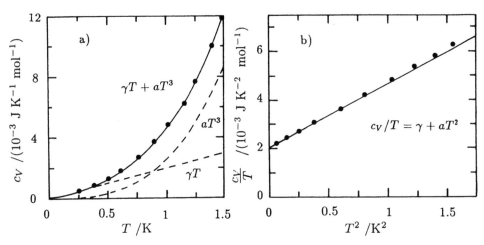

Figure 10.4: Heat capacity of potassium at low temperature. a) Plot of $c_V(T)$. The dashed curves are the contributions of the electrons (γT) and of the lattice (aT^3). b) Plot of c_V/T as a function of T^2. The intercept with the vertical axis yields $\gamma = 2.08 \times 10^{-3}$ SI and the slope is $a = 2.57 \times 10^{-3}$ SI [W.H. Lien and N.E. Phillips, *Phys. Rev. A* **133**, 1370 (1964)].

γ (10.37) for several metals, as well as the Debye temperatures Θ_D, which are related to the coefficient a by $a = 12\pi^4 R/5\Theta_D^3$ (8.60). The difference between the theoretical and experimental values of γ is due to the approximations made in the model of the free electron gas in metals. The difference can be accounted for by assigning to the electrons a "thermal effective mass" m^* such that the electron gas model yields the observed value of γ. This mass is therefore defined as

$$\frac{m^*}{m} = \frac{\gamma \text{ (experimental)}}{\gamma \text{ (theoretical)}} \ . \tag{10.39}$$

Application to ^3He

Figure 10.5 shows the heat capacity of liquid helium ^{3}He for $T < 0.3$ K at a pressure $P = 0.12$ atm ($v \simeq 37$ cm^3 mol^{-1}). For $T \lesssim 50$ mK, it displays a linear region in which $c_V \simeq 2.9$ RT. If the non-interacting fermion gas model is applied to liquid helium ^{3}He with $J = 1/2$ and $v = 37$ cm^3 mol^{-1}, we find from (10.37) that $c_V = 1.0$ RT ($\epsilon_F^0 = 4.3 \times 10^{-4}$ eV and $T_F = \epsilon_F^0/k = 5.0$ K). The theoretical value of γ thus found (1.0 R) differs from the experimental value (2.9 R). This difference can be explained by the presence of interactions. As in the case of metals, these interactions can be accounted for by introducing a thermal effective mass for the helium atoms ^{3}He, $m^* = 2.9$ m. On replacing m by m^* in equation (10.24) for ϵ_F^0, we find an effective Fermi energy $\epsilon_F^* = 1.5 \times 10^{-4}$ eV, which corresponds to a temperature $T_F^* = \epsilon_F^*/k = 1.7$ K. It can be seen that the temperature range for which the behaviour is linear is indeed $T \ll T_F^*$.

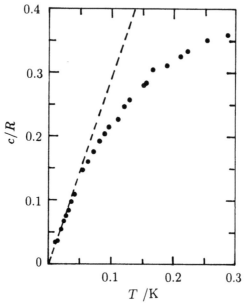

Figure 10.5: Molar heat capacity of helium ^{3}He at $P = 0.12$ atm. The dashed line is $c = 2.89$ RT [A.C. Anderson, W. Reese and J.C. Wheatley, *Phys. Rev.* **130**, 495 (1963)].

10.3.5 Ideal Electron Gas with Arbitrary Density of States

Up till now we have been looking at ideal gases of free fermions for which the dispersion relation is $\epsilon = p^2/2m$. In a metal, the conduction electrons can still be considered as having no interactions among each other, but their interaction with the lattice modifies the dispersion relation (Ch. 11). We therefore consider the general case of arbitrary dispersion $\epsilon = \epsilon(\mathbf{p})$. It is then preferable to use the variable ϵ instead of $\mathbf{p}$, and thus to introduce the density of states $g(\epsilon)$.

The number of electronic states with energy less than a given value ϵ is

$$G(\epsilon) = \int_{\epsilon(\mathbf{p}) < \epsilon} \frac{2d^3\mathbf{r}\, d^3\mathbf{p}}{h^3} = \frac{2V}{h^3} \int_{\epsilon(\mathbf{p}) < \epsilon} d^3\mathbf{p} \ , \tag{10.40}$$

where the integration over $\mathbf{p}$ is taken inside the surface $\epsilon(\mathbf{p}) = \epsilon$. The density of energy states is then

$$\begin{aligned} g(\epsilon) &= \frac{G(\epsilon + d\epsilon) - G(\epsilon)}{d\epsilon} = G'(\epsilon) \\ &= \frac{2V}{h^3} \int d^3\mathbf{p}\, \delta(\epsilon - \epsilon(\mathbf{p})) \ , \end{aligned} \tag{10.41}$$

where δ is the Dirac distribution having the general property

$$\int f(x)\, \delta(g(x))\, dx = \frac{f(a)}{|g'(a)|} \qquad \text{where} \qquad g(a) = 0 \ . \tag{10.42}$$

In the particular case of free electrons, this density of states is equal to (Fig. 10.6)

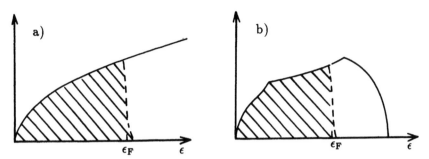

Figure 10.6: Density of energy states $g(\epsilon)$ (full lines) and population density $n(\epsilon)g(\epsilon)$ at a temperature T $(kT \ll \epsilon_F)$ (dashed lines): a) free electron gas; b) arbitrary case.

$$g(\epsilon) = \frac{2V}{h^3} \int 4\pi p^2\, dp\, \delta\left(\epsilon - \frac{p^2}{2m}\right) = \frac{4\pi V}{h^3}(2m)^{3/2}\epsilon^{1/2} \ , \qquad (10.43)$$

in agreement with equation (1.9), which was obtained directly from $g(\mathbf{r},\, \mathbf{p}) = 2/h^3$. The electronic population density is then

$$\frac{dN}{d\epsilon} = \frac{g(\epsilon)}{\exp\left[\beta(\epsilon - \epsilon_F)\right] + 1} = n(\epsilon)\, g(\epsilon) \ .$$

The form of the Fermi function leads to the following consequences:
- at absolute zero, all electronic states below ϵ_F^0 are occupied. We then have

$$\frac{dN}{d\epsilon} = \begin{cases} g(\epsilon) & \text{for } \epsilon < \epsilon_F^0 \\ 0 & \text{for } \epsilon > \epsilon_F^0 \end{cases} \qquad (10.44)$$

the Fermi energy at absolute zero being defined by

$$N = \int_0^{\epsilon_F^0} g(\epsilon)\, d\epsilon = G(\epsilon_F^0) \ . \qquad (10.45)$$

- at an arbitrary temperature $(T \ll \epsilon_F^0/k)$, the shape of the population density is very similar to that at absolute zero (Fig. 10.6).

The thermodynamic functions are found from the grand potential

$$\Omega = -kT \int g(\epsilon) \ln\left[1 + \exp\beta(\epsilon_F - \epsilon)\right]\, d\epsilon = -\int \frac{G(\epsilon)\, d\epsilon}{\exp\beta(\epsilon - \epsilon_F) + 1}$$

where the second equation is obtained on integrating by parts. With the Sommerfeld asymptotic expansion (10.29), this potential becomes

$$\Omega(T,\, V,\, \epsilon_F) = -\int_0^{\epsilon_F} G(\epsilon)\, d\epsilon - \frac{\pi^2}{6}(kT)^2 g(\epsilon_F) + \ldots \qquad (10.46)$$

From this equation, the following equalities can be derived,

$$N = -\left(\frac{\partial \Omega}{\partial \epsilon_F}\right)_{T,\, V} = G(\epsilon_F) + \frac{\pi^2}{6}(kT)^2 g'(\epsilon_F) + \ldots$$

$$S = -\left(\frac{\partial \Omega}{\partial T}\right)_{V,\, \epsilon_F} = \frac{\pi^2}{3}k^2 T g(\epsilon_F) + \ldots$$

At absolute zero, the first of these equations reduces to (10.45) and, on expanding $G(\epsilon_F)$ in the form $G(\epsilon_F^0) + (\epsilon_F - \epsilon_F^0)g(\epsilon_F^0)$, we get for the chemical potential

$$\mu \equiv \epsilon_F = \epsilon_F^0 - \frac{\pi^2}{6}\frac{g'(\epsilon_F^0)}{g(\epsilon_F^0)}(kT)^2 + \ldots \tag{10.47}$$

In the same approximation, the second equation

$$S = \frac{\pi^2}{3}k^2 T g(\epsilon_F^0)$$

yields the heat capacity

$$C_V = T\left(\frac{\partial S}{\partial T}\right)_{V,\ N} = \frac{\pi^2}{3}k^2 g(\epsilon_F^0)T = \gamma T \ , \tag{10.48}$$

which, as in (10.37), depends linearly on T.

The above expressions are valid for an arbitrary density of states. They yield the same results as in paragraph 10.3.4 for a free electron gas, in which

$$g(\epsilon_F^0) = \frac{4\pi V}{h^3}(2m)^{3/2}\epsilon_F^{0\ 1/2} = \frac{3}{2}\frac{N}{\epsilon_F^0} \ , \tag{10.49}$$

where we have used expression (10.24) for ϵ_F^0.

Exercise 10.3 *Internal energy of an electron gas*

Determine the internal energy of an electron gas from the relation $U = \sum N_i \epsilon_i$ using the Sommerfeld expansion for an arbitrary density of states. Then rederive equation (10.48) for the heat capacity C_V.

Solution

The internal energy is

$$U = \int_0^\infty \epsilon \times n(\epsilon)g(\epsilon)\ d\epsilon = \int \frac{\epsilon g(\epsilon)\ d\epsilon}{\exp \beta(\epsilon - \epsilon_F) + 1} \ .$$

In the Sommerfeld approximation, we have

$$U = \int_0^{\epsilon_F} \epsilon g(\epsilon)\ d\epsilon + \frac{\pi^2}{6}(kT)^2 \frac{d}{d\epsilon}[\epsilon g(\epsilon)]_{\epsilon=\epsilon_F} + \ldots$$

Expanding around ϵ_F^0, we get

$$U = \int_0^{\epsilon_F^0} \epsilon g(\epsilon)\ d\epsilon + (\epsilon_F - \epsilon_F^0)\epsilon_F^0 g(\epsilon_F^0) + \frac{\pi^2}{6}(kT)^2\left[g(\epsilon_F^0) + \epsilon_F^0 g'(\epsilon_F^0)\right] + \ldots$$

and substituting equation (10.47) for ϵ_F,

$$U = \int_0^{\epsilon_F^0} \epsilon g(\epsilon)\ d\epsilon + \frac{\pi^2}{6}(kT)^2 g(\epsilon_F^0) = U_0 + \frac{\pi^2}{6}(kT)^2 g(\epsilon_F^0) \ . \tag{10.50}$$

As will be recognized from its integral form, the term U_0 is the zero point energy. Furthermore, $U - U_0$ varies as T^2, which gives a heat capacity that is linear in T, as found above (10.48).

10.4 PROPERTIES OF FERMION GASES IN A MAGNETIC FIELD

In this paragraph we deal with the magnetization properties of a perfect gas of fermions. Comparisons with experiment can be found in helium ^{3}He and in the conduction electrons of metals. In these examples the particles have a spin $J = 1/2$ and the gyromagnetic ratio is $g = 2$. The magnetic properties of fermion gases originate from two causes:

- orientation of the spin magnetic moments (Pauli paramagnetism, 1926),
- changes in the trajectories of charged particles (Landau diamagnetism for $\mu_B B \ll kT$, de Haas-van Alphen effect for $kT \ll \mu_B B$).

10.4.1 Pauli paramagnetism

When a particle of spin $J = 1/2$ and $g = 2$ is placed in a magnetic flux density $\mathbf{B}$, it acquires a magnetic energy (4.28)

$$\epsilon_\pm = \pm\frac{1}{2}g\mu_B B = \pm\mu_B B \tag{10.51}$$

where the sign $\pm$ describes the spin state (parallel or antiparallel to $\mathbf{B}$). This energy adds to the kinetic energy, which we continue to write as ϵ_i. The thermodynamic functions are then the sum of two terms, one for the particles with their spin parallel to the field with energy $\epsilon_i + \mu_B B$ and the other with spin antiparallel having energy $\epsilon_i - \mu_B B$ (the spin degeneracy is thus lifted). In particular, the grand potential is written $\Omega = \Omega_+ + \Omega_-$ and the total magnetic moment of the substance is found by differentiation, i.e.,

$$\mathcal{M} = -\left(\frac{\partial\Omega}{\partial B}\right)_{T,\,V,\,\mu} \quad ,$$

from which, to permit comparison with experiment, the variable μ must be eliminated and replaced by N.

Absolute zero

Because of the simplicity of the method, we start by examining the case $T = 0$ K. At this temperature, all energy levels lower than the Fermi energy ϵ_F are full, while higher levels are empty. As in (10.21), the constraint $N = \sum N_i$ can then be written

$$\begin{aligned}
N &= \sum_{\epsilon_i + \mu_B B < \epsilon_F} g_i + \sum_{\epsilon_i - \mu_B B < \epsilon_F} g_i \\
&= \sum_{\epsilon_i < \epsilon_F - \mu_B B} g_i + \sum_{\epsilon_i < \epsilon_F + \mu_B B} g_i \quad .
\end{aligned} \tag{10.52}$$

This expression sums over each spin state (Fig. 10.7). On introducing the notations

$$p_F^\pm = [2m(\epsilon_F \mp \mu_B B)]^{1/2} = (2m\epsilon_F)^{1/2}\left[1 \mp \frac{\mu_B B}{2\epsilon_F} - \frac{1}{8}\left(\frac{\mu_B B}{\epsilon_F}\right)^2 + \dots\right], \tag{10.53}$$

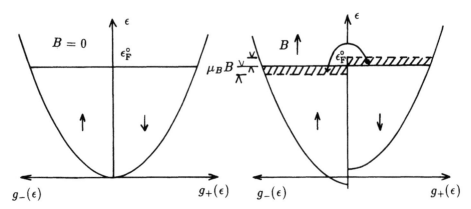

Figure 10.7: Population of the energy levels in zero or non zero field. The right (left) half of each diagram corresponds to particles with spin parallel (antiparallel) to the field. The arrows indicate the direction of the magnetic moment of the particles, which is in the opposite direction to their spin. Note that $g(\epsilon) = g_+(\epsilon) + g_-(\epsilon)$.

and, on summing over continuous variables as in (10.21), we get

$$
\begin{aligned}
N &= \frac{4\pi V}{3h^3}\left(p_F^{+3} + p_F^{-3}\right) \\
&= \frac{8\pi V}{3h^3}(2m)^{3/2}\epsilon_F^{5/2}\left[1 + \frac{3}{8}\left(\frac{\mu_B B}{\epsilon_F}\right)^2 + \dots\right].
\end{aligned}
\tag{10.54}
$$

For $\mu_B B \ll \epsilon_F$, the terms in B^2 are small and hence the Fermi energy ϵ_F is always given by its value in zero field ϵ_F^0 (10.24). We thus see (Fig. 10.7) that applying a magnetic field readjusts the populations in such a way that ϵ_F^0 remains constant.

Next, the magnetic moment $\mathcal{M}$ of the substance can be found by using the relation $\mathcal{M} = \sum N_i \mu_i$, as for N (10.52). Thus

$$
\mathcal{M} = \sum_{\epsilon_i < \epsilon_F^0 - \mu_B B} g_i \times (-\mu_B) + \sum_{\epsilon_i < \epsilon_F^0 + \mu_B B} g_i \times \mu_B .
$$

Transforming to continuous notations gives, as in (10.54)

$$
\begin{aligned}
\mathcal{M} &= \frac{4\pi V}{3h^3}\mu_B\left(-p_F^{+3} + p_F^{-3}\right) \\
&= \frac{8\pi V}{3h^3}(2m)^{3/2}\mu_B\left(\frac{3}{2}\mu_B B \epsilon_F^{0\,1/2} + \dots\right).
\end{aligned}
\tag{10.55}
$$

Using equation (10.24) to replace V by the extensive variable N (or else by taking the ratio of (10.55) and (10.54)), we find for the magnetic moment of the free fermion gas

$$
\mathcal{M} = \frac{3}{2}N\mu_B\frac{\mu_B B}{\epsilon_F^0} = \frac{3}{2}N\frac{\mu_B^2 B}{\epsilon_F^0} .
\tag{10.56}
$$

It can be seen that $\mathcal{M}$ is proportional to B, which shows the existence of paramagnetism connected with the orientation of the magnetic moment of the particles.
In the framework of Fermi-Dirac statistics, this phenomenon is called Pauli paramagnetism, in distinction to transition metal paramagnetism which is described by
Maxwell-Boltzmann statistics (Brillouin model). The Pauli magnetic susceptibility

$$\chi = \frac{\mathcal{M}}{V}\frac{\mu_0}{B} = \frac{3}{2}\mu_0\frac{N}{V}\frac{\mu_B^2}{\epsilon_F^0} \tag{10.57}$$

is much smaller than that found in the Brillouin model (4.49) for weak fields, since
$kT \ll \epsilon_F^0$.

The value of the magnetic moment $\mathcal{M}$ can be found by inspecting Figure 10.7.
The magnetic moment arises from reversal of the electron spins occupying energy
states close to the Fermi level (hatched area on the right), whose number is $g(\epsilon_F^0)/2\times$
$\mu_B B$ (area of the zone). We then have

$$\mathcal{M} = \frac{1}{2}g(\epsilon_F^0)\mu_B B \times 2\mu_B = g(\epsilon_F^0)\mu_B^2 B \ . \tag{10.58}$$

This is equivalent to equation (10.56), taking into account expression (10.49) for
the density of states of a gas of free fermions. Note that equation (10.58) is valid
for arbitrary densities.

From the above discussion it can be seen that the smallness of the Pauli susceptibility is a consequence of the small fraction of electrons that participate (only those
in the neighbourhood of the Fermi level).

Temperatures above absolute zero

We now calculate the thermodynamic potential $\Omega(T, V, \mu, B)$ defined by equation
(2.34). It is a sum of two terms Ω_+ and Ω_- for the two possible spin orientations
of the particles. Using the notations (10.5 and 6) of paragraph 10.3.1, and taking
into account the magnetic energy, these terms can be written

$$\Omega_\pm = -kT\frac{4\pi V}{h^3}\int_0^\infty p^2\,dp\ln\left[1 + \exp\beta\left(\mu - \frac{p^2}{2m} \mp \mu_B B\right)\right]$$
$$= -kT\frac{Z(T, V)}{2}\,f\,(\beta\mu \mp \beta\mu_B B) \ , \tag{10.59a}$$

and therefore

$$\Omega(T, V, \mu, B) = -kT\frac{Z}{2}\left[f\left(\nu - \frac{\mu_B B}{kT}\right) + f\left(\nu + \frac{\mu_B B}{kT}\right)\right]$$
$$= -kTZ(T, V)\left[f(\nu) + \frac{f''(\nu)}{2}\left(\frac{\mu_B B}{kT}\right)^2 + \ldots\right] \tag{10.59b}$$

The magnetic moment of the substance is then

$$\mathcal{M}(T, V, \mu, B) = -\left(\frac{\partial\Omega}{\partial B}\right)_{T, V, \mu} = \frac{\mu_B^2 B}{kT}Z(T, V)\left[f''(\nu) + \ldots\right] \ , \tag{10.60}$$

where the chemical potential $\mu = \epsilon_F = kT\nu$ is obtained from the relation

$$N = -\left(\frac{\partial \Omega}{\partial \mu}\right)_{T, V, B} = Z(T, V)\left[f'(\nu) + \frac{1}{2}f^{(3)}(\nu)\left(\frac{\mu_B B}{kT}\right)^2 + \dots\right] . \quad (10.61)$$

For weak fields, $\mu_B B \ll kT$, and the above equation reduces to $N = Zf'(\nu)$, as found in (10.10). This therefore shows that $\mu \equiv \epsilon_F$ does not vary significantly with the magnetic field and has the same form as in zero field. On taking the ratio of equations (10.60 and 61), it can be seen that the magnetic moment

$$\mathcal{M} = N\frac{\mu_B^2 B}{kT}\frac{f''(\nu)}{f'(\nu)} , \qquad \nu = \mu/kT , \quad (10.62)$$

is proportional to B, with a coefficient that we now examine for the two limiting cases $\nu \ll -1$ and $\nu \gg 1$.

The situation $\nu \ll -1$ arises for molecular gases to which the limit of corrected Maxwell-Boltzmann statistics can be applied. The function $f(\nu)$ is then practically equal to e^ν, so that $f'(\nu) \simeq f''(\nu)$. The magnetic moment (10.62) then becomes

$$\mathcal{M} = N\frac{\mu_B^2 B}{kT} \qquad (\mu \ll -kT) \quad (10.63a)$$

which is the same as the Curie law (4.48) for a spin $J = 1/2$ with $g = 2$. Note that this relation was established for Maxwell-Boltzmann statistics, but it is also valid in corrected Maxwell-Boltzmann statistics since the kinetic and magnetic energies are additive and separable.

The situation for which $\nu \gg 1$ corresponds to free electrons in a metal. The function $f(\nu)$ then takes the limiting form (10.28), which gives

$$\frac{f''(\nu)}{f'(\nu)} = \frac{d}{d\nu}\ln f'(\nu) = \frac{3}{2\nu}\left(1 - \frac{\pi^2}{6\nu^2}\right) .$$

With the limiting form (10.33) for ν, the magnetic moment (10.31) becomes

$$\mathcal{M} = \frac{3N\mu_B^2}{2\epsilon_F^0}\left[1 - \frac{\pi^2}{12}\left(\frac{kT}{\epsilon_F^0}\right)^2\right]B . \quad (10.63b)$$

At absolute zero this yields equation (10.56), and since $kT \ll \epsilon_F^0$ for electrons in a metal, the magnetic moment depends only weakly upon the temperature.

In the general case, for arbitrary values of ν, the magnetic moment is found from the general formula (10.62); ν is given by (10.13), $f'(\nu) = N/Z$, written in the form

$$f'(\nu) = \frac{4}{3\sqrt{\pi}}\left(\frac{\epsilon_F^0}{kT}\right)^{3/2} = \frac{4}{3\sqrt{\pi}}\left(\frac{T_F}{T}\right)^{3/2} . \quad (10.64)$$

For each value of $x = kT/\epsilon_F^0 = T/T_F$, $f'(\nu)$ can be determined and a graphic solution found for $f''(\nu)$ from Figure 10.2. The function

$$y = \frac{\mathcal{M}kT}{N\mu_B^2 B} = \frac{f''(\nu)}{f'(\nu)} , \quad (10.65)$$

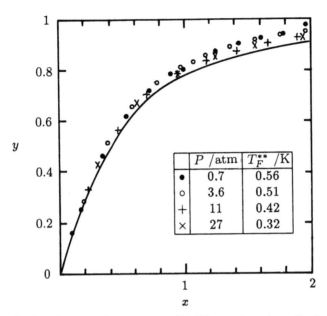

Figure 10.8: Reduced magnetic moment y (10.65) as a function of reduced temperature $x = T/T_F$. The curve shows the behaviour of an ideal gas of fermions with spin 1/2. The experimental points are for liquid helium ^{3}He. The experimental values of T_F^{**} used for the variable $x = T/T_F^{**}$ are indicated in the figure for different pressures [B.T. Beal and J. Hatton, *Phys. Rev. A* **139**, 1751 (1965)].

obtained point by point, is shown in Figure 10.8 in terms of the variable x. The two limiting cases discussed above appear again here. The first, ($\nu \ll -1$), occurs at high temperatures ($x \gg 1$), for which $y \to 1$, in accordance with Curie's law (10.63a); the second ($\nu \gg 1$) corresponds to the case $x \ll 1$ and gives the linear dependence $y = 3x/2$ found from (10.63b).

Application to ^{3}He

The magnetic properties of ^{3}He have been extensively investigated as a function of temperature and pressure [J. Wilks, *op. cit.*]. For temperatures above 2 K ($T > 2$ K), Curie's law (10.63a) is obeyed both in the liquid and in the gas phase, which is consistent with a model of independent particles of spin 1/2. Below 0.2 K, the magnetic susceptibility of helium is described by a relation of the form $\chi = \chi_0(1 - bT^2)$ in agreement with the theoretical expression (10.63b). If the theoretical expression (10.24) for ϵ_F^0 is used, however, neither does the susceptibility extrapolated to absolute zero χ_0 take the value (10.57), nor is the constant b equal to $\pi^2 k^2/12\epsilon_F^0{}^2$. This is due to the fact that interactions between atoms of the liquid and between their spins can no longer be neglected.

It is, however, possible to explain all the experimental results using the independent particle model by introducing a phenomenological constant

$$\epsilon_F^{**} = kT_F^{**} = \frac{3}{2}\frac{N}{V}\frac{\mu_0 \, \mu_B^2}{\chi_0} \tag{10.66}$$

defined through relation (10.57) with the value of the experimental magnetic susceptibility χ_0 extrapolated to absolute zero. The experimental values of the reduced magnetization y (10.65) are then plotted as a function of $x = T/T_F^{**}$ for different pressures (Fig. 10.8). It can be seen that the experimental points lie on a single curve that is independent of the pressure (it follows a scaling law) and which is close to the theoretical curve of the independent particle model.

In summary, for a gas of independent particles, $T_F = \epsilon_F^0/k$ can be found from three independent measurements extrapolated to absolute zero, namely the density, using relation (10.24), the heat capacity (10.37), and the magnetic susceptibility (10.57). In liquid ^{3}He, these three temperatures, denoted by T_F, T_F^* and T_F^{**}, have different values on account of the different interactions between atoms. As we have just shown, one or other of these parameters is used, depending on the phenomenon investigated.

Lastly, we note that for temperatures lower than 3 mK, ^{3}He exhibits superfluid phases, and the simple model above is no longer valid.

10.4.2 Electron Gas

Landau levels

In classical mechanics, an electron placed in a uniform magnetic field describes a helical trajectory. The projection of its motion perpendicular to the field (xOy plane) describes a circle with constant angular velocity

$$\omega = \frac{eB}{m} \tag{10.67}$$

and in the direction parallel to the field Oz its motion is uniform. In quantum mechanics, the energy of the circular motion is quantized

$$\epsilon_\perp = \frac{p_\perp^2}{2m} = \frac{p_x^2 + p_y^2}{2m} \rightarrow \left(j + \frac{1}{2}\right)\hbar\omega = \left(j + \frac{1}{2}\right)\frac{e\hbar}{m}B = 2\left(j + \frac{1}{2}\right)\mu_B B, \tag{10.68}$$

where j is zero or a positive integer such that the kinetic energy of the electron becomes

$$\epsilon_i = \frac{1}{2m}\left(p_x^2 + p_y^2 + p_z^2\right) \rightarrow \epsilon = 2\left(j + \frac{1}{2}\right)\mu_B B + \frac{p_z^2}{2m}. \tag{10.69}$$

States corresponding to the same value of j constitute a *Landau level* (Fig. 10.9a).

To determine the degeneracy connected with each value of j, we observe that in the (p_x, p_y) plane, the curves j =constant are circles $p_x^2 + p_y^2 = 4m\mu_B B(j + 1/2)$ (10.68) whose areas increase in arithmetic progression with steps of ehB. In phase

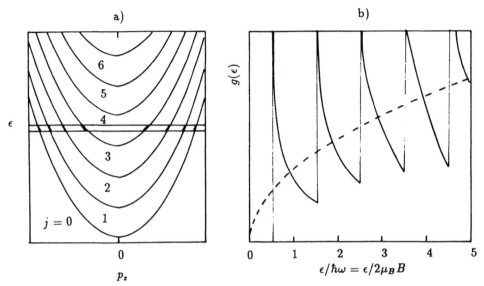

Figure 10.9: a) Diagram of Landau levels (j =const.) as a function of p_z. The two horizontal straight lines are separated by $d\epsilon$ and the thick parts of the curves correspond to energy states lying between ϵ and $\epsilon + d\epsilon$. b) Density of states of the electron motion (10.71). The dashed curve is the density of states in zero field (1.9).

space (x, y, p_x, p_y), the surfaces j =constant are therefore cylinders that bound domains of size $4\pi m \mu_B B \, S_{xy}$, where S_{xy} is the cross-sectional area of the container perpendicular to B. This domain corresponds to quantum states whose number, equal to $4\pi m \mu_B B \, S_{xy}/h^2$, is characteristic of the degeneracy associated with the value of j. As the degeneracy associated with translation along the z axis is $L_z dp_z/h$ (where L_z is the length of the container), that associated with the energy (10.69) is therefore

$$g_i = \frac{V \, d^3 p}{h^3} \quad \rightarrow \quad \frac{L_z \, dp_z}{h} \times \frac{4\pi m \mu_B B \, S_{xy}}{h^2} = \frac{4\pi V}{h^3} m \mu_B B \, dp_z \quad . \qquad (10.70)$$

The density of states of motion $g(\epsilon)$ is found by considering the states with energy lying between ϵ and $\epsilon + d\epsilon$ (Fig. 10.9a). The contributions from each value of j to this density are such that the expression obtained from (10.69)

$$p_z = (2m)^{1/2} \left[\epsilon - 2 \left(j + \frac{1}{2} \right) \mu_B B \right]^{1/2} \qquad (10.71)$$

is real. Then, taking into account the two possible signs for p_z,

$$
\begin{aligned}
g(\epsilon) &= \frac{4\pi V}{h^3} m \mu_B B \sum_j \frac{2 dp_z}{d\epsilon} \\
&= \frac{2\pi V}{h^3} (2m)^{3/2} \mu_B B \sum_j \left[\epsilon - 2 \left(j + \frac{1}{2} \right) \mu_B B \right]^{-1/2} . \qquad (10.72)
\end{aligned}
$$

This density, shown in Figure 10.9b, has discontinuities at $\epsilon = 2(j + 1/2)\mu_B B$. Note, however, that as $B \to 0$, the width and spacing of the discontinuities tend to zero.

We can now write the total energy of an electron in a magnetic field by adding its kinetic energy (10.69) to that of its magnetic moment (10.51), i.e.,

$$\epsilon = 2\left(j + \frac{1}{2}\right)\mu_B B + \frac{p_z^2}{2m} \pm \mu_B B \quad .$$

It follows that the grand potential $\Omega(T, V, \epsilon_F, B)$ is the sum of two terms Ω_+ and Ω_- for the spin states parallel and antiparallel to the field B. These are given by

$$\begin{aligned}
\Omega_\pm &= -kT \int \frac{4\pi V}{h^3} m\mu_B B \, dp_z \\
&\times \sum_j \ln\left[1 + \exp\beta\left(\epsilon_F - 2\left(j + \frac{1}{2}\right)\mu_B B - \frac{p_z^2}{2m} \mp \mu_B B\right)\right]. \quad (10.73)
\end{aligned}$$

For electrons in a metal, kT and $\mu_B B$ are both always very much smaller than the Fermi energy ϵ_F. We now continue this investigation for the two limiting cases $kT \gg \mu_B B$ and $kT \ll \mu_B B$.

Electronic paramagnetism

We first consider the physical case of electrons in a metal at ordinary temperatures, where $kT \sim 3 \times 10^{-2}$ eV$\gg \mu_B B \sim 6 \times 10^{-5}$ eV $(B \sim 1$ T$)$. To calculate the sum over j appearing in (10.73), we make use of the approximation

$$\sum_{j=0}^{\infty} f\left(j + \frac{1}{2}\right) = \int_0^\infty f(j) \, dj + \frac{1}{24} f'(0) + \dots \quad (10.74)$$

which is valid if f varies slowly in the vicinity of 0. Each term $\Omega_\pm$ then has the form $\Omega_\pm = \Omega_\pm^{(0)} + \Omega_\pm^{(1)} + \dots$ with

$$\begin{aligned}
\Omega_\pm^{(0)} &= -kT\frac{4\pi V}{h^3} m\mu_B B \\
&\times \int dp_z \, dj \ln\left[1 + \exp\beta\left(\epsilon_F - 2j\mu_B B - \frac{p_z^2}{2m} \mp \mu_B B\right)\right] \quad (10.75) \\
\Omega_\pm^{(1)} &= \frac{4\pi V}{h^3} m \cdot \frac{(\mu_B B)^2}{12} \int \frac{dp_z}{\exp\beta\left((p_z^2/2m) - \epsilon_F \pm \mu_B B\right) + 1} \quad .
\end{aligned}$$

With the change of variable $j \to p_\perp^2 = 4m\mu_B Bj$ inspired by (10.68), the $\Omega_\pm^{(0)}$ take the form (10.59a). Furthermore, changing the variable $\epsilon = p_z^2/2m$ in the equation for $\Omega_\pm^{(1)}$ yields

$$\Omega_\pm^{(1)} = \frac{4\pi V}{h^3} \frac{\mu_B^2 B^2}{24} (2m)^{3/2} \int_0^\infty \frac{\epsilon^{-1/2} \, d\epsilon}{\exp\beta\left(\epsilon - \epsilon_F \pm \mu_B B\right) + 1}$$

which gives, with the Sommerfeld series (10.29),

$$\Omega_{\pm}^{(1)} = \frac{4\pi V}{h^3} (2m)^{3/2} \frac{\mu_B^2 B^2}{12} \epsilon_F^{1/2} \left[1 - \frac{\pi^2}{24} \left(\frac{kT}{\epsilon_F} \right)^2 + \ldots \right] . \tag{10.76a}$$

To determine the chemical potential $\mu \equiv \epsilon_F$, we may go to the limit $B = 0$ ($\mu_B B \ll kT$), for which ϵ_F is given by (10.34). Using this Fermi energy, we can write for the terms $\Omega_{\pm}^{(1)}$

$$\Omega_{\pm}^{(1)} = \frac{N}{8} \frac{\mu_B^2 B^2}{\epsilon_F^0} \left[1 - \frac{\pi^2}{12} \left(\frac{kT}{\epsilon_F^0} \right)^2 \right] . \tag{10.76b}$$

The magnetization $\mathcal{M}$ of the electron gas is then a sum of two terms, $\mathcal{M}^{(0)}$ and $\mathcal{M}^{(1)}$. The term $\mathcal{M}^{(0)}$, found from the function $\Omega^{(0)} = \Omega_+^{(0)} + \Omega_-^{(0)}$, takes the form (10.63b), which we already encountered in Pauli paramagnetism due to the orientation of spin magnetic moments. The term $\mathcal{M}^{(1)}$ obtained from $\Omega^{(1)} = \Omega_+^{(1)} + \Omega_-^{(1)}$ is given by

$$\mathcal{M}^{(1)} = -\frac{\partial \Omega^{(1)}}{\partial B} = -\frac{N}{2} \frac{\mu_B^2}{\epsilon_F^0} \left[1 - \frac{\pi^2}{12} \left(\frac{kT}{\epsilon_F^0} \right)^2 \right] B .$$

This contribution to the magnetic moment, which is linear in B and antiparallel to the field, corresponds to the phenomenon of Landau diamagnetism in which the electron trajectories are wrapped around the lines of field so as to counteract its effect.

The total magnetic moment $\mathcal{M} = \mathcal{M}^{(0)} + \mathcal{M}^{(1)}$ of the gas of free electrons and its susceptibility are then given by

$$\mathcal{M} = \frac{\chi V B}{\mu_0} \quad \text{where} \quad \chi = \frac{\mu_0 N \mu_B^2}{V \epsilon_F^0} \left[1 - \frac{\pi^2}{12} \left(\frac{kT}{\epsilon_F^0} \right)^2 \right]^2 , \tag{10.77}$$

which describes the general phenomenon of electronic paramagnetism. Note that the ratio between the contributions of Landau diamagnetism and Pauli paramagnetism, equal to $-1/3$ for a free electron gas, may be different if interactions are taken into account.

Electronic paramagnetic susceptibility depends only weakly on temperature and is very much smaller than that of ionic paramagnetism (Brillouin model)

$$\frac{\chi \text{ (electrons)}}{\chi \text{ (Brillouin)}} \sim \frac{kT}{\epsilon_F^0} .$$

For this reason, electronic paramagnetism is masked by that of the ions in the case of the transition metals. The phenomenon is therefore observed in metals that contain non magnetic ions. It is, however, of the same order of magnitude as ionic diamagnetism, and for this reason electronic magnetic susceptibilities are not obtained directly.

Table 10.2: Magnetic properties of monovalent metals. Free electron density $n = N/V$, theoretical and experimental magnetic susceptibility χ_{theor} (10.77) and χ_{exp}, ratio of magnetic and thermal effective masses m^{**} and m^*, and magnetic susceptibility χ_{ion} due to ions [W.D. Knight, *Solid State Physics* **2**, 93 (1956)].

Metal	n $/10^{22}$ cm^{-3}	χ_{theor} $/10^{-6}$ SI	χ_{exp} $/10^{-6}$ SI	m^{**}/m^*	χ_{ion} $/10^{-6}$ SI
Li	4.60	6.60	24.2	1.63	− 0.6
Na	2.54	5.35	10.7	1.60	− 2.2
K	1.32	4.45	9.8	1.85	− 4.0
Rb	1.08	4.04	10.1	1.95	− 5.8
Cs	0.86	3.86	11.9	2.05	− 7.2
Cu	8.46	8.15	25.8	2.31	− 34.8
Ag	5.86	7.18	26.4	3.59	− 51.5
Au	5.90	7.23	38.8	4.61	− 72.8

Table 10.2 compares measured electronic magnetic susceptibilities of monovalent metals with the theoretical values (10.77) for a free electron gas. Agreement is found only in the order of magnitude. The theory can then be corrected by introducing a magnetic effective mass m^{**} of the electron such that

$$\frac{m^{**}}{m} = \frac{\chi \text{ (experimental)}}{\chi \text{ (theoretical)}} \ . \tag{10.78}$$

The ratio of the effective mass determined in this way to the thermal effective mass m^* (10.39), taking into account the theoretical expressions (10.37) and (10.77) for γ and χ,

$$\frac{m^{**}}{m^*} = \frac{\pi^2}{2} \frac{Rk}{\mu_0 \mu_B^2} \frac{V}{N} \frac{\chi}{\gamma} \text{ (experimental)} \ , \tag{10.79}$$

is in general different from 1. The corrections introduced in the form of effective masses are therefore different in thermal and magnetic phenomena, which can be explained by the fact that the interactions act in a different way.

de Haas-van Alphen effect

We now consider the case of electrons in a metal at low temperatures ($T \sim 1$ K) in a strong field ($B \sim 10$ T) where $kT \ll \mu_B B$. In equation (10.73) for $\Omega_{\pm}$, the function of j varies rapidly in the neighbourhood of $j = 0$, and hence approximation (10.74) is no longer valid. We therefore use the exact Poisson formula

$$\sum_{j=0}^{\infty} f\left(j + \frac{1}{2}\right) = \int_0^{\infty} f(j) \, dj + 2\sum_{n=1}^{\infty}(-1)^n \int f(j) \cos 2\pi n j \, dj \ , \tag{10.80}$$

so that we have

$$\Omega_\pm = \Omega_\pm^{(0)} - \frac{4\pi V}{h^3} 2m\mu_B B \sum_{n=1}^{\infty}(-1)^n \int dp_z \ dj \ f_\pm(j, \ p_z)\cos 2\pi nj \quad (10.81)$$

with $f_\pm(j, \ p_z) = kT \ln\left[1 + \exp\beta\left(\epsilon_F - 2j\mu_B B - \frac{p_z^2}{2m} \mp \mu_B B\right)\right]$.

The $\Omega_\pm^{(0)}$ are those appearing in (10.75). Since their sum $\Omega^{(0)} = \Omega_+^{(0)} + \Omega_-^{(0)}$ is the grand potential (10.59), it includes the orientation effect of the magnetic moments of the electrons. To calculate the integrals in (10.81), we use the approximation $T = 0$. A somewhat tedious calculation then gives

$$\int dp_z \ dj \ f_\pm(j, \ p_z)\cos 2\pi nj = \frac{\mu_B B}{\pi^2}\frac{p_F^\pm}{n^2} - \frac{(-1)^n}{2\pi^2}\frac{(2m)^{1/2}(\mu_B B)^{3/2}}{n^{5/2}}\cos\left(\frac{\pi n\epsilon_F}{\mu_B B} - \frac{\pi}{4}\right)$$

where $p_F^\pm$ are the momenta introduced in (10.53). Given that

$$\sum_{n=1}^{\infty}\frac{(-1)^n}{n^2} = -\frac{\pi^2}{12} \ ,$$

the terms $\Omega_\pm$ (10.81) can be written

$$\Omega_\pm = \Omega_\pm^{(0)} + \frac{4\pi V}{h^3}m\mu_B B\left[\frac{\mu_B B}{6}p_F^\pm + \frac{(2m)^{1/2}(\mu_B B)^{3/2}}{\pi^2}\sum_n\frac{1}{n^{5/2}}\cos\left(\frac{\pi n\epsilon_F}{\mu_B B} - \frac{\pi}{4}\right)\right].$$

The second term in this expression,

$$\Omega_\pm^{(1)} = \frac{4\pi V}{h^3}\frac{mp_F^\pm}{6}\mu_B^2 B^2 \simeq \frac{4\pi V}{h^3}(2m)^{3/2}\frac{\mu_B^2 B^2}{12}\epsilon_F^{1/2},$$

is identified with the term $\Omega_\pm^{(1)}$ found in (10.76) for Landau diamagnetism at absolute zero.

The grand potential Ω of a free electron gas at absolute zero is then given by

$$\Omega(T = 0, \ V, \ \mu \equiv \epsilon_F, \ B) = \Omega^{(0)} + \Omega^{(1)}$$
$$+ \ \frac{4V}{\pi h^3}(2m)^{3/2}(\mu_B B)^{5/2}\sum_n\frac{1}{n^{5/2}}\cos\left(\frac{\pi n\epsilon_F}{\mu_B B} - \frac{\pi}{4}\right). \quad (10.82)$$

In conclusion, Ω is a sum of three terms,
 • $\Omega^{(0)}$, defined in (10.59), is the contribution from the orientation of the electron spins;
 • $\Omega^{(1)}$, defined in (10.76), is the contribution from the helical trajectories of the electrons;
 • a term that oscillates as B varies, which is of the order of $(\mu_B B/\epsilon_F)^{1/2}$ with respect to $\Omega^{(0)}$ and $\Omega^{(1)}$.
The reason for this oscillation is as follows. At absolute zero, all the energy levels

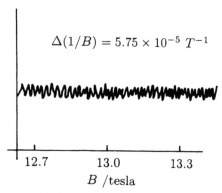

$$\Delta(1/B) = 5.75 \times 10^{-5}\ T^{-1}$$

12.7   13.0   13.3

B /tesla

Figure 10.10: de Haas-van Alphen effect in potassium at $T = 1.07$ K. The ordinate axis shows the value of the oscillatory component of the magnetic susceptibility [A.C. Thorsen and T.G. Berlincourt, *Phys. Rev. Lett.* **6**, 617 (1961)].

up to the Fermi level ϵ_F are populated. As this level depends only weakly on the magnetic field, we call it ϵ_F^0. The density of states (10.72) in the vicinity of ϵ_F^0 then becomes infinite (Fig. 10.9b) each time that $1/B$ is equal to $(2j + 1)\mu_B/\epsilon_F^0$, that is, with a periodicity

$$\Delta(1/B) = 2\mu_B/\epsilon_F^0 \ . \tag{10.83}$$

This period is just that of the oscillating term in (10.82). If kT is of the order of $\mu_B B$, the transition zone of the Fermi function is spread over several discontinuities of $g(\epsilon)$ and, on average, the oscillations disappear.

The magnetic moment $\mathcal{M}$ is found by differentiating expression (10.82) for Ω with respect to B. This moment contains several terms of different magnitudes. Conserving only the main term, which comes from the oscillatory contribution in Ω, we finally get

$$\mathcal{M} = -\frac{3}{2\pi} N\mu_B \left(\frac{\mu_B B}{\epsilon_F^0}\right)^{1/2} \sum_n \frac{1}{n^{3/2}} \sin\left(\frac{\pi n \epsilon_F^0}{\mu_B B} - \frac{\pi}{4}\right) \ . \tag{10.84a}$$

This magnetic moment is of the order of $(\epsilon_F^0/\mu_B B)^{1/2} \sim 30$ times larger than the magnetic moment of the electron gas at room temperature (10.77). Its most remarkable property is its oscillatory behaviour as a function of B with the periodicity (10.83). This effect, observed by W.J. de Haas and P.M. van Alphen in 1930, bears their names and was explained qualitatively by R. Peierls in 1933 and in detail by L.D. Landau in 1939. Figure 10.10 shows the oscillations of the magnetization in potassium at 1.07 K in a magnetic field of the order of 13 T. The measured periodicity of the oscillations, $\Delta(1/B) = 5.75 \times 10^{-5}\ T^{-1}$, is in good agreement with the theoretical value $\Delta(1/B) = 5.52 \times 10^{-5}\ T^{-1}$ calculated from (10.83) together with the numerical values in Table 10.1.

Note that the period (10.83) can be written in the form

$$\Delta(1/B) = \frac{e\hbar}{m\epsilon_F} = \frac{eh}{\pi p_F^2} = \frac{eh}{A_F} \tag{10.84b}$$

where $A_F = \pi p_F^2$ is the area of a great circle on the Fermi sphere.

10.5   ELEMENTARY THEORY OF CONDUCTION IN METALS

10.5.1   Drude Model

In 1900 P. Drude proposed a simple model for metallic conduction. He supposed that it is caused by free electrons, and made the following assumptions:
i)   The free electrons do not interact at a distance, neither among each other nor with the ions of the crystal lattice. Consequently, in the absence of an external field, they move in a straight line.
ii)   Each electron experiences collisions with the ions or with the other electrons at a rate $\nu = 1/\tau$. In analogy with (6.30), the distribution law for the delay between two successive collisions is given by

$$dP = e^{-t/\tau} \, \frac{dt}{\tau} \; . \tag{10.85}$$

iii)   After each collision, assumed to be instantaneous, the electron is "thermalized", that is, its velocity is distributed randomly according to the equilibrium distribution of velocities at the point of collision.

These assumptions are sufficient to construct an elementary theory for conduction that is in good agreement with experiment.

10.5.2   Free Electron Velocity Distribution

For his model, P. Drude used the Maxwell velocity distribution (6.5). A. Sommerfeld (1928) corrected Drude's results by introducing the velocity distribution of Fermi-Dirac statistics. The required relationship is found by making the following changes of notation in the Fermi-Dirac distribution (10.1)

$$g_i \; \to \; \frac{2d^3\mathbf{r} \, d^3\mathbf{p}}{h^3} \qquad \epsilon_i \; \to \; \frac{\mathbf{p}^2}{2m} \qquad \mathbf{p} \; \to \; m\mathbf{v}. \tag{10.86}$$

This gives the number of electrons contained in the elementary volume $d^3\mathbf{r}$ having a velocity $\mathbf{v}$ to within $d^3\mathbf{v}$

$$d^6 N_{\mathbf{r}, \, \mathbf{v}} = \frac{2m^3}{h^3} \left[\exp\left(\frac{m\mathbf{v}^2}{2kT} - \frac{\epsilon_F}{kT} \right) + 1 \right]^{-1} d^3\mathbf{r} \, d^3\mathbf{v} \tag{10.87}$$

where ϵ_F is the Fermi level (10.34). As we have seen, this distribution depends only weakly on temperature, and from now on we shall consider it in the approximation $T = 0$, for which

$$d^6 N_{\mathbf{r}, \, \mathbf{v}} = \begin{cases} 2m^3/h^3 \;\; d^3\mathbf{r} \, d^3\mathbf{v} & |\mathbf{v}| < v_F^0 \\ 0 & |\mathbf{v}| > v_F^0 \end{cases}$$

where $v_F^0 = (2\epsilon_F^0/m)^{1/2}$ is a velocity of the order of 10^6 m s^{-1} for $\epsilon_F^0 \sim 5$ eV (§10.3.3). It can thus be seen that all velocities up to the Fermi velocity v_F^0 are equally probable. It is then easily verified that

$$\bar{v} = \frac{1}{N}\int d^6 N_{\mathbf{r},\,\mathbf{v}}\,|\mathbf{v}| = \frac{8\pi m}{h^3}\frac{V}{N}\epsilon_F^{0\,2} = \frac{3}{4}v_F^0\ ,$$

$$\overline{v^2} = \frac{1}{N}\int d^6 N_{\mathbf{r},\,\mathbf{v}}\,v^2 = \frac{8\pi m^{1/2}}{5h^3}\frac{V}{N}\epsilon_F^{0\,5/2} = \frac{3}{5}v_F^{0\,2}\ , \tag{10.88}$$

$$\bar{v}_x = 0 \qquad \text{and} \qquad \overline{v_x^2} = \frac{1}{3}\overline{v^2} = \frac{1}{5}v_F^{0\,2}\ .$$

10.5.3   Ohm's Law

Consider a metal subjected to a constant uniform electric field $\mathbf{E}$. A free electron experiences a collision at $t = 0$ at the point $\mathbf{r}_0$ and moves off again with a velocity $\mathbf{v}_0$. This electron then uniformly accelerates in the electric field according to

$$m\frac{d\mathbf{v}}{dt} = -e\mathbf{E}\ ,$$

so that its velocity and position at a later time t prior to another collision are

$$\frac{d\mathbf{r}}{dt} = \mathbf{v} = \mathbf{v}_0 - \frac{e}{m}\mathbf{E}t \qquad \text{and} \qquad \mathbf{r} = \mathbf{r}_0 + \mathbf{v}_0 t - \frac{e}{m}\mathbf{E}\frac{t^2}{2}\ . \tag{10.89}$$

We see that the displacement of the electron $\mathbf{d} = \mathbf{r} - \mathbf{r}_0$ contains a "disordered" term $\mathbf{v}_0 t$ corresponding to thermal agitation and an "ordered" term $-e\mathbf{E}t^2/2m$ corresponding to the effect of the electric field. The mean displacement between two collisions is then found by performing two averages, one over the distribution of velocities $\mathbf{v}_0$, and the other over the distribution of time intervals t between two collisions. Since $< \mathbf{v}_0 >= 0$ and, from (6.31), $< t^2 >= 2\tau^2$, the mean displacement of an electron between two collisions is

$$< \mathbf{d} >= -\frac{e}{m}\mathbf{E}\tau^2\ .$$

We can calculate the average "drift" velocity of the electron by considering a large number of successive free paths

$$< \mathbf{v} >= \frac{\sum_i \mathbf{d}_i}{\sum_i t_i} = \frac{N < \mathbf{d} >}{N\tau} = -\frac{e}{m}\tau\mathbf{E}\ , \tag{10.90}$$

where the average time between two successive collisions is τ. The electrons thus move up the electric field with a velocity that is proportional to E, where the proportionality factor $\mu = e\tau/m$ is called the electron *mobility*. An electric current is thus generated of density

$$\mathbf{j} = n\times(-e)\times < \mathbf{v} >= \frac{ne^2\tau}{m}\mathbf{E} = \sigma\mathbf{E} \tag{10.91}$$

where n is the number density of free electrons. This is the microscopic form of Ohm's law, which yields the expression for the *electrical conductivity*

$$\sigma = \frac{ne^2}{m}\tau = \frac{ne^2}{m}\frac{\lambda}{\bar{v}} = ne\mu\ . \tag{10.92}$$

Table 10.3: Conduction properties of monovalent metals at $0°C$: electrical resistivity $\rho \equiv 1/\sigma$ /(10^{-8} Ω m), free electron density n /(10^{22} cm^{-3}), mean free time of flight τ /(10^{-14} s), Hall coefficient R_H /(10^{-10} m^3 C^{-1}) and thermal conductivity K /(W m^{-1} K^{-1}). The ratio $K/\sigma T$ is in 10^{-8} SI.

Metal	ρ	n	τ	R_H	$-neR_H$	K	$K/\sigma T$
Li	8.55	4.60	0.9	− 1.7	1.25	71	2.22
Na	4.3	2.54	3.3	− 2.5	1.02	135	2.13
K	6.1	1.32	4.4	− 4.2	0.89	99	2.21
Rb	11.6	1.08	2.8				
Cs	19	0.86	2.2	− 7.8	1.07		
Cu	1.55	8.46	2.7	− 0.55	0.75	400	2.27
Ag	1.47	5.86	4.1	− 0.84	0.79	418	2.25
Au	2.01	5.90	3.0	− 0.72	0.68	311	2.29

As Ohm's law is found to be true experimentally, measurement of the conductivity σ (or the resistivity $\rho = 1/\sigma$) allows τ to be calculated. Table 10.3 shows that this time is of the order of 10^{-14} s. This value shows, firstly, that the drift velocity of the electrons (current velocity), given by (10.90) is of the order of one metre per second for electric fields of the order of a few volts per centimetre. Secondly, it shows that the mean free path of the electrons $\lambda = \bar{v}\tau$, of the order of 100 Å, is much larger than the interatomic spacing.

Note that the kinetic energy acquired by the electron between two consecutive collisions

$$\Delta\epsilon_c = \frac{1}{2}m\left(v^2 - v_0^2\right) = \frac{e^2}{2m}\mathbf{E}^2 t^2 - e\mathbf{v}_0.\mathbf{E}t$$

is surrendered to the lattice after each collision. The average value of this energy is equal to $<\Delta\epsilon_c> = e^2 E^2 \tau^2 / m$ and the power per unit volume transferred to the lattice,

$$\mathcal{P} = \frac{N}{V}\frac{<\Delta\epsilon_c>}{\tau} = \frac{ne^2\tau}{m}E^2$$

thus satisfies Joule's law

$$\mathcal{P} = \mathbf{j}.\mathbf{E} = \sigma E^2 \quad . \tag{10.93}$$

Finally, we recall that the results of this paragraph make no appeal to the Fermi-Dirac distribution. They are therefore independent of the statistics under consideration.

10.5.4 Electrical Conductivity of Metals

In the above discussion we saw that the mean free path of the electrons is much larger than the distance between atoms. This fact, which may appear surprising, is

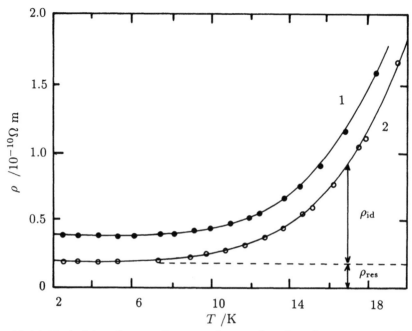

Figure 10.11: Resistivity of two sodium samples as a function of temperature. Sample 2 was subjected to prolonged annealing which improved its crystallinity. The decomposition into residual and ideal resistivity is shown [D.K.C. MacDonald and K. Mendelssohn, *Proc. R. Soc. A* (London) **202**, 103, (1950)].

the result of several factors. First, owing to the Pauli exclusion principle, almost all the electrons occupy states located below the Fermi energy. For this reason, after a collision, changes in the quantum state of an electron are possible only close to the Fermi surface and hence only a small fraction of the electrons is involved. Second, the metal ions occupy only a small part of the crystal volume (§8.1). Lastly, in a perfect lattice, the motion of the electrons adjusts itself to match the periodic potential of the crystal (Ch.11).

Collisions experienced by the electrons are then mainly due to deviations from a perfectly periodic crystal structure caused by

• phonons, or lattice vibrations;

• crystal defects (vacancies, interstitial ions, dislocations, polycrystalline structure, ...);

• chemical impurities.

The total resistivity of the metal can then be considered as being the sum of several terms corresponding to each collision mechanism. As the phonon contribution, which is zero at absolute zero, is the only one that is temperature dependent, we can write (Matthiessen's rule)

$$\rho = \rho_{\text{res}} + \rho_{\text{id}}(T) \tag{10.94}$$

where ρ_{res} is the residual resistivity at absolute zero due to the presence of physical and chemical defects in the crystal, and $\rho_{\text{id}}(T)$, called the ideal resistivity, is that

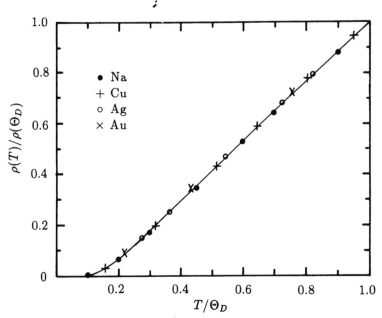

Figure 10.12: Resistivity ρ of several metals as a function of temperature T in reduced variables. The Debye temperatures Θ_D are taken to be 202 (Na), 310 (Cu), 220 (Ag) and 185 K (Au) and the corresponding resistivities are 2.27 (Na), 1.80 (Cu), 1.16 (Ag) and 1.32×10^{-8} Ω m (Au) [G.K. White and S.B. Woods, *Philos. Trans. R. Soc. A* **251**, 273 (1959)]. The continuous line is the Bloch function (10.95).

due to collisions with phonons. Figure 10.11 shows the resistivity of sodium at low temperature, where it can be seen that the residual value ρ_{res} depends on the quality of the sample, while the ideal value ρ_{id}, depends only on the temperature. Note that the residual resistivity is negligible outside the low temperature region.

Figure 10.12 shows the temperature dependence of ρ for several metals. At high temperatures, the dependence is linear owing to the fact that the mean free path of the electrons λ is inversely proportional to the phonon density, i.e., in this region, to the temperature (cf. Comprehension Exercise 8.13). At low temperatures, the ideal resistivity varies as T^n where $n \sim 5$. Using a simple model, F. Bloch (1929) proposed the semi-empirical interpolation

$$\rho_{id} \propto \left(\frac{T}{\Theta_D}\right)^5 \int_0^{\Theta_D/T} \frac{x^5 \, dx}{(e^x - 1)(1 - e^{-x})} \tag{10.95}$$

where Θ_D is the Debye temperature of the solid. This relation satisfactorily describes the experimental results for a large number of metals (Fig. 10.12), where the value of Θ_D may de different from that found in heat capacity measurements.

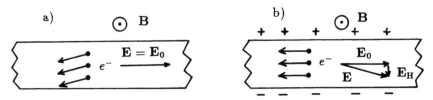

Figure 10.13: Diagram of the Hall effect. Electric and magnetic field prevailing a) at the start of the transient regime and b) in the permanent regime. The mean velocity of the electrons makes a constant angle with the direction of the total electric field **E**.

10.5.5   Hall Effect

When a magnetic flux density **B** is applied to a conductor that is already subjected to an electric field **E**, the combined effect on the electrons is given by

$$m\frac{d\mathbf{v}}{dt} = -e\left(\mathbf{E} + \mathbf{v} \times \mathbf{B}\right)\ \ .$$

Since the frequency of collisions is much greater than the cyclotron frequency $\omega = eB/m$ (10.67), the magnetic field has very little effect between two collisions, and **v** in the second term on the right hand side can be replaced by expression (10.89), which is valid in zero field. Integration of the above equation of motion gives

$$\mathbf{r} = \mathbf{r}_0 + \mathbf{v}_0 t - \frac{e}{m}(\mathbf{E} + \mathbf{v}_0 \times \mathbf{B})\frac{t^2}{2} + \frac{e^2}{m^2}(\mathbf{E} \times \mathbf{B})\frac{t^3}{6}\ \ .$$

The mean displacement between two collisions is then

$$
\begin{aligned}
< \mathbf{d} > &= < \mathbf{r} - \mathbf{r}_0 > = -\frac{e}{m}\mathbf{E}\frac{< t^2 >}{2} + \frac{e^2}{m^2}(\mathbf{E} \times \mathbf{B})\frac{< t^3 >}{6} \\
&= -\frac{e\tau^2}{m}\left[\mathbf{E} - \frac{e\tau}{m}(\mathbf{E} \times \mathbf{B})\right]\ \ .
\end{aligned}
\tag{10.96}
$$

This result shows that the electrons no longer follow the electric field lines. The corresponding current density is therefore

$$\mathbf{j} = -\frac{ne}{\tau} < \mathbf{d} > = \sigma\left[\mathbf{E} - \frac{\sigma}{ne}(\mathbf{E} \times \mathbf{B})\right] \tag{10.97}$$

where use has been made of equation (10.92) for σ.

In a wire-shaped conductor, the applied electric field $\mathbf{E}_0$ is parallel to the wire. During a transitory period, the electrons acquire a transverse velocity owing to the term in $\mathbf{E} \times \mathbf{B}$ and charges accumulate at the surface of the wire in this direction (Fig. 10.13). This accumulated charge generates an electric field $\mathbf{E}_H$ orthogonal to $\mathbf{E}_0$ and to **B**, called the Hall field (E.H. Hall, 1879), which compensates the effect of the magnetic field. In the permanent regime, the electrons then move along the direction of the wire ($\mathbf{j} \parallel \mathbf{E}_0$). On writing $\mathbf{E} = \mathbf{E}_0 + \mathbf{E}_H$ in (10.97) and separating the parallel and perpendicular terms, we find

$$\mathbf{E}_H = \frac{\sigma}{ne}(\mathbf{E}_0 \times \mathbf{B}) \quad \text{and} \quad \mathbf{j} = \sigma\left(\mathbf{E}_0 - \frac{\sigma}{ne}(\mathbf{E}_H \times \mathbf{B})\right) \simeq \sigma\mathbf{E}_0$$

$$\text{that is,} \quad \mathbf{E}_H = -\frac{1}{ne}\mathbf{B} \times \mathbf{j}. \tag{10.98}$$

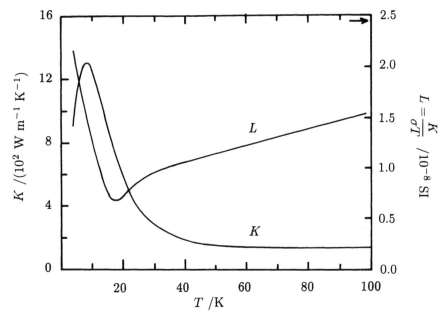

Figure 10.14: Thermal conductivity K and Lorenz number L of a sample of sodium at low temperature. The arrow indicates the position of the theoretical value of $L = \pi^2 k^2/3e^2$ [R. Berman and D.K.C. MacDonald, *Proc. Roy. Soc. A* (London) **209**, 368 (1951)].

The Hall field is perpendicular and proportional both to the current density **j** and to the magnetic field **B**, where the proportionality constant, called the Hall coefficient, is

$$R_H = -\frac{1}{ne} \; . \tag{10.99}$$

The minus sign $-$ corresponds to the sign of the charge carriers. Positive charge carriers would generate a Hall field in the opposite direction. The value of the coefficient R_H for several monovalent metals is given in Table 10.3. Since equation (10.99) involves only the electron density, it is in fairly good agreement with the experimental value for these metals. Note that the Hall field is frequently used to measure the value of magnetic fields (semiconductor Hall effect probe).

10.5.6   Thermal Conductivity

At room temperature, the thermal conductivity of metals is of the order of 10 to 100 times higher than that of dielectrics. This difference is due to the fact that transport of energy by electrons in a metal is greater than that carried by phonons.

The thermal conductivity of the electrons can be estimated from equation (6.42), which also applies to electrons

$$K = \frac{1}{3} \, n\lambda\bar{v} \, \frac{c_V}{\mathcal{N}} = \frac{1}{3} \, n\tau(\bar{v})^2 \, \frac{c_V}{\mathcal{N}}$$

where c_V is the electronic molar heat capacity (10.37). On inserting the expressions for c_V and $\bar{v}$ (10.88), we obtain

$$K = \frac{3\pi^2}{32}\, n\tau v_F^{02}\, \frac{k^2 T}{\epsilon_F^0} = \frac{3\pi^2}{16}\, \frac{n\tau k^2}{m}\, T \qquad \text{(approximately)} \ .$$

An exact calculation using Boltzmann's equation yields a result that differs from the above expression merely by a numerical factor close to 2,

$$K = \frac{\pi^2}{3}\, \frac{n\tau k^2}{m} T \qquad \text{(exactly)} \ . \tag{10.100}$$

Since thermal conduction by phonons is negligible, this expression gives the total thermal conductivity of the metal.

It is of interest to compare the thermal and electrical conductivities K and σ (10.92) through the ratio $L = K/\sigma T$, called the Lorenz number. The present model gives

$$L = \frac{K}{\sigma T} = \frac{\pi^2}{3}\, \frac{k^2}{e^2} = 2.44 \times 10^{-8} \ \text{SI} \ . \tag{10.101}$$

This result is in agreement with the law of Wiedemann and Franz (1853) who noted that good heat conductors are also good electrical conductors, and who showed that at a given temperature the ratio K/σ is the same for different metals. Table 10.3 lists the experimental values of the Lorenz number for various metals at 0°C. Good agreement is found with the model. At lower temperatures, the Lorenz number falls below the value (10.101) (Fig. 10.14). This is due to the fact that electron-phonon scattering at small angles generates a lower resistance for the transport of electric charge than for the transport of energy. At the lowest temperatures the Lorenz number recovers its value (10.101), while σ and K/T reach their residual values.

BIBLIOGRAPHY

J.G. Daunt, The Electronic Specific Heat in Metals, *Prog. in Low Temp. Phys.* **1**, 202 (1955).

W.D. Knight, Electron paramagnetism and nuclear resonance in metals, *Solid State Physics* **2**, 93 (1956).

D.K.C. MacDonald, Electrical conductivity of metals and alloys at low temperatures, *Handbuch der Physik* **14**, 137 (1956).

K. Mendelssohn and H.M. Rosenberg, The thermal conductivity of metals at low temperatures, *Solid State Physics*, **12**, 223 (1961).

W.D. Nottingham, *Handbuch der Physik* **21**, 1 (1956).

D. Schoenberg, The de Haas-van Alphen effect, *Prog. in Low Temp. Phys.* **2**, 226 (1957).

J. Wilks, *The Properties of Liquid and Solid Helium*, Clarendon Press, Oxford (1967).

Historical References

F. Bloch, *Z. Phys.* **52**, 555 (1929); **59**, 208 (1930).

W. J. de Haas and P.W. van Alphen, *Leiden Comm.* 208d, 212a (1930) and 220d (1932).
P. Drude, *Ann. Physik* **3**, 369 (1900).
E.H. Hall, *Phil. Mag.* **9**, 225 (1879) and **10**, 301 (1880).
L.D. Landau, appendix by D. Schoenberg, *Proc. Roy. Soc. A* (London) **170**, 341 (1939).
W. Pauli, *Z. Phys.* **41**, 81 (1926).
R. Peierls, *Z. Phys.* **81** 186 (1933).
A. Sommerfeld, *Z. Phys.* **47**, 1 (1928).
G. Wiedemann and R. Franz, *Ann. Physik* **89**, 497 (1853).

COMPREHENSION EXERCISES

10.1 Determine the Fermi energy at absolute zero for an electron gas in one dimension confined to a segment of length L. Same question for a two-dimensional gas. [Ans.: $h^2 N^2/8mL^2$; $h^2 N/4\pi mS$].

10.2 Calculate the wavelength associated with an electron of energy equal to the Fermi energy of copper ($N/V = 8.46 \times 10^{22}$ cm^{-3}). Compare this to the distance between two nearest neighbour atoms ($d = 2.55$ Å). [Ans.: $\lambda = 4.63$ Å].

10.3 Calculate the mean kinetic energy of an electron in copper at absolute zero ($N/V = 8.46 \times 10^{22}$ cm^{-3}). At what temperature does a gas molecule have the same average kinetic energy? [Ans.: 4.22 eV; 32 600K].

10.4 Calculate the value of the pressure P of the electron gas in copper at absolute zero ($N/V = 8.46 \times 10^{22}$ cm^{-3}) [Ans.: 3.81 $\times$ 10^{10} Pa $\doteq$ 376 000 atm].

10.5 Using Figures 10.3 or 6, show by a qualitative argument that the difference between the internal energies $U - U_0$ at $T \neq 0$ and $T = 0$ is equal to $(kT)^2 g(\epsilon_F^0)$, to within a numerical factor.

10.6 Calculate and plot the derivative of the Fermi function $n(\epsilon)$. What is its limit at $T = 0$?

10.7 For a free electron gas in two dimensions, determine the Fermi energy at an arbitrary temperature. Discuss its value, making use of the Sommerfeld expansion (10.29) [Ans.: $\epsilon_F = kT \ln(\exp(\beta\epsilon_F^0) - 1)$ equal to ϵ_F^0 to within an exponentially small term].

10.8 Determine the molar density of states $g(\epsilon_F^0)$ for a two-dimensional electron gas as a function of S/N, where S is the surface area of the gas. Calculate this density for graphite, where $S/N = 2.62 \times 10^{-20}$ m^2. Hence deduce the value of the coefficient γ of the electronic molar heat capacity ($c_V = \gamma T$) for graphite. This value is very different from that found experimentally, $\gamma = 13.8 \times 10^{-6}$ J K^{-2} mol^{-1}, since the conduction electrons remain closely bound to the carbon atoms [B.J.C. van der Hoeven Jr. and P.H. Keesom, *Phys. Rev.* **130**, 1318 (1963)] [Ans.: 4.1 $\times$ 10^{41} J^{-1} mol^{-1}; 2.6 $\times$ 10^{-6} JK^{-2} mol^{-1}].

10.9 Show that as B tends to 0, the density of states (10.72) of an electron gas in a magnetic field yields the normal density of states (1.9) in the absence of field. (Hint: replace the sum by an integral.)

10.10 What would be the mean free path of the free electrons in a metal if their only interaction were their contact with the ions of the network? Take the example of sodium where the ionic radius is $r = 0.98$ Å and the electron density is $n = 2.5 \times 10^{22}$ cm^{-3}. Why is this value much smaller than the experimental value $\lambda \sim 300$ Å ? [Ans.: $\lambda \simeq 1/\pi r^2 n = 13$ Å].

10.11 Estimate the electric field in a copper wire of cross-section 1 mm^2 carrying a current of 1 A ($\rho \simeq 10^{-8}$ Ω m) [Ans.: 10^{-2} Vm^{-1}].

10.12 Show that the ideal electrical resistivity given by the Bloch equation (10.95) varies as T for $T \gg \Theta_D$ and as T^5 for $T \ll \Theta_D$.

10.13 A sample has the shape of a rectangular parallelepiped of sides a, b, c. A current is made to flow parallel to a by applying a potential difference U. A magnetic field B is established parallel to b. Show that a potential difference $V_H = R_H \sigma B U c/a$ is generated. Calculate V_H for a sample having $a = 1$ cm, $c = 0.01$ cm subjected to a voltage $U = 0.5$ V and a field $B = 1$ T. Take the examples of copper ($R_H \sigma = 3.5 \times 10^{-3}$ SI) and a semiconductor ($R_H \sigma \sim 0.1$ SI) [Ans.: 17.5 μV; 0.5 mV].

PROBLEM 10.1 WHITE DWARF STARS

1 Relativistic electron gas

1.1a Write down the expression for the equilibrium distribution of a free electron gas and make a graph of this distribution at $T = 0$ K. The Fermi energy is ϵ_F^0.

1.1b Recall the equation for the translational density of states and hence deduce expressions for p_F^0 and $x_F = p_F^0/mc$, where p_F^0 is the radius of the Fermi sphere in momentum space.

1.1c The energy of a particle with momentum p is given in relativity by $\epsilon = (p^2 c^2 + m^2 c^4)^{1/2}$. In the classical limit $p/mc \ll 1$, this gives $\epsilon \simeq mc^2 + p^2/2m$ and, in the ultrarelativistic limit $p/mc \gg 1$, it yields $\epsilon \simeq pc$. Use the relativistic expression for ϵ to write the Fermi energy ϵ_F^0 corresponding to the momentum p_F^0 as a function of x_F. Discuss the two limits $x_F \ll 1$ and $x_F \gg 1$.

1.2a Find an expression for the internal energy of the electron gas at absolute zero, $U_0 = \sum N_i \epsilon_i$, in the variables V and x_F. Express this result in terms of the function $h(x) = \int_0^x t^2(t^2 + 1)^{1/2} \, dt$, which will be left undetermined. The limiting forms of $h(x)$ for $x \ll 1$ and $x \gg 1$ are respectively

$$h(x) = \frac{x^3}{3} + \frac{x^5}{10} + \dots \quad (x \ll 1) \qquad \text{and} \qquad h(x) = \frac{x^4}{4} \quad (x \gg 1) \ .$$

1.2b Hence derive an expression for the kinetic pressure P_0 of the electrons as a function of x_F, of $h(x_F)$ and of its derivative $h'(x_F)$.

1.2c Show that in the classical limit $x_F \ll 1$, $P_0 V = 2(U_0 - Nmc^2)/3$.

1.2d Show also that in the ultrarelativistic limit $x_F \gg 1$, $P_0 V = U_0/3$. Comment on these results.

2 Application to white dwarves

A white dwarf is an old star, consisting essentially of ^{4}He, with a mass M, similar to that of the Sun ($M_S = 2.0 \times 10^{30}$ kg), and with extremely high density $\rho \sim 10^{10}$ kg

$m^{-3} \sim 10^7 \rho_S$, where ρ_S is the mean density of the Sun. The internal temperature T of the star is of the order of 10^7 K. At this temperature, helium is completely ionized and it can be assumed that the star is composed of N free electrons and $N/2$ helium nuclei whose effect is to neutralize the electron charge and to allow cohesion of the star by gravitational interaction.

2.1a Calculate the volume V, then the radius R of a typical white dwarf having $M = M_S$ and $\rho = 10^{10}$ kg m^{-3}. Compare R with the radius of the Earth, equal to 6400 km.

2.1b Calculate the number of electrons N and the density N/V of this star (atomic mass of helium $M(^4\text{He}) = 4.0$ g mol^{-1}).

2.1c Calculate the value of p_F^0 for the electron gas in the typical white dwarf, as well as that of x_F. Hence deduce that relativistic kinematics must be used to calculate the energy of the electrons.

2.1d Calculate ϵ_F^0 and hence show that the properties of the electron gas in the white dwarf can be described by the approximation $T = 0$ K.

2.2 The total energy of a white dwarf is the sum of the kinetic energy of the electrons and the gravitational energy of the star, since the kinetic energy of the nuclei and the electrostatic energy are both negligible. A dimensional argument shows that the gravitational energy of a star of mass M and radius R has the form

$$E = -\alpha G \frac{M^2}{R}$$

where G is the gravitational constant and α is a number that depends on the distribution of matter inside the star ($\alpha = 3/5$ if the distribution is uniform).

2.2a Express the volume V and the radius R of the star as functions of N and x_F, then of M and x_F. Assume that the mass of a helium atom is equal to $4m_p$, where m_p is the mass of a proton.

2.2b Write the kinetic energy U_0 of the electrons as a function of M and x_F.

2.2c Express the gravitational energy of the star in the form

$$E = -\frac{3}{8}\alpha\gamma mc^2 \left(\frac{M}{m_p}\right)^{5/3} x_F$$

where γ is a dimensionless constant to be determined. For the numerical calculations, take $\gamma = 1.03 \times 10^{-38}$.

2.3a Write down the condition that yields the equilibrium radius R_0 for a star of given mass M.

2.3b State this equilibrium condition and express it in the form

$$f(x_F) = \alpha\gamma \left(\frac{M}{m_p}\right)^{2/3} .$$

Express $f(x)$ as a function of $h(x)$ and of $h'(x)$.

2.3c The function $f(x)$ is shown in Figure 10.15. Use a graphical method to determine the equilibrium radius R_0 of a white dwarf of mass $M = M_S$. Take $\alpha = 3/5$.

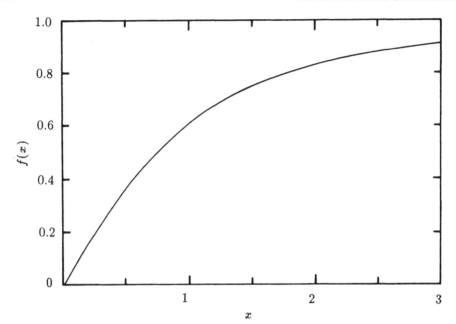

Figure 10.15: Function f defined in text

2.3d Show that white dwarves cannot have a mass greater than a limiting mass M_l (Chandrasekhar limit) and give this as a function of M_S.

Numerical data: $hc = 1.99 \times 10^{-25}$ J m; $mc^2 = 8.19 \times 10^{-14}$ J; $k = 1.38 \times 10^{-23}$ J K^{-1}; $m_p = 1.67 \times 10^{-27}$ kg.

SOLUTION
1 Relativistic electron gas
1.1a See §10.2. The Fermi energy is identical to the chemical potential μ.

1.1b The density of states for electrons of spin $1/2$ is $2d^3\mathbf{r}d^3\mathbf{p}/h^3$. On integrating over $\mathbf{r}$ and $\mathbf{p}$, we get (§10.3.3)

$$N = 2\frac{V}{h^3} \times \frac{4}{3}\pi p_F^{03}$$

$$\text{or} \quad p_F^0 = \frac{h}{2}\left(\frac{3N}{\pi V}\right)^{1/3} \quad \text{and} \quad x_F = \frac{h}{2mc}\left(\frac{3N}{\pi V}\right)^{1/3} \quad . \quad (10.102)$$

1.1c We have

$$\epsilon_F^0 = \left(p_F^{0\,2}c^2 + m^2c^4\right)^{1/2} = mc^2\left(x_F^2 + 1\right)^{1/2} \quad .$$

For $x_F \ll 1$, the Fermi energy is given by

$$\epsilon_F^0 \simeq mc^2\left(1 + \frac{x_F^2}{2}\right) = mc^2 + \frac{h^2}{8m}\left(\frac{3N}{\pi V}\right)^{2/3} \quad .$$

Except for the rest mass energy term, this yields equation (10.24) for the Fermi energy of a non-relativistic electron gas. For $x_F \gg 1$, we have

$$\epsilon_F^0 \simeq mc^2 x_F = \frac{hc}{2} \left(\frac{3N}{\pi V} \right)^{1/3} .$$

1.2a The internal energy at absolute zero is

$$U_0 = \sum_{\epsilon_i < \epsilon_F^0} g_i \epsilon_i = \int_0^{p_F^0} 2 \frac{V}{h^3} \times 4\pi p^2 \, dp \left[p^2 c^2 + m^2 c^4 \right]^{1/2} .$$

Changing the variable $t = p/mc$ gives

$$U_0 = \frac{8\pi V}{h^3} m^4 c^5 h(x_F) . \qquad (10.103)$$

1.2b The pressure of the star is given by

$$P_0 = - \left(\frac{\partial F_0}{\partial V} \right)_N = - \left(\frac{\partial U_0}{\partial V} \right)_N = \frac{8\pi}{h^3} m^4 c^5 \left[-h(x_F) - V h'(x_F) \left(\frac{\partial x_F}{\partial V} \right)_N \right]$$

since $F_0 = U_0 - T_0 S_0 = U_0$. Using equation (10.102) for x_F, we find

$$P_0 = \frac{8\pi}{h^3} m^4 c^5 \left[\frac{x_F}{3} h'(x_F) - h(x_F) \right] .$$

1.2c In the limit of $h(x)$ when $x_F \ll 1$, U_0 and P_0 become

$$U_0 = \frac{8\pi V}{h^3} m^4 c^5 \left(\frac{x_F^3}{3} + \frac{x_F^5}{10} \right) = \frac{8\pi V}{3h^3} m^4 c^5 x_F^3 + \frac{4\pi V}{5h^3} m^4 c^5 x_F^5$$

and $$P_0 = \frac{8\pi}{h^3} m^4 c^5 \left[\frac{x_F}{3} \left(x_F^2 + \frac{x_F^4}{2} \right) - \left(\frac{x_F^3}{3} + \frac{x_F^5}{10} \right) \right] = \frac{8\pi}{15h^3} m^4 c^5 x_F^5 .$$

On substituting (10.102) for x_F, it is seen that the first term in U_0 is equal to Nmc^2, which, in the classical limit, corresponds to the rest mass energy of the electrons. On comparing U_0 and P_0, it is then found that

$$P_0 V = \frac{2}{3} \left(U_0 - Nmc^2 \right) .$$

This result is simply the general relation $PV = 2U/3$ that applies to non-relativistic gases, where U now denotes the internal energy of the gas (excluding the rest mass energy).

1.2d With the limiting form of $h(x)$ as $x \gg 1$, U_0 and P_0 become

$$U_0 = \frac{8\pi V}{h^3} m^4 c^5 \times \frac{x_F^4}{4}$$

and $$P_0 = \frac{8\pi}{h^3} m^4 c^5 \left[\frac{x_F}{3} \times x_F^3 - \frac{x_F^4}{4} \right] = \frac{8\pi}{h^3} m^4 c^5 \times \frac{x_F^4}{12} .$$

It follows that $P_0 V = U_0/3$. This is the general relation $PV = U/3$ that applies to an ultrarelativistic gas ($\epsilon = pc$).

2 Application to white dwarves

2.1a The volume and radius of the star are given by

$$V = \frac{M_S}{\rho} = 2.0 \times 10^{20} \text{ m}^3 \qquad \text{and} \qquad R = \left(\frac{3V}{4\pi}\right)^{1/3} = 3.6 \times 10^6 \text{ m} \quad .$$

The radius of this white dwarf is comparable to that of the Earth.

2.1b The number of helium atoms in the star is

$$\frac{N}{2} = \frac{M_S}{M(^4\text{He})} \times \mathcal{N} = 3.0 \times 10^{56} \quad .$$

There are therefore 3×10^{56} helium nuclei and $N = 6 \times 10^{56}$ electrons in the star. The electron density is thus $N/V = 3 \times 10^{36} \text{ m}^{-3}$, which is approximately 10^8 greater than in metals.

2.1c The value of the density N/V calculated above gives $p_F^0 = 4.7 \times 10^{-22}$ kg m s^{-1} and $x_F = p_F^0/mc = 1.72$. This result shows that electrons lying close to the surface of the Fermi sphere are relativistic. Relativistic kinematics must therefore be used.

2.1d The value of the Fermi energy is

$$\epsilon_F^0 = mc^2 \left(x_F^2 + 1\right)^{1/2} = 1.63 \times 10^{-13} \text{ J} = 1.02 \times 10^6 \text{ eV} \quad .$$

Since the temperature of the star is $T = 10^7$ K, it follows that $kT/\epsilon_F^0 = 8 \times 10^{-4} \ll 1$. The approximation of absolute zero is therefore acceptable.

2.2a From equation (10.102) for x_F, we find

$$V = \frac{3N}{\pi} \left(\frac{h}{2mc}\right)^3 \frac{1}{x_F^3} \qquad \text{and} \qquad R = \left(\frac{3V}{4\pi}\right)^{1/3} = \frac{h}{2mc} \left(\frac{9N}{4\pi^2}\right)^{1/3} \frac{1}{x_F} \quad .$$

Since the mass of the star is $M = (N/2) \times 4m_p = 2Nm_p$, we get finally

$$V = \frac{3}{2\pi} \left(\frac{h}{2mc}\right)^3 \frac{M}{m_p} \frac{1}{x_F^3} \qquad \text{and} \qquad R = \frac{h}{4mc} \left(\frac{3}{\pi}\right)^{2/3} \left(\frac{M}{m_p}\right)^{1/3} \frac{1}{x_F} \quad . \quad (10.104)$$

2.2b Substituting the above expression for V into equation (10.103) for U_0 yields

$$U_0 = \frac{3}{2} mc^2 \frac{M}{m_p} \frac{h(x_F)}{x_F^3} \quad . \quad (10.105)$$

2.2c Similarly, substituting expression (10.104) for R into that for E yields

$$
\begin{aligned}
E &= -\alpha G M^2 \frac{4mc}{h} \left(\frac{\pi}{3}\right)^{2/3} \left(\frac{m_p}{M}\right)^{1/3} x_F \\
&= -\alpha \frac{4G m_p^2}{hc} \left(\frac{\pi}{3}\right)^{2/3} mc^2 \left(\frac{M}{m_p}\right)^{5/3} x_F \quad .
\end{aligned}
\quad (10.106)
$$

This expression has the required form, where the dimensionless constant is

$$\gamma = \frac{32}{3} \left(\frac{\pi}{3}\right)^{2/3} \frac{Gm_p^2}{hc} \ .$$

2.3a The equilibrium radius of the star is given by the condition of minimum total energy with respect to R, with M being kept constant, i.e.,

$$\left(\frac{dU}{dR}\right)_{R=R_0} = 0 \ .$$

Since the variables R and x_F are related through equation (10.104), this condition can be written $dU/dx_F = 0$.

2.3b With equations (10.105,106) for U_0 and E, the equilibrium condition gives

$$\frac{dU}{dx_F} = \frac{3}{2}mc^2 \frac{M}{m_p} \frac{x_F h'(x_F) - 3h(x_F)}{x_F^4} - \frac{3}{8}\alpha\gamma mc^2 \left(\frac{M}{m_p}\right)^{5/3} = 0 \ ,$$

which reduces to

$$f(x_F) = 4\frac{x_F h'(x_F) - 3h(x_F)}{x_F^4} = \alpha\gamma \left(\frac{M}{m_p}\right)^{2/3} \ .$$

2.3c Taking $M = M_S = 2 \times 10^{30}$ kg, we must find the graphical solution of $f(x_F) = 0.70$. Reading off from Figure 10.15, we get $x_F = 1.25$. From (10.104)

$$R_0 = 5.00 \times 10^6 \text{ m} = 5000 \text{ km} \ .$$

This confirms the order of magnitude taken in 2.1 for the radius of a typical star.

2.3d When x_F tends to infinity, the limiting form of $h(x)$ indicates that $f(x_F) \to 1$. There are therefore no solutions to the equation of equilibrium when the mass is greater than

$$M_l = m_p \times \left(\frac{1}{\alpha\gamma}\right)^{3/2} = 3.44 \times 10^{30} \text{ kg} = 1.72 \ M_S \ .$$

For masses greater than M_l, no equilibrium state exists in which the kinetic pressure of the electronic gas can counterbalance the gravitational attraction. The star thus contracts and, as the temperature rises, new nuclear reactions occur and the star enters a different regime.

A more complete theory taking account of variations in density with the depth in the star gives $M_l = 1.44 \ M_S$ (Chandrasekhar limit). The fact that the elementary model yields results close to those of the general model is a consequence of the low compressibility of the electron gas in the star, and stems from the Pauli exclusion principle.

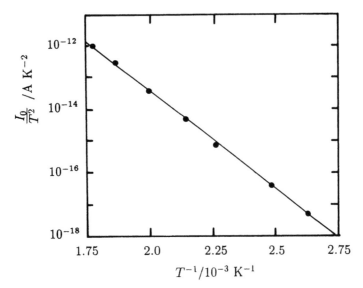

Figure 10.16: Thermionic emission from a tungsten filament covered by an oxide layer. I_0 is the total electric current [C.S. Hung, *J. App. Phys.* **21**, 37 (1950)].

PROBLEM 10.2 EMISSION OF ELECTRONS FROM METALS

To gain an understanding of electron emission from metals, we adopt a model in which the free electrons constitute a gas of non-interacting particles and the ions of the crystal generate a uniform potential well of energy $-\epsilon_A$ ($\epsilon_A > 0$) inside the crystal (ϵ_A is called the affinity of the metal). Electrons that have sufficient kinetic energy can leave the metal, giving rise to the phenomenon of thermionic emission, which is important at high temperatures.

1 What is the density of electrons $d^3n_{\mathbf{p}}$ with momentum equal to $\mathbf{p}$ to within $d^3\mathbf{p}$, the Fermi energy of the electrons being ϵ_F? Hence find an expression for $d^3n_{\mathbf{v}}$, where $\mathbf{v}$ is the velocity.

2 Calculate the density of electrons dn_{v_z} with velocity in a direction Oz lying between v_z and $v_z + dv_z$. Set $u = m(v_x^2 + v_y^2)/2kT$ and observe that

$$\frac{1}{Ce^u + 1} = \frac{e^{-u}}{C + e^{-u}} \quad .$$

3 What is the number of electrons $d^6N_{\mathbf{v}}'$ with velocity $\mathbf{v}$ to within $d^3\mathbf{v}$ that cross a surface perpendicular to Oz of area dS in the interval dt? Hence deduce an expression for the elementary current density dj_z associated with the electrons whose normal component of velocity lies between v_z and $v_z + dv_z$. Set $\epsilon_z = mv_z^2/2$.

4 Assume that the only electrons that can leave the metal are those for which $\epsilon_z > \epsilon_A$. Find an expression for the total current density $j_0 \equiv j_z$ of electrons leaving the metal knowing that the work function $w = \epsilon_A - \epsilon_F$ is much greater than kT (Richardson equation).

5 Figure 10.16 shows the current I_0 emitted by a tungsten cathode covered with an oxide layer, as a function of temperature T. Show that the Richardson equation is consistent with the data and determine the value of w.

6 If a potential difference V^* ($V^* < 0$) is applied between a cathode and a collecting anode of the same metal (Schottky effect), only electrons with an energy $\epsilon_z > \epsilon_A - eV^*$ can reach the anode and contribute to the current.

6a What is the ratio I/I_0 of the currents measured with and without the potential difference V^*?

6b Measurements carried out with the tungsten cathode described in question 5 and a tantalum anode show that the current I is of the form

$$\log_{10} \frac{I}{I_0} \simeq 4650 \frac{V^*}{T} \ ,$$

where $V^* = V - V_0$ is an effective potential difference related to the applied potential difference V and $V_0 = 3.2$ volts is a constant. Is this empirical relation in agreement with the theoretical equation found in 6a? What is the significance of the quantity $\Delta w = eV_0$?

SOLUTION

1 Referring to paragraph 10.5.2 , we have

$$d^3 n_{\mathbf{v}} = \frac{2m^3}{h^3} \left[\exp\left(\frac{m\mathbf{v}^2}{2kT} - \frac{\epsilon_F}{kT} \right) + 1 \right]^{-1} d^3\mathbf{v} \ .$$

2 Transforming to polar co-ordinates (ρ, ϕ) in the (v_x, v_y) plane, and then setting $u = m\left(v_x^2 + v_y^2 \right)/2kT = m\rho^2/2kT$, we find

$$dn_{v_z} = \frac{2m^3}{h^3} dv_z \int\int \frac{kT}{m} \, du \, d\phi \, \frac{1}{Ce^u + 1} \ , \qquad C = \exp\left(\frac{mv_z^2}{2kT} - \frac{\epsilon_F}{kT} \right) \ .$$

Integration over ϕ ($d\phi \rightarrow 2\pi$), together with the identity stated in the question, yields

$$
\begin{aligned}
dn_{v_z} &= \frac{4\pi m^2}{h^3} kT \left[-\ln(C + e^{-u}) \right]_0^\infty \, dv_z \\
&= \frac{4\pi m^2}{h^3} kT \ln\left[1 + \exp\left(\frac{\epsilon_F}{kT} - \frac{mv_z^2}{2kT} \right) \right] \, dv_z \ .
\end{aligned}
$$

3 As already seen in paragraph 6.3.1, $d^6 N'_{\mathbf{v}} = d^3 n_{\mathbf{v}} v_z \, dt dS$. Integration over v_x and v_y yields for the number of electrons whose velocity normal to the surface is equal to v_z, to within dv_z,

$$d^4 N'_{v_z} = dn_{v_z} v_z \, dt dS.$$

The associated current density $dj_z = ed^4 N'_{v_z}/dtdS$ is

$$dj_z = \frac{4\pi me}{h^3} kT \ln\left[1 + \exp\left(\frac{\epsilon_F - \epsilon_z}{kT} \right) \right] \, d\epsilon_z \qquad (10.107)$$

where $d\epsilon_z = mv_z \, dv_z$.

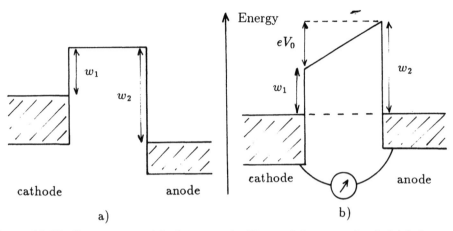

Figure 10.17: Contact potential of two metals. Shape of the energy levels (a) before and (b) after contact. The contact potential V_0 is given by the difference in work functions of the cathode and the anode, $eV_0 = w_2 - w_1$.

4 The required current density is found by integrating dj_z over all values of ϵ_z greater than ϵ_A. As $\epsilon_z - \epsilon_F > \epsilon_A - \epsilon_F \gg kT$, the exponential in dj_z is much smaller than 1 and we have

$$
\begin{aligned}
j_0 &= \frac{4\pi m e}{h^3} kT \exp\left(\frac{\epsilon_F}{kT}\right) \int_{\epsilon_A}^{\infty} \exp\left(-\frac{\epsilon_z}{kT}\right) d\epsilon_z , \\
&= \frac{4\pi m e}{h^3} (kT)^2 \exp\left(-\frac{w}{kT}\right) .
\end{aligned}
\tag{10.108}
$$

This is the Richardson equation. Note that, in the range of integration, $N_i/g_i \ll 1$, and approximation (10.3) for the Fermi function has been used implicitly.

5 Figure 10.16 shows that the current I_0 varies as $T^2 \exp(-B/T)$ in agreement with the Richardson equation. The numerical value of the slope B determined from the figure is $B = 14\ 400$ K. From this the work function can be calculated, i.e.

$$
w = \frac{kB}{e} = 1.24 \text{ eV} .
$$

Note that the precision obtained from the experimental data for the value of the exponent of T is poor. The model of Fowler, which takes into account the oxide layer, leads to a factor $T^{5/4}$ instead of T^2 and fits the data equally well. Furthermore, the theoretical normalization factor is larger than the experimental factor. One reason for this discrepancy is the existence of a quantum mechanical reflection factor on crossing a potential barrier.

6a We need only recalculate 4, replacing ϵ_A by $\epsilon_A - eV^*$, and therefore w by $w - eV^*$. The current density (10.108) then becomes

$$
j = j_0 \exp\left(\frac{eV^*}{kT}\right) ,
$$

and a similar relation exists between I and I_0.

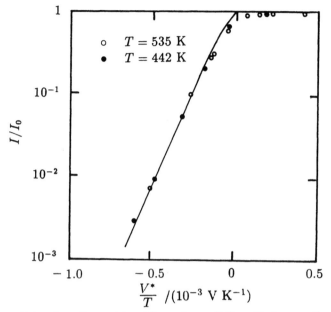

Figure 10.18: Schottky effect with a bias voltage ($V^* < 0$) for a cylindrical tungsten cathode covered with oxide. I and I_0 are the currents with and without a potential difference. The full curve shows the Schottky relation for cylindrical geometry [C.S. Hung. *J. Appl. Phys.* **21**, 37 (1950)].

6b Numerically, the theoretical relation between I and I_0 is therefore

$$\log_{10}\frac{I}{I_0} = 0.43 \times \frac{eV^*}{kT} = 5040\frac{V^*}{T} \quad ,$$

in satisfactory agreement with experiment.

The experimentally observed term V_0, called the contact potential, comes from the fact that the cathode and the anode are made of different metals. When cathode and anode are connected by a conductor, electrons flow from one metal to the other and become localized on opposite facing surfaces. This creates an electrostatic field between the metals and a potential difference that increases until the Fermi energies become equal (Fig. 10.17). The quantity

$$\Delta w \equiv eV_0 = w_2 - w_1$$

is the difference between the work function of the anode w_2 and that of the cathode w_1. It follows that the work function of tantalum is $w_2 = w_1 + eV_0 = 4.4$ eV. Figure 10.18 shows experimental results of the Schottky effect. The equation stated in the question is in fact only approximate. The theoretical Schottky relation, obtained by taking into account the kinetics of the electrons between a coaxial cathode and anode, is shown as a full line. For V^*/T less than about -2.5×10^{-4} V K^{-1}, this relation is linear with a slope equal to that stated in the question. Above -2.5×10^{-4} V K^{-1}, the current is lower than that predicted by the Schottky relation, owing to surface defects on the cathode. It follows that, for the current to reach its ideal value in zero field (saturation current), a slightly positive potential difference V^* (~ 50 mV) must be applied. For larger values of V^*, the current increases again since the reduction of the potential barrier makes it easier to remove electrons.

Chapter 11

Electronic Properties of Solids

11.1 INTRODUCTION

The success of the free electron model in metals (Ch. 10) led to a more detailed investigation of the electronic levels in solids and to the discovery that these levels cluster into energy bands. In particular, these bands provide an explanation for the physical differences between metals and insulators.

As a first example of a metal we take sodium. The free atom possesses 11 electrons with the configuration $1s^2\,2s^2\,2p^6\,3s$. The first 10 electrons occupy full layers, and have a spatial range of about 2Å (Fig. 11.1), while the eleventh electron has a $3s$ orbital that extends beyond 5 Å. If two sodium atoms come close to each other, their $3s$ orbitals start to overlap when their separation is about 10 Å and the degeneracy of the $3s$ levels of each atom is lifted. In the crystal the separation between atoms is 3.7 Å and the number N of $3s$ levels is equal to the number of atoms. These levels are distributed continuously in an energy band called the $3s$ band (Fig. 11.2). If the spin of the electrons is taken into account, the $3s$ band in fact contains $2N$ electron states, although it contains only N electrons. It is therefore

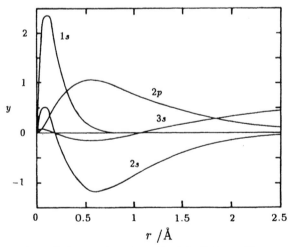

Figure 11.1: Electronic radial wave functions $R_{nl}(r)$ of the sodium atom in its ground state. The quantity plotted is $y = rR_{nl}$ [D.R. Hartree and W. Hartree, *Proc. Roy. Soc. A* (London) **193**, 299 (1948)].

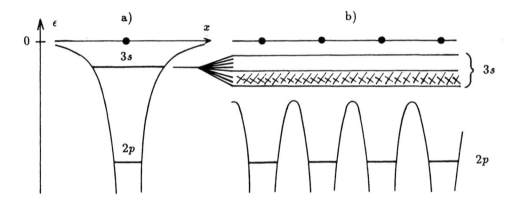

Figure 11.2: Electronic energy levels in sodium: a) free atom, b) crystal. The vertical axis shows the energy, and the abscissa indicates the spatial position. In addition, the curves in this figure show the electrostatic potential, and the filled circles on the x axis indicate the position of the nuclei. In the solid, the $2p$ levels remain degenerate and the $3s$ levels form a band that is half-filled with free electrons.

half full, and the electrons preferentially occupy the lowest levels (Fig. 11.3a). The electrostatic potential to which the electrons are subjected, being the sum of the potentials generated by each atom, is lower than that of the free atom (Fig. 11.2). Consequently the $3s$ electrons are free, and sodium is a metal.

We now consider the case of an insulator, silicon. The electronic configuration of the free atom is $1s^2\ 2s^2\ 2p^6\ 3s^2\ 3p^2$, and it possesses 4 external electrons, the $3p$ sub-layer being incomplete. In the solid, hybridization of the electronic states occurs, and this leads to the formation of two separate energy bands, the $3s - 3p$ band and the $3p$ band, each containing 4 electronic states per atom (Fig. 11.3b). The external electrons (4 per atom) then completely fill the (lower) $3s - 3p$ band, while the $3p$ band remains empty. Electrical conduction cannot occur and silicon is an insulator.

Lastly, we take the case of magnesium, whose electronic configuration is $1s^2\ 2s^2\ 2p^6\ 3s^2$. In the solid, the N atoms liberate $2N$ $3s$ electrons which, if there were no overlap of the $3s$ and $3p$ bands (Fig. 11.3c), would have completely filled the $3s$ band. Some electrons do however occupy the bottom of the $3p$ band, leaving vacant states at the top of the $3s$ band. It follows that magnesium is metallic. As a general rule, elements whose free atoms have their outer electrons in the s, d and f sub-layers are metallic (alkali metals, alkali-earths, transition metals ...). Those whose outer electrons are in the p sub-layer tend to be insulators if they are light, and metallic if they are heavy.

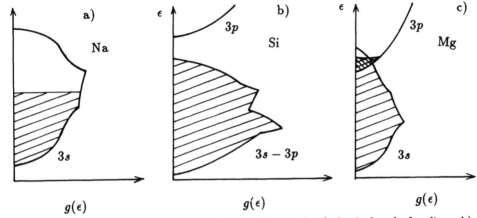

Figure 11.3: Diagram of the electronic density of states in a) the $3s$ band of sodium, b) the $3s - 3p$ and $3p$ bands of silicon, c) the $3s$ and $3p$ bands of magnesium. When the levels occupied by the electrons (hatched areas) do not completely fill the band, the substance is metallic (a and c). In the opposite case b), the material is an insulator.

11.2   ELECTRONIC STATES IN A PERIODIC POTENTIAL

In this paragraph, we investigate the electronic states of a crystal using a one-dimensional model and generalize the results to the three-dimensional case.

11.2.1   Bloch Theorem

The periodic structure of crystals imposes spatial periodicity on the electrostatic potential $\phi(\mathbf{x})$ that an electron experiences from the nuclei and from the other electrons. Thus, for a one-dimensional solid with period a, we have $\phi(x+a) = \phi(x)$.

Let $\psi(x)$ be the electronic wave function, which is a solution of the Schrödinger equation

$$-\frac{\hbar^2}{2m}\frac{d^2\psi}{dx^2} + \phi(x)\psi = \epsilon\,\psi(x) \ , \tag{11.1}$$

associated with the energy ϵ. Clearly the function $\psi(x + a)$ is also a solution of the Schrödinger equation with the same energy, and the two functions $\psi(x)$ and $\psi(x + a)$ therefore differ only by a phase factor, i.e.

$$\psi(x + a) = e^{i\varphi}\psi(x) \tag{11.2}$$

where φ is a phase that can be restricted to the interval $[-\pi, +\pi[$. Introducing now the quantity $k = \varphi/a$ comprised between $-\pi/a$ and π/a, we can write $\psi(x)$ in the form

$$\psi(x) = u(x)e^{ikx}. \tag{11.3}$$

It can be verified that $u(x)$ is a function that allows for the periodicity of the network and depends on k. This result is the Bloch theorem. It shows that the electronic

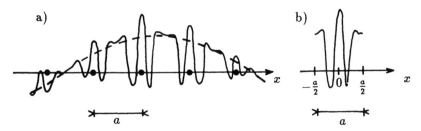

Figure 11.4: a) Sketch of a function $\psi(x)$ with $k = \pi/4a$. The exponential $\exp(ikx)$ is drawn as a dashed curve. b) The periodic function $u(x)$ is shown diagrammatically for one of its periods $[-a/2,\ a/2[$.

wave functions look like sine waves that are perturbed in the neighbourhood of the ions (Fig. 11.4).

k is then interpreted as the wave vector associated with the electron, its values being confined to the interval

$$-\frac{\pi}{a} \le k < \frac{\pi}{a} \ . \tag{11.4}$$

Bloch's theorem can be generalized to 3 dimensions as

$$\psi(\mathbf{x}) = u(\mathbf{x})\exp(i\mathbf{k}.\mathbf{x}). \tag{11.5}$$

Here, $u(\mathbf{x})$ is a function of $\mathbf{x}$, having the same periodicity as the crystal lattice, and depends on $\mathbf{k}$. The wave vector $\mathbf{k}$ is limited to one unit cell in reciprocal space, or the first Brillouin zone in general (§8.3.4). Note that, as for phonons (§8.4.1), the vector $\mathbf{p} = \hbar\mathbf{k}$ is not the momentum of the electron. This vector $\mathbf{p}$, however, plays the role of an effective momentum (§11.3.1).

11.2.2   Classification of Electronic States

Bloch's theorem provides a means of classifying the electronic states, as we shall show for the simple one-dimensional case of a constant potential, which is treated as having a period a. This is the situation of free electrons. We take the potential to be zero by shifting the origin of the energy.

Each electronic state is a plane wave characterized by the arbitrary parameter k' and whose wave function and associated energy have the form

$$\psi(x) = \exp(ik'x) \qquad \text{and} \qquad \epsilon = \frac{\hbar^2 k'^2}{2m} \ . \tag{11.6}$$

We introduce the wave vector k, restricted to the interval $[-\pi/a,\ \pi/a[$, and the integer n', defined by

$$k' = \frac{2n'\pi}{a} + k \ . \tag{11.7}$$

Equations (11.6) can then be written as

$$\psi_{n'k}(x) = \exp\left(i\frac{2n'\pi x}{a}\right)\exp(ikx) \quad \text{and} \quad \epsilon_{n'}(k) = \frac{\hbar^2}{2m}\left(\frac{2n'\pi}{a} + k\right)^2 \ , \tag{11.8}$$

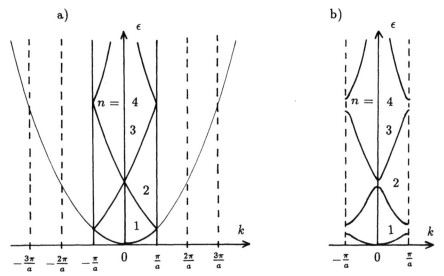

Figure 11.5: Electronic dispersion curves $\epsilon_n(k)$ in a one-dimensional crystal of length a: a) free electrons and b) Kronig-Penney model with $P = 1$. The Brillouin zone number n defines the energy band. In figure a) is also shown the curve $\epsilon = \hbar^2 k^2 / 2m$, the limits of the first 3 Brillouin zones being indicated.

where we have replaced $\psi(x)$ by the Bloch theorem result (11.3). We thus see that each electronic state is characterized by a wave vector k located in the first Brillouin zone (11.4) and by an arbitrary integer n'. The resulting dispersion curves $\epsilon_{n'}(k)$ are shown in Figure 11.5a.

As will be seen in the next paragraph, the appearance of energy bands means that instead of using the number n' defined in (11.7), it is preferable to use another number n such that

$$(n - 1)\frac{\pi}{a} \leq |k'| \leq \frac{n\pi}{a} \quad . \tag{11.9}$$

n is the Brillouin zone number, with the first zone (11.4) corresponding to $n = 1$.

These results can be generalized to the case of free electrons in 3-dimensional space. The vector $\mathbf{k}'$ decomposes into a sum of the wave vector $\mathbf{k}$ belonging to the first Brillouin zone and a vector $\mathbf{K}$ of the reciprocal lattice. The Brillouin zone number n to which $\mathbf{k}'$ belongs is determined as follows (Fig. 11.6):

- perpendicular bisector planes are drawn for all the vectors of the reciprocal lattice (Bragg planes);
- the zone number n is the number of Bragg planes cut by the vector $\mathbf{k}'$, plus 1.

As an illustration, Figure 11.7 shows the first Brillouin zone and the dispersion curves of a free electron for a face-centred cubic lattice (cf. Fig. 8.11).

11.2.3   Kronig-Penney Model

The Kronig-Penney model (1930) is one-dimensional. Through simple calculations, it yields dispersion relations that have the general characteristics found in real

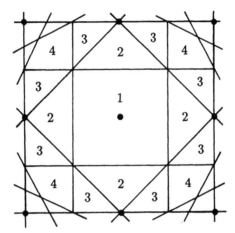

Figure 11.6: Brillouin zones for a two-dimensional square lattice (square direct and reciprocal lattices). In the figure, zones $n = 1$, 2 and 3 are complete, while zones $n = 4$, 5 and 6 appear partially.

substances, i.e. energy bands separated by gaps, and the general shape of the dispersion curves.

In this model, the periodic form of the potential $\phi(x)$ is

$$\phi(x) = \alpha \sum_{p=-\infty}^{\infty} \delta(x - pa) - \frac{\alpha}{a} \ .$$

Here Dirac δ functions are used to simulate the strongly repulsive regions in the immediate vicinity of the ions, while the constant term $-\alpha/a$ describes the attractive regions between the ions (the value of the constant term is chosen so that the mean value of ϕ is zero over a lattice period).

The solutions of the Schrödinger equation are defined piece-wise between two consecutive Dirac peaks and are given by

$$\psi(x) = A_p e^{ik'x} + B_p e^{-ik'x} \tag{11.10a}$$

in the interval $pa \leq x < (p+1)a$. The parameter k' that characterizes a solution can take an arbitrary value related to the energy by

$$\epsilon = -\frac{\alpha}{a} + \frac{\hbar^2 k'^2}{2m} \ . \tag{11.10b}$$

First, the matching conditions for the wave function and its derivative for a given k' are specified at a point $x = pa$ at which there is a discontinuity in ϕ. Second, the wave vector k is introduced through Bloch's theorem (11.2,3) in the form

$$\psi\left((p+1)a\right) = e^{ika}\psi(pa) \ .$$

This gives the following implicit equation that determines k', and hence ϵ, as a function of k

$$\cos k'a + P\frac{\sin k'a}{k'a} = \cos ka , \qquad \text{with} \qquad P = \frac{ma\alpha}{\hbar^2} \ . \tag{11.11}$$

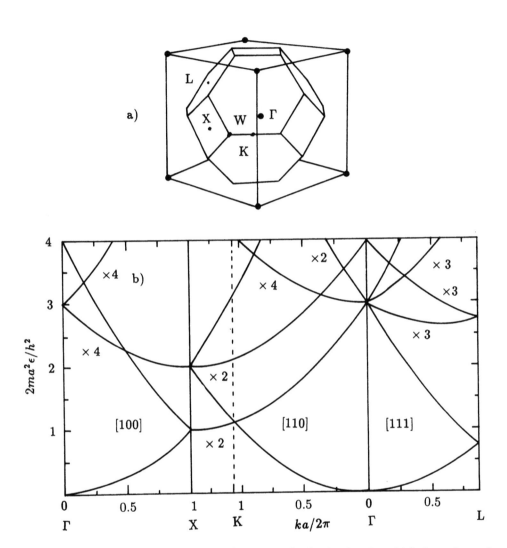

Figure 11.7: a) First Brillouin zone of a face-centred cubic lattice, for which the reciprocal lattice is cubic centred. The particular points that repeat by symmetry are: Γ = centre of the cuboctahedron, X = centre of the square faces, L = centre of the hexagonal faces, K = mid-point of an arbitrary edge, W = vertex of the polyhedron. b) Electronic dispersion curves $\epsilon(\mathbf{k})$ in the three directions indicated for free electrons in this lattice with side a. The number of superimposed branches is indicated by ×2, ×3 or ×4.

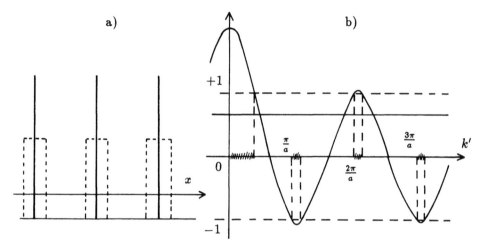

Figure 11.8: a) Potential energy $\phi(x)$ in the Kronig-Penney model. The Dirac peaks are the limits of the rectangular functions (dashes) having constant surface area α and whose width tends to zero. b) Graph of the function on the left hand side of equation (11.11) with $P = 1$. The horizontal straight line with height lying between -1 and $+1$ yields a graphic solution of (11.11). The forbidden values of k' are shown hatched.

Graphical solution of this equation (Fig. 11.8b) shows that
- for each value of k, there is an infinite number of discrete values of k', and therefore of energies. The corresponding dispersion relations $\epsilon_n(k)$ are shown in Figure 11.5b.
- since $\cos ka$ lies between -1 and $+1$, not all values of k' and hence not all energies are accessible. It can thus be seen that the energy levels are grouped into allowed bands that are separated by forbidden gaps.
- at the bottom of each allowed band, alternately at $k_1 = 0$ and at $k_1 = \pm\pi/a$, the energy takes the quadratic form

$$\epsilon_n(k) \simeq \epsilon_{1n} + \frac{\hbar^2}{2m_{1n}^*}(k - k_1)^2 \qquad (11.12a)$$

where m_{1n}^* is called the effective mass of the electron at the bottom of the band n.
- at the top of each allowed band, both at $k_2 = \pm\pi/a$ and at $k_2 = 0$, the energy has the quadratic form

$$\epsilon_n(k) \simeq \epsilon_{2n} - \frac{\hbar^2}{2m_{2n}^*}(k - k_2)^2 \qquad (11.12b)$$

where m_{2n}^* is called the effective mass of the "hole" at the top of the band n.

11.2.4   General Results

11.For three-dimensional solids, the potential is approximated by functions with highly complex forms. Solving the Schrödinger equation by numerical methods yields the dispersion relations $\epsilon_n(\mathbf{k})$ for the electrons. The resulting pattern of energy bands is intermediate between that found by broadening the electronic levels of

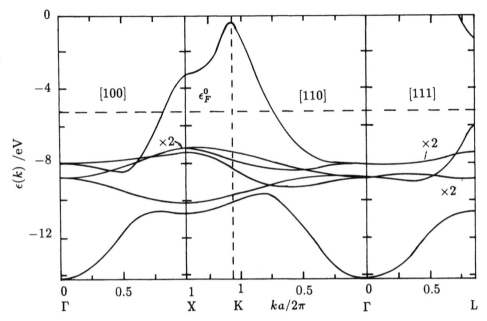

Figure 11.9: Calculated electronic dispersion curves $\epsilon_n(\mathbf{k})$ for copper in the directions indicated. The horizontal line shows the Fermi energy ϵ_F^0, where zero energy corresponds to an electron outside the metal. The lattice parameter of copper is $a = 3.61$ Å . Compare with Figure 11.7 [G.A. Burdick, *Phys. Rev.* **129**, 138 (1963)].

the free atom (Fig. 11.2) and that for a free electron (Fig. 11.7). Figure 11.9 shows the diagram found by one of these methods for copper. The electronic configuration of the free atom is $3d^{10}\ 4s$, and the inner electrons have the highly stable configuration of argon. It can be seen that for almost every value of $\mathbf{k}$, there are 5 states having energy close to -8 eV that make up the "3d band", and one state whose energy varies between -15 and 0 eV that forms the "4s band". These designations are merely a convention, since, for certain values of $\mathbf{k}$, the 6 levels are very close, and the distinction $3d - 4s$ is therefore arbitrary. Note, however, that the shape of the dispersion curve of the 4s band is closely similar to that of the lowest curve for a free electron (Fig. 11.7).

We observe that the dispersion curves in copper have the following general characteristics: they all display an extremum at the centre of the first Brillouin zone ($k = 0$, point Γ) as well as at the centre of the faces of this zone (points X and L), i.e. at points such that $\mathbf{k} = \mathbf{K}/2$, where $\mathbf{K}$ is a vector of the reciprocal lattice. This property follows from the translational invariance of the lattice and its symmetry, which require that

$$\epsilon(-\mathbf{k}) = \epsilon(\mathbf{k}) \quad \text{and} \quad \epsilon(\mathbf{k} + \mathbf{K}) = \epsilon(\mathbf{k}) \ .$$

Finally, we point out that experimental information on band structure can be obtained from investigations of the absorption of soft X-rays and from photo-emission of electrons.

11.3   FERMI SURFACE

11.3.1   Density of Electronic States

In solving the Schrödinger equation, the boundary conditions at the limits of the crystal must be respected. One suitable approach is the *periodic boundary condition* of Born-von Kármán. In the one-dimensional case, this means that the wave function must satisfy

$$\psi(x + L) = \psi(x) \ ,$$

where $L = Na$ is the length of the crystal. From Bloch's theorem (11.3) it then follows that $\exp(ikL) = 1$. The wave vector k, lying between $-\pi/a$ and $+\pi/a$, is given by

$$k = 2n\pi/L \qquad \text{with} \qquad n = -\frac{N}{2}, \ \dots \ , \ \frac{N}{2} \ .$$

The N values of k are then regularly spaced by $2\pi/L$ with a density $g(k)$ given by

$$g(k) \ dk = 2 \times \frac{L}{2\pi} \ dk \qquad \text{with} \qquad \int_{-\pi/a}^{\pi/a} g(k) \ dk = 2N \ .$$

The factor 2 is included here to account for the 2 spin orientations of the electron for each value of k.

In three dimensions, the periodic boundary condition gives for the density $g(\mathbf{k})$

$$g(\mathbf{k}) \ d^3\mathbf{k} = 2 \times \frac{V}{(2\pi)^3} \ d^3\mathbf{k} \qquad \text{with} \qquad \int_{BZ} g(\mathbf{k}) \ d^3\mathbf{k} = 2N \ , \qquad (11.13)$$

where BZ denotes the region of the first Brillouin zone. This density can also be found from the density in phase space $(\mathbf{x}, \ \mathbf{p})$, such that

$$g(\mathbf{x}, \ \mathbf{p}) \ d^3\mathbf{x} \ d^3\mathbf{p} = 2 \times \frac{d^3\mathbf{x} \ d^3\mathbf{p}}{h^3}$$

where we have set $\mathbf{p} = \hbar\mathbf{k}$. This is one of the reasons for which the vector $\mathbf{p}$ plays the role of the momentum of the electron.

As in (8.41), the density of energy states in the band n is then given by

$$g_n(\epsilon) = 2 \times \frac{V}{(2\pi)^3} \int d^3\mathbf{k} \ \delta\left(\epsilon - \epsilon_n(\mathbf{k})\right) \qquad (11.14)$$

with

$$\int_{\epsilon_{1n}}^{\epsilon_{2n}} g_n(\epsilon) \ d\epsilon = 2N \qquad (11.15)$$

where ϵ_{1n} and ϵ_{2n} are the minimum and maximum energies in the band n. For free electrons, the total density of states takes the form (10.43)

$$g(\epsilon) = \sum_n g_n(\epsilon) = \frac{4\pi V}{h^3}(2m)^{3/2}\epsilon^{1/2} \ . \qquad (11.16)$$

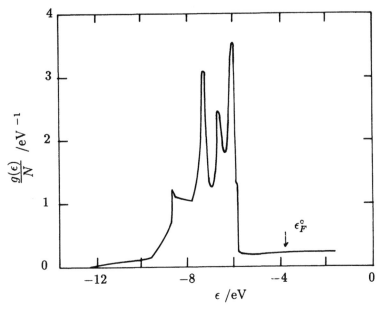

Figure 11.10: Calculated density of states per atom in copper. Note the accumulation of states near -8 eV corresponding to the d band. The position of the Fermi energy ϵ_F^0 is indicated [F.M. Mueller, *Phys. Rev.* **153**, 150 (1967) and E.C. Snow, *Phys. Rev.* **171**, 785 (1968)].

Figure 11.10 shows a real density of states. It displays van Hove discontinuities (§8.4.2) for values of ϵ at which the vector $\mathbf{k}$ is equal to zero or is directed towards the centre of a face in the first Brillouin zone ($\mathbf{k} = \mathbf{K}/2$). Moreover, near the bottom or the top of the band n, this density respectively takes the form

$$g_n(\epsilon) \;=\; \frac{4\pi V}{h^3}\,(2m_{1n}^*)^{3/2}\,(\epsilon - \epsilon_{1n})^{1/2} \tag{11.17a}$$

$$\text{and}\qquad g_n(\epsilon) \;=\; \frac{4\pi V}{h^3}\,(2m_{2n}^*)^{3/2}\,(\epsilon_{2n} - \epsilon)^{1/2} \;. \tag{11.17b}$$

These expressions have the same form as those for free electrons (11.16), and the parameters m_{1n}^* and m_{2n}^* are the effective masses encountered in (11.12).

11.3.2   Fermi Surface

At absolute zero, the electrons occupy all states with energy smaller than a value ϵ_F^0 called the Fermi energy. The surface in reciprocal space for which $\epsilon(\mathbf{k}) = \epsilon_F^0$ is called the Fermi surface and this plays a fundamental role in the detailed understanding of the electronic properties of solids. For free electrons, this surface is a sphere, called the Fermi sphere. For the alkali metals, the Fermi surface is almost spherical, which explains why the free electron model is satisfactory for these metals. For other metals, this surface is complex and can even define a boundary between separate regions containing occupied quantum states (Fig. 11.11).

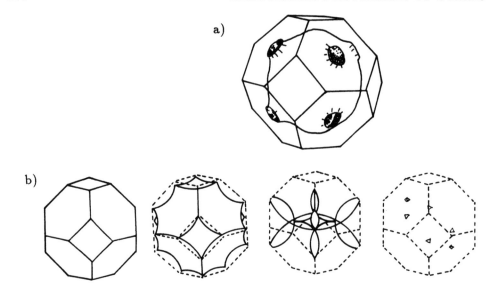

Figure 11.11: Fermi surfaces of two face-centred cubic metals. a) Monovalent metal: copper. b) Trivalent metal: aluminium. For the latter, the Fermi surface has 4 sheets. The first is identical to the first Brillouin zone. This zone and the second sheet limit the occupied states, and the enclosed region is empty. The third and fourth sheets contain the remaining occupied states [W.A. Harrison, *Phys. Rev.* **118**, 1183 (1960)].

The shape of the Fermi surface can be determined by various methods, of which by far the most powerful is that based on the de Haas-van Alphen effect (§10.4.2). L. Onsager (1952) established that, in the general case, the magnetic moment of a substance oscillates as a function of the inverse of the magnetic field according to equation (10.84). This particular expression was derived for a free electron gas, namely

$$\Delta\left(\frac{1}{B}\right) = \frac{eh}{A_F} \qquad (11.18)$$

where A_F is the maximum cross section of the Fermi surface cut by a plane normal to the magnetic field.

By measuring the oscillation period for various magnetic field orientations in the crystal, the shape and size of the Fermi surface can be found.

11.3.3 Electrical Conductivity

Except for superconductors (Ch. 15), no substance carries electrical currents without the presence of an electric field. In non zero field, certain materials conduct electricity, while others do not. The latter, called insulators, are those whose energy bands at absolute zero are either completely occupied by electrons (full band) or completely empty (empty band). In a full band, states occupied by the (indistinguishable) electrons remain unchanged whether a field is applied or not. It follows that the electrons in full bands do not participate in the electric current.

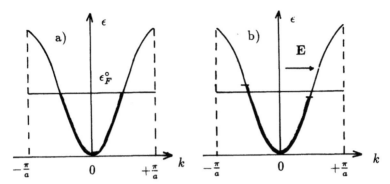

Figure 11.12: Occupied electronic states (thick lines) in a band of a monovalent metal in one dimension where the dispersion relation is $\epsilon = \epsilon_0 \sin^2(ka/2)$, a) in zero field and b) in an electric field. The electrons ascend the electric field and give rise to a current in the direction of the field.

For electrons in a partially filled band, the conduction model that was used in paragraph 10.5 for a free electron gas can still be employed. The results are a generalization of those obtained earlier. It can thus be shown that the electric field **E** causes the time average of the momentum of an electron to increase from zero to

$$< \mathbf{p} > = \hbar < \mathbf{k} > \qquad \text{with} \qquad < \mathbf{k} > = -\frac{e\tau}{\hbar}\mathbf{E} \qquad (11.19)$$

where τ is the mean time between two collisions (cf. (10.90)). Accordingly, the occupied electronic states of the band are modified as shown in Figure 11.12, and a current is generated of density

$$\mathbf{j} = \frac{1}{V}\int 2\frac{d^3x\, d^3p}{h^3}(-e) < \mathbf{v} > = -e\int \frac{d^3k}{4\pi^3} < \mathbf{v} > \qquad (11.20)$$

where $< \mathbf{v} >$ is the drift velocity of an electron of wave vector **k**. Note that the magnitude of $< \mathbf{k} >$ is much smaller than the size of the first Brillouin zone and therefore the region of integration can be taken to be the same as the region of occupied states in zero field.

To relate $< \mathbf{v} >$ to $< \mathbf{k} >$, we observe that the velocity **v** of an electron with momentum $\mathbf{p} = \hbar\mathbf{k}$ is found from the dispersion relation by

$$\mathbf{v}(\mathbf{k}) = \frac{\partial\epsilon}{\partial\mathbf{p}} = \frac{1}{\hbar}\frac{\partial\epsilon}{\partial\mathbf{k}} \qquad . \qquad (11.21)$$

This relation gives the group velocity of the wave packet describing an electron. If the solid is isotropic, the drift velocity of an electron is found by writing

$$< \mathbf{v} > \; = \; \mathbf{v}\,(\mathbf{k}+ < \mathbf{k} >) - \mathbf{v}(\mathbf{k}) \simeq \frac{1}{\hbar}\frac{\partial^2\epsilon}{\partial k^2} < \mathbf{k} >$$

$$= \; -e\tau\frac{1}{\hbar^2}\frac{\partial^2\epsilon}{\partial k^2}\,\mathbf{E} \qquad .$$

For a free electron, $\epsilon = \hbar^2 k^2/2m$, and we recover equation (10.90). Introducing now the notation

$$\frac{1}{m^*(\mathbf{k})} = \frac{1}{\hbar^2}\frac{\partial^2\epsilon}{\partial k^2} = \frac{\partial^2\epsilon}{\partial p^2} \tag{11.22}$$

the current density becomes

$$\mathbf{j} = e^2\tau \int \frac{d^3\mathbf{k}}{4\pi^3}\frac{1}{m^*(\mathbf{k})}\mathbf{E} \quad . \tag{11.23}$$

The quantity $m^*(\mathbf{k})$ defines the curvature of the dispersion curve of the band n, and has the dimensions of a mass. It is called the effective mass of an electron with wave vector $\mathbf{k}$. With free electrons, the effective mass is identical to that of the electron m. At the bottom of the band n, this effective mass is given by (cf. (11.12a))

$$m^*(\mathbf{k}_1) = m_{1n}^* \quad ,$$

while, at the top of the band, the effective mass is negative and is given by (11.12b)

$$m^*(\mathbf{k}_2) = -m_{2n}^* < 0 \quad .$$

The conductivity σ_n of the electrons in band n is then

$$\sigma_n = \frac{\mathbf{j}}{\mathbf{E}} = e^2\tau \int \frac{d^3\mathbf{k}}{4\pi^3}\frac{1}{m^*(\mathbf{k})} \quad . \tag{11.24}$$

For a full band, the integral is zero. Note that the conductivity of the substance is the sum of the conductivities of the partially filled bands, and that for a free electron gas ($m^* = m$), this gives equation (10.92).

Since solids are anisotropic, the effective mass and the conductivity are in fact second order tensors m_{ij}^* and σ_{ij}, such that

$$(m^{*-1})_{ij} = \frac{\partial^2\epsilon}{\partial p_i\,\partial p_j} = \frac{1}{\hbar^2}\frac{\partial^2\epsilon}{\partial k_i\,\partial k_j}$$

$$\text{and} \qquad \sigma_{ij} = e^2\tau \int \frac{d^3\mathbf{k}}{4\pi^3}(m^{*-1})_{ij} \quad . \tag{11.25}$$

For isotropic substances, these tensors take the form of the unit tensor.

Exercise 11.1 *Conductivity of a metal in one dimension*

Consider a one-dimensional crystal of length a for which the only band that can conduct electricity has the dispersion relation $\epsilon = \epsilon_0 \sin^2(ka/2)$ (Fig. 11.12).

1 Show that at the bottom and at the top of the band the dispersion relation has the form (11.12). Determine the parameters m_1^* and m_2^*.

2 Determine the effective mass $m^*(k)$ and hence derive an expression for the conductivity σ as a function of the Fermi wave vector k_F.

3 Calculate k_F and σ for a crystal in which there are z free electrons per atom. Discuss the cases $z = 1$ and $z = 2$.

Solution

1 On making series expansions around $k_1 = 0$ (bottom of the band) and $k_2 = +\pi/a$ (top of the band), we get

$$\epsilon(k) \simeq \epsilon_0 \frac{a^2}{4} k^2 \qquad (k \simeq k_1 = 0)$$

$$\epsilon(k) = \epsilon_0 \left[1 - \sin^2\left(\frac{ka}{2} - \frac{\pi}{2} \right) \right] \simeq \epsilon_0 \left[1 - \frac{a^2}{4}(k - k_2)^2 \right] \qquad \left(k \simeq k_2 = \frac{\pi}{a} \right) .$$

The effective masses of the electron and the hole are then

$$m_1^* = m_2^* = \frac{2\hbar^2}{\epsilon_0 a^2} .$$

2 The effective mass $m^*(\mathbf{k})$, given by (11.22), is equal to

$$m^*(k) = \frac{2\hbar^2}{\epsilon_0 a^2} \times \frac{1}{\cos ka} = \frac{m_1^*}{\cos ka} .$$

It can be checked that $m^*(0) = m_1^*$ and $m^*(\pi/a) = -m_2^*$. For a one-dimensional model, the electrical conductivity from (11.24) is

$$\sigma = e^2\tau \int \frac{dk}{\pi} \times \frac{1}{m^*(k)} = \frac{e^2\tau}{\pi m_1^*} \int dk \cos ka .$$

On integrating between $-k_F$ and $+k_F$, we find

$$\sigma = \frac{e^2\tau}{m_1^*} \frac{2 \sin k_F a}{\pi a} .$$

3 The electronic density of states of the one-dimensional metal is

$$g(k) \, dk = 2 \times \frac{L \, dp}{h} = \frac{L}{\pi} dk .$$

Since all states for which $|k| < k_F$ are populated, we have

$$zN = \frac{L}{\pi} \times 2k_F , \qquad \text{i.e.} \qquad k_F = z\frac{\pi}{2} \times \frac{N}{L} = z\frac{\pi}{2a}$$

where zN is the total number of free electrons.

For $z = 1$, we have $k_F = \pi/2a$ and the band is half full (Fig. 11.12); the crystal is metallic and its conductivity is

$$\sigma = \frac{2e^2\tau}{\pi a m_1^*} .$$

Remembering that the linear density of electrons is $n = N/L = 1/a$, we see that this expression is of the form (10.92), with the substitution $m \to \pi m_1^*/2$.

For $z = 2$, we have $k_F = \pi/a$ and the band is full. The crystal is insulating, and its conductivity is zero. Real divalent metals are conductors owing to overlap of the s and p bands (Fig. 11.3).

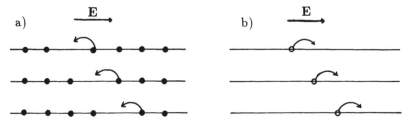

Figure 11.13: Diagram of successive displacements in space a) of an electron in an almost full band and b) of the corresponding hole. The electron moves up the electric field, while the hole moves down. In both cases, the current, which is directed along **E**, has the same value.

11.3.4 Holes

Imagine a substance that has a energy band that is full except for a state located at the top of the band. As we shall see, its properties can be understood more easily by considering a fictitious particle called a hole.

In the presence of an electric field, a current flows whose density is given by (11.20), where the integral is extended to all states of the band excluding the one that is empty. Since the integral for a full band is zero, we have

$$\mathbf{j} = \mathbf{0} \text{ (full band)} - \mathbf{j} \text{ (vacant state)}$$
$$= +e < \mathbf{v} > ,$$

which is equivalent to there being a *hole* with charge $+e$ that replaces all the electrons in the band (Fig. 11.13). A similar argument, with equation (11.23) for the density, leads to

$$\mathbf{j} = -\frac{e^2\tau}{m^*(\mathbf{k})} \mathbf{E} \quad ,$$

by which it is understood that the hole has a mass $m_h = -m^*(\mathbf{k}) > 0$. This result explains the term effective mass of the hole, used earlier.

The energy of the electrons in the band can also be written

$$U = \sum N_i \epsilon_i = U_0 - \epsilon \text{ (vacant state)}$$

where U_0 is the internal energy of the full band. In terms of holes, this equality becomes

$$U = U_0 + \epsilon \text{ (hole)} \qquad \text{with} \qquad \epsilon \text{ (hole)} = -\epsilon \text{ (vacant state)} \quad .$$

Thus the energy of the holes varies in the opposite way to that of the electrons (Fig. 11.14).

In real situations, if the number of vacant states in a band is fairly small, it is worthwhile to describe the properties of the band in terms of holes rather than electrons. It should be emphasized that, in the same band, these two representations are mutually exclusive. In cases where several bands are involved, one band may be

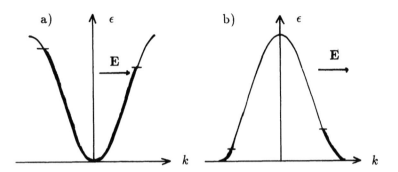

Figure 11.14: Interpretation of the electronic properties of a band in terms of a) electrons or b) holes. The thick lines show the occupied states. In b), the dispersion curve a) is inverted ($\epsilon \rightarrow -\epsilon$) and the charge of the particles has changed sign ($-e \rightarrow +e$).

represented by holes and another by electrons. This is the case respectively for the $3s$ and $3p$ bands in magnesium (Fig. 11.3) and also for the valence and conduction bands in semiconductors (§11.4). The total conductivity of the metal is then the sum of two conductivities, that of the holes in one band and of the electrons in the other.

In a magnetic field, holes create a Hall field in the direction opposite to that of the electrons (Fig. 11.15). Many metals therefore have a positive Hall constant, while the free electron model predicts a negative constant (§10.5.5). Band theory and the notion of holes have alone been able to provide an explanation for this phenomenon, called the anomalous Hall effect.

In the hole representation, the distribution of holes in the different energy levels must be considered at a given temperature T. The electron distribution is given by the Fermi function $n(\epsilon)$ (10.2). The distribution of holes is therefore

$$n_h(\epsilon) = 1 - n(\epsilon) = \frac{1}{\exp \beta(\epsilon_h - \mu_h) + 1} \tag{11.26}$$

where $\mu_h = -\mu$ is the chemical potential of the holes and $\epsilon_h = -\epsilon$ their energy. The distribution relation for the holes is also a Fermi distribution.

11.4 SEMICONDUCTORS

11.4.1 Introduction

Materials in which the energy bands are either entirely full or entirely empty at absolute zero are insulators. At non-zero temperatures, however, some electrons leave the top of the first full band (valence band) and enter the bottom of the first empty band (conduction band), making the substance weakly conducting. The conductivity varies with ϵ_g/kT, where ϵ_g is the width of the gap between the valence and the conduction bands (Fig. 11.16). In some materials, the value of ϵ_g is quite large and the conductivity is negligible; this is the case in diamond ($\epsilon_g = 5.5$ eV).

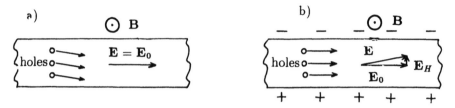

Figure 11.15: Anomalous Hall effect. This phenomenon is caused by holes, in the same way as the normal Hall effect is produced by electrons (cf. caption Fig. 10.13).

In other substances, where ϵ_g is smaller, there is a measurable conductivity that varies strongly with the temperature. These are called intrinsic semiconductors, examples of which are silicon ($\epsilon_g = 1.1$ eV) and germanium ($\epsilon_g = 0.7$ eV), which lie below diamond in the Periodic Table of the elements. Their atoms have 4 $ns^2 \, np^2$ valence electrons and form covalent crystals by sharing electrons in pairs between a central atom and its four surrounding atoms (Fig. 11.17a). When a valence electron moves into the conduction band, a hole is generated that is free to move. The conductivity is then the sum of two contributions, one for electrons in the conduction band and the other for holes in the valence band. Composite covalent semiconductor materials also exist, for example Ga As ($\epsilon_g = 1.4$ eV), In Sb ($\epsilon_g = 0.16$ eV), composed of trivalent (Ga, In) and pentavalent atoms (As, Sb).

If atoms of valence 3 or 5 are substituted in small amounts (one in 10^3 to 10^9 atoms) into an intrinsic (e.g. tetravalent) semiconductor, doped, or extrinsic semiconductors are obtained. If the semiconductor is doped with pentavalent atoms (for example, germanium doped with arsenic), an energy level (donor level) appears near the conduction band (Fig. 11.17b). At absolute zero, the first four electrons of each doping atom, which are occupied in valence bonds, populate the valence band, while the fifth is in the donor level. It only needs a weak thermal excitation to liberate the latter and to send it into the conduction band. The resulting conduction is essentially due to these electrons and is insensitive to the temperature. This type of doping is called negative doping or n-type.

When an intrinsic tetravalent semiconductor is doped with trivalent atoms (for example germanium doped with gallium), an energy level (acceptor level) detaches itself from the valence band (Fig. 11.17c). At absolute zero, the three valence electrons of each doping atom populate the valence band and an incomplete bond is established corresponding to a hole in the acceptor level. In the presence of weak thermal excitation a valence electron will move into this level and thus create a hole in the valence band. The resulting conduction is essentially due to these (positive) holes and is insensitive to the temperature. This is called positive doping or p-type.

11.4.2   Statistics of Intrinsic Semiconductors

We begin by calculating the electron density n_c in the conduction band and the hole density p_v in the valence band as a function of the Fermi energy ϵ_F. These are

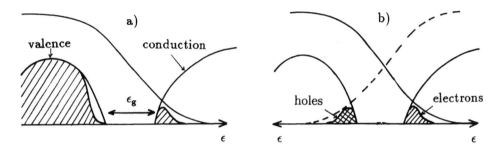

Figure 11.16: a) Fermi function, density of states and population density of electrons (hatched zones) in an intrinsic semiconductor; ϵ_g is the width of the gap. b) Same diagram, in terms of holes in the valence band.

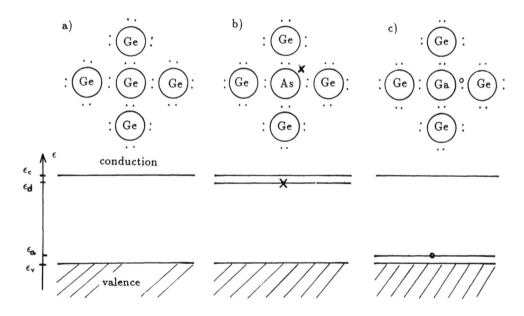

Figure 11.17: Two dimensional diagram of the covalent bonds and energy levels in a) intrinsic, b) n-type and c) p-type semiconductors. In the n-type semiconductor b), a donor level ϵ_d, which has the same number of states as doping atoms, appears near the conduction band. In the p-type semiconductor c), an acceptor level ϵ_a, which contains as many states as doping atoms, separates from the valence band.

given by

$$n_c = \frac{1}{V} \int_{\epsilon_c}^{\infty} g(\epsilon) n(\epsilon) \, d\epsilon \qquad \text{and} \qquad p_v = \frac{1}{V} \int_{-\infty}^{\epsilon_v} g(\epsilon) n_h(\epsilon) \, d\epsilon$$

where $n(\epsilon)$ and $n_h(\epsilon)$ are respectively the Fermi functions for electrons (10.2) and for holes (11.26). As the main contribution to these integrals comes from the edges of the bands (Fig. 11.16), the general shape of the bands is of little consequence and expressions (11.17a and b) can be used, since they are valid at the band edges. The electron density is thus

$$n_c = \frac{4\pi}{h^3} (2m_c)^{3/2} \int_{\epsilon_c}^{\infty} \frac{(\epsilon - \epsilon_c)^{1/2} d\epsilon}{e^{\beta(\epsilon - \epsilon_F)} + 1} \quad .$$

When $\epsilon_c - \epsilon_F$ is much larger than kT, Fermi's equation can be replaced by its Boltzmann limit (10.3) and, with the change of variable $x^2 = \beta(\epsilon - \epsilon_c)$, it follows that

$$n_c = \frac{4\pi}{h^3} (2m_c kT)^{3/2} \exp \beta(\epsilon_F - \epsilon_c) \int_0^{\infty} 2x^2 \, e^{-x^2} \, dx \quad .$$

Since the integral is equal to $\pi^{1/2}/2$ (Table 6.1), we finally get

$$n_c = g_c(T) \exp \beta(\epsilon_F - \epsilon_c) \qquad \text{with} \qquad g_c(T) = 2\frac{(2\pi m_c kT)^{3/2}}{h^3} \quad . \qquad (11.27)$$

The quantity $g_c(T)$ is called the effective conduction density of states per unit volume, and equation (11.27) has the form of the Maxwell-Boltzmann distribution for a single conduction level with energy ϵ_c.

Similarly, for the density of holes,

$$p_v = \frac{4\pi}{h^3} (2m_v)^{3/2} \int_{-\infty}^{\epsilon_v} (\epsilon_v - \epsilon)^{1/2} \frac{d\epsilon}{e^{\beta(\epsilon_F - \epsilon)} + 1} \quad .$$

When $\epsilon_F - \epsilon_v$ is much greater than kT, an analogous calculation gives

$$p_v = g_v(T) \exp \beta(\epsilon_v - \epsilon_F) \qquad \text{with} \qquad g_v(T) = 2\frac{(2\pi m_v kT)^{3/2}}{h^3} \quad , \qquad (11.28)$$

where $g_v(T)$ is called the effective valence density of states per unit volume.

Note that the product $n_c p_v$ is independent of ϵ_F and is given by

$$n_c p_v = g_c(T) g_v(T) \exp(-\beta \epsilon_g) \qquad (11.29)$$

where we have introduced the gap width $\epsilon_g = \epsilon_c - \epsilon_v$. Equations (11.27-29) are also valid for extrinsic semiconductors (§11.4.3), provided that ϵ_F is separated by several kT from the edges of the band.

For an intrinsic semiconductor, the condition that determines the Fermi level, namely $N = \sum N_i$, can be written

$$p_v = n_c \equiv n_i \qquad (11.30)$$

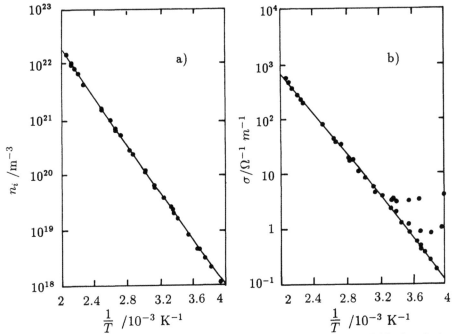

Figure 11.18: a) Number of electron-hole pairs per unit volume and b) electrical conductivity of germanium in its intrinsic region. The curve a) has the form (11.32). The points that deviate from the curve in b) correspond to samples in their extrinsic region [F.J. Morin and J.P. Maita, *Phys. Rev.* **94**, 1525 (1954)]. Note that the density of atoms in germanium is 4.4×10^{28} m^{-3}.

since the holes in the valence band are created by removing electrons that now populate the conduction band. On equating (11.27 and 28), we find

$$g_c \exp \beta(\epsilon_F - \epsilon_c) = g_v \exp \beta(\epsilon_v - \epsilon_F)$$

which gives

$$\epsilon_F = \frac{\epsilon_c + \epsilon_v}{2} + \frac{kT}{2} \ln \frac{g_v}{g_c} = \frac{\epsilon_c + \epsilon_v}{2} + \frac{3kT}{4} \ln \frac{m_v}{m_c} \quad . \qquad (11.31)$$

The Fermi level thus lies near the middle of the gap $(\epsilon_c + \epsilon_v)/2$, since kT is always smaller than the width ϵ_g of the gap.

Lastly, substituting ϵ_F into (11.27,28), or by using (11.29,30), we find for the electron and hole densities

$$n_i \equiv n_c = p_v = 2 \frac{(2\pi kT)^{3/2}}{h^3} (m_c m_v)^{3/4} \exp\left(-\frac{\epsilon_g}{2kT}\right) \quad . \qquad (11.32)$$

Owing to the exponential term, these densities depend strongly on T. For germanium, it is found experimentally (Fig. 11.18a) that

$$n_i \text{ /m}^{-3} = 1.76 \times 10^{22} [T \text{ /K}]^{3/2} \exp(-\frac{4550}{T \text{ /K}})$$

which is consistent with equation (11.32) for $\epsilon_g/2k = 4550$ K, i.e. $\epsilon_g = 0.784$ eV. For germanium, however, $m_c = m_v = 0.6\ m$ and the experimental numerical factor is therefore 7.9 times larger than the theoretical value. This discrepancy can be explained by taking into account the temperature variation of the gap. If we assume a linear dependence of the form

$$\epsilon_g(T) = \epsilon_g(0) - \alpha T \ ,$$

the theoretical expression (11.32) then contains an extra factor $\exp(\alpha/2k)$. Agreement is found on setting $\alpha = 3.6 \times 10^{-4}$ eV K^{-1}, i.e.

$$\epsilon_g(T) \ /\text{eV} = 0.784 - 3.6 \times 10^{-4}[T \ /\text{K}] \ .$$

Since the electrical conductivity of semiconductors is the sum of the conductivities of the electrons and the holes as in (10.92), for the intrinsic case we have

$$\sigma = n_c e\mu_c + p_v e\mu_v = n_i e(\mu_c + \mu_v) \tag{11.33}$$

where μ_c and μ_v are the mobilities of the electrons in the conduction band and of the holes in the valence band, respectively. The mobility, being related to the collision rate of the charge carriers (electrons or holes) with phonons, decreases with increasing temperature. Since the number of carriers contains an exponential factor, however, this factor dominates in the expression for σ and for this reason, the conductivity increases with temperature (Fig. 11.18b).

11.4.3   Statistics of Extrinsic Semiconductors

In this discussion, we consider an n-type semiconductor with a single donor level (Fig. 11.17b). We wish to determine both its Fermi level and the electron density in the conduction band as a function of T. For this, we observe that at absolute zero the donor level and the valence band are both full, while the conduction band is empty. At higher temperatures, the number of electrons in the conduction band n_c is equal to the number of holes in the donor level plus that in the valence band, i.e., from (11.27,28)

$$n_c \equiv g_c(T) \exp \beta(\epsilon_F - \epsilon_c) = \frac{g_d}{\exp \beta(\epsilon_F - \epsilon_d) + 1} + g_v(T) \exp \beta(\epsilon_v - \epsilon_F) \ . \tag{11.34}$$

To solve this equation for ϵ_F, we consider in turn three temperature domains (Fig. 11.19):

a) For $kT \ll \epsilon_c - \epsilon_d$, the contribution from the valence band can be neglected and the Boltzmann approximation can be used for the donor level, i.e.,

$$g_c(T) \exp \beta(\epsilon_F - \epsilon_c) = g_d \exp \beta(\epsilon_d - \epsilon_F) \ ,$$

from which it follows that

$$\epsilon_F = \frac{\epsilon_c + \epsilon_d}{2} - \frac{kT}{2} \ln \frac{g_c}{g_d} \quad \text{and} \quad n_c = (g_d g_c)^{1/2} \exp \frac{\beta(\epsilon_d - \epsilon_c)}{2} \ . \tag{11.35a}$$

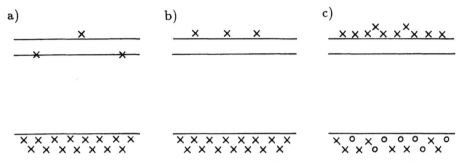

Figure 11.19: Diagram of the three temperature domains in an n-type semiconductor: a) low temperature, b) extrinsic region, c) intrinsic region.

At absolute zero, we find, as previously, that the conduction band is empty, and the Fermi level lies mid-way between the donor level and the conduction band.

b) for $\epsilon_c - \epsilon_d < kT \ll \epsilon_c - \epsilon_v$, the contribution from the valence band can still be neglected, and the term $\exp \beta(\epsilon_F - \epsilon_d)$ is negligible. It thus follows that $g_c \exp \beta(\epsilon_F - \epsilon_c) = g_d$, which yields

$$\epsilon_F = \epsilon_c - kT \ln \frac{g_c}{g_d} \qquad \text{and} \qquad n_c = g_d \; . \tag{11.35b}$$

We see that all the electrons have left the donor level for the conduction band. The semiconductor is said to be in its extrinsic region.

c) When the valence band contribution becomes greater than that of the donor level, the latter can be neglected, and the semiconductor is in its intrinsic domain for which ϵ_F and n_c are given by (11.31 and 32).

The density of conduction electrons n_c is shown in Figure 11.20 for a hypothetical case resembling that of germanium doped with arsenic. The density of valence holes p_v, given by the law of mass action (11.29), is large only in the intrinsic region where it is equal to n_c. The Fermi level, which is close to the donor level in the extrinsic region, shifts quickly towards the middle of the gap as we move out of the extrinsic into the intrinsic region.

Similar results are obtained for p-type doping, by permuting the roles of the electrons and holes and of the conduction and valence bands. In practice, a semiconductor is never pure and it always contains various impurities of both p and n type. There is, however, partial compensation, with electrons from the donor levels filling the acceptor levels. In its extrinsic region, the semiconductor behaves as if it were n- or p-type only if there is an excess of donor or acceptor atoms. What is more, the intrinsic behaviour appears at temperatures that become lower with increasing purity.

In the extrinsic region of a semiconductor (case b), conduction is provided by the majority carriers (electrons in n-type and holes in p-type), and the electrical conductivity

$$\sigma = n_c e \mu_c \qquad \text{or} \qquad p_v e \mu_v \tag{11.36}$$

remains practically constant (Fig. 11.21a). Its variation is due only to the temperature dependence of the majority carrier mobility, μ_c or μ_v. From this property μ_c

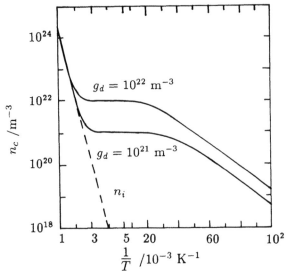

Figure 11.20: Density of conduction electrons in an n-type semiconductor. The full lines are the solutions of equation (11.34) for two different types of doping. We take $m_c = m_v = 0.6\ m$, $\epsilon_g = \epsilon_c - \epsilon_v = 0.68$ eV and $\epsilon_c - \epsilon_d = 0.0127$ eV. The dashed curve shows the electron density in an intrinsic semiconductor. Note the change of scale for $1/T = 5 \times 10^{-3}$ K^{-1}.

and μ_v can be calculated as a function of T if the doping levels have been measured by chemical analysis.

Further information on semiconductors can be obtained by measuring the Hall constant R_H (Fig. 11.21b). When two types of charge carrier are present, this quantity is given by

$$R_H = \frac{1}{e} \frac{p_v \mu_v^2 - n_c \mu_c^2}{(p_v \mu_v + n_c \mu_c)^2} \quad . \tag{11.37}$$

This equation can be also deduced by generalizing the equations of paragraph 10.5.5. In particular, at the points where the Hall constant changes sign, $p_v/n_c = (\mu_c/\mu_v)^2$. In the extrinsic region ($p_v = 0$ or $n_c = 0$), the Hall constant simplifies to

$$R_H = -\frac{1}{n_c e} \qquad \text{or} \qquad \frac{1}{p_v e}$$

and from the product

$$\mu_H \equiv |R_H|\sigma = \mu_c \quad \text{or} \quad \mu_v \tag{11.38}$$

the mobilities can in principle be determined. In fact, expression (11.37) for R_H, which comes from the Drude model, is approximate and the quantity μ_H, called the Hall mobility, differs from the carrier mobility by a factor of the order of unity. Note that the electron mobility μ_c is generally greater than that of holes, μ_v. Both decrease with increasing temperature, and in germanium at 300 K for example,

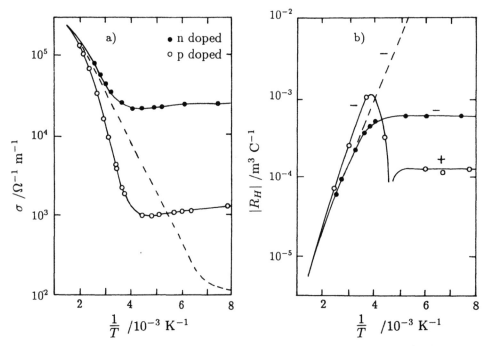

Figure 11.21: Electrical conductivity σ and Hall constant R_H of samples of indium anti-monide In Sb of n- or p-type. The sign of the Hall constant is indicated beside the curves. The dashed curves are for a pure sample [O. Madelung and H. Weiss, *Z. Naturforsch. A* **9**, 527 (1954)].

their values are $\mu_c \simeq 0.4$ m^2 V^{-1} s^{-1} and $\mu_v \simeq 0.2$ m^2 V^{-1} s^{-1}.

Exercise 11.2 *Compensated semiconductors*

Consider a semiconductor with a donor level and an acceptor level whose respective degeneracies are such that $g_d \gg g_a$, in the temperature range where the valence band has no influence.

1 Describe the system at absolute zero.

2 Write down the relation defining ϵ_F. Hence derive an expression giving the number of conduction electrons, n_c, as a function of temperature.

3.1 Calculate n_c in the lowest temperature range. State the limits of this range in a sample for which $\epsilon_c - \epsilon_d = 0.0127$ eV, $m_c = 0.6\, m$, $g_d = 10^{22}$ m^{-3} and $g_a = 10^{20}$ m^{-3}.

3.2 What is the value of n_c at the highest temperature?

Solution

1 At absolute zero, the valence and conduction bands are respectively full and empty. Excess electrons from donor atoms are "trapped" in the holes of the acceptor atoms. There are therefore $g_d - g_a$ electrons or g_a holes in the donor level, and the acceptor level is saturated.

2 At all temperatures, the number of electrons in the conduction band is equal to the number of holes in the donor level in excess of g_a. As in (11.34), we find

$$n_c = \frac{g_d}{\exp \beta(\epsilon_F - \epsilon_d) + 1} - g_a \quad .$$

On replacing $\exp \beta \epsilon_F$ by $n_c \exp \beta \epsilon_c / g_c(T)$ in (11.25), we find for the equation defining n_c

$$\frac{n_c(g_a + n_c)}{g_d - g_a - n_c} = g_c \exp \beta (\epsilon_d - \epsilon_c) \quad . \tag{11.39}$$

3.1 When $T \to 0$, the right hand side of (11.39), and hence n_c, is very small, and we have

$$n_c = \frac{g_d - g_a}{g_a} g_c \, \exp \beta (\epsilon_d - \epsilon_c) \quad .$$

This equality is in fact valid as long as $n_c \ll g_a$, that is

$$g_c(T) \exp \beta (\epsilon_d - \epsilon_c) \ll \frac{g_a^2}{g_d} \quad .$$

In numerical terms, it follows that the range of validity is $T < 20$ K.

3.2 Above 20 K, n_c becomes large compared with g_a. The influence of the acceptor level is then negligible ($g_a \ll n_c$, g_d) and the compensation effect ceases. Equations (11.35a and b) become applicable again.

11.4.4 p-n Junctions

The interest of semiconductors lies in the properties of p-n junctions. These junctions are created by doping two neighbouring regions of a semiconductor crystal, one with type p and the other with type n (Fig. 11.22a). Since the chemical potential (or Fermi energy) is higher in the n region, the electrons leave this region for the p region of lower chemical potential, where they are trapped by acceptor atoms (electron-hole recombination). The n region then takes on a positive charge by reason of the electron deficit and the p region becomes negatively charged (Fig. 11.22c). An electric field therefore develops which, at equilibrium, hinders further electrons from crossing the junction. This field corresponds to an electrostatic potential step (Fig. 11.22d), which shifts the energy levels as shown in Figure 11.22b.

Typically, the height of the potential step is about the same as the width of the gap ϵ_g, i.e. between 0.5 and 1 V, and the width of the transition region is of the order of 1 to 0.01 μm for doping levels between 10^{20} and 10^{24} m^{-3}.

If a positive voltage is now applied to the p region, the potential step is decreased and the Fermi level drops in the p region. The conduction electrons in the n region move towards the p region, i.e. a current develops from the p to the n region. This current adds to that of the valence holes moving in the opposite direction. However, if a negative voltage is applied to the p region, the potential step increases and only the few electrons that populate the conduction band of the p region move to the n region, generating only a very weak current from the n to the p region. The same is true for the valence holes.

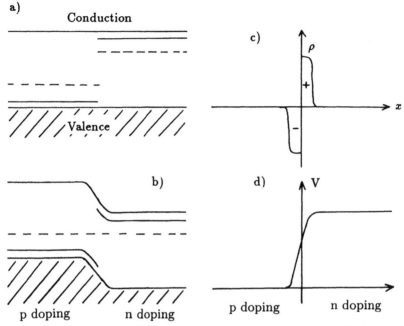

Figure 11.22: a) and b) Bands and energy levels of a *p-n* junction before and after reaching equilibrium. The dashed lines are the Fermi levels. c) and d) Charge density and electrostatic potential in the vicinity of a *p-n* junction. Note that the electrostatic energy of an electron is $-eV$.

It can be shown that the current density crossing the junction is of the form

$$j = j_0 \left\{ \exp \left(\frac{eV}{kT} \right) - 1 \right\} \, ,$$

where j_0 depends on the nature of the diode and on its temperature. Roughly speaking, it can be said that the *p-n* junction conducts current in only one direction (diode effect), and rectifies an alternating current into a direct current. It is nonetheless noteworthy that if the bias voltage becomes too large ($V < 0$), the diode again starts to conduct, on the one hand because of an avalanche effect, and on the other hand because of direct tunnelling of conduction electrons from the n region towards the valence band in the p region. This phenomenon is used as a voltage regulator in Zener diodes.

The junction transistor is composed of three contiguous regions of the same semiconductor, successively doped *n-p-n* or *p-n-p*. The two outer regions are called the emitter and collector, and the central region, which is very thin ($\sim 0.1 \ \mu$m), is the base (Fig. 11.23). The majority carriers of the emitter and collector cross this region easily as long as the base voltage does not raise the potential barrier too high. By this means the transistor acts as an electronic gate. Furthermore, the majority carriers of the emitter and collector become minority carriers in the base, and so have a certain probability ($\sim 10^{-2}$) of recombination, thereby giving rise to a base current. By the same token, a current in the base I_b produces an

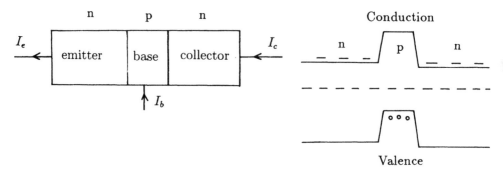

Figure 11.23: n-p-n junction transistor and diagram of the energy bands. The p-type base forms a potential barrier, which, depending on the applied voltage, controls the flow of electrons from the emitter to the collector. The dashed line indicates the Fermi level.

amplified collector current I_c. By this property the transistor acts as an amplifier of variations in current with a gain $\beta \sim 50$.

The properties of diodes and transistors and their use in microelectronics confer on semiconductor materials a place of utmost importance.

BIBLIOGRAPHY

J. Auvray, *Circuits et composants électroniques*, Hermann (1975).
J. Callaway, Electron energy bands in solid, *Solid State Physics* **7**, 99 (1958).
H.Y. Fan, Valence semi-conductors, germanium and silicon, *Solid State Physics* **1**, 283 (1955).
W.A. Harrison, *Solid State Theory*, McGraw-Hill (1970).
Historical References
R. de L. Kronig and W.G. Penney, *Proc. Roy. Soc. A* (London) **130**, 499 (1930).

COMPREHENSION EXERCISES

11.1 Consider a square two-dimensional reciprocal lattice (Fig. 11.6) of side $b = 2\pi/a$. The components of the wave vector $\mathbf{k}'$ of a free electron are 0.6 b and $-0.8\, b$. To which Brillouin zone does it belong, and what is the vector $\mathbf{k}$ corresponding to the first Brillouin zone? [Ans.: $n = 4$ and $\mathbf{k} = -0.4\, b$; 0.2 b].

11.2 Consider a square two-dimensional crystal of side a in which each atom liberates z electrons, taken to be free. Show that $k_F = (z/2\pi)^{1/2}b$, where $b = 2\pi/a$ is the side of the reciprocal lattice (Fig. 11.6). How many different Brillouin zones are intersected by the Fermi circle for $z = 1$, 2 and 4 [Ans.: $n = 1$, 2, 4]. Draw the curve $\epsilon = \epsilon_F^0$ in the first Brillouin zone for each of the three cases and indicate the regions where $\epsilon < \epsilon_F^0$.

11.3 Draw the first four Brillouin zones for a rectangular reciprocal lattice of sides b and 2 b.

11.4 Show that the radius of the Fermi sphere of a free electron gas in a face-centred cubic lattice (4 atoms per cell of side a) is $k_F = b(3z/2\pi)^{1/3}$ with $b =$

$2\pi/a$. For what values of z a) does the Fermi sphere completely enclose the first Brillouin zone, and b) does the first Brillouin zone completely enclose the Fermi sphere (Fig. 11.7a)? Take $\Gamma L = (3/16)^{1/2}b$ and $\Gamma N = (5/16)^{1/2}b$ [Ans.: a) $z \geq 3$ and b) $z \leq 2$].

11.5 Calculate the density of states $g(\epsilon)$ for the one-dimensional model of Exercise 11.1 with $z = 1$. Hence, using (10.48) deduce the electronic heat capacity c_V [Ans.: $c_V/R = 2\pi kT/3\epsilon_0$].

11.6 Determine the wave vector and the velocity of a hole resulting from the absence of an electron of wave vector $\mathbf{k}$ and velocity $\mathbf{v}$. Hint: consult Figure 11.3 [Ans.: $-\mathbf{k}$ and $-\mathbf{v}$].

11.7 Compare the density of conduction electrons and the conductivity of germanium (Fig.11.18) with the same quantities in the metals of Table 10.3.

11.8 Show that the internal energy per unit volume of an intrinsic semiconductor is given by $U = U(T = 0) + \epsilon_g n_i(T)$ where n_i is given by (11.32). Hence deduce the expression for the heat capacity per unit volume c_V of the sample. Answer the same question for an extrinsic semiconductor.

11.9 For an n-type extrinsic semiconductor, solve equation (11.34) for the Fermi energy ϵ_F in the temperature range where the valence band contribution is negligible. Hence derive an expression for n_c, the density of conduction electrons. Find the limiting cases (11.35a and b).

11.10 Show that the potential difference at a p-n junction is

$$V \equiv V_n - V_p = \frac{\epsilon_g}{e} - \frac{kT}{e} \ln \frac{g_c g_v}{g_d g_a} \quad .$$

Calculate this value at 300 K for $g_a = g_d = 10^{22}$ m^{-3}, $g_c = g_v = 2 \times 10^{21}$ $(T/K)^{3/2}$ m^{-3} and $\epsilon_g = 0.68$ eV [Ans.: $V = 0.3$ V].

Chapter 12

Gibbs Theory of Ensembles

12.1 SYSTEMS OF INTERACTING PARTICLES

The statistical method expounded in the first part of this book successfully explains a large number of physical phenomena. Until now, however, we have only considered problems that can be reduced to models of non-interacting particles, in which the Hamiltonian operator $\hat{H}$ is a sum of single-particle terms $\hat{h}_i$. Thus

$$\hat{H} = \sum_{i=1}^{N} \hat{h}_i \ , \tag{12.1}$$

that is to say, $U = \sum N_i \epsilon_i$. If interactions are to be included, the elementary method must be extended. In particular, we must re-examine the way in which the microscopic (or quantum) state of a system is described.

12.1.1 Gas of Particles

Let us consider a gas of N particles that possess only degrees of translational freedom. In classical mechanics, the state of this gas at any instant can be described by the $6N$ variables $\mathbf{r}_1$, $\mathbf{p}_1$, $\ldots$, $\mathbf{r}_N$, $\mathbf{p}_N$ that define the positions and momenta of the particles. The corresponding energy is

$$E = \sum_{i=1}^{N} \frac{\mathbf{p}_i^2}{2m} + V\left(\mathbf{r}_1, \ \ldots \ \mathbf{r}_N\right) \ , \tag{12.2}$$

where the term V is the interaction energy. In the course of time, the co-ordinates and momenta of the particles change, and the characteristic point describing the state of the system moves in a $6N$-dimensional phase space. The density of quantum states g in this phase space is given by a generalization of equation (1.6),

$$g\left(\mathbf{r}_1, \ \mathbf{p}_1, \ \ldots \ \mathbf{r}_N, \ \mathbf{p}_N\right) \prod_{i=1}^{N} d^3\mathbf{r}_i \ d^3\mathbf{p}_i = \frac{1}{N!} \prod_{i=1}^{N} \frac{d^3\mathbf{r}_i \ d^3\mathbf{p}_i}{h^3} \ . \tag{12.3}$$

In g it can be seen that for each particle there is a factor $1/h^3$, in addition to the indistinguishability factor coming from the $N!$ identical physical states that are obtained by permuting the co-ordinates and momenta of the N particles.

This "semi-classical" description will be used when the effects connected with the quantum mechanical character of the particles (fermions or bosons) do not arise. If this is not the case, we must apply the general method for describing quantum systems (§12.1.3).

12.1.2   System of Localized Magnetic Moments

We consider a system of N magnetic ions of spin S located on the sites of a lattice. The set of N quantum numbers m_i ($i = 1, \ldots N$ and $-S \leq m_i \leq S$) characterizing the projection of the spin of each atom on a quantization axis defines a state of the system

$$|r >\equiv |m_1, \ldots m_i, \ldots m_N > \quad . \tag{12.4}$$

The $(2S + 1)^N$ states of this type form a basis in the space of quantum states for this system. At each instant, the system is in a state $|\Psi >$, which is a superposition of the basis states

$$|\Psi >= \sum_r c_r|r > \quad \text{with} \quad c_r =< r|\Psi > \quad \text{and} \quad \sum_r |c_r|^2 = 1 \ . \tag{12.5}$$

In the course of time, it evolves in accordance with Schrödinger's equation

$$i\hbar \frac{\partial}{\partial t}|\Psi >= \hat{H}|\Psi > \tag{12.6}$$

where the Hamiltonian operator $\hat{H}$ has the form

$$\hat{H} = g\mu_B \sum_{i=1}^{N} \hat{\mathbf{S}}_i.\mathbf{B} + \hat{V}\left(\hat{\mathbf{S}}_1, \ldots \hat{\mathbf{S}}_i, \ldots \hat{\mathbf{S}}_N\right) \ . \tag{12.7}$$

In this Hamiltonian, the first term represents the interaction between the magnetic moments and the magnetic field $\mathbf{B}$, and the second describes the interactions between moments. The vector operators $\hat{\mathbf{S}}_i$ are defined from their components by

$$\hat{S}_{zi} |m_1, \ldots m_i, \ldots m_N > \ = \ m_i|m_1, \ldots m_i, \ldots m_N > \tag{12.8a}$$

$$\hat{S}_{\pm i} |m_1, \ldots m_i, \ldots m_N > \ =$$

$$\sqrt{J(J + 1) - m_i(m_i \pm 1)} \qquad |m_1, \ldots (m_i \pm 1), \ldots m_N > \tag{12.8b}$$

$$\text{with} \qquad \hat{S}_{\pm i} \ = \ \hat{S}_{xi} \pm i\hat{S}_{yi} \ . \tag{12.8c}$$

We recall that for a spin $S = 1/2$, with Pauli matrices

$$\hat{\sigma}_x = \begin{pmatrix} 0 & 1 \\ 1 & 0 \end{pmatrix} \qquad \hat{\sigma}_y = \begin{pmatrix} 0 & -i \\ i & 0 \end{pmatrix} \qquad \hat{\sigma}_z = \begin{pmatrix} 1 & 0 \\ 0 & -1 \end{pmatrix} \tag{12.9}$$

the operators $\hat{\mathbf{S}}$ can be written as

$$\hat{\mathbf{S}} = \frac{1}{2}\,\hat{\sigma} \ . \tag{12.10}$$

12.1.3 Arbitrary Quantum System

In the general case, the quantum state $|\Psi>$ of an arbitrary system can be decomposed into a basis of states $|r>$ $(r = 1, 2, \ldots)$ as in (12.5), and its time dependence is governed by the Schrödinger equation (12.6). The choice of basis states depends on the system, and is generally taken to be the eigenstate basis of some operator $\hat{A}$ (or of several operators). For the system of magnetic moments (§12.1.2), the eigenstate basis chosen is the one that is common to the N operators $\hat{S}_{z1}, \ldots \hat{S}_{zi}, \ldots \hat{S}_{zN}$. In the following, we shall often refer to the eigenstate basis of the Hamiltonian $\hat{H}$ defined by

$$\hat{H}|r> = E_r|r> \quad . \tag{12.11}$$

This is the basis of stationary states whose energy is E_r.

In the formalism of quantum mechanics, all the information about a system is contained in a Hermitian operator $\hat{\rho}$ called the density matrix (or operator) which satisfies the normalization condition

$$\text{Tr } \hat{\rho} \equiv \sum_r <r|\hat{\rho}|r> = 1 \quad . \tag{12.12}$$

In problems for which a wave function can be defined (isolated systems, for example), the density operator is the projector on the state $|\Psi>$ that characterizes the system, and is given by

$$\hat{\rho} = |\Psi><\Psi| \quad . \tag{12.13a}$$

In this case, and in this case alone, the density operator satisfies $\hat{\rho}^2 = \hat{\rho}$ and we have

$$\rho_{rs} \equiv <s|\hat{\rho}|r> = c_s^* c_r \quad . \tag{12.13b}$$

Later we shall see that the density operator plays an important role in statistical thermodynamics.

12.2 MEASUREMENT OF A QUANTITY

When a quantity A is measured in a quantum system which at a given instant is in the state $|\Psi>$, any one of the eigenvalues A_r of the operator $\hat{A}$ associated with A will be found, with a probability $|c_r|^2 = |<r|\Psi>|^2$. Here $|r>$ is the eigenstate of $\hat{A}$ corresponding to the eigenvalue A_r

$$\hat{A}|r> = A_r|r> \quad .$$

The measurement of a quantity is therefore probabilistic, and is quantum mechanical in nature. The mean value $<A>$ of the resulting measurement is

$$\begin{aligned}
<A> &= \sum_r |<r|\Psi>|^2 A_r = \sum_r <r|\Psi><\Psi|r> A_r \\
&= \sum_r <r|\hat{\rho}A_r|r> = \sum_r <r|\hat{\rho}\hat{A}|r> \\
&= \text{Tr }(\hat{\rho}\hat{A}) \tag{12.14a}
\end{aligned}$$

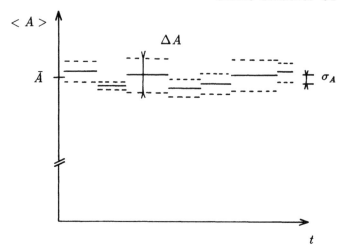

Figure 12.1: Schematic diagram of the time dependence of $< A >$, the mean quantum value of an operator $\hat{A}$ associated with the quantity A (full line), and its standard deviation ΔA (dashes). The standard deviation σ_A including the time fluctuations is shown.

and its standard deviation ΔA is given by

$$\Delta A = \left[< A^2 > - < A >^2 \right]^{1/2} \quad \text{with} \quad < A^2 >= \text{Tr} \left(\hat{\rho} \hat{A}^2 \right) . \quad (12.14b)$$

We recall that we also have

$$< A >=< \Psi | \hat{A} | \Psi > . \quad (12.14c)$$

Furthermore, a system at equilibrium that contains a large number of particles is continually changing its quantum state owing to interactions with the walls of the container, with the thermostat,... The outcome of measuring a quantity A, which is already probabilistic for quantum mechanical reasons, also has a probability distribution due to the variation of the system in time (Fig. 12.1). The process of measurement, even when it is described as being instantaneous, in fact lasts for a time τ that is extremely long on the molecular scale. The mean value of the result is given by

$$\overline{A} \equiv \frac{1}{\tau} \int_t^{t+\tau} < A > \, dt \quad (12.15a)$$

and its standard deviation is

$$\sigma_A = \left[\overline{A^2} - \overline{A}^2 \right]^{1/2} . \quad (12.15b)$$

Later, however, we shall show (§12.5.3 and §12.6.5) that the ratio $\sigma_A / \overline{A}$ varies as $1/\sqrt{N}$. In the thermodynamic limit ($N \to \infty$), the fluctuations in the measured quantities are therefore negligible, which is in agreement with experimental observation.

The result of measuring a quantity A thus involves two averaging processes, one being quantum mechanical, the other connected with the evolution in time. This justifies the use of statistical methods for investigating systems that contain large numbers of particles. Such systems can then be described only by the statistical probabilities P_r of finding the system in the state $|r>$. The outcome of measuring a quantity A then takes the form

$$\overline{A} = \sum_r P_r A_r \ . \tag{12.16}$$

The internal energy U of a system is thus given by

$$U \equiv \overline{E} = \sum_r P_r E_r \tag{12.17}$$

and the entropy is defined by

$$S = -k \sum_r P_r \ln P_r \ , \tag{12.18}$$

a formula which will be proved later (§12.4).

There is no method that allows the probability distribution P_r to be derived from the Schrödinger equation, since the systems under consideration possess a very large number of degrees of freedom. Following J.W.Gibbs, we are therefore obliged to postulate the form for the probability distribution P_r (§12.3).

We note that systems are in general described by a statistical density operator (or matrix) $\hat{\rho}$, which is a generalization of the quantum density operator and is such that

$$\overline{A} = \ \mathrm{Tr} \ (\hat{\rho}\hat{A}) \qquad \text{with} \qquad \hat{\rho}^+ = \hat{\rho} \quad \text{and} \quad \mathrm{Tr} \ \hat{\rho} = 1 \ . \tag{12.19}$$

Equation (12.16) is valid only if the statistical density operator is diagonal in the eigenvector basis of the operator $\hat{A}$. In this case $\rho_{rr} = P_r$ and $\rho_{rs} = 0$ $(r \neq s)$.

Exercise 12.1 *Quantum density matrix of one particle with spin 1/2 selected out of two*

We consider an isolated system of two (identical or not) particles with spin 1/2 in one of the triplet or singlet states $|++>$, $(|+->+|-+>)/\sqrt{2}$, $|-->$ or $(|+->-|-+>)/\sqrt{2}$. We are particularly interested in the state of the first of these particles.

1 For each of the 4 states calculate the mean quantum value and the quantum standard deviation of the three spin operators $\hat{\mathbf{S}} \equiv \hat{\sigma}/2$ for the first particle.

2 For the mean value of the measurements of the spin components of the first particle, confirm that the quantum mechanical density matrix is equal respectively to $(1+\hat{\sigma}_z)/2$, $1/2$, $(1-\hat{\sigma}_z)/2$ and $1/2$ for the 4 states considered. Here, 1 represents the 2×2 unit matrix.

3 When $\hat{\rho}^2 = \hat{\rho}$, show that the state of the first particle can be described by a wave function, which is to be specified.

Table 12.1: Expectation values of spin components for two spin 1/2 particles

State	$\|++>$	$\frac{\|+->+\|-+>}{\sqrt{2}}$	$\|-->$	$\frac{\|+->-\|-+>}{\sqrt{2}}$
$< S_x >=< S_y >$	0	0	0	0
$\Delta S_x = \Delta S_y$	0	0	0	0
$< S_z >$	1/2	0	−1/2	0
$< S_z^2 >$	1/4	1/4	1/4	1/4
ΔS_z	0	1/2	0	1/2

Solution

1 The property (12.8b) of the operators $S_\pm = S_x \pm iS_y$ shows that the mean values $< S_\pm >$ and $< S_\pm^2 >$ calculated from equation (12.14c) are zero. $< S_z >$ and $< S_z^2 >$ can be calculated from (12.8a). This yields Table 12.1.

2 By observing that

$$\text{Tr }\hat{\sigma}_i = 0 \qquad \text{and} \qquad \text{Tr }\hat{\sigma}_i\hat{\sigma}_j = 2\delta_{ij} \ ,$$

it is verified that $< S_i >= \text{Tr } (\hat{\rho}\hat{S}_i) \equiv \text{Tr } (\hat{\rho}\hat{\sigma}_i)/2$ for the four states. We show how the expression for the proposed density matrices $\hat{\rho}$ is obtained. The Hermitian condition and the normalization condition (12.19) restrict the form of the density operator to

$$\hat{\rho} = (1 + a_x\hat{\sigma}_x + a_y\hat{\sigma}_y + a_z\hat{\sigma}_z)/2 \equiv (1 + \mathbf{a}. \ \hat{\sigma})/2 \qquad \text{(a real)} \ .$$

Since $< S_x >=< S_y >= 0$, it follows that $a_x = a_y = 0$ for each of the 4 states, by virtue of (12.14a) and the properties of the traces of Pauli matrices. From the value of $< S_z >$ it follows that $a_z = 1$, 0, −1 and 0, respectively, thereby giving rise to the 4 forms of the density matrix.

3 It is seen that $\hat{\rho}^2 = \hat{\rho}$ for the two states $|++>$ and $|-->$, for which the first particle is respectively in the definite state $|+>$ and $|->$ ($\Delta S = 0$). In the two other cases for which $\hat{\rho}^2 \neq \hat{\rho}$, particle 1 is in each of the two states $|+>$ and $|->$ with probability 1/2. However, no wave function of the type $|\Psi >= \alpha|+> + \beta|->$ can account for the values of $< S >$ and ΔS. The density matrix must then be used.

12.3   POSTULATE OF STATISTICAL THERMODYNAMICS

To investigate the form of the probability distribution, J.W. Gibbs introduced the notion of statistical ensembles. These are sets containing a huge number of elements that are identical to the system investigated and which are subject to the same constraints and external parameters. In these statistical ensembles, it is assumed that all the quantum states of the system being considered are attained at all times by a number of elements that is proportional to P_r. The outcome of a measurement of a quantity for the system can then be identified with the mean value of that quantity over the elements of a Gibbs ensemble.

The study of these ensembles allowed certain properties of the probability distribution P_r to be defined, and gave rise to the fundamental postulate of statistical thermodynamics: *all states accessible to a system that have the same number of particles, the same energy, the same external parameters and are subject to the same constraints occur with the same probability.* This postulate is a generalization to arbitrary systems of the first postulate in the elementary theory (§1.4). In particular it states that the probability P_r is a function only of the number of particles N in the system, of the energy E_r, of the external parameters (volume, magnetic field, ...) and of the nature of the constraints (isolated system or not, closed system or not, ...).

In what follows, on the basis of this postulate, we shall determine the probability distributions P_r in three kinds of ensemble:

• the *microcanonical* ensemble, for an isolated closed system (N and U fixed);

• the *canonical* ensemble, for a closed system in contact with a thermostat at temperature T (N and T fixed);

• the *grand canonical* ensemble, for an open system that can exchange particles with a reservoir of particles with chemical potential μ and in contact with a thermostat at temperature T (μ and T fixed).

The thermodynamic properties of the systems investigated do not depend on the nature of the constraints (T or U fixed, N or μ) and therefore on the ensemble used. For reasons of simplicity, the choice of ensemble is therefore made on the basis of the physical problem being investigated.

12.4 MICROCANONICAL ENSEMBLE

In the microcanonical ensemble, we consider an isolated closed system, that is, one where the energy E and the number of particles N are constant, and where the external parameters V, B, etc., are fixed. According to the postulate stated above, all the quantum (microscopic) states are equally probable.

Let $W(E, N, V, B, \ldots)$, or, for brevity, $W(E)$, be the total number of quantum states that are accessible to the system under the conditions imposed. The probability of occurrence of each of the states is then

$$P_r = 1/W(E) \qquad \text{with} \qquad \sum_r P_r = 1 \; , \tag{12.20}$$

and the value measured for a quantity A is given by

$$\overline{A} = \sum_r A_r/W(E) \; . \tag{12.21}$$

In particular, from (12.17,18), the internal energy and the entropy are equal to

$$U \;=\; \sum_r P_r E_r = E \times \sum_r P_r = E \tag{12.22}$$

$$\text{and} \quad S \;=\; -k \sum_r P_r \ln P_r = k \sum_r P_r \ln W(E) = k \ln W(E) \; . \tag{12.23}$$

The last expression is similar to the Boltzmann relation (1.39). It differs, however, in that equation (12.23) gives the entropy of the system at equilibrium as a function of the total number of accessible states of energy E, i.e. $W(E)$, while the Boltzmann relation involves the thermodynamic probability W of an arbitrary macroscopic state (even out of equilibrium). $W(E)$ is in fact the sum of all the thermodynamic probabilities W. Nevertheless, as we saw in paragraph 1.4, in the thermodynamic limit $W(E)$ is almost the same as $W_{\max}$, the thermodynamic probability at equilibrium. This similarity provides an a *posteriori* justification for the definition of the entropy (12.18).

Thus, if $W(E) \equiv W(E, N, V, B, \dots)$ is known, equation (12.23) yields the entropy of a system at equilibrium as a function of the energy E (which, for an isolated system, is identical to the internal energy U), of the number of particles N and of the external parameters $V, B, \dots$, i.e. $S(U, N, V, B, \dots)$. As we have already seen (§1.6), when expressed in its natural variables, this function is sufficient to define the properties of the substance. However, the equation for $W(E)$, and hence that for the entropy, is extremely difficult to establish once the particles begin to interact to any significant extent. For this reason the microcanonical ensemble is used only for non-interacting particles. The results are identical to those found by elementary methods in the first part of this book. When interactions between particles cannot be neglected, other ensembles are used, for which a large number of calculation techniques have been developed.

Finally, we note that, since the energy of a system is quantized, it is natural to consider for $W(E)$ the number of states having energies lying between E and $E + \delta E$, where δE is of the order of a quantum of energy. In the thermodynamic limit, the physical results are not affected by the value of δE, as will be seen in the following exercise.

Exercise 12.2 *Ideal gas in the microcanonical ensemble*
1 Calculate the size τ of the region in $6N$ dimensional phase space that delimits the states accessible to an ideal gas with N identical particles contained in a vessel of volume V and having energy less than E. The volume of a sphere of radius r in d dimensions is equal to $\pi^{d/2}r^d/\Gamma(d/2+1)$ where the gamma function, defined by

$$\Gamma(x) = \int_0^\infty t^{x-1}e^{-t}\, dt \quad, \tag{12.24a}$$

has the property that $\Gamma(x+1) = x!$ when x is an integer and

$$\ln\Gamma(x+1) = x\ln x - x + \frac{1}{2}\ln 2\pi x + O\left(\frac{1}{x}\right) \quad. \tag{12.24b}$$

2a Write down the number $\Phi(E)$ of different quantum states of the gas enclosed in this region and hence derive $W(E)$, the number of quantum states of the gas with energy lying between E and $E + \delta E$, where the energy δE is very much smaller than E, and close to the quantum of kinetic energy $h^2/8mV^{2/3}$ (§1.2.1).
2b Calculate $\Phi(E)$ and $W(E)$ numerically for a monatomic gas with molar mass $M = 4$ g mol^{-1}, containing $N = 2.5 \times 10^{22}$ atoms, with energy $E = 150$ J and occupying a volume $V = 1$ L (approximately 1 litre of helium at NTP).

3 Express the entropy S of the system in the thermodynamic limit $N \to \infty$, $V \to \infty$, $E \to \infty$ when V/N and E/N are constant.

4 Hence derive an expression for the functions U and S in the variables T, V and N.

Solution

1 If $\mathbf{r}_1$, $\mathbf{p}_1$, ... $\mathbf{r}_N$, $\mathbf{p}_N$ denote the $6N$ dimensions in phase space, the required region is defined by

$$\mathbf{r}_i = (x_i, \, y_i, \, z_i) \in V \text{ for } i \; = \; 1, \dots, N$$

$$\text{and} \quad \frac{\mathbf{p}_1^2}{2m} + \dots + \frac{\mathbf{p}_N^2}{2m} = \frac{1}{2m} \sum_i \mathbf{p}_i^2 \; \leq \; E \; .$$

Its size is therefore

$$\tau = \int \prod_{i=1}^{N} d^3\mathbf{r}_i \; d^3\mathbf{p}_i = \prod_i \int d^3\mathbf{r}_i \times \int \prod_i d^3\mathbf{p}_i \; .$$

In this expression, the first integral is equal to V and the second is the volume of a sphere of radius $(2mE)^{1/2}$ in N dimensions. We therefore have

$$\tau = \frac{\pi^{3N/2} V^N (2mE)^{3N/2}}{\Gamma\left(\frac{3N}{2} + 1\right)} \; .$$

2a According to (12.3), in the phase space of N particles, the density of states is equal to $1/h^{3N}N!$. The term $N!$ accounts for the indistinguishability of the particles. The number of states $\Phi(E)$ is then

$$\Phi(E) = \frac{\tau}{h^{3N} N!} = \frac{V^N (2\pi mE)^{3N/2}}{h^{3N} N! \Gamma\left(\frac{3N}{2} + 1\right)} \; ,$$

and $W(E)$ is therefore

$$W(E) = \Phi(E + \delta E) - \Phi(E) = \frac{\partial \Phi}{\partial E} \, \delta E = \frac{3N}{2E} \, \delta E \Phi(E) \; .$$

Note that $W(E)$ depends strongly upon E: $W(E) \sim E^{3N/2}$.

2b The numerical values of $\Phi(E)$ and $W(E)$ are extremely large. We therefore calculate their logarithm. Using Stirling's formula (1.32) and (12.24b), we have

$$
\begin{aligned}
\ln \Phi(E) \; = \; & N \ln \frac{V}{h^3} (2\pi mE)^{3/2} - N \ln N + N - \frac{1}{2} \ln 2\pi N \\
& - \frac{3N}{2} \ln \frac{3N}{2} + \frac{3N}{2} - \frac{1}{2} \ln 3\pi N \\
= \; & N \ln \frac{V}{Nh^3} \left(\frac{4\pi mE}{3N}\right)^{3/2} + \frac{5N}{2} - \frac{1}{2} \ln 6\pi^2 N^2 \; .
\end{aligned}
$$

Term by term, we have numerically

$$\ln \Phi(E) = 3.15 \times 10^{23} + 6.25 \times 10^{22} - 53.6 \ .$$

Clearly, the last term, which enters into the value of $\Phi(E)$ as a factor of 5.2×10^{-24}, can be neglected in this expression. Similarly, we have

$$
\begin{aligned}
\ln W(E) \ &= \ \ln \Phi(E) + \ln \left(\frac{3N}{2E} \delta E \right) \\
&= \ 3.78 \times 10^{23} + 219 \ .
\end{aligned}
$$

The second term is negligible compared to the first, and this shows that either $W(E)$ or $\Phi(E)$ can be used equally well.

3 From equation (12.23), we have

$$\frac{S}{Nk} = \frac{1}{N} \ln \left(\frac{3N}{2E} \delta E \right) + \frac{1}{N} \ln \Phi(E) \simeq \frac{1}{N} \ln \Phi(E) \ ,$$

and hence, from the result of 2b

$$\frac{S}{k} = N \ln \frac{V}{Nh^3} \left(\frac{4\pi m E}{3N} \right)^{3/2} + \frac{5N}{2} \ . \tag{12.25}$$

Note that S/Nk is intensive in form, as all the neglected terms are in $(\ln V)/N$ or $(\ln N)/N$ at most.

4 Since the system is isolated, $U = E$, and equation (12.25) yields S in terms of the variables U, N and V. The temperature is introduced through the equation of classical thermodynamics

$$\frac{1}{T} = \left(\frac{\partial S}{\partial U} \right)_{N, \, V} \ , \quad \text{i.e., here} \quad \frac{1}{T} = \frac{\partial S}{\partial E} \, (E = U) = \frac{3}{2} Nk \times \frac{1}{U} \ .$$

From this we derive the expression for the internal energy of an ideal (monatomic) gas, $U = 3NkT/2$, which was already found in (5.23). On substituting into (12.25), the same expression for the entropy $S(T, \, V, \, N)$ is recovered as was obtained in (5.26) by the elementary method. From these two thermodynamic functions the free energy of the gas $F(T, \, V, \, N)$ can therefore be found.

12.5   CANONICAL ENSEMBLE

12.5.1   Canonical Distribution

Systems that do not exchange work with the outside (constant V, B, ...), but can exchange heat with a thermostat at temperature T, come into the domain of the canonical ensemble. According to Gibbs, the elements of this ensemble are all identical systems that are subject to the same constraints of temperature T, number of particles N, volume V, ..., but for which the energy E, chemical potential, pressure, ... change in the course of time.

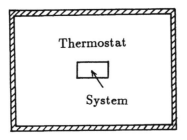

Figure 12.2: Canonical ensemble

We wish to determine the probability distribution P_r for the occurrence of a quantum state r. This is called the canonical distribution. For this, we use the results of the microcanonical ensemble by considering the isolated closed system composed of the system investigated plus the thermostat, which is assumed to be sufficiently large for its energy always to be much greater than that of the system (Fig. 12.2). As the total system is isolated, its energy E_0 must be constant and equal to the sum of the energy of the system investigated and that of the thermostat, plus the interaction energy between the particles in each of them. For sufficiently large systems, these interactions, which are localized at the contact surface with the thermostat, are negligible in comparison with the intrinsic energies of the thermostat and of the system. Therefore, if the latter is in the state r with energy E_r, then the energy of the thermostat is $E' = E_0 - E_r$, and it can occupy one of the $W'(E') = W'(E_0 - E_r)$ accessible states for this energy. The total system therefore has a total number of accessible states equal to $W_0(E_0) = \sum_r W'(E_0 - E_r)$. Since the total system is isolated, all states have the same probability and the probability that the system considered is in the state r is

$$P_r = \frac{W'(E_0 - E_r)}{W_0(E_0)} = \frac{W'(E_0 - E_r)}{\sum_r W'(E_0 - E_r)} \quad . \tag{12.26}$$

To define this probability, we perform a series expansion, making use of the fact that E_r is much smaller than E_0. However, since the function $W'(E')$ varies very strongly with E', we expand its logarithm to first order, i.e.

$$\ln W'(E_0 - E_r) = \ln W'(E_0) - E_r \left(\frac{\partial \ln W'}{\partial E'} \right)_{E'=E_0} + \ldots \tag{12.27}$$

According to equation (12.23), the entropy $S'(E_0)$ of the thermostat of energy E_0, divided by k, is $\ln W'(E_0)$. This is therefore a constant independent of E_r. Similarly, we have

$$\left(\frac{\partial \ln W'}{\partial E'} \right)_{E'=E_0} = \frac{1}{k} \left(\frac{\partial S'}{\partial E'} \right)_{E'=E_0} = \frac{1}{kT} \equiv \beta$$

where T is the temperature of the thermostat. Finally, we get

$$\ln W'(E_0 - E_r) = \ln W'(E_0) - \beta E_r$$

i.e., $\qquad W'(E_0 - E_r) = W'(E_0)e^{-\beta E_r} \quad .$

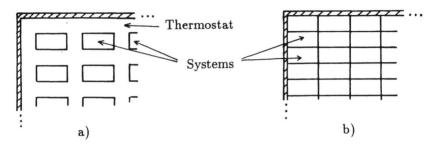

Figure 12.3: Two equivalent representations of the canonical ensemble.

On substituting into equation (12.26), we obtain the canonical distribution for the probability P_r of occurrence of a state r with energy E_r

$$P_r = \frac{e^{-\beta E_r}}{Q_N} \qquad \text{with} \qquad Q_N = \sum_r e^{-\beta E_r} \ . \tag{12.28a}$$

The quantity Q_N, called the canonical partition function, acts as the normalization factor for the canonical distribution. It depends on the parameters defining the state of the system, i.e., number of particles N, temperature T, external parameters $(V, B, \ldots)$.

The canonical distribution has the same form as the Maxwell-Boltzmann distribution (3.1,2). The following argument explains why. The canonical ensemble contains a large number of identical systems, each of which is in contact with a thermostat at temperature T (Fig. 12.3a). It is clear, however, that for each of the systems, the role of thermostat can equally well be played by the set of all the other systems (Fig. 12.3b). It is therefore possible to consider the canonical ensemble as a closed isolated thermodynamic system, composed of the previous systems each of which in turn is considered as a localized macroscopic "particle" having no interaction. Maxwell-Boltzmann statistics, applied to these "particles", allows the canonical distribution to be recovered, since the energy states of the "particles" considered here are the energy states of the systems $\epsilon_i \to E_r$.

Lastly, we note that the statistical density matrix is

$$\hat{\rho} = \frac{e^{-\beta \hat{H}}}{Q_N} \qquad \text{with} \qquad Q_N = \text{Tr} \ e^{-\beta \hat{H}} \ . \tag{12.28b}$$

This expression is a generalization of the probability distribution (12.28a).

12.5.2   Thermodynamic Functions

The internal energy U and entropy S of a system can now be formulated using the basic equations (12.17,18). We have respectively

$$U \equiv \overline{E} = \sum_r P_r E_r = \frac{1}{Q_N} \sum_r E_r e^{-\beta E_r} = -\frac{\partial \ln Q_N}{\partial \beta} \tag{12.29}$$

and $\quad S = -k \sum_r P_r \ln P_r = k \sum_r P_r \left(\beta E_r + \ln Q_N \right) = \dfrac{U}{T} + k \ln Q_N$. $\quad$ (12.30)

Using both these functions, the free energy F of the system can be written in its natural variables $T,\ N,\ V,\ \ldots$ as

$$F = U - TS = -kT \ln Q_N \quad . \tag{12.31}$$

We then know that all the thermodynamic properties of the system can be obtained from the differential relation

$$dF = -S\ dT + \mu\ dN - P\ dV - M\ dB + \ldots \tag{12.32}$$

Thus

$$P = -\frac{\partial F}{\partial V} = kT \frac{\partial}{\partial V} \left(\ln Q_N \right) \qquad \text{and} \qquad M = -\frac{\partial F}{\partial B} = kT \frac{\partial}{\partial B} \left(\ln Q_N \right) \quad .$$

Using (12.28a) to define the canonical partition function, we get for the pressure

$$P = \frac{kT}{Q_N} \sum_r -\beta \frac{\partial E_r}{\partial V} \exp(-\beta E_r) = -\sum_r P_r \frac{\partial E_r}{\partial V} = -\overline{\frac{\partial E_r}{\partial V}} \tag{12.33}$$

and, similarly, for the magnetic moment

$$M = -\overline{\frac{\partial E_r}{\partial B}} \quad . \tag{12.34}$$

These two expressions reveal the averaged character of the pressure and the magnetic moment. Finally, the heat capacity at constant volume is given by

$$C_V = \left(\frac{\partial U}{\partial T} \right)_V = k\beta^2 \frac{\partial^2 \ln Q_N}{\partial \beta^2} \quad . \tag{12.35}$$

Exercise 12.3 Maxwell-Boltzmann statistics from the canonical distribution
$\quad$ Consider a system of N identical distinguishable particles having no interactions. Calculate the canonical partition function Q_N as a function of the partition function for one particle Z and rederive expression (2.20) for the free energy in Maxwell-Boltzmann statistics.

Solution
$\quad$ A state r of the system is characterized by the states $i_1,\ i_2,\ \ldots\ i_N$ of each of the particles. Denoting by $\epsilon(i) \equiv \epsilon_i$ the energy of a particle in the state i, the energy E_r of the state r of the system is given by

$$E_r = \epsilon(i_1) + \epsilon(i_2) + \ldots \epsilon(i_N) \quad .$$

The canonical partition function then becomes

$$Q_N = \sum_{i_1} \sum_{i_2} \cdots \sum_{i_N} \exp(-\beta E_r) \quad .$$

As the energy E_r is additive, we have

$$Q_N = \sum_{i_1} \exp\left[-\beta \epsilon(i_1)\right] \times \ldots \times \sum_{i_N} \exp\left[-\beta \epsilon(i_N)\right] \quad .$$

The N sums appearing in Q_N are all equal and they are the one-particle partition function Z. Hence

$$Q_N = Z^N \quad . \tag{12.36}$$

The free energy then becomes

$$F = -kT \ln Q_N = -NkT \ln Z$$

in agreement with equation (2.20), which was found using the elementary theory in Maxwell-Boltzmann statistics.

Exercise 12.4 *Free energy of an ideal monatomic gas*
Calculate the canonical partition function Q_N for an ideal monatomic gas, neglecting the excited electronic states. Hence derive the equation for the free energy of the gas.

Solution
The partition function is given by

$$Q_N = \text{Tr}\, e^{-\beta \hat{H}} = \int g\left(\mathbf{r}_1,\, \mathbf{p}_1,\, \ldots\right) \prod_{i=1}^{N} d^3\mathbf{r}_i\, d^3\mathbf{p}_i\, e^{-\beta E(\mathbf{r}_1,\, \mathbf{p}_1,\, \ldots\,)}$$

where the energy E and the degeneracy g are given by (12.2,3) with $V\left(\mathbf{r}_1,\, \ldots\, \mathbf{r}_N\right) = 0$ (ideal gas). As E is a sum, we have

$$Q_N = \frac{V^N}{N!\, h^{3N}} \prod_{i=1}^{N} \int d^3\mathbf{p}_i\, e^{-\beta \mathbf{p}_i^2/2m} \quad .$$

Since the integrals are all identical and equal to $(2\pi mkT)^{3/2}$, we find

$$Q_N = \frac{Z^N}{N!} \qquad \text{with} \qquad Z = \frac{V}{h^3}\, (2\pi mkT)^{3/2} \quad . \tag{12.37}$$

It will be recognized that Z is the translational partition function for one particle (5.18). The equation $Q_N = Z^N/N!$ applies to any ideal gas.
The free energy F is found from equation (12.31). We thus obtain

$$F = -NkT \left[\ln \frac{V}{N} \frac{(2\pi mkT)^{3/2}}{h^3} + 1\right] \quad ,$$

which is identical to expression (5.21) found by the elementary method.

Exercise 12.5 *Magnetic moment and susceptibility*
  Using expression (12.7) for the Hamiltonian, show that the magnetic moment
and the molar susceptibility of a system of N spins are given by

$$\mathcal{M}_M = -\mathcal{N}g\mu_B \frac{1}{N} \sum_i \overline{S_i} \quad \text{and} \quad \chi_M = \mathcal{N}\mu_0 \frac{g^2\mu_B^2}{kT} \times \frac{1}{N} \sum_{ij} \left(\overline{S_iS_j} - \overline{S_i}\,\overline{S_j} \right) \quad (12.38)$$

Solution
  From equation (12.34) for $\mathcal{M}$ and from the Hamiltonian, it follows that

$$\mathcal{M}_M = -\frac{\mathcal{N}}{N} \frac{\partial \overline{E_r}}{\partial B} = -\mathcal{N}g\mu_B \frac{1}{N} \sum_i \overline{S_i}$$

where, according to (12.28b), we have

$$\overline{S_i} = \text{Tr } S_i\hat{\rho} = \frac{1}{Q_N} \text{Tr } S_i e^{-\beta\hat{H}} \quad .$$

The susceptibility is given by

$$\chi_M = \mu_0 \frac{\partial \mathcal{M}_M}{\partial B} = -\frac{\mathcal{N}}{N}\mu_0 g\mu_B \sum_i \frac{\partial \overline{S_i}}{\partial B}$$

$$\text{with} \quad \frac{\partial \overline{S_i}}{\partial B} = -\frac{\beta}{Q_N} \text{Tr } S_i \frac{\partial \hat{H}}{\partial B} e^{-\beta\hat{H}} - \frac{1}{Q_N^2} \frac{\partial Q_N}{\partial B} \text{Tr } S_i e^{-\beta\hat{H}}$$

$$= -\frac{g\mu_B}{kT} \sum_j \text{Tr } S_iS_j\hat{\rho} - \frac{\partial \ln Q_N}{\partial B} \overline{S_i}$$

$$= -\frac{g\mu_B}{kT} \sum_j \overline{S_iS_j} + \frac{g\mu_B}{kT} \sum_j \overline{S_j}\,\overline{S_i} \quad .$$

We thus find the same result as in (12.38). Note that $\overline{S_i} = \overline{S_j}$ and that the quantity
$\overline{S_iS_j} - \overline{S_i}\,\overline{S_j}$ is the correlation function of the spins i and j, which depends only
on their relative position. For substances that are not isotropic, equations (12.38)
apply to each component, and χ then has a tensor character.

12.5.3 Fluctuations

A measurement of a thermodynamic quantity is the result of an averaging process.
We can now calculate the fluctuation of some of these quantities, in particular that
of the internal energy. To do this, we calculate $\overline{E_r^2}$. Thus

$$\overline{E_r^2} = \frac{1}{Q_N} \sum_r E_r^2 e^{-\beta E_r} = \frac{1}{Q_N} \frac{\partial^2 Q_N}{\partial \beta^2}$$

and from this we derive the standard deviation δU in a measurement of the energy

$$(\delta U)^2 = \overline{E_r^2} - \overline{E_r}^2 = \frac{1}{Q_N}\frac{\partial^2 Q_N}{\partial\beta^2} - \frac{1}{Q_N^2}\left(\frac{\partial Q_N}{\partial\beta}\right)^2 = \frac{\partial^2 \ln Q_N}{\partial\beta^2} \quad .$$

Using equation (12.29) for the internal energy, this standard deviation becomes

$$(\delta U)^2 = -\frac{\partial U}{\partial\beta} = kT^2\frac{\partial U}{\partial T} = kT^2 C_V \tag{12.39}$$

where C_V is the heat capacity of the system. The relative fluctuation in the internal energy is therefore equal to

$$\frac{\delta U}{\overline{E_r}} \equiv \frac{\delta U}{U} = \frac{(kT^2 C_V)^{1/2}}{U} = N^{-1/2}\left(\frac{NkT^2 C_V}{U^2}\right)^{1/2} = N^{-1/2}\left(\frac{RT^2 c_V}{u^2}\right)^{1/2}$$

where u and c_V are the molar internal energy and molar heat capacity of the system. The intensive ratio $RT^2 c_V/u^2$ is generally of the order of unity (for an ideal monatomic gas with $c_V = 3R/2$, this ratio is equal to 2/3), which shows that the relative fluctuation is of the order of $1/\sqrt{N}$. This is extremely small for macroscopic systems, which contain $N \sim 10^{23}$ particles.

Exercise 12.6 *Fluctuation of pressure in an ideal monatomic gas*
    In an ideal monatomic gas, the translational energy levels (1.2) of the atoms vary with the volume V as $V^{-2/3}$. Using (12.33), rederive the equation of state of a perfect gas and determine the fluctuation in the pressure of the gas. Are the results valid for a perfect di- or polyatomic gas?

Solution
    Since the energy of the gas is the sum of the translational energies of the atoms, the energies E_r also vary as $V^{-2/3}$, and hence $\partial E_r/\partial V = -2E_r/3V$. It then follows that

$$P \equiv \overline{P} = -\frac{\overline{\partial E_r}}{\partial V} = \frac{2}{3V}\overline{E_r} = \frac{2}{3V}U = \frac{NkT}{V} \quad ,$$

since the internal energy of a perfect monatomic gas is $U = 3NkT/2$. To calculate the fluctuation in pressure δP, we calculate $\overline{P^2}$. We have

$$\overline{P^2} = \overline{\left(\frac{\partial E_r}{\partial V}\right)^2} = \frac{4}{9V^2}\overline{E_r^2}$$

and, consequently, from (12.39)

$$(\delta P)^2 = \overline{P^2} - \overline{P}^2 = \frac{4}{9V^2}\left(\overline{E_r^2} - \overline{E_r}^2\right) = \frac{4}{9V^2}(\delta U)^2 = \frac{4}{9V^2}kT^2 C_V \quad .$$

The relative fluctuation in pressure is therefore equal to

$$\frac{\delta P}{P} = \frac{2T}{3V}(kC_V)^{1/2} \times \frac{V}{NkT} = \left(\frac{2}{3N}\right)^{1/2} \quad . \tag{12.40}$$

We see that the relative fluctuation in pressure is also of the order of $1/\sqrt{N}$. The validity of this result extends to di- and polyatomic gases, since the contributions from the energy of rotation and vibration are independent of V and vanish from $\partial E_r / \partial V$.

Exercise 12.7 *Fluctuation of the magnetic moment*
   Determine the fluctuation of the magnetic moment in a system of spins.

Solution
   Since, apart from a factor $-g\mu_B$, $\mathcal{M}$ is equal to the mean value of $\sum_i S_i$, we have by definition

$$\frac{\overline{(\delta\mathcal{M})^2}}{g^2\mu_B^2} = \overline{\left(\sum_i S_i\right)^2} - \left(\overline{\sum_i S_i}\right)^2 = \overline{\sum_{ij} S_i S_j} - \overline{\sum_i S_i}\ \overline{\sum_j S_j}$$

$$= \sum_{ij}\left(\overline{S_i S_j} - \overline{S_i}\ \overline{S_j}\right)\ .$$

Using (12.38), we find

$$\overline{(\delta\mathcal{M})^2} = \frac{kT\chi}{\mu_0}\ , \tag{12.41}$$

or, for the molar magnetic moment $(M_M = \mathcal{M}\, \mathcal{N}/N)$,

$$\overline{(\delta M_M)^2} = \frac{RT\chi_M}{N\mu_0}\ .$$

For perfect paramagnetic substances, χ_M is given by (4.49) and thus

$$\overline{(\delta M_M)^2} = \frac{1}{N}\frac{\mathcal{N}^2\mu^2}{3}J(J+1) \quad \text{or} \quad \frac{\delta M_M}{M_{Ms}} = \sqrt{\frac{J+1}{3J}}\ \frac{1}{\sqrt{N}}$$

which shows that the fluctuation is again in $1/\sqrt{N}$. For an arbitrary substance, fluctuations are always negligible except when χ_M diverges, i.e. when $\partial M/\partial B$ becomes infinite, which is the case at the Curie point of ferromagnetic substances (Chap. 13).

12.6   GRAND CANONICAL ENSEMBLE

12.6.1   Grand Canonical Distribution

Systems are often encountered in which the number of particles is not constant. This occurs for example for a gaseous phase lying above a liquid, or also for a black body in which photons continually appear and disappear. To study such systems, which are generally composed of indistinguishable particles, the framework of the grand canonical ensemble is adopted. Here, systems are considered for which the external parameters $(V,\ B,\ \dots)$ are constant, and which are in contact both with a thermostat at temperature T and with a reservoir of particles each of chemical

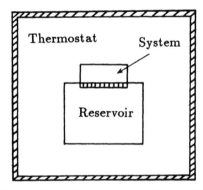

Figure 12.4: Grand canonical ensemble

potential μ. The state of the system is now defined by the number of particles N and also by a quantum number r of the same kind as that used for canonical ensembles, in which N is held constant.

To find the grand canonical distribution giving the probability P_{Nr} of finding the system in a state N, r, of energy E_{Nr}, we consider the closed isolated system composed of the system investigated plus the thermostat and the reservoir. The latter are both assumed to be large compared with the system (Fig. 12.4). This total system therefore has an energy E_0, and a number N_0 of particles of the same kind as those in the reservoir. It follows that, if the system investigated contains N particles and is in a state r, the energy of the thermostat plus the reservoir is $E' = E_0 - E_{Nr}$ and the number of particles is $N' = N_0 - N$. It then occupies one of the $W'(E', N')$ accessible states. As the total system is isolated, it can therefore with equal probability occupy any of the

$$W_0(E_0) = \sum_N \sum_r W'(E_0 - E_{Nr}, N_0 - N)$$

states. The probability that the system is in the state N, r is

$$P_{Nr} = \frac{W'(E_0 - E_{Nr}, N_0 - N)}{W_0(E_0)} \quad .$$

As with canonical ensembles, on performing a series expansion of $\ln W'$, we find

$$\ln W'(E_0 - E_{Nr}, N_0 - N) = \ln W'(E_0, N_0) - E_{Nr} \left(\frac{\partial \ln W'}{\partial E'} \right)_0$$
$$- N \left(\frac{\partial \ln W'}{\partial N'} \right)_0 ,$$

where the subscript zero means $E' = E_0$ and $N' = N_0$. As $\ln W'(E_0, N_0)$ is the entropy, divided by k, of the thermostat plus the reservoir, the thermodynamic relation

$$dU = T\, dS + \mu\, dN - P\, dV\, \ldots \quad \text{or} \quad dS = \frac{dU}{T} - \frac{\mu}{T} dN + \frac{P}{T} dV + \ldots$$

yields

$$\frac{\partial \ln W'}{\partial E'} = \frac{1}{k}\frac{\partial S'}{\partial E'} = \frac{1}{kT} \qquad \text{and} \qquad \frac{\partial \ln W'}{\partial N'} = \frac{1}{k}\frac{\partial S'}{\partial N'} = -\frac{\mu}{kT}$$

where T and μ are the temperature of the thermostat and the chemical potential of a particle of the reservoir. We therefore have

$$W'\left(E_0 - E_{Nr}, \; N_0 - N\right) = W'\left(E_0, \; N_0\right)e^{-\beta E_{Nr} + \beta N\mu}$$

and we find for the grand canonical distribution

$$P_{Nr} = \frac{e^{\beta N\mu - \beta E_{Nr}}}{\Xi} \qquad \text{with} \qquad \Xi = \sum_{N,r} e^{\beta N\mu - \beta E_{Nr}} \qquad (12.42)$$

where Ξ, called the grand canonical partition function, is a function of T, μ and the external parameters $(V, B, \dots)$. The grand canonical partition function is related to the canonical partition functions of the system by

$$\Xi(T, \; \mu, \; V, \; \dots) = \sum_{N} e^{\beta N\mu} \sum_{r} e^{-\beta E_{Nr}} = \sum_{N=0}^{\infty} e^{\beta N\mu} \, Q_N(T, \; V, \; \dots) \quad . \quad (12.43)$$

12.6.2   Thermodynamic Functions

Let us now calculate the internal energy U and the entropy S of the system by inserting the appropriate expressions into equations (12.17,18), and also the mean number of particles $\overline{N}$ in the system. To do this, we observe that the logarithmic derivatives of the partition function Ξ with respect to β and μ are given by

$$\frac{\partial \ln \Xi}{\partial \beta} = \frac{1}{\Xi} \sum_{N,r} (N\mu - E_{Nr})\, e^{\beta N\mu - \beta E_{Nr}} = \mu \sum_{N,r} P_{Nr} N - \sum_{N,r} P_{Nr} E_{Nr}$$

and

$$\frac{\partial \ln \Xi}{\partial \mu} = \frac{1}{\Xi} \sum_{N,r} \beta N e^{\beta N\mu - \beta E_{Nr}} = \frac{1}{kT} \sum_{N,r} P_{Nr} N \quad .$$

The second of these equations yields the mean number of particles, namely

$$\overline{N} = \sum_{N,r} P_{Nr} N = kT \, \frac{\partial \ln \Xi}{\partial \mu} \qquad (12.44)$$

and the first gives

$$U = \sum_{N,r} P_{Nr} E_{Nr} = \overline{N}\mu - \frac{\partial \ln \Xi}{\partial \beta} \quad . \qquad (12.45)$$

Lastly, the entropy (12.18) is given here by

$$S = -k \sum_{N,r} P_{Nr} \ln P_{Nr} = -k \sum_{N,r} P_{Nr}\left[\beta N\mu - \beta E_{Nr} - \ln \Xi\right]$$

$$= -\frac{\overline{N}\mu}{T} + \frac{U}{T} + k \ln \Xi \quad . \qquad (12.46)$$

These three quantities are expressed in the variables T, μ, V, ... for which the appropriate thermodynamic potential is

$$\Omega(T, \mu, V, \ldots) = U - TS - \overline{N}\mu , \tag{12.47a}$$

the differential of which is

$$d\Omega = -S \, dT - \overline{N} \, d\mu - P \, dV + \ldots \qquad . \tag{12.47b}$$

Rearranging the terms in (12.46), we find for the statistical definition of the grand potential Ω

$$\Omega(T, \mu, V, \ldots) = -kT \ln \Xi(T, \mu, V, \ldots) . \tag{12.48}$$

This equation for Ω shows that if we know the grand canonical partition function Ξ, then all the properties of the system can be determined. Thus

$$\overline{N} = -\frac{\partial \Omega}{\partial \mu} = kT\frac{\partial \ln \Xi}{\partial \mu} \qquad \text{and} \qquad P = -\frac{\partial \Omega}{\partial V} = kT\frac{\partial \ln \Xi}{\partial V} \qquad . \tag{12.49}$$

On substituting (12.42) for Ξ, these equations become

$$\overline{N} = \frac{1}{\Xi}\sum_{N,r} Ne^{\beta N\mu - \beta E_{Nr}} = \sum_{N,r} N P_{Nr} \tag{12.50}$$

$$\overline{P} = \frac{1}{\Xi}\sum_{N,r} -\frac{\partial E_{Nr}}{\partial V}e^{\beta N\mu - \beta E_{Nr}} = \sum_{N,r}\left(-\frac{\partial E_{Nr}}{\partial V}\right) P_{Nr}, \tag{12.51}$$

which reveals the averaged character of these physical quantities.

Exercise 12.8 *Grand potential of an ideal gas*
    Derive the expression for the grand canonical partition function of an ideal gas as a function of the single particle partition function Z. Hence deduce the equation for the grand potential $\Omega(T, \mu, V)$.

Solution
    From equations (12.43 and 37), we have

$$\Xi = \sum_{N=0}^{\infty} e^{\beta N\mu}\frac{Z^N}{N!} = \exp\left(Ze^{\beta\mu}\right) \quad ,$$

from which it follows that

$$\Omega(T, \mu, V) = -kT \ln \Xi = -kT Z(T, V)e^{\beta\mu} \quad .$$

This expression is identical to that found by elementary methods (2.37).

12.6.3   Systems of Identical Particles

Fock space

When the system being investigated consists of identical particles, it is not correct to say that its state is specified by the state of each of the particles, as well as by the number N, because it is impossible to distinguish one particle from another. The correct description defines the number of particles occupying each single-particle quantum state. If we then choose a basis of one-particle states with index i, then the basis of the states of the system can be taken to be

$$|n_1, \ ... \ n_i, \ ... \ > \equiv |N, r > \quad , \tag{12.52}$$

where the n_i are positive integers or zero such that their sum is equal to N. This basis is thus particularly suited to open systems if the numbers n_i are considered to be independent variables. The state $|0, \ 0, \ ... \ >$ thus corresponds to a system having no particles, the state $|0, \ ... \ 0, \ 1, \ 0, \ ... \ >$ to a system containing 1 particle that is in the state $i, ...$ The space encompassed by the states (12.52) is called Fock space. In this space, a set of operators $\hat{n}_i$ is introduced defined by

$$\hat{n}_i \ |n_1, \ ... \ n_i, \ ... \ > = n_i \ |n_1, \ ... \ n_i, \ ... \ > \tag{12.53}$$

These measure the number of particles in state i when the system is in a state $|N, r >$. Such operators, which commute, in turn allow the particle number operator to be defined by

$$\hat{N} = \sum_i \hat{n}_i \ . \tag{12.54}$$

We can therefore now rewrite the grand canonical partition function (12.42) in the form

$$\Xi = \sum_{n_1, ... n_i, ...} \exp\left[\beta\left(\sum_i n_i\right)\mu - \beta E\right] \equiv \mathrm{Tr} \ e^{\beta \hat{N}\mu - \beta \hat{H}} \tag{12.55}$$

and give the equation for the statistical density operator

$$\hat{\rho} = \frac{1}{\Xi}e^{\beta \hat{N}\mu - \beta \hat{H}} \quad . \tag{12.56}$$

12.6.4   Creation and Annihilation Operators

In the formalism of second quantization, for each one-particle state a non Hermitian operator $\hat{a}_i$ is introduced, called the annihilation operator for the state i. This transforms a state having n_i particles in state i into a state with $n_i - 1$ particles

$$\hat{a}_i|... \ n_i, ... \ > = \sqrt{n_i}|... \ n_i - 1, \ ... \ > \quad . \tag{12.57a}$$

The adjoint operator $\hat{a}_i^+$, called the creation operator for the state i, then transforms from a state with n_i particles in i into a state with $n_i + 1$ particles, i.e.

$$\hat{a}_i^+|... \ n_i, ... \ > = \sqrt{n_i + 1}|... \ n_i + 1, \ ... \ > \quad . \tag{12.57b}$$

Starting from a state in Fock space, with these operators all the other states can be constructed by iteration. They are connected to the operators $\hat{n}_i$ through the basic relation

$$\hat{n}_i = \hat{a}_i^+ \hat{a}_i \ . \tag{12.58}$$

Furthermore, it can be shown that any Fock space operator can always be expressed as a superposition of products of annihilation and creation operators.

The properties of these operators are, however, different according to the quantum mechanical character (fermion or boson) of the particles. As a consequence of the Pauli exclusion principle, systems of fermions can exist only in states of Fock space for which $n_i = 0$ or 1 for all i, while boson systems can occupy any state ($n_i = 0,\ 1,\ 2,\ \ldots$). It can be shown that the annihilation and creation operators are completely specified by imposing the following rules

$$\hat{a}_i \hat{a}_j + \hat{a}_j \hat{a}_i = 0 \quad \text{and} \quad \hat{a}_i \hat{a}_j^+ + \hat{a}_j^+ \hat{a}_i = \delta_{ij} \quad \text{(fermions)} \tag{12.59a}$$

$$\hat{a}_i \hat{a}_j - \hat{a}_j \hat{a}_i = 0 \quad \text{and} \quad \hat{a}_i \hat{a}_j^+ - \hat{a}_j^+ \hat{a}_i = \delta_{ij} \quad \text{(bosons)} \ . \tag{12.59b}$$

Investigations into quantum systems of interacting particles rely on the use of these operators.

Ideal gas

When the interaction energy of the particles is negligible, the grand potential adopts a more compact form. The energy of the system can be written $E = \sum_i n_i \epsilon_i$, corresponding to the Hamiltonian

$$\hat{H} = \sum_i \epsilon_i \hat{n}_i \tag{12.60}$$

and the partition function (12.55) factorizes

$$\Xi = \sum_{n_1} e^{\beta n_1 (\mu - \epsilon_1)} \times \ldots \sum_{n_i} e^{\beta n_i (\mu - \epsilon_i)} \times \ldots$$

so that the grand potential (12.48) becomes

$$\Omega = -kT \sum_i \ln \left[\sum_{n_i} e^{\beta n_i (\mu - \epsilon_i)} \right] \ .$$

We now consider a gas of fermions. Here, the sum over n_i contains only the two terms $n_i = 0$ and 1 (Pauli exclusion principle) and the grand potential takes the form

$$\Omega \equiv \Omega_{FD} = -kT \sum_i \ln \left[1 + e^{\beta (\mu - \epsilon_i)} \right]$$

which is identical to equation (2.34) found by the elementary method in Fermi-Dirac statistics. For bosons, the sum over n_i, taken over all integer values, is an infinite geometrical series of progression ratio $\exp \beta(\epsilon_i - \mu)$, and we have

$$\Omega \equiv \Omega_{BE} = kT \sum_i \ln \left[1 - e^{\beta(\mu - \epsilon_i)} \right] .$$

This is identical to equation (2.34) found in Bose-Einstein statistics.

It is thus clear that the method of grand canonical ensembles yields the same results as the elementary method.

12.6.5   Fluctuations

In a grand canonical ensemble, as in a canonical ensemble, a quantity in the system may vary through exchange of energy with the thermostat. It can also vary as a result of exchange of matter with the reservoir. The fluctuation in the outcome of a measurement is therefore different from that found for canonical ensembles.

We first calculate the fluctuation δN in the number of particles in the system. This fluctuation does not exist in the canonical ensemble. We have

$$\overline{N^2} = \sum_{N,r} N^2 P_{Nr} = \frac{1}{\Xi} \sum_{N,r} N^2 e^{N\beta\mu - \beta E_{Nr}} = \frac{k^2 T^2}{\Xi} \frac{\partial^2 \Xi}{\partial \mu^2} \quad ,$$

which gives for the square of the fluctuation

$$(\delta N)^2 = \overline{N^2} - \overline{N}^2 = \frac{k^2 T^2}{\Xi} \frac{\partial^2 \Xi}{\partial \mu^2} - \frac{k^2 T^2}{\Xi^2} \left(\frac{\partial \Xi}{\partial \mu} \right)^2 = k^2 T^2 \frac{\partial^2 \ln \Xi}{\partial \mu^2} = kT \frac{\partial \overline{N}}{\partial \mu} \quad ,$$

where we have used equation (12.49). To specify $(\delta N)^2$, we calculate $(\partial \mu / \partial \overline{N})_{T, V}$, making use of the Gibbs-Duhem equation from classical thermodynamics

$$\overline{N} \, d\mu = V \, dP - S \, dT \quad .$$

Using the equation of state $P = P(v, T) = P(\mathcal{N}V/N, T)$, we transform to the variables T, V, N. Thus

$$N \, d\mu = \left(V \frac{\partial P}{\partial T} - S \right) dT + \frac{N}{N} V \frac{\partial P}{\partial v} \, dV - \frac{\mathcal{N}V^2}{N^2} \frac{\partial P}{\partial v} \, dN$$

and hence

$$\left(\frac{\partial \mu}{\partial N} \right)_{T,V} = -\frac{\mathcal{N}V^2}{N^3} \left(\frac{\partial P}{\partial v} \right)_T = -\frac{V}{N^2} v \left(\frac{\partial P}{\partial v} \right)_T = \frac{V}{N^2 \chi_T}$$

where χ_T is the isothermal compressibility of the gas. Therefore

$$(\delta N)^2 = \frac{kT}{V} \chi_T \overline{N}^2 \quad ,$$

and the relative fluctuation in the number of particles, which is equal to that of the particle density $n = \overline{N}/V$, is

$$\frac{\delta N}{\overline{N}} = \frac{\delta n}{n} = \left(\frac{kT}{V}\chi_T\right)^{1/2} . \tag{12.61}$$

For ideal gases, where $\chi_T = 1/P$, this fluctuation is equal to $1/\sqrt{N}$ and is therefore negligible. For a real gas close to its critical point, however, the isothermal compressibility becomes infinite (§14.4.1) and extremely large density fluctuations occur, which give rise to the phenomenon of critical opalescence.

Let us now calculate the fluctuation of the internal energy in the same way as the canonical fluctuation (§12.5.3). Setting $\nu = \beta\mu$, we have

$$
\begin{aligned}
U &\equiv \overline{E_{Nr}} = \sum_{N,r} P_{Nr}\, E_{Nr} = \frac{1}{\Xi}\sum_{N,r} E_{Nr}\, e^{N\nu - \beta E_{Nr}} \\
&= -\frac{1}{\Xi}\left(\frac{\partial\Xi}{\partial\beta}\right)_\nu = -\left(\frac{\partial\ln\Xi}{\partial\beta}\right)_\nu \\
\overline{E_{Nr}^2} &= \sum_{N,r} P_{Nr}\, E_{Nr}^2 = \frac{1}{\Xi}\sum_{N,r} E_{Nr}^2\, e^{N\nu - \beta E_{Nr}} = \frac{1}{\Xi}\left(\frac{\partial^2\Xi}{\partial\beta^2}\right)_\nu ,
\end{aligned}
\tag{12.62}
$$

which yields

$$(\delta U)^2 = \overline{E_{Nr}^2} - \overline{E_{Nr}}^2 = \frac{1}{\Xi}\frac{\partial^2\Xi}{\partial\beta^2} - \frac{1}{\Xi^2}\left(\frac{\partial\Xi}{\partial\beta}\right)^2 = \left(\frac{\partial^2\ln\Xi}{\partial\beta^2}\right)_\nu . \tag{12.63}$$

Using equation (12.62) for the internal energy, we finally find

$$(\delta U)^2 = -\left(\frac{\partial U}{\partial\beta}\right)_\nu = kT^2\left(\frac{\partial U}{\partial T}\right)_\nu .$$

This expression for the fluctuation in energy resembles that for the canonical fluctuation (12.39), but in the present case the derivative is taken at constant $\nu = \beta\mu$ instead of at constant N. A tedious calculation involving thermodynamic relations gives for $(\delta U)^2$

$$(\delta U)^2 = kT^2 C_V + \left(\frac{\partial U}{\partial N}\right)_{T,V}^2 (\delta N)^2. \tag{12.64}$$

This shows that the canonical fluctuation $(kT^2 C_V)^{1/2}$ and the fluctuation in energy, $(\partial U/\partial\overline{N})\,\delta N$, due the variation in the number of particles, add quadratically to give the grand canonical fluctuation.

Exercise 12.9 *Grand canonical fluctuation of energy for an ideal monatomic gas*

The grand potential of an ideal monatomic gas is given by equation (2.37), $\Omega(T, \nu, V) = -kTZe^\nu$, where $Z(T, V)$ is the partition function for one particle

(5.18). Using (12.62,63), find the internal energy U and its mean square fluctuation $(\delta U)^2$ as a function of ν and of Z. Hence derive an expression for the relative fluctuation $\delta U/U$ and compare with the canonical form.

Solution

The equation for Ω shows that $\ln \Xi = Ze^{\nu}$. Since Z varies as $T^{3/2}$, U and $(\delta U)^2$ are given by

$$U = -\left(\frac{\partial \ln \Xi}{\partial \beta}\right)_{\nu} = kT^2 \left(\frac{\partial \ln \Xi}{\partial T}\right)_{\nu} = \frac{3}{2} kT Ze^{\nu}$$

$$\text{and} \quad (\delta U)^2 = \left(\frac{\partial^2 \ln \Xi}{\partial \beta^2}\right)_{\nu} = -\left(\frac{\partial U}{\partial \beta}\right)_{\nu} = kT^2 \left(\frac{\partial U}{\partial T}\right)_{\nu} = \frac{15}{4} k^2 T^2 Ze^{\nu} .$$

Therefore

$$\frac{(\delta U)^2}{U^2} = \frac{5}{3} \frac{1}{Ze^{\nu}} = -\frac{5}{3} \frac{kT}{\Omega} = \frac{5}{3} \frac{kT}{PV} = \frac{5}{3N} .$$

The square of the relative canonical fluctuation is just $2/3N$ (§12.5.3), the difference being due to the relative fluctuation in the number of particles, which is equal to $1/N$.

BIBLIOGRAPHY

Historical references: density matrix
P.A.M. Dirac, *The Principles of Quantum Mechanics*, Oxford (1935).
J. von Neumann, *Mathematische Grundlagen der Quantenmechanik*, Berlin (1932).
Historical references: ensembles
W. Gibbs, *Collected works* **2** *Elementary Principles in Statistical Mechanics*, Longmans (1928).
W. Pauli, *Z. Physik* **41**, 81 (1927).

COMPREHENSION EXERCISES

12.1 Consider a system of N non interacting spins $1/2$, placed in a constant magnetic flux density B. The energy of this isolated system is E. Calculate the entropy of the system $S(N, E, B)$ [Ans.: Equation (1.60)].

12.2 For a certain thermostat the number of states is given by $W' = AE'^{\alpha N'}$ where α is a numerical factor close to 1. Make a series expansion of $W'(E_0 - E_r)$ and of its logarithm in the neighbourhood of E_0, and verify that the expansion can be restricted to order 1 only for the logarithm.

PROBLEM 12.1 MODELS OF ADSORPTION OF GASES BY A SOLID

Molecules of a gas can be trapped at the surface of a solid. This is the phenomenon of adsorption. Two types of adsorption are to be distinguished, namely chemisorption, in which the adsorbed molecules form a chemical bond with the atoms of the surface, and physical adsorption, in which the gas molecules are bound to the surface by van der Waals forces. In the former the molecules are adsorbed at a finite number of sites and thus form a monomolecular layer. In the latter the molecules can be deposited in several layers on the surface. We propose to investigate a simple model for each type of adsorption.

1 Gas phase

We first consider the gaseous phase which, for simplicity, we take to be an ideal monatomic gas of molar mass M.

1.1 Give the general equations for the grand canonical partition function Ξ and the density operator $\hat{\rho}$ as a function of the operators $\hat{H}$ (Hamiltonian) and $\hat{N}$ (number of particles).

1.2 Calculate the function $\Xi(T, V, \mu)$ for the gas, where T, V and μ are respectively the temperature, volume and chemical potential for one gas particle. Neglect degrees of freedom other than that of translation, and assume that the molecules are indistinguishable. Hence derive the equation for the grand potential $\Omega(T, V, \mu)$.

1.3a Express the chemical potential μ as a function of T, V and $\overline{N}$, where $\overline{N}$ is the mean number of molecules at equilibrium.

1.3b Show that μ can be written in the variables T and P as

$$\mu = kT \ln \frac{P}{P_0(T)} \qquad \text{with} \qquad P_0(T) \propto T^{5/2} \quad .$$

2 Langmuir model

In the Langmuir model, it is assumed that the adsorbing solid possesses N_0 sites at which the gas molecules can be trapped with an energy - ϵ. To each site is attributed a quantum number n_i $(i = 1, \ldots N_0)$ equal to 0 if the site is unoccupied, and to 1 if the site contains a molecule. In this model, the site can be occupied by one molecule at most. Since the number of molecules in the adsorbed phase fluctuates, grand canonical ensemble methods may be applied to this phase.

2.1a Express the number of molecules N of the adsorbed phase as a function of the n_i.

2.1b Write down the Hamiltonian H of the adsorbed phase as a function of n_i and of ϵ.

2.1c Hence derive an expression for the Hamiltonian $H' = H - N\mu$.

2.2a Give the grand canonical partition function Ξ for the adsorbed phase.

2.2b Show that Ξ may be written in the form Z^{N_0}, where Z can be taken to be the partition function of a site. Find an expression for $Z(T, \mu)$.

2.2c Calculate the mean number N_A of molecules in the adsorbed phase as a function of T and μ.

2.3a What is the condition for equilibrium between the adsorbed phase and the gas phase? Use this condition to express N_A as a function of the temperature T and the pressure P of the gas (Langmuir adsorption isotherm). Sketch an isotherm $N_A(P)$.

2.3b Adsorption measurements of nitrogen on wood charcoal show that the isotherms obey an equation of the form

$$\frac{P}{x} = a(T) + b(T)P \ ,$$

where x is the ratio of the mass of adsorbed gas to that of the adsorbent. Show that these results are compatible with the Langmuir equation.

2.3c In this investigation, the measurements gave

$$a\ (20\ ^\circ C) = 85\ \text{atm} \qquad \text{and} \qquad b\ (20\ ^\circ C) = 6.6$$
$$a\ (-77\ ^\circ C) = 15\ \text{atm} \qquad \text{and} \qquad b\ (-77\ ^\circ C) = 4.4 \quad .$$

From these data, derive the trapping energy for one mole, $\mathcal{N}\epsilon$.

3 B.E.T. model

In the B.E.T. (Brunauer, Emmett, Teller) model, it is assumed that each of the N_0 sites on the surface can trap an unlimited number of molecules, the first having an energy $-\epsilon_1$, and the others with energies $-\epsilon_2$ ($\epsilon_1 > \epsilon_2 > 0$).

3.1 Calculate the grand canonical partition function $Z(T, \mu)$ for one site and hence, as in 2.2b, derive the grand partition function Ξ for the adsorbed phase. Set $z = \exp(\beta\mu)$, $z_1 = \exp(\beta\epsilon_1)$ and $z_2 = \exp(\beta\epsilon_2)$.

3.2a Calculate the mean number N_A of molecules in the adsorbed phase as a function of z.

3.2b Use the condition of equilibrium with the gas phase to write N_A as a function of $\xi = Pe^{\beta\epsilon_2}/P_0(T)$. Set $c = \exp\beta(\epsilon_1 - \epsilon_2) = z_1/z_2$.

3.2c Sketch $N_A(x)$. What is the physical meaning of the divergence of the function? What phenomenon does the expression $P_e = P_0(T)e^{-\beta\epsilon_2}$ describe?

3.3a Investigations into the adsorption of various gases by different catalysts show that over a wide range of pressure the isotherms obey an equation of the form

$$\frac{P}{(P_e - P)V} = A(T) + B(T)\frac{P}{P_e} \ ,$$

where V is the volume of adsorbed gas expressed in normal conditions of temperature and pressure. Are these results compatible with the B.E.T. model?

3.3b The adsorption of nitrogen by 50.4 g of an iron-based catalyst, measured at 77.3 K ($P_e = 1$ atm), gave the results $A = 0.25 \times 10^{-4}$ cm^{-3} and $B = 7.7 \times 10^{-3}$ cm^{-3}. Calculate the value of c. For nitrogen, assumed to be a symmetric diatomic ideal gas, statistical thermodynamics yields P_0 (77.3 K)$= 2.7 \times 10^6$ atm. Hence derive the values of ϵ_2 and ϵ_1 for one mole. Discuss, and compare with the molar latent heat of evaporation of nitrogen at atmospheric pressure $\Delta H = 5.6$ kJ mol^{-1}.

Numerical data: Gas constant $R = 8.315$ kJ mol^{-1}.

SOLUTION

1 The gas phase

1.1 The equations for Ξ and $\hat{\rho}$ are given in (12.55, 56).

1.2 Ξ and Ω are given in exercise (12.8) with

$$Z = \frac{V}{h^3} \, (2\pi mkT)^{3/2} \quad .$$

1.3a We have

$$N \equiv \overline{N} = -\frac{\partial \Omega}{\partial \mu} = Z(T, V)e^{\beta \mu} \quad ,$$

from which it follows that

$$\mu = -kT \ln \frac{Z}{N} = -kT \ln \left(\frac{V}{N} \frac{(2\pi mkT)^{3/2}}{h^3} \right) \quad ,$$

which is identical to (5.27).

1.3b On changing to the variable $P = NkT/V$ as in paragraph 5.2.4, we find, as in (5.34)

$$\mu = kT \ln \frac{P}{P_0(T)} \qquad \text{with} \qquad P_0(T) = e^{i_0} M^{3/2} T^{5/2} \quad .$$

2 Langmuir model

2.1 Clearly we have

$$N = \sum_{i=1}^{N_0} n_i \quad \text{et} \quad H = -N\epsilon = -\epsilon \sum_{i=1}^{N_0} n_i \quad , \quad \text{and hence} \quad H' = -(\mu + \epsilon) \sum_{i=1}^{N_0} n_i \quad .$$

2.2a Expression (12.55) for Ξ here becomes

$$\Xi = \sum_{n_1=0}^{1} \cdots \sum_{n_{N_0}=0}^{1} e^{-\beta H'}$$

since all the states of the adsorbed phase (2^{N_0} in number) are found by varying the n_i independently.

2.2b Since H' is a sum over i, we may write

$$\Xi = \sum_{n_1=0}^{1} e^{\beta(\mu+\epsilon)n_1} \cdots \sum_{n_{N_0}=0}^{1} e^{\beta(\mu+\epsilon)n_{N_0}} = Z^{N_0}$$

$$\text{with} \qquad Z(T, \mu) = \sum_{n=0}^{1} e^{\beta(\mu+\epsilon)n} = 1 + e^{\beta(\mu+\epsilon)} \quad .$$

2.2c We have

$$N_A = -\frac{\partial \Omega}{\partial \mu} = kT \frac{\partial \ln \Xi}{\partial \mu} = N_0 kT \frac{\partial \ln Z}{\partial \mu} \quad .$$

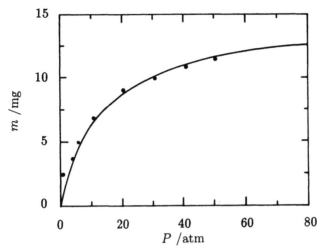

Figure 12.5: Mass of nitrogen adsorbed by 96.4 mg of charcoal at 20 °C. The curve is the Langmuir isotherm for $a = 85$ atm and $b = 6.6$ [J.W. McBain and G.T. Britton, *J. Am. Chem. Soc.* **52**, 2198 (1930)].

From the above equation for Z, it follows that

$$N_A = \frac{N_0}{1 + e^{-\beta(\epsilon+\mu)}} \quad .$$

2.3a At equilibrium, equality of the chemical potential in the two phases means that $e^{-\beta\mu} = P_0(T)/P$, and hence

$$N_A = \frac{N_0}{1 + e^{-\beta\epsilon} P_0(T)/P} \quad .$$

An isotherm is shown in Figure 12.5.

2.3b N_A is proportional to the dimensionless variable x, i.e., $N_A = Kx$. The Langmuir equation above can then be written as

$$\frac{P}{x} = \frac{K}{N_0}(P + e^{-\beta\epsilon} P_0) \quad .$$

The fit of the experimental data to this equation gives $a = Ke^{-\beta\epsilon} P_0/N_0$ and $b = K/N_0$.

2.3c $a/b = e^{-\beta\epsilon} P_0$, and since P_0 is proportional to $T^{5/2}$, we have

$$\frac{a/b\ (20\ °C)}{a/b\ (-77\ °C)} = \frac{P_0\ (20\ °C)}{P_0\ (-77\ °C)} \frac{e^{-\beta_1\epsilon}}{e^{-\beta_2\epsilon}} = \left(\frac{T_1}{T_2}\right)^{5/2} \exp\left[\frac{\epsilon}{k}\left(\frac{1}{T_2} - \frac{1}{T_1}\right)\right] = 3.8$$

with $T_1 = 293$ K and $T_2 = 196$ K. It follows that $\epsilon/k = 192$ K, and thus $\mathcal{N}\epsilon = R \times \epsilon/k \simeq 1600$ J mol^{-1}.

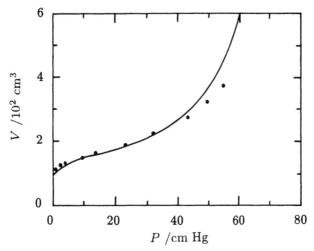

Figure 12.6: Volume (at N.T.P.) of nitrogen adsorbed on 50.4 g of an iron-based catalyst at 77.3 K. The curve shows the B.E.T. isotherm for $A = 0.25 \times 10^{-4}$ cm^{-3} and $B = 7.7 \times 10^{-3}$ cm^{-3} [S. Brunauer, P.H. Emmet and E. Teller, *J. Am. Chem. Soc.* **60**, 309 (1938)].

3 B.E.T. model

3.1 The grand canonical partition function for a site is given by

$$Z = \sum_{N=0}^{\infty} e^{\beta(N\mu - \epsilon_N)} = \sum_{N=0}^{\infty} z^N e^{-\beta \epsilon_N}$$

where ϵ_N is the energy of N molecules adsorbed on the site. Therefore

$$
\begin{aligned}
Z &= 1 + z e^{\beta \epsilon_1} + z^2 e^{\beta(\epsilon_1 + \epsilon_2)} + z^3 e^{\beta(\epsilon_1 + 2\epsilon_2)} + \dots \\
&= 1 + z e^{\beta \epsilon_1} \left[1 + z e^{\beta \epsilon_2} + z^2 e^{2\beta \epsilon_2} + \dots \right] \\
&= 1 + \frac{z e^{\beta \epsilon_1}}{1 - z e^{\beta \epsilon_2}} = \frac{1 + z(z_1 - z_2)}{1 - z z_2}
\end{aligned}
$$

As in 2.2b, the grand canonical partition function of the adsorbed phase Ξ is equal to Z^{N_0}. As particles on different sites do not interact, the N_0 sites are independent.

3.2a As in 2.2c, we have

$$\frac{N_A}{N_0} = kT \frac{\partial \ln Z}{\partial \mu} = z \frac{\partial \ln Z}{\partial z} \quad .$$

On substituting for Z, we find

$$\frac{N_A}{N_0} = \frac{z z_1}{[1 + z(z_1 - z_2)][1 - z z_2]} \quad .$$

3.2b The condition of equilibrium gives $z = P/P_0(T)$, i.e.,

$$\frac{N_A}{N_0} = \frac{c\xi}{[1 + (c - 1)\xi][1 - \xi]} \quad .$$

This relation is the B.E.T. equation.

3.2c The curve of $N_A(\xi)$ is shown in Figure 12.6, where the axes are $P \propto \xi$ and $V \propto N_A$.

The number of adsorbed atoms diverges for $\xi = 1$, i.e. for $P = P_e(T) \equiv P_0(T)e^{-\beta \epsilon_2}$. $P_e(T)$ can be pictured as the saturated vapour pressure at temperature T (the pressure at which the gas liquefies).

3.3a Since N_A is proportional to V and ξ is equal to P/P_e, the B.E.T. equation can be written in the form

$$\frac{\xi}{(1 - \xi)} \frac{N_0}{N_A} = \frac{1 + (c - 1)\xi}{c} \qquad \text{or} \qquad \frac{P}{(P_e - P)V} = A + B\frac{P}{P_e} \quad .$$

In this form, the B.E.T. equation is similar to the variation found experimentally, with $B/A = c - 1$.

3.3b From the numerical values of A and B, we find $c = 310$, or

$$\mathcal{N}(\epsilon_1 - \epsilon_2) = 3.7 \text{ kJ mol}^{-1} \quad ,$$

and from the values of P_e and P_0, we get

$$\mathcal{N}\epsilon_2 = 9.5 \text{ kJ mol}^{-1} \quad ,$$
$$\text{which gives} \qquad \mathcal{N}\epsilon_1 = 13.2 \text{ kJ mol}^{-1} \quad .$$

This confirms that the first adsorbed "layer" is bound more strongly than the subsequent ones ($\epsilon_1 > \epsilon_2$).

The B.E.T. isotherm provides a good fit to the data for nitrogen up to about 40 cm of mercury, which corresponds to adsorption of the first layers. At higher pressures the B.E.T. curve lies above the experimental points, which can be attributed to the fact that the following layers are even less strongly bound. The energy $\mathcal{N}\epsilon_i$ then tends to the value in the liquid $\mathcal{N}\epsilon_L = \Delta H$, equal to 5.6 kJ mol^{-1}.

Chapter 13

Magnetic Materials

13.1   INTRODUCTION

In Chapter 4 we examined the properties of magnetic substances in a temperature range where interactions between magnetic ions can be neglected (ideal paramagnetism). We now focus our attention on these interactions in order to interpret all the magnetic properties of solids. At sufficiently low temperatures, the existence of different types of ordered phases arises from the fact that these interactions tend to align the spins of two neighbouring ions parallel or antiparallel. The principal phases are:

- ferromagnetic phases, in which the spins of all the magnetic ions are parallel;

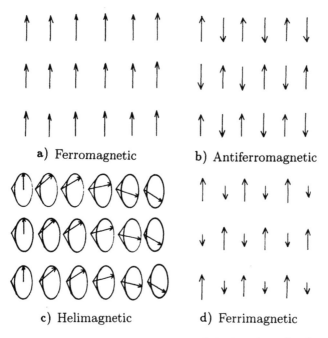

a) Ferromagnetic   b) Antiferromagnetic

c) Helimagnetic   d) Ferrimagnetic

Figure 13.1: Diagram of the spin arrangements of the ions in ordered magnetic phases.

427

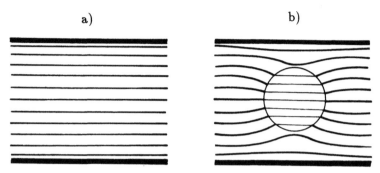

Figure 13.2: Field lines in an infinite solenoid (a) *in vacuo* and b) with a homogeneous spherical sample of an isotropic magnetic material.

these exhibit spontaneous magnetization (Fig. 13.1a);
• antiferromagnetic phases, in which the spins of two nearest neighbour ions are antiparallel; these display no spontaneous magnetization (Fig. 13.1b);
• helimagnetic phases, in which the spins of two neighbouring ions lie at a given angle to each other (Fig. 13.1c);
• ferrimagnetic phases in which there are several different magnetic ions with various arrangements; these display spontaneous magnetization (Fig. 13.1d).

In what follows, we shall devote our attention mainly to substances that exhibit a ferromagnetic phase. We shall refer to them simply as ferromagnetic substances.

First, we refer the reader to paragraph 4.2, and we briefly review magnetostatics. When a current passes through a sufficiently long solenoid containing n turns per unit length, a uniform magnetic field $\mathbf{H_0}$ is established *in vacuo*, with flux density $\mathbf{B_0}$ (Fig. 13.2a), where

$$\mathbf{B_0} = \mu_0 \mathbf{H_0} , \qquad \text{with} \quad H_0 = nI . \tag{13.1}$$

If a sample of magnetic material is present in the solenoid, it becomes magnetized, and the field and the magnetic flux density are no longer uniform (Fig. 13.2b). If, however, the sample, which is homogeneous and isotropic (e.g., a polycrystal), is an ellipsoid with one axis parallel to that of the solenoid, then the magnetization $\mathbf{M}$, the field $\mathbf{H}$ and the flux density $\mathbf{B}$ are parallel and uniform inside the material and they are governed by the algebraic relation

$$B = \mu_0(H + M) . \tag{13.2}$$

The field H inside the substance is related to that *in vacuo*, H_0, by

$$H = H_0 - H_d \qquad \text{with} \quad H_d = DM , \tag{13.3}$$

where H_d is the demagnetizing field generated by the magnetization of the substance, and D is the demagnetizing factor, the value of which lies between 0 and 1 ($D = 0$ for an infinite cylinder, $D = 1/3$ for a sphere and $D = 1$ for a plate perpendicular to the field).

Measurements of the magnetization yield the relation $M = M(H_0, T)$. From equations (13.3 and 2) the magnetic equation of state, $M = M(B, T)$, can be

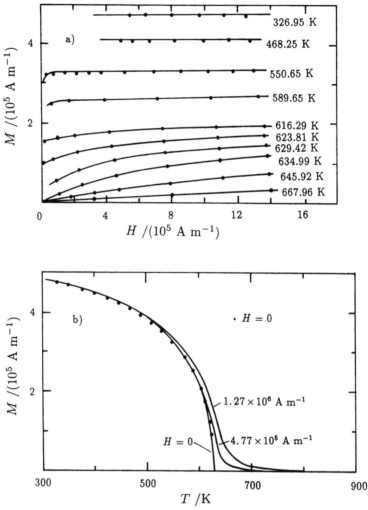

Figure 13.3: Magnetic equation of state of nickel. a) $M(H, T_i)$ isotherms; b) Isofield lines $M(H_i, T)$. The field H inside the sample is equal to the external field corrected for the demagnetizing field (13.3). The curves $M(B, T_i)$ and $M(B_i, T)$, obtained from $B = \mu_0(H + M)$, have the same shape. The Curie temperature is $T_C = 630.75$ K [P. Weiss and R. Forrer, *Ann. Phys.* (Paris) **5**, 153 (1926)].

obtained. Note that the equation of state cannot account for hysteresis phenomena that occur in ordered phases with small fields (§13.6.3).

13.2   PROPERTIES OF FERROMAGNETIC MATERIALS

13.2.1   Magnetic Equation of State

The experimentally observed magnetic equation of state for a ferromagnetic substance is plotted in Figure 13.3. It displays a characteristic critical temperature T_C,

Table 13.1: Physical constants of ferromagnetic substances. X: CuK_2 Cl_4, $2H_2O$; M: molar mass$/10^{-3}$ kg mol^{-1}; ρ: density$/10^3$ kg m^{-3}; T_C: Curie temperature/K; θ: Weiss temperature /K; C_M: Curie constant$/(4\pi\ 10^{-6}$ m^3 K mol$^{-1})$; n_{eff}: effective ferromagnetic number of magnetons; α, β, γ, δ: critical exponents.

	Fe	Ni	Gd	CrBr₃	EuO	GdCl₃	X
M	55.85	58.71	157.26	291.76	167.96	263.27	319.57
ρ	7.86	8.90	7.95	7.20	8.21	4.52	1.16
T_C	1041.5[a]	631.4[b]	293[c]	37[c]	69.4[b]	2.2[c]	0.88[d]
θ	1093[e]	650[e]	302.5[e]				
C_M	1.26[e]	0.32[e]	7.8[e]		7.9[f]		
n_{eff}	2.23[c]	0.606[c]	7.02[c]	1.96[c]	6.9[f]	5.74[c]	1[d]
α	−0.140[a]	−0.091[f]	−0.20[g]		−0.045[b]		~ −0.1[d]
β	0.34[h]	0.390[i]	0.399[g]	0.368[j]	0.368[b]		
γ	1.33[h]	1.315[i]	1.3[h]	1.22[j]	1.31[b]		1.36[d]
δ		4.22[j]	4.0[d]	4.3[j]	4.46[k]		

[a] G. Ahlers et al., *Phys. Rev. B* **12**, 1938 (1975).
[b] M. Barmatz et al., *Phys. Rev. B* **12**, 1947 (1975).
[c] N.W. Ashcroft and N.D. Mermin, *op. cit.*, p. 697.
[d] L.P. Kadanoff et al., *Rev. Mod. Phys.* **39**, 395 (1967).
[e] A.H. Morrish, *op. cit.*, p. 270.
[f] B.T. Matthias et al., *Phys. Rev. Lett.* **7**, 160 (1961).
[g] A.R. Chowdhury et al., *Phys. Rev. B* **33**, 6231 (1986).
[h] C. Kittel, *op. cit.*, p. 461.
[i] N. Stüsser et al., *Phys. Rev. B* **33**, 6423 (1986).
[j] H.E. Stanley, *op. cit.*
[k] N. Menyuk et al., *Phys. Rev. B* **3**, 1689 (1971).lp

called the Curie temperature, that separates the paramagnetic phase $(T > T_C)$, in which the magnetization is zero in zero field, from the ferromagnetic phase $(T < T_C)$, in which spontaneous magnetization can occur in zero field. The Curie temperatures of several ferromagnetic substances are listed in Table 13.1. In the paramagnetic phase, the low field magnetization is proportional to the field H (§4.2.3)

$$\mathbf{M} = [\chi]\mathbf{H} \qquad \text{or} \qquad M_i = \chi_{ij}H_j \ , \tag{13.4}$$

where the magnetic susceptibility is a tensor for single crystals and a scalar for polycrystalline solids. Typical temperature variations of the susceptibility are shown in Figure 13.4. At high temperatures $(T \gg T_C)$, the susceptibility obeys the Curie-

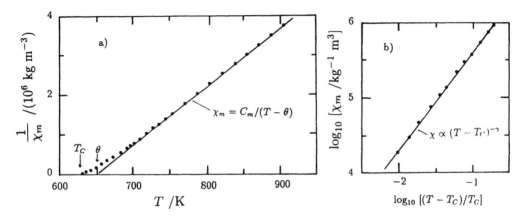

Figure 13.4: Specific magnetic susceptibility per unit mass $\chi_m = \chi/\rho$ of the paramagnetic phase of nickel as a function of temperature. a) The straight line shows the Curie-Weiss law (13.5) with $\theta = 650$ K and $C_m = C/\rho = 6.8 \times 10^{-5}$ m^3 K kg^{-1}. b) The straight line is the power law (13.6) in a double logarithmic plot with $T_C = 627.2$ K and $\gamma = 1.31$ [P. Weiss and R. Forrer, *Ann. Phys.* (Paris) **5**, 153 (1926)].

Weiss law

$$\chi = \frac{C}{T - \theta} \qquad \text{or} \qquad \frac{1}{\chi} = \frac{T - \theta}{C} \ , \tag{13.5}$$

where C and θ are positive constants called the Curie constant and the Weiss temperature respectively. In the neighbourhood of T_C, χ follows a power law

$$\chi \propto \left(\frac{T - T_C}{T_C}\right)^{-\gamma} \tag{13.6}$$

where γ, called the critical exponent, takes a value close to 1.3 (Table 13.1).

In the ferromagnetic phase at temperatures $T \ll T_C$, the molar magnetic moment M_M depends weakly on the field strength. It tends to a saturation value M_{Ms} at absolute zero, corresponding to the maximum orientation of the elementary moments μ . The maximum value of μ_z can then be determined (Table 13.1) from the relation

$$M_{Ms} = \mathcal{N} \left(\mu_z\right)_{max} = \mathcal{N}\mu_B n_{\text{eff}} \tag{13.7}$$

where n_{eff} is the effective ferromagnetic number of Bohr magnetons, not to be confused with the effective paramagnetic number of magnetons p (4.53). Close to absolute zero (Fig. 13.5a), the reduced magnetization takes the form

$$R \equiv \frac{M_M}{M_{Ms}} = 1 - A_1 T^{3/2} - A_2 T^{5/2} \ . \tag{13.8}$$

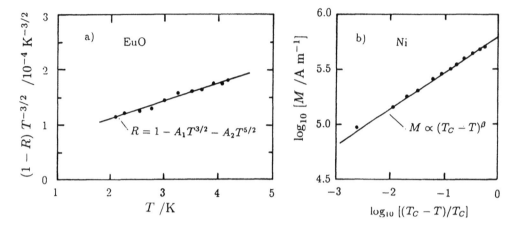

Figure 13.5: Reduced spontaneous magnetization R in the ferromagnetic phase as a function of temperature. a) Magnetization of europium oxide EuO at low temperatures. The straight line is equation (13.8) with $A_1 = 0.45 \times 10^{-4}$ /K$^{-3/2}$ and $A_2 = 0.33 \times 10^{-4}$ /K$^{-5/2}$ [E.L. Boyd, *Phys. Rev.* **145**, 174 (1966)]. b) Magnetization of nickel near the critical point. The straight line in the double logarithmic plot shows the power law (13.9) with $T_C = 627.2$ K and $\beta = 0.33$ [P. Weiss and R. Forrer, *Ann. Phys.* (Paris) **5**, 153 (1926)].

Finally, near the Curie temperature, the reduced spontaneous magnetization decreases strongly (Fig. 13.5b), following a power law

$$R \propto \left(\frac{T_C - T}{T_C} \right)^{\beta} \quad , \tag{13.9}$$

where the value of the critical exponent β is about 0.35 (Table 13.1).

At the critical temperature, the reduced magnetization of the substance varies with the magnetic field (Fig. 13.6) as

$$R \propto H^{1/\delta} \quad , \tag{13.10}$$

where the critical exponent δ is approximately equal to 4 (Table 13.1).

13.2.2   Heat Capacity

The heat capacity at constant volume C_V of a substance is obtained by measuring the heat capacity at constant pressure C_P in zero field and applying equation (8.48). Depending on the material, it is the sum of several contributions:

• the lattice heat capacity, investigated in Chapter 8, which varies as T^3 at low temperatures;

• an electronic heat capacity, linear in T, which occurs in metals (Chap. 10);

• a Schottky heat capacity coming from the splitting of the electronic ground state

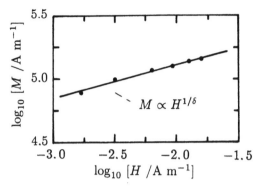

Figure 13.6: Magnetization of nickel as a function of H at the critical temperature $T_C = 627.2$ K. The straight line in the logarithmic plot is the power law (13.10) with $\delta = 4$ [P. Weiss and R. Forrer, *Ann. Phys.* (Paris) **5**, 153 (1926)].

of the magnetic ion, and which contributes a term in A/T^2 (§3.4.3);

- the magnetic heat capacity $C_V^{(m)}$ due to interactions between magnetic ions.

The magnetic heat capacity (Fig. 13.7), found by subtraction, has a characteristic maximum at the critical point. Near this point, it is described by the continuous finite expression

$$C_V^{(m)} = K - A \left| \frac{T - T_C}{T_C} \right|^{-\alpha} , \qquad (13.11)$$

where the value of the critical exponent α is negative and of the order of -0.1 (Table 13.1). At low temperatures (Fig. 13.8) the heat capacity takes the form

$$\frac{c_V^{(m)}}{R} = B_1 T^{3/2} + B_2 T^{5/2} . \qquad (13.12)$$

13.3 INTERACTIONS BETWEEN MAGNETIC IONS

Interactions between magnetic ions are of two kinds: magnetic interactions between dipoles, and electrostatic exchange interactions between the electrons of two neighbouring ions resulting from the symmetrization rule. The latter are called exchange interactions.

The magnetic interaction energy between two dipoles $\boldsymbol{\mu}_i$ and $\boldsymbol{\mu}_j$ associated with the magnetic ions i and j in the crystal is given by

$$\epsilon_{ij} = \frac{\mu_0}{4\pi} \left[\frac{\boldsymbol{\mu}_i \cdot \boldsymbol{\mu}_j}{r_{ij}^3} - \frac{3(\boldsymbol{\mu}_i \cdot \mathbf{r}_{ij})(\boldsymbol{\mu}_j \cdot \mathbf{r}_{ij})}{r_{ij}^5} \right] \sim \frac{\mu_0}{4\pi} \frac{\mu_i \mu_j}{r_{ij}^3} . \qquad (13.13)$$

This energy can be evaluated by assuming r_{ij} to be of the same order of magnitude as the distance between nearest neighbour magnetic ions (for example 2 Å), and that μ_i and μ_j are of the order of μ_B. We thus find $\epsilon_{ij} = 1.1 \times 10^{-24}$ J $= 6.7 \times 10^{-6}$ eV. The magnetic interaction produces significant effects only at temperatures of the order of

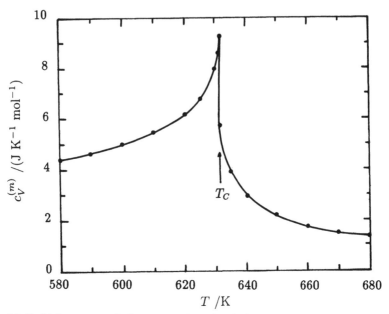

Figure 13.7: Molar magnetic heat capacity of nickel near the critical point $T_C = 631.5$ K [D.L. Connelly et al., *Phys. Rev. B* **3**, 924 (1971)].

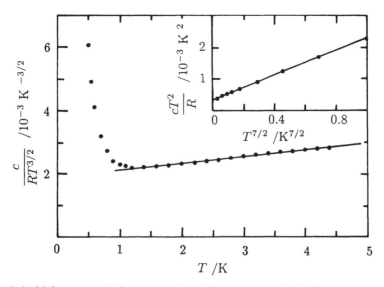

Figure 13.8: Molar magnetic heat capacity c of europium oxide EuO at low temperatures. The straight line is equation (13.12) with $B_1 = 1.92 \times 10^{-3}$ /K$^{-3/2}$ and $B_2 = 0.23 \times 10^{-3}$ /K$^{-5/2}$. The upturn at the lowest temperatures is due to a Schottky anomaly in A/T^2, as shown in the inset ($A/R = 0.38$ /K^2) [O.W. Dietrich et al., *Phys. Rev. B* **12**, 2844 (1975)].

$\epsilon_{ij}/k \simeq 0.1$ K. In substances that exhibit ferromagnetism in the temperature range above 1 K, it is much weaker than the exchange interaction. However, because it decreases as $1/r^3$ while the exchange interaction decreases exponentially, it gives rise to ferromagnetism at very low temperatures in highly dilute magnetic substances, such as chrome potassium alum $CrK(SO_4)_2$, $12H_2O$ ($T_C = 0.004$ K).

The exchange interaction between electrons of neighbouring ions is due to Coulomb forces and depends on the spin state of the electrons by virtue of the symmetrization rule (W. Heisenberg, 1926). Consider two electrons belonging to two neighbouring ions, whose orbital wave functions are respectively $\phi_1(r)$ and $\phi_2(r)$. Since the total wave function ψ must be antisymmetric under exchange of the electrons, either the spin wave function χ or the orbital wave function ϕ must be symmetric, while the other is antisymmetric. The total wave function with symmetric spin corresponds to one of the three states

$$| + + >, \qquad \frac{1}{\sqrt{2}}(| + - > + | - + >), \qquad | - - >$$

and to a total spin $S = 1$ (triplet state). The antisymmetric wave function

$$\frac{1}{\sqrt{2}}(| + - > - | - + >) \quad ,$$

has a total spin $S = 0$ (singlet state). The total orbital wave function is one of the two functions

$$\phi = \frac{1}{\sqrt{2}}[\phi_1(r_1) \phi_2(r_2) \mp \phi_1(r_2) \phi_2(r_1)] \quad ,$$

where the sign $-$ is associated with the triplet states ($S = 1$) and the sign $+$ with the singlet state ($S = 0$).

The Coulomb interaction potential between the electrons $V = e^2/r_{12}$ gives rise to an interaction energy

$$\epsilon = < \phi\,|V|\,\phi > = < \phi\,|\frac{e^2}{r_{12}}|\,\phi > \quad ,$$

which depends on the spins only through the symmetry of ϕ. On expanding as a series, we find

$$\epsilon = K \mp J = K + [1 - S(S+1)]\,J \tag{13.14}$$

where K and J are the integrals

$$K = \int \frac{e^2}{r_{12}}\,|\phi_1(r_1)|^2\,|\phi_2(r_2)|^2\,d^3r_1\,d^3r_2$$

$$J = \int \frac{e^2}{r_{12}}\,\phi_1(r_1)\phi_1^*(r_2)\phi_2(r_2)\phi_2^*(r_1)\,d^3r_1\,d^3r_2 \quad . \tag{13.15}$$

Since $\hat{S}^2 = S(S+1)$ and also

$$\hat{S}^2 = (\hat{s}_1 + \hat{s}_2)^2 = \hat{s}_1^2 + \hat{s}_2^2 + 2\hat{s}_1 \cdot \hat{s}_2 = \frac{3}{2} + 2\hat{s}_1 \cdot \hat{s}_2 \quad ,$$

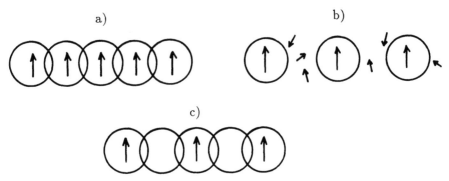

Figure 13.9: Diagrammatic illustration of exchange interactions; a) direct, b) indirect, through conduction electrons, and c) superexchange, through electrons of non magnetic ions.

the interaction energy (13.14) between the two electrons becomes

$$\epsilon = K - \frac{J}{2} - 2J\mathbf{s_1}.\mathbf{s_2} \quad . \tag{13.16}$$

This energy, from which the term $K - J/2$ is generally omitted, is called the exchange energy, and J is called the exchange integral, since expression (13.15) involves an exchange of the co-ordinates of the two electrons. We emphasize that the interaction is purely of Coulomb type, even though the electron spin enters the expression for the energy through the symmetrization rule. Exchange energy is a quantum mechanical notion that has no classical equivalent.

As magnetic ions generally contain several electrons in incomplete sub-layers, equation (13.16) is generalized in the form

$$\epsilon = -2J \, \mathbf{S_1}.\mathbf{S_2} \quad . \tag{13.17}$$

The exchange integral J has the dimensions of energy. It depends strongly on the separation between the ions. At short distances it is negative, then it turns positive. At large distances, it tends exponentially to zero. In metallic manganese, the separation between ions is $d = 2.2$ Å, and the exchange integral is thus negative, while for the ions in metallic cobalt and nickel ($d = 2.5$ Å) the exchange integral is positive. Note that positive values of J favour a parallel arrangement of the spins of the ions, which, as we shall show, leads to ferromagnetism. Negative values of J favour an antiparallel arrangement and lead to the phenomenon of antiferromagnetism.

Theoretical calculations, however, show that the model in which electrons are localized in the ions (Fig. 13.9a) is not in principle sufficient to explain the order of magnitude of J. In metals, the dominant exchange interaction occurs through conduction electrons (RKKY theory) (Fig. 13.9b). Similarly, in ionic compounds, magnetic ions exhibit a so-called superexchange interaction, which is transmitted by the electrons of the non magnetic ions (such as O^{--}, Cl^-, F^-) that separate them (Fig. 13.9c). Equation (13.17), however, provides a good phenomenological description which turns out to be very useful in interpreting the magnetic properties of materials.

13.4 MODELS OF FERROMAGNETISM

13.4.1 Heisenberg Models

On the basis of equation (13.17) for the exchange energy between two ions, W. Heisenberg (1928) proposed the following Hamiltonian for a system of interacting spins (§12.1.2)

$$
\begin{aligned}
\hat{H} &= g\mu_B \mathbf{B} \cdot \sum_i \hat{\mathbf{S}}_i - 2 \sum_{<ij>} J_{ij} \hat{\mathbf{S}}_i \cdot \hat{\mathbf{S}}_j \\
&= g\mu_B \mathbf{B} \cdot \sum_i \hat{\mathbf{S}}_i - \sum_{i,\, j \neq i} J_{ij} \hat{\mathbf{S}}_i \cdot \hat{\mathbf{S}}_j \; . \tag{13.18}
\end{aligned}
$$

The first term in this Hamiltonian is the coupling between the magnetic field and the ions, and takes the form

$$
-\mathcal{M} \cdot \mathbf{B} \qquad \text{with} \qquad \mathcal{M} = -g\mu_B \sum_i \mathbf{S}_i \; . \tag{13.19}
$$

The second term describes the exchange interaction between the ions, in which the exchange integral $J_{ij} = J_{ji}$ depends only on the distance between the ions i and j. It is the sum over all ion pairs $< ij >$ in the term (13.17), which is also half the sum over all pairs i and j (the factor $1/2$ corrects for double counting in the same ion pair, ij and ji).

Note that the Hamiltonian $\hat{H}$ is an operator that can be expressed as a function of the N vector operators $\hat{\mathbf{S}}_i$ defined in (12.8).

In the framework of canonical ensembles, the canonical partition function Q_N of the spin system is given by the Hamiltonian as follows

$$
Q_N = \text{Tr } e^{-\beta \hat{H}} = \sum_r < r \, | e^{-\beta \hat{H}} | \, r > \; , \tag{13.20}
$$

where $| r > \equiv | m_1, \ldots m_N >$ is one of the $(2S+1)^N$ possible spin states, and where the sum over r means

$$
\sum_r \quad \to \quad \sum_{m_1 = -S}^{S} \cdots \sum_{m_N = -S}^{S} \; . \tag{13.21}
$$

With a knowledge of this partition function we can then derive the magnetic properties of the substance, as we saw earlier in paragraph 12.5.2. However, an exact determination of the partition function with the Hamiltonian (13.18) has not so far proved possible. Only approximate methods and simplified models have been found to be of use. Comparisons of these results with experiments confirm the validity of the Heisenberg Hamiltonian.

At this point we introduce notations that will be useful later on. As in (4.47) the saturation magnetic moment is defined by

$$
\mathcal{M}_S = N S g \mu_B \tag{13.22a}
$$

and its molar value by

$$M_{MS} = \mathcal{N}Sg\mu_B = 5.585 \; Sg \; \text{SI} \; . \tag{13.22b}$$

In addition, we denote by w the sum of the exchange integrals of one ion i with all the others

$$w = \sum_{j \neq i} J_{ij} \; , \tag{13.23a}$$

this quantity being independent of the position of the ion i. If we consider the interactions of an ion only with its z equidistant nearest neighbours, we have

$$w = zJ \; . \tag{13.23b}$$

Lastly, we define a temperature θ such that

$$k\theta = \frac{2}{3} S(S+1)w \; . \tag{13.24}$$

We shall see through several examples that in the Heisenberg model this temperature is the Weiss temperature.

13.4.2 Molecular Field Method

Molecular field

One of the difficulties of working with the Hamiltonian (13.18) resides in the fact that it is not linear in S_i. One way of linearizing it is to make the substitution

$$\mathbf{S}_i.\mathbf{S}_j \; \rightarrow \; \mathbf{S}_i.\bar{\mathbf{S}} + \bar{\mathbf{S}}.\mathbf{S}_j - \bar{\mathbf{S}}.\bar{\mathbf{S}} \tag{13.25}$$

where $\bar{\mathbf{S}}$ is the mean canonical value of $\mathbf{S}_i$ (and of $\mathbf{S}_j$) and is such that, from (13.19)

$$|\bar{\mathbf{S}}| = \frac{M}{Ng\mu_B} = S\frac{M}{M_S} = SR \; . \tag{13.26}$$

It has been shown that linearization becomes increasingly valid if the number z of neighbours of an ion is large. The simpler transformation $\mathbf{S}_i \cdot \mathbf{S}_j \; \rightarrow \; \mathbf{S}_i.\bar{\mathbf{S}}$ is also often used. It carries the same physical meaning but in certain cases it can lead to double counting.

With the transformation (13.25), the Heisenberg Hamiltonian can be written as

$$\hat{H} = g\mu_B\mathbf{B} \cdot \sum_i \hat{\mathbf{S}}_i - 2\bar{\mathbf{S}} \cdot \sum_{i,\, j \neq i} J_{ij}\hat{\mathbf{S}}_i + \left(\bar{\mathbf{S}}\right)^2 \sum_{i,\, j \neq i} J_{ij} \; ,$$

where the dummy indices i and j can be interchanged. With equation (13.23a), the Hamiltonian becomes

$$\hat{H} = g\mu_B \left(\mathbf{B} - \frac{2w}{g\mu_B}\bar{\mathbf{S}}\right) \sum_i \hat{\mathbf{S}}_i + Nw \left(\bar{\mathbf{S}}\right)^2 \; . \tag{13.27}$$

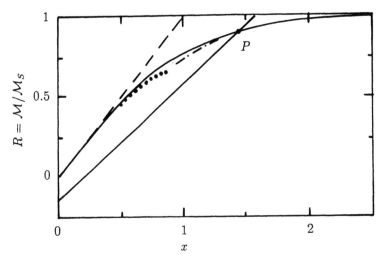

Figure 13.10: Graphical solution of equation (13.34) for $S = 1/2$. The continuous curve is the Brillouin function $B_{1/2}(x) = \tanh x$, and the full straight line is equation (13.35). These intersect at P, the ordinate of which gives the value of $R = \mathcal{M}/\mathcal{M}_S$. The dashed straight line is the tangent at the origin of the Brillouin function, the dotted curve is the approximation using $x - x^3/3$ for $\tanh x$, and the dashed-dotted line is the approximation $1 - \exp(-2x)$. In zero magnetic field, the straight line (13.35) passes through the origin. Brillouin curves for various values of S are shown in Figure (4.8).

Apart from a constant term, this Hamiltonian corresponds to a system of non-interacting spins in an effective field $\mathbf{B}_{\text{eff}} = \mathbf{B} + \mathbf{B}_W$, which is the sum of the magnetic flux density $\mathbf{B}$ and a fictitious field. The latter is, from (13.26)

$$B_W = \frac{2w}{g\mu_B} \, |\bar{\mathbf{S}}| = \frac{2wS}{g\mu_B} \, R \ . \qquad (13.28a)$$

A similar fictitious field, proportional to the reduced magnetization R, was introduced in 1907 by P. Weiss under the name of molecular field. It was used to explain numerous properties of magnetic materials. We shall introduce the quantity

$$q = \frac{2wS}{g\mu_B} \ , \qquad (13.28b)$$

which is characteristic of the substance and has the same dimensions as a magnetic flux density.

The Hamiltonian that we end up with is given by

$$\hat{H} = g\mu_B B_{\text{eff}} \sum_i \hat{S}_{zi} + NS^2 wR^2 \qquad \text{with} \qquad B_{\text{eff}} = B + qR \ , \qquad (13.29)$$

where we have chosen the z axis parallel to B.

Magnetic moment

When the term independent of the $\hat{S}_{zi}$ is omitted, the corresponding canonical partition function is

$$Q_N = \text{Tr}\left[\exp\left(-\frac{x}{S}\sum_i \hat{S}_{zi}\right)\right], \quad \text{with} \quad x = Sg\frac{\mu_B B_{\text{eff}}}{kT} = \frac{M_S B_{\text{eff}}}{NkT} \quad (13.30)$$

As the Hamiltonian is diagonal in the basis of the states $|\, r >$ (12.4), the partition function can be written

$$\begin{aligned}
Q_N &= \sum_{m_1=-S}^{S} \cdots \sum_{m_N=-S}^{S} \exp(-xm_1/S) \,\cdots\, \exp(-xm_N/S) \\
&= \left[\sum_{m=-S}^{S} \exp(-xm/S)\right]^N = Z^N \ , \quad (13.31)
\end{aligned}$$

where Z is the single particle partition function (4.36,37) encountered in connection with the Brillouin model of paramagnetism. The free energy is then given by

$$F = -kT\ln Q_N = -NkT\ln Z \ ,$$

or, on reinserting the constant term in $\hat{H}$ and using (4.37) for Z,

$$F = -NkT\ln\frac{\sinh\frac{2S+1}{2S}x}{\sinh\frac{1}{2S}x} + NS^2 wR^2 \ , \quad (13.32)$$

with

$$x = \frac{M_S(B+qR)}{NkT} \ . \quad (13.33)$$

In this equation, in addition to the variables N, T and B, there is the reduced magnetization $R = M/M_S$, which can be found with the help of the relation $M = -(\partial F/\partial B)$. Thus

$$M = NkT\frac{\partial\ln Z}{\partial B} = M_S\frac{d\ln Z}{dx} \ ,$$

which, on introducing the Brillouin function (4.41), yields

$$R = \frac{M}{M_S} = B_S(x) = B_S\left[\frac{M_S(B+qR)}{NkT}\right] \ . \quad (13.34)$$

This equation defines R, and hence M, implicitly as a function of B and T. Since it is not possible to find an analytical expression for M, we are obliged to solve graphically, or else use approximate methods.

To solve equation (13.34) graphically, we observe that R can be expressed as a function of x, either through (13.34) or by the relation

$$R = \frac{NkT}{qM_S}x - \frac{B}{q} \quad (13.35)$$

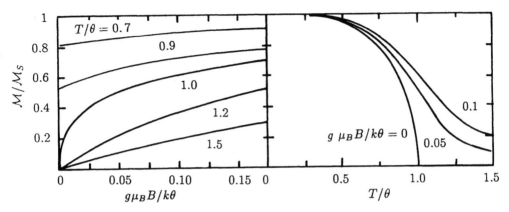

Figure 13.11: $\mathcal{M}(B, T_i)$ isotherms and $\mathcal{M}(B_i, T)$ isofield curves found by the molecular field method for $S = 1/2$. The temperature θ is given by (13.24). Spontaneous magnetization arises for $T < \theta$, and the Curie temperature is therefore $T_C = \theta$ in this model.

which comes from (13.33). The ordinate of the intersection between $R = B_S(x)$ and the straight line (13.35) yields the value of $R = \mathcal{M}/\mathcal{M}_S$ for a given B and T (Fig. 13.10). The resulting isotherms $\mathcal{M}(B, T_i)$ and isofield curves $\mathcal{M}(T, B_i)$ are shown in Figure 13.11. Clearly, there is qualitative agreement with the experimental curves (Fig. 13.3).

The solution in zero field ($B = 0$) requires a separate discussion. In this case, the straight line (13.35) goes through the origin. At sufficiently high temperatures, the slope of this straight line, which is proportional to the temperature, is greater than the initial slope of the Brillouin function, and the solution of equation (13.34) is $\mathcal{M} = 0$. At sufficiently low temperatures, when the slope becomes smaller than the initial slope of the Brillouin function, a new solution appears, $\mathcal{M} \neq 0$, corresponding to spontaneous magnetization (the state $\mathcal{M} = 0$ is unstable). The transition from the paramagnetic to the ferromagnetic phase occurs at a temperature T_C at which the straight line and the Brillouin function have the same slope at the origin, or, bearing in mind (4.45),

$$\frac{NkT_C}{q\mathcal{M}_S} = \frac{S+1}{3S} \quad .$$

The molecular field model yields a Curie temperature, which, from (13.22 and 28), is equal to

$$T_C = \frac{(S+1)}{3S} \frac{q\mathcal{M}_S}{Nk} = \frac{2}{3} \frac{S(S+1)w}{k} \equiv \theta \quad . \tag{13.36}$$

In the paramagnetic phase in a weak field, the magnetic moment $\mathcal{M}$ is much smaller than the saturation moment $\mathcal{M}_S$ ($R \ll 1$) and in (13.34) the Brillouin function can be replaced by its tangent at the origin, $(S + 1)x/3S$, found from (4.45). This gives

$$\frac{\mathcal{M}}{\mathcal{M}_S} = \frac{S+1}{3S} \frac{\mathcal{M}_S B}{NkT} + \frac{\theta}{T} \frac{\mathcal{M}}{\mathcal{M}_S} \quad ,$$

and hence

$$M = \frac{S+1}{3S}\frac{\mathcal{M}_S^2}{Nk}\frac{B}{T-\theta} = N\frac{S(S+1)g^2\mu_B^2}{3k}\frac{B}{T-\theta} \ . \tag{13.37}$$

This relation is a Curie-Weiss law in which the Weiss temperature is equal to θ, defined in (13.24), and the Curie constant has the same form as in equation (4.48). It agrees with the experimental results for $T \gg T_C$. However, it yields the same values for the Weiss temperature θ and the Curie temperature T_C, which is contrary to experimental observations ($T_C < \theta$). In addition, equation (13.37), when extended to the whole temperature range, yields a critical exponent γ (13.6) equal to 1, in disagreement with the experimental data.

In the ferromagnetic phase, solving graphically for various values of S yields the temperature dependence of the spontaneous magnetization that is shown in Figure 13.12. Agreement with experiment is satisfactory, but only qualitatively. In particular for $\mathcal{M}/\mathcal{M}_S < 0.5$ (Fig. 13.10), the magnetization can be found from (13.34) by replacing the Brillouin function by its series expansion (4.45). Equation (13.34) then becomes

$$\frac{\mathcal{M}}{\mathcal{M}_S} = \frac{\theta}{T}\frac{\mathcal{M}}{\mathcal{M}_S} - \frac{(2S+1)^4 - 1}{45(2S)^4}\left(\frac{2S^2 w}{kT}\frac{\mathcal{M}}{\mathcal{M}_S}\right)^3 \ .$$

On simplification, this equation leads to $\mathcal{M} \propto T(\theta - T)^{1/2}$. Near the Curie point $T_C = \theta$, therefore, the magnetic moment varies as

$$\mathcal{M} \propto (\theta - T)^\beta \qquad \text{with} \qquad \beta = 1/2 \ . \tag{13.38}$$

This value for the critical exponent β is in disagreement with the experimental values $\beta \simeq 0.35$.

Likewise, for $\mathcal{M}/\mathcal{M}_S > 0.8$ (Fig. 13.10), the Brillouin function can be replaced by its limit (4.45) as x tends to infinity. It then follows that

$$\frac{\mathcal{M}}{\mathcal{M}_S} = 1 - \frac{1}{S}\exp\left(-2S\frac{w}{kT}\frac{\mathcal{M}}{\mathcal{M}_S}\right) \simeq 1 - \frac{1}{S}\exp\left(-2S\frac{w}{kT}\right) \ .$$

In this expression the magnetic moment at absolute zero tends to its saturation value $\mathcal{M}_S$ much faster than the experimentally observed power law (13.8) (Fig. 13.12).

Lastly, for $T = T_C$, a simple but tedious calculation shows that the magnetic moment varies as

$$\mathcal{M} \propto B^{1/\delta} \qquad \text{with} \qquad \delta = 3 \ . \tag{13.39}$$

This result is roughly in agreement with experiment ($\delta \simeq 4$).

Heat capacity

The internal energy can be found either from $U = -\partial \ln Q_N/\partial\beta$, or by taking the mean value of the Hamiltonian. The latter method is more direct, and from (13.26

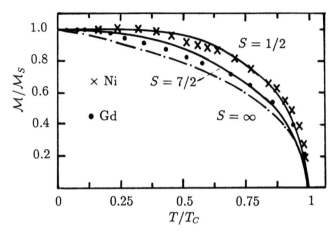

Figure 13.12: Spontaneous magnetization of nickel and gadolinium as a function of temperature. The full lines are given by the molecular field model with $S = 1/2$ and $S = 7/2$ corresponding to the effective spins of the Ni^{2+} and Gd^{3+} ions. The dashed curve is the limit $S = \infty$ in this model [P. Weiss and R. Forrer, *Ann. Phys.* (Paris) **5**, 153 (1926) and **12**, 1279 (1929); H.E. Nigh et al., *Phys. Rev.* **132**, 1092 (1963)].

and 27) gives

$$U = g\mu_B \left(\mathbf{B} - \frac{2w}{g\mu_B} \mathbf{\bar{S}} \right) N\mathbf{\bar{S}} + Nw(\mathbf{\bar{S}})^2$$

$$= -\mathcal{M} \cdot \mathbf{B} - NS^2 w \left(\frac{\mathcal{M}}{\mathcal{M}_S} \right)^2 . \tag{13.40}$$

The first term in U is the energy of the substance in the external field, and the second term is the exchange energy of the ions.

The magnetic heat capacity in zero field is then

$$C_V^{(m)} = \frac{\partial U}{\partial T} = -\frac{2NS^2 w}{\mathcal{M}_S^2} \mathcal{M} \frac{d\mathcal{M}}{dT} . \tag{13.41}$$

In the paramagnetic phase, since the magnetic moment is zero, this heat capacity vanishes. In the ferromagnetic phase, the magnetic moment is a decreasing function of the temperature and the heat capacity is positive. It is plotted in Figure 13.13. Agreement with experiment (Fig. 13.7) is qualitative. In contradiction with the measurements, the model predicts a discontinuity at the Curie temperature equal to

$$\Delta c_V^{(m)} = -\frac{5S(S+1)}{2S(S+1)+1} R , \tag{13.42}$$

which, by definition, corresponds to a critical exponent $\alpha = 0$. Similarly, at low temperatures, the model predicts an exponential decrease instead of the power law (13.12) that is observed experimentally.

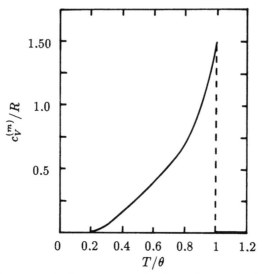

Figure 13.13: Molar magnetic heat capacity in the molecular field model for a material with spin $S = 1/2$.

Discussion and improvements

The molecular field method provides an explanation of ferromagnetic phenomena. In particular, at high temperatures it is exact, and it predicts the appearance of spontaneous magnetization, with the establishment of long range order below a critical temperature. It does not, however, give a good description of the behaviour near the critical point or at low temperatures.

Studies of this method in physical space of various dimensions d show that it yields the same results at the critical point as the Heisenberg model for $d > 4$. This is related to the fact that as the number of nearest neighbours increases, approximation (13.25) becomes more accurate. The deficiency of the molecular field method is that it neglects correlations between spins, the importance of which increases as the number of neighbours decreases. These correlations show up as short range order, even above the critical point where long range order has vanished. In the Bethe - Peierls - Weiss method, the interaction between neighbouring ions is treated exactly and that between more distant ions is handled by the molecular field method. This improves the results. In particular, they predict $T_C \neq \theta$ and a non-zero heat capacity above the Curie point. This method can be generalized by considering successive layers of neighbours.

Close to absolute zero, the discrepancy stems from the fact that the molecular field method is equivalent to taking the spins to be independent. When spin waves and magnons (§13.4.5) are introduced, agreement is re-established, just as the introduction of elastic waves and phonons gave agreement with the Einstein model.

13.4.3   Simplified Models

Introduction

The molecular field method is an approximation that starts from an exact Hamiltonian. It is possible to simplify the Hamiltonian and to employ exact methods, for example, by including interactions between nearest neighbours only. The dimension d of space can also be reduced (or increased). Lastly, it is possible to consider spins having a number of components D that is different from 3. If the spin has only one component $(D = 1)$ we get the Ising model; spins with two components $(D = 2)$ yield the XY model, and an infinite number of components $(D = \infty)$ gives the spherical model. These models have led to a broader understanding of ferromagnetism, and it has been possible to apply them to certain physical examples.

Among the models employed, the simplest is that of Ising $(D = 1)$ with spins $S = 1/2$ and interactions between nearest neighbours only. The Hamiltonian (13.18) for this system becomes

$$\hat{H} = g\mu_B B \sum_i \hat{S}_{zi} - 2J \sum_{<ij>} \hat{S}_{zi}\hat{S}_{zj} \; .$$

It is therefore diagonal in the basis of the states $|\; r >$ (13.20) and its eigenvalues are

$$
\begin{aligned}
E_r &= g\mu_B B \sum_i m_i - 2J \sum_{<ij>} m_i m_j \\
&= \frac{g\mu_B B}{2} \sum_i \alpha_i - \frac{J}{2} \sum_{<ij>} \alpha_i \alpha_j \qquad (13.43)
\end{aligned}
$$

where the $\alpha_i = 2m_i$ are quantum numbers that take the values $+1$ or -1. Note that the sum over $< ij >$ in (13.43) is restricted to neighbouring pairs of ions and therefore has $zN/2$ terms.

One-dimensional Ising model

A model with spin $S = 1/2$ having only one component $(D = 1)$ was first investigated by V. Lenz (1920) and solved by E. Ising (1925) in one-dimensional space $(d = 1)$ for interactions between nearest neighbour ions. In this model, where each ion has two neighbours $(z = 2)$, the energy of the state $|\; r >$ is

$$E_r = \frac{g\mu_B B}{2} \sum_{i=1}^{N} \alpha_i - \frac{J}{2} \sum_{i=1}^{N} \alpha_i \alpha_{i+1} \; . \qquad (13.44)$$

Note that we have added the term $\alpha_N \alpha_{N+1}$. By identifying α_{N+1} with α_1 we eliminate the boundary effects since it is assumed that the line of spins forms a closed circle.

The canonical partition function is then given by

$$Q_N = \sum_r e^{-\beta E_r} = \sum_{\alpha_1 = \pm 1} \cdots \sum_{\alpha_N = \pm 1} K(\alpha_1, \alpha_2)\, K(\alpha_2, \alpha_3) \; \cdots \; K(\alpha_N, \alpha_1)$$

where $K(\alpha, \alpha')$ is a set of four numbers defined by

$$K(\alpha, \alpha') = \exp\left(-\frac{\beta g \mu_B B}{2} \frac{\alpha + \alpha'}{2} + \frac{\beta J}{2} \alpha\alpha'\right) . \tag{13.45}$$

If K is taken to be a 2×2 matrix, the above expression for Q_N is the trace of the matrix K^N. As the trace of a matrix is the sum of its eigenvalues, we have

$$Q_N = \text{Tr } K^N = \lambda_1^N + \lambda_2^N = \lambda_1^N \left[1 + \left(\frac{\lambda_2}{\lambda_1}\right)^N\right] .$$

where λ_1 and $\lambda_2 < \lambda_1$ are the eigenvalues of the matrix K. In the thermodynamic limit $N \to \infty$, $(\lambda_2/\lambda_1)^N$ tends to zero and we have

$$Q_N = \lambda_1^N . \tag{13.46}$$

From (13.45), the matrix K is given by

$$K = \begin{bmatrix} \exp(-x + a) & \exp(-a) \\ \exp(-a) & \exp(x + a) \end{bmatrix} ,$$

with

$$x = \frac{g \mu_B B}{kT} \qquad \text{and} \qquad a = \frac{J}{2kT} . \tag{13.47}$$

Since the characteristic equation is $\lambda^2 - 2\lambda e^a \cosh x + 2 \sinh 2a = 0$, the eigenvalues of K, which are solutions of this equation, are

$$\begin{aligned} \lambda_{1,2} &= e^a \cosh x \pm \left[e^{2a} \cosh^2 x - 2 \sinh 2a\right]^{1/2} \\ &= e^a \left[\cosh x \pm \left(\sinh^2 x + e^{-4a}\right)^{1/2}\right] . \end{aligned} \tag{13.48}$$

The free energy is then given by

$$F = -kT \ln Q_N = -NkT \ln \lambda_1 ,$$

and the magnetic moment of the substance is

$$\begin{aligned} \mathcal{M} &= -\frac{\partial F}{\partial B} = \frac{NkT}{\lambda_1} \frac{d\lambda_1}{dB} = \frac{N g \mu_B B}{2\lambda_1} \frac{d\lambda_1}{dx} \\ &= \frac{N g \mu_B}{2} \frac{\sinh(g \mu_B B/2kT)}{\left[\sinh^2(g \mu_B B/2kT) + \exp(-2J/kT)\right]^{1/2}} . \end{aligned} \tag{13.49}$$

It can be seen that the magnetic moment is zero in zero magnetic field. This one-dimensional model therefore has no ferromagnetic phase. This is related to the fact that the number of near neighbours is small ($z = 2$). In weak fields, the magnetization has the form

$$\mathcal{M} = N \frac{g^2 \mu_B^2}{4k} \frac{e^{J/kT}}{T} B , \tag{13.50}$$

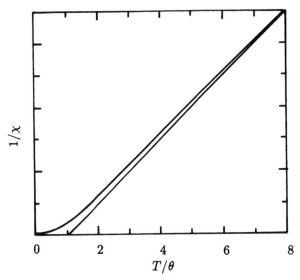

Figure 13.14: Magnetic susceptibility in the one-dimensional Ising model. The straight line is the Curie-Weiss law (13.5).

which corresponds to the susceptibility shown in Figure 13.14. Note that the susceptibility is given by

$$\frac{1}{\chi} \propto T e^{-J/kT} = T \left(1 - \frac{J}{kT} + \dots \right) = T - \frac{J}{k} + \dots \quad . \tag{13.51}$$

This expression shows that at high temperatures the substance obeys a Curie-Weiss law with $\theta = J/k$, in agreement with equation (13.24) with $z = 2$ and $S = 1/2$.

In zero field the internal energy is given by

$$U = -\frac{d \ln Q_N}{d\beta} = -N \frac{d}{d\beta} \ln \left(2 \cosh \frac{\beta J}{2}\right) = -N \frac{J}{2} \tanh \frac{J}{2kT} \quad . \tag{13.52}$$

Thus at absolute zero the energy of the substance is $U_0 = -NJ/2$. All the spins are therefore parallel, which means that the substance becomes ferromagnetic at absolute zero. Finally, we note that the molar heat capacity is given by

$$\frac{c_V}{R} = \left(\frac{J}{2kT}\right)^2 \frac{1}{\cosh^2 (J/2kT)} \quad , \tag{13.53}$$

which corresponds to a Schottky heat capacity (3.24), where $\epsilon = J$ is the energy required to reverse a spin.

Other models

Other models can be solved exactly. One example is the Ising model for a square lattice (two dimensions). This was solved by L. Onsager (1944) with interactions between nearest neighbours and for a spin $1/2$. It exhibits a phase transition at T_C where $\sinh \theta / 2T_C = 1$, i.e. $T_C = 0.567 \, \theta$. The critical exponents obtained are

$$\alpha = 0 \ , \qquad \beta = \frac{1}{8} \ , \qquad \gamma = \frac{7}{4} \qquad \text{and} \qquad \delta = 15 \ . \qquad (13.54)$$

The critical exponent $\alpha = 0$ corresponds to a heat capacity that varies as $\ln |T - T_C|$. Other types of lattice (triangular or hexagonal) yield the same critical exponents. Only the value of T_C / θ changes.

For the spherical model $(D = \infty)$ in three dimensions $(d = 3)$ the critical exponents are

$$\alpha = -1 \ , \qquad \beta = \frac{1}{2} \ , \qquad \gamma = 2 \qquad \text{and} \qquad \delta = 5 \ . \qquad (13.55)$$

The value of α corresponds to a heat capacity (13.11) that is continuous at the critical temperature but has a discontinuity in its slope.

13.4.4   High Temperature Expansion

We consider the simple Ising model of paragraph 13.4.3 whose energy is given by (13.43). The partition function has the form

$$Q_N = \sum_{\alpha_1 = \pm 1} \cdots \sum_{\alpha_N = \pm 1} \prod_k \exp x \alpha_k \prod_{<ij>} \exp \left(-a \alpha_i \alpha_j \right) \ ,$$

where x and a are given in (13.47). As the terms α_k and $\alpha_i \alpha_j$ are equal to $+1$ or -1, with the identity

$$\exp(\pm u) = \cosh u \pm \sinh u = \cosh u \, (1 \pm \tanh u) \ ,$$

we can transcribe Q_N into the form

$$Q_N \ = \ (\cosh x)^N \, (\cosh a)^{zN/2}$$
$$\sum_{\alpha_1, \, \ldots \, , \alpha_N} \prod_k (1 + \alpha_k \tanh x) \prod_{<ij>} (1 + \alpha_i \alpha_j \tanh a) \ . \qquad (13.56)$$

When T tends to infinity, x and a tend to zero and the sum appearing in (13.56) yields a series expansion for Q_N as a function of $1/T$ via $\tanh x$ and $\tanh a$. In this sum the term in $(\tanh x)^p \, (\tanh a)^q$ contains p factors of type α_k and q factors of type $\alpha_i \alpha_j$. It can be represented by a graph (Fig. 13.15) that comprises p crosses symbolizing the sites k, and q segments joining two neighbouring sites $< ij >$. Since we have the relationships

$$\sum_{\alpha = \pm 1} \alpha^{2r+1} = 0 \qquad \text{and} \qquad \sum_{\alpha = \pm 1} \alpha^{2r} = 2 \ ,$$

all graphs containing one of the α_i raised to an odd power are zero, while the others contribute 2^N. The coefficient of the term in $(\tanh x)^p \, (\tanh a)^q$ is therefore equal

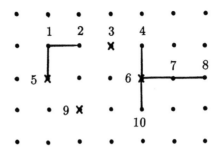

Figure 13.15: Graph corresponding to one term of the expansion (13.56) of Q_N on a square lattice. This term is in $(\tanh x)^4 \, (\tanh a)^6$ since it contains 4 crosses and 6 segments. Sites 1, 5, 6, 7 as well as the unnumbered sites contribute to this term by a factor 2, and the others (2, 3, 4, 8, 9, 10) each contribute by a factor zero. The graph is therefore zero.

to 2^N times the number of non-zero graphs containing p crosses and q segments. The two remaining terms of lowest order correspond to the graphs

$$p = 2 \, , \quad q = 1 \, , \qquad \text{and} \qquad p = 2 \, , \quad q = 2 \, .$$

The number of graphs is respectively $zN/2$ and $z(z-1)N/2$. The succeeding terms depend on the type of lattice.

The first terms in the expansion of Q_N are thus

$$Q_N = (2 \cosh x)^N (\cosh a)^{zN/2}$$
$$\left[1 + \frac{Nz}{2} \tanh^2 x \tanh a + \frac{Nz(z-1)}{2} \tanh^2 x \tanh^2 a + \ldots \right]$$

and the magnetic moment of the substance is given by

$$\mathcal{M} = kT \frac{\partial \ln Q_N}{\partial B} = g\mu_B \frac{\partial \ln Q_N}{\partial x}$$
$$= \frac{Ng\mu_B}{2} \tanh x \left[1 + z \frac{\tanh a}{\cosh^2 x} + z(z-1) \frac{\tanh^2 a}{\cosh^2 x} + \ldots \right] \, .$$

For weak fields ($x \to 0$) it can be seen that $\mathcal{M}$ is proportional to B, and the series expansion of the magnetic susceptibility is

$$\chi = \frac{C}{T} \left[1 + z \tanh \frac{J}{2kT} + z(z-1) \tanh^2 \frac{J}{2kT} + \ldots \right]$$
$$= \frac{C}{T} \left[1 + \frac{\theta}{T} + \frac{z-1}{z} \frac{\theta^2}{T^2} + \ldots \right] \, , \tag{13.57}$$

where C is the Curie constant (4.50) and $\theta = Jz/2k$ is the Weiss temperature (13.24). Note that the first two terms in this expansion (and also the third term when $z \to \infty$) are identical to those in the expansion of the Curie-Weiss relation (13.5).

By means of term-by-term identification of the high temperature expansion of quantities such as χ (13.57) with that of functions containing singular factors like $(1 - T_C/T)^{-\gamma}$, the critical temperature T_C and the various critical exponents can be determined. Thus equation (13.57) corresponds to an expansion to second order in θ/T of the function

$$\chi = \frac{C}{T} \left(\frac{T}{T - T_C} \right)^{\gamma} \qquad \text{with} \qquad \frac{T_C}{\theta} = \frac{z - 2}{z} \qquad \text{and} \qquad \gamma = \frac{z}{z - 2} \ .$$

This approximate result is in reasonable agreement with the values listed in Table (13.1) for $z = 12$ (face-centred cubic lattice).

The choice of the function for identification is not unique, but when the number of terms in the expansion increases, the values of T_C and of the critical exponents become independent of the choice.

Other expansions exist, e.g. low temperature expansions. The most powerful method is that based on group renormalization in quantum field theory, where the expansion is performed on the dimension d of physical space, for example as a function of $\epsilon = 4 - d$. This method shows that the critical exponents do not depend on the type of lattice. For an Ising model in three dimensions ($D = 1$, $d = 3$), it is found that

$$\alpha = 0.107 \quad ; \quad \beta = 0.327 \quad ; \quad \gamma = 1.239 \quad ; \quad \delta = 4.789 \ ,$$

where the error is in the last significant figure. For a three dimensional Heisenberg model ($D = 3$, $d = 3$)

$$\alpha = -0.115 \quad ; \quad \beta = 0.365 \quad ; \quad \gamma = 1.386 \quad ; \quad \delta = 4.795$$

[J.C. Le Guillou and J. Zinn-Justin, *Phys. Rev. B* **21**, 3976 (1980) and *J. Phys. (Paris)* **48**, 19 (1987)]. The last results are in good agreement with the experimental values listed in Table 13.1.

13.4.5   Spin Waves and Magnons

One-dimensional model

At absolute zero, a ferromagnetic substance is in its ground state, in which all the spins are aligned to the maximum degree. To investigate this substance close to absolute zero, the first excited states must be found. The properties of these states can be found by considering a simple one-dimensional model consisting of a row of N ions of spin $S = 1/2$ with spacing a which interact only with nearest neighbours. In zero field the Hamiltonian of this system is given by

$$\hat{H} = -2J \sum_{i=1}^{N} \hat{\mathbf{S}}_i \cdot \hat{\mathbf{S}}_{i+1}$$

$$= -2J \sum_{i=1}^{N} \left(\hat{S}_{x,i} \hat{S}_{x,i+1} + \hat{S}_{y,i} \hat{S}_{y,i+1} + \hat{S}_{z,i} \hat{S}_{z,i+1} \right)$$

$$= -J \sum_{i=1}^{N} \left(\hat{S}_{+,i}\hat{S}_{-,i+1} + \hat{S}_{-,i}\hat{S}_{+,i+1} + 2\hat{S}_{z,i}\hat{S}_{z,i+1} \right) \tag{13.58}$$

where the operators $\hat{S}$ are defined in (12.8).

In the ground state $| 0 > \equiv | + + \ldots + >$ of this system, the projection of all the spins is $+1/2$. In this state, the energy E_0, obtained by operating $\hat{H}$ on the ket $| 0 >$, is

$$E_0 = -N \frac{J}{2}, \tag{13.59}$$

since the S_+ operators give a zero result, and only the S_z operators contribute, each with a factor $1/2$. This result can be explained by attributing an interaction energy $-J/2$ to two parallel neighbouring spins.

Now let us consider an excited state of the row of spins, denoted $| n >$, in which the n-th spin is reversed, i.e.

$$| n > \equiv | + + \ldots + - + \ldots + > . \tag{13.60}$$

The N states of this type are not eigenstates of the Hamiltonian, since

$$\hat{H}| n > = (E_0 + 2J)| n > -J | n+1 > -J| n-1 > . \tag{13.61}$$

This result is obtained using the relations

$$\hat{S}_{+,i}| n > = \delta_{in} | 0 > , \qquad \hat{S}_{-,i} | 0 > = | i > ,$$

$$\hat{S}_{z,i}| n > = \frac{1}{2} | n > \quad (n \neq i) , \qquad \hat{S}_{z,n} | n > = -\frac{1}{2} | n > .$$

The state $| n >$ is therefore not stationary and the position of the reversed spin will propagate. By contrast, states of type

$$| k > = \frac{1}{\sqrt{N}} \sum_{n=1}^{N} e^{ikna} | n > \tag{13.62}$$

are stationary, since

$$\hat{H}| k > = \frac{1}{\sqrt{N}} \sum e^{ikna} \hat{H}| n >$$

$$= \frac{1}{\sqrt{N}} \left[(E_0 + 2J) \sum e^{ikna}| n > \right.$$

$$\left. -J \sum e^{ikna}| n+1 > -J \sum e^{ikna}| n-1 > \right]$$

i.e., on setting $n' = n + 1$ and $n' = n - 1$ respectively in the last two sums,

$$\hat{H} | k > = \left(E_0 + 2J - Je^{-ika} - Je^{ika} \right) | k > . \tag{13.63}$$

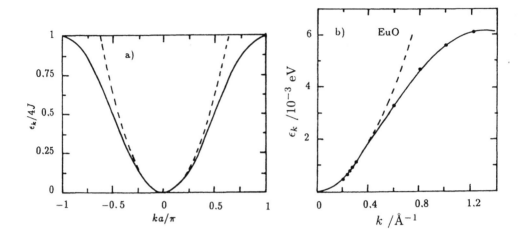

Figure 13.16: Dispersion curves for spin waves. The parabolic approximation is shown as a dashed line. a) Row of spins 1/2 with nearest neighbour interactions (Eq. (13.64)). b) Europium oxide powder EuO at 5.5 K. The experimental points were obtained by neutron diffraction. The coefficient in the parabolic relation is $D = 10.8 \times 10^{-3}$ eV Å^2. The curve is the result of a model with interactions between nearest neighbours ($J_1 = 5.2 \times 10^{-5}$ eV) and second nearest neighbours ($J_2 = 1.3 \times 10^{-5}$ eV) that takes account of the anisotropy [L. Passell et al., *Phys. Rev.* B **14**, 4897 (1976)].

If E_0 is taken as the origin, the energy of the state $\mid k >$ is then

$$\epsilon_k = J\left(2 - e^{-ika} - e^{ika}\right)$$
$$= 2J(1 - \cos ka) = 4J\sin^2\frac{ka}{2} \quad . \tag{13.64}$$

This dispersion relation is shown in Figure 13.16a.

A state $\mid k >$ can be visualized as follows. Since

$$< k \mid \hat{S}_{z,n} \mid k > = \frac{N-2}{2N} \qquad \text{and} \qquad < k \mid \hat{S}_{\pm,n} \mid k > = 0 \; ,$$

it can be seen that the projection of the spin n on the z axis is practically equal to 1/2, and the mean value of the other projections is zero. Moreover, since the scalar product of the projections of the spins n and n' in the xy plane is

$$< k \mid S_{x,n}\, S_{x,n'} + S_{y,n}\, S_{y,n'} \mid k > = < k \mid S_{+,n}\, S_{-,n'} + S_{-,n}\, S_{+,n'} \mid k >$$
$$= \frac{1}{N}\cos(n - n')ka \; ,$$

each spin may be considered as having a transverse projection of modulus $1/\sqrt{N}$ and shifted by an angle ka with respect to that of its neighbours (Fig. 13.17).

Note that, just as for elastic waves in solids (§8.4.1), the values of k can be restricted to the interval $-\pi/a$, $+\pi/a$, since the ket $\mid k + 2\pi/a >$ is the same as

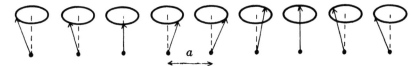

Figure 13.17: Diagram of the instantaneous orientation of the row of spins through which a spin wave with wave vector $k = \pi/4a$ is travelling. With advancing time, the spins precess with angular velocity $\omega = \epsilon_k/\hbar$. The phase difference between two neighbouring spins is equal to ka.

$| k >$. Furthermore, there are only N independent states $| k >$, which are linear functions of the N states $| n >$. Taking regularly spaced values

$$k = \frac{2\pi}{Na} p \qquad \text{with} \qquad -\frac{N}{2} \leq p \leq \frac{N}{2} \ ,$$

satisfies the condition $\exp iNka = 1$ corresponding to the periodic boundary condition of Born-von Kármán (or that of a closed loop).

Generalization to three dimensions

The above discussion can be generalized to three dimensions. The spin waves are then described by a wave vector $\mathbf{k}$ that can take N values uniformly distributed in the first Brillouin zone. The dispersion relation $\epsilon(\mathbf{k})$ of these waves is anisotropic. For small values of $\mathbf{k}$ it is quadratic and, in cubic crystals, its form is isotropic

$$\epsilon(\mathbf{k}) = Dk^2 \ , \tag{13.65}$$

where D is a constant.

Other excited states can be constructed by reversing two or more spins. They can, as a first approximation, be considered as being a superposition of non-interacting spin waves. It can then be shown that quantization of these waves gives rise to particles called magnons, which obey Bose-Einstein statistics and whose dispersion relation is $\epsilon = \epsilon(\mathbf{k})$. The magnetic state of a substance can be considered as a superposition of the ground state and a gas composed of an indeterminate number of magnons N_{mag} at temperature T. In particular the magnetic moment is

$$\mathcal{M} = \mathcal{M}_s - N_{mag} g \mu_B \ , \tag{13.66}$$

since the presence of each magnon corresponds to a spin reversal (for $S = 1/2$) and its magnetic energy is equal to the magnetic energy at absolute zero, plus that of the gas of magnons. Thus

$$N_{mag} = \int \frac{d^3r \, d^3p}{h^3} \frac{1}{e^{\beta\epsilon} - 1} \qquad \text{and} \qquad U^{(m)} - E_0 = \int \frac{d^3r \, d^3p}{h^3} \frac{\epsilon}{e^{\beta\epsilon} - 1} \ .$$

At low temperatures, only magnons with small values of $\mathbf{k} = \mathbf{p}/\hbar$ are created, and the above integrals can be calculated on replacing ϵ by

$$\epsilon = Dk^2 = Dp^2/\hbar^2$$

and then integrating over all values of p. With the change of variable $x = \beta D p^2 / \hbar^2$, the number of magnons and the energy become

$$N_{mag} = \frac{V}{4\pi^2} \left(\frac{kT}{D}\right)^{3/2} \int_0^\infty \frac{x^{1/2} dx}{e^x - 1}$$

$$\text{and} \quad U^{(m)} - E_0 = \frac{V}{4\pi^2} \left(\frac{kT}{D}\right)^{3/2} kT \int_0^\infty \frac{x^{3/2} dx}{e^x - 1} .$$

As the values of the numerical integrals are respectively 2.315 and 1.784 we get finally

$$N_{mag} = 2.315 \frac{N}{4\pi^2} \left(\frac{T}{\Theta}\right)^{3/2} \quad \text{and} \quad U^{(m)} - E_0 = 1.784 \frac{N}{4\pi^2} \left(\frac{T}{\Theta}\right)^{3/2} kT , \quad (13.67)$$

where $\Theta = (N/V)^{2/3} D/k$ is a characteristic temperature. The reduced magnetization of the substance is then

$$R = \frac{\mathcal{M}}{\mathcal{M}_S} = 1 - N_{mag} \frac{g\mu_B}{\mathcal{M}_S} = 1 - \frac{5.86 \times 10^{-2}}{S} \left(\frac{T}{\Theta}\right)^{3/2} . \qquad (13.68)$$

This result is in agreement with the first terms of the experimentally observed relation (13.8), where

$$A_1 = \frac{5.86 \times 10^{-2}}{S} \frac{1}{\Theta^{3/2}} . \qquad (13.69)$$

Another measurable quantity, the molar magnetic heat capacity, is given by

$$c_V^{(m)} = \frac{du^{(m)}}{dT} = \frac{5}{2} \frac{1.784}{4\pi^2} \left(\frac{T}{\Theta}\right)^{3/2} \mathcal{N}k = 0.113 \, R \left(\frac{T}{\Theta}\right)^{3/2} , \qquad (13.70)$$

which is in agreement with the first term of the experimental relation (13.12), with

$$B_1 = \frac{0.113}{\Theta^{3/2}} . \qquad (13.71)$$

For a quantitative comparison, we consider europium oxide, for which the constant D in the dispersion relation (13.65) is equal to $D = 10.8 \times 10^{-3}$ eV Å^2 (Fig. 13.16). With the density and molar mass given in Table 13.1, this yields $\Theta = 11.9$ K, $A_1 = 4.1 \times 10^{-4}$ /K$^{-3/2}$ and $B_1 = 2.7 \times 10^{-3}$ /K$^{-3/2}$. The value of B_1 is in fairly good agreement with that found experimentally, 1.92×10^{-3} /K$^{-3/2}$ (Fig. 13.8), while that of A_1 differs by an order of magnitude ($A_1 = 0.45 \times 10^{-4}$ /K$^{-3/2}$, Fig. 13.5). Better agreement is found when magnetic dipolar interactions (13.13) are taken into account. Note that the model predicts that the ratio B_1/A_1 is equal to 1.93 S, and is independent of Θ and of D, i.e. 6.76 for europium salts Eu^{2+} ($S = 7/2$). For the oxide EuO the experimental value uncorrected for the dipolar interactions is 6.5 times too high, while for the sulphide EuS the uncorrected value is only about two times too large [O.W. Dietrich et al., *Phys. Rev.* A **133**, 811 (1964)].

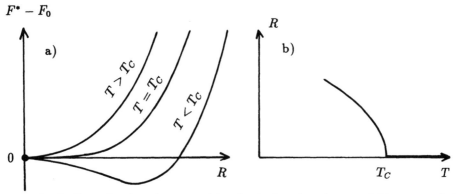

Figure 13.18: Second order phase transition in the Landau theory. a) Variation of the potential F^* (13.72) as a function of the order parameter R near the critical temperature T_C. b) Variation with temperature of the order parameter. The latter remains continuous at $T = T_C$.

When account is taken of the discrepancy between the experimental dispersion curve and the parabolic approximation (13.64), the terms in $T^{5/2}$, $T^{7/2}$, ... appear, as observed experimentally. Lastly, interactions between magnons, which have been neglected until now, contribute through an additional term in T^4. This is unimportant at low temperatures.

Finally, we note that magnons exist only in the Heisenberg model. In an Ising model, the states $\mid n >$ (13.60) are degenerate eigenstates of the Hamiltonian of energy $E_0 + 2J$ (Eq. (13.61)). At low temperatures the Ising model is equivalent to a two-level model and the observable quantities thus vary as $\exp(-A/T)$.

13.5   THEORIES OF CRITICAL PHENOMENA

13.5.1   Landau Theory

In 1937, L. Landau proposed a general phenomenological description for phase transitions and critical phenomena. It consists in expanding a thermodynamic potential of the substance away from equilibrium in powers of an order parameter R. In magnetism, if variations in volume are neglected, this potential is the function $F^*(T, \mathbf{B}, \mathcal{M}) = U(T, \mathbf{B}, \mathcal{M}) - TS(T, \mathbf{B}, \mathcal{M})$, which, when $\mathcal{M}$ takes its equilibrium value that minimizes F^*, is just the free energy $F(T, \mathbf{B})$ of the substance. The reduced magnetization $R = \mathcal{M}/\mathcal{M}_S$ here plays the role of an order parameter.

We first consider the substance in zero magnetic field. If it is assumed that the material is isotropic, the potential F^* is an even function of $\mathcal{M} = \mid \mathcal{M} \mid = R\mathcal{M}_S$. For small values of R, we can then expand F^*,

$$F^*(T, R) = F_0(T) + \frac{a(T)}{2} R^2 + \frac{b(T)}{4} R^4 + \dots \tag{13.72}$$

The shape of this function is shown in Figure 13.18a for $b > 0$. We see that two cases arise depending on the sign of a: if a is positive, the function has a minimum at $R = 0$; if a is negative, the minimum occurs at $R = (-a/b)^{1/2}$. Let T_C be the temperature at which $a(T)$ is equal to zero. Close to this temperature, $a(T)$ has the form

$$a(T) = A(T - T_C) + \dots \ . \tag{13.73}$$

Spontaneous reduced magnetization

$$R = \left[\frac{A}{b} (T_C - T) \right]^{1/2} \qquad (T < T_C) \tag{13.74}$$

then appears below T_C if A (as well as b) is positive, while the magnetization is zero for $T > T_C$. This result, which is identical to that found using the molecular field method (13.38), predicts a value for the critical exponent $\beta = 1/2$.

The free energy F of the substance at equilibrium is then given by

$$F(T) = \begin{cases} F_0(T) - A^2 (T_C - T)^2/4b & (T < T_C) \\ F_0(T) & (T > T_C) \end{cases} . \tag{13.75}$$

It can be seen that $F(T)$ is continuous at the critical point, as also is its derivative $-S(T)$. The heat capacity, however, exhibits a discontinuity

$$C_V(T) = T \frac{dS}{dT} = \begin{cases} C_V^0(T) + A^2 T/2b & (T < T_C) \\ C_V^0(T) & (T > T_C) \end{cases} . \tag{13.76}$$

The jump in heat capacity found in this way is similar to that predicted by molecular field theory (§13.4.2).

In a magnetic field, the potential F^* contains an extra term $-\mathcal{M}B = -\mathcal{M}_S BR$, such that in the paramagnetic phase, $(a > 0)$, we have

$$F^* = F_0(T) - \mathcal{M}_S BR + \frac{a(T)}{2} R^2 + \dots \ .$$

This function exhibits a minimum for

$$R = \frac{\mathcal{M}_S B}{a(T)} = \frac{\mathcal{M}_S B}{A(T - T_C)} . \tag{13.77}$$

We thus recover the Curie-Weiss law. Landau's theory, applied to magnetism, predicts a second order para-ferromagnetic phase transition (continuous first derivatives of F, i.e. $\mathcal{M}$ and S, and discontinuous second derivatives, i.e. C_V). Although this result is incorrect, the Landau theory provides a general phenomenological interpretation of phase transitions in the neighbourhood of the critical point.

The Landau theory also provides an interpretation of first order phase transitions. If the coefficient b is negative, the term in R^6 must be included in the series expansion of F^*

$$F^*(T, R) = F_0(T) + \frac{a}{2} R^2 + \frac{b}{4} R^4 + \frac{c}{6} R^6 + \dots \ . \tag{13.78}$$

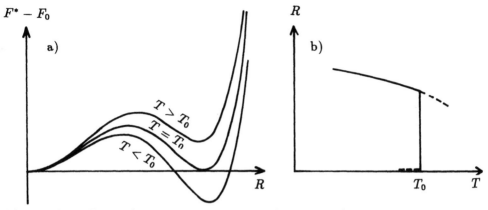

Figure 13.19: First order phase change in the Landau theory. a) Variation of the potential F^* (13.78) as a function of the order parameter R close to the phase transition temperature T_0. b) Variation of the order parameter as a function of temperature. For $T = T_0$ the order parameter is discontinuous and two phases can coexist. The dashed extrapolations of the two curves $R(T)$ represent metastable phases corresponding to the secondary minima.

This function is shown in Figure 13.19a for $c > 0$. As long as the equation $F^*(T, R) = F_0(T)$ admits of only one solution, the order parameter at equilibrium is zero. When other solutions appear, the order parameter that minimizes F^* suddenly becomes positive (Fig. 13.19b). At the transition temperature T_0, two minima occur corresponding to the coexistence of two phases, namely a disordered phase $R = 0$, and an ordered phase $R \neq 0$. This is characteristic of a first order liquid-vapour phase transition. The presence of secondary minima gives rise to the possibility of metastable phases.

Landau's theory introduced the notion that a universal description for phase transition phenomena may be possible.

13.5.2 Scale Invariance

The Landau theory predicts that magnetic phenomena have second order phase transitions. This is in disagreement with experiment, where the results show that close to the critical point the potential F^* cannot be expanded as in equation (13.72), since it does not have an analytical form.

With the assumption of scale invariance, the free energy F in the neighbourhood of the critical temperature is taken to be a generalized homogeneous function. This means that there are two exponents a and b such that for all positive numbers λ

$$F(\lambda^a t, \ \lambda^b B) = \lambda \ F(t, \ B) \ , \tag{13.79}$$

where t is the reduced variable

$$t = \frac{T - T_C}{T_C} \ . \tag{13.80}$$

On differentiating, it is found that the other quantities are also homogeneous. Thus,

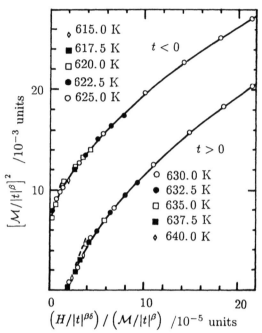

Figure 13.20: Reduced equation of state of nickel. The experimental points confirm the validity of the form of equation (13.87). The values chosen for the parameters are $T_C = 627.4$ K, $\beta = 0.378$ and $\delta = 4.54$. Here, H is expressed in oersteds (1 Oe$= 10^3/4\pi$ A m^{-1}), and $\mathcal{M}$ in cgs emu per gramme (1 cgs emu $= 10^{-3}$ A m^2) [J.S. Kouvel and J.B. Comly, *Phys. Rev. Lett.*, **20**, 1237, (1968)].

the derivative of (13.79) with respect to B yields

$$\lambda^b \, \mathcal{M}(\lambda^a t, \, \lambda^b B) = \lambda \, \mathcal{M}(t, \, B) \ ,$$

or, on replacing λ by $\lambda^{1/(1-b)}$

$$\mathcal{M}(\lambda^{a/(1-b)} t, \, \lambda^{b/(1-b)} B) = \lambda \, \mathcal{M}(t, \, B) \ . \tag{13.81}$$

This equation shows than in zero field, if λ takes the particular value $\lambda = |t|^{(b-1)/a}$, then

$$\mathcal{M}(t, \, 0) = |t|^{(1-b)/a} \, \mathcal{M}(\pm 1, \, 0) \ . \tag{13.82}$$

For the magnetization to be zero in zero field above the critical temperature ($t/|t| = +1$), we must set $\mathcal{M}(+1, \, 0) = 0$. Below T_C ($t/|t| = -1$), equation (13.82) indicates that the spontaneous magnetization obeys the power law (13.9), where the critical exponent is

$$\beta = \frac{1-b}{a} \ . \tag{13.83}$$

Similarly, on taking $\lambda = B^{(b-1)/b}$, equation (13.81) indicates that at the critical temperature

$$\mathcal{M}(0, \, B) = B^{(1-b)/b} \, \mathcal{M}(0, \, 1) \tag{13.84}$$

in agreement with the power law (13.10), where the critical exponent is

$$\delta = \frac{b}{1-b} \quad . \tag{13.85}$$

More generally, equation (13.81) can be written in the form

$$\mathcal{M}(t,\ B) = |t|^{\ (1-b)/a}\ \mathcal{M}(\pm 1,\ |t|^{\ -b/a}B) \tag{13.86}$$

which is a generalization of (13.82) to the case of non zero fields. Replacement of b and a by the critical exponents β and δ yields the reduced form of this relation

$$\frac{\mathcal{M}}{|t|^{\ \beta}} = m_{\pm}\left(\frac{B}{|t|^{\ \beta\delta}}\right) \tag{13.87}$$

which is confirmed remarkably by experiment (Fig. 13.20).

The magnetic susceptibility is found by differentiating expression (13.87) for $\mathcal{M}$ with respect to B and then letting B tend to zero. This gives

$$\chi(t) = |t|^{\ \beta(1-\delta)}\ m'_{\pm}(0) \quad .$$

In the paramagnetic phase this expression is a power law of the form (13.6) where the critical exponent is

$$\gamma = \beta(\delta - 1) \quad . \tag{13.88}$$

This relation between critical exponents is found to be valid both in experiment (Table 13.1) and in the various theoretical models.

On differentiating the free energy with respect to temperature, the entropy and the heat capacity are successively obtained. Thus

$$\begin{aligned}
S(\lambda^{a/(1-a)}t,\ \lambda^{b/(1-a)}B) &= \lambda\, S(t,\ B) \\
C_{V,B}(\lambda^{a/(1-2a)}t,\ \lambda^{b/(1-2a)}B) &= \lambda\, C_{V,B}(t,\ B) \quad .
\end{aligned}$$

In particular, in zero field the latter relation shows that the heat capacity takes the form

$$C_V(t) = |t|^{(1-2a)/a}\ C_{V,B}(\pm 1,\ 0) \quad ,$$

and therefore obeys the power law (13.11) with

$$\alpha = 2 - \frac{1}{a} \quad .$$

Eliminating b in (13.83 and 85) gives

$$\frac{1}{a} = \beta(\delta + 1) = \gamma + 2\beta$$

and the critical exponents α, β and γ are subject to

$$\alpha + 2\beta + \gamma = 2 \quad . \tag{13.89}$$

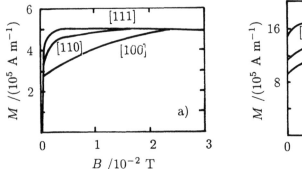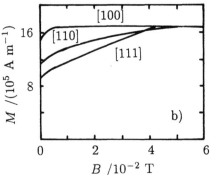

Figure 13.21: Magnetization curves of single crystals of nickel (a) and iron (b) at room temperature, with a field applied along the directions of the symmetry axes of the cube [from K. Honda et al., *Sci. Rep. Tôhoku, Imp. Univ.* **24**, 391 (1935) and **17**, 111 (1928)].

This relation between critical exponents is verified experimentally (Table 13.1) and is also found in the various models.

The assumption of scale invariance is thus confirmed by experiment. In particular, it states that there are only two independent critical exponents. This hypothesis, introduced heuristically, was proved by L.P. Kadanoff (1966) and subsequently demonstrated by K.G. Wilson (1972) using renormalization group methods of quantum field theory. It was thus shown that classes of systems exist that have the same critical exponents (universality classes) and are characterized only by the dimensions of space (d) and of the order parameter (D). It is striking that the exponents depend neither on the exchange integrals, nor on the magnitude of the spins, nor on the type of lattice, ...

13.6   ANISOTROPY AND FERROMAGNETIC DOMAINS

13.6.1   Magnetocrystalline Anisotropy

In a ferromagnetic crystal, the magnetic ions are subjected to a crystal field (§4.5.2). This causes the spontaneous magnetization to align parallel to preferred directions known as the easy directions of magnetization (Fig. 13.21). For nickel and cubic iron these are respectively the directions of a [111] diagonal and a [100] edge of the cube. Anisotropy can be taken into account in the Landau theory by assuming that the potential F^* depends on the direction of the magnetic moment $\mathcal{M}$, with direction cosines α_1, α_2, α_3. In an expansion restricted to terms in $\mathcal{M}^4$, expression (13.72) for F^* in a cubic lattice must contain an additional anisotropy term

$$F_a^* = \frac{b'}{2}\,\mathcal{M}^4\left(\alpha_1^2\alpha_2^2 + \alpha_2^2\alpha_3^2 + \alpha_3^2\alpha_1^2\right) = \frac{b'}{4}\,\mathcal{M}^4\left(1 - \alpha_1^4 - \alpha_2^4 - \alpha_3^4\right), \quad (13.90)$$

which is the only term up to order 4 in α_i that is allowed by the symmetry.

For a given value of $\mathcal{M}$, the anisotropy term of the potential exhibits extrema in directions that, for reasons of symmetry, are parallel to the symmetry axes of the cube. Their values are respectively

$$F_a^* = 0 \quad \text{for} \quad \alpha_1 = 1, \ \alpha_2 = \alpha_3 = 0 \quad ([100]\text{direction}) \ ,$$

$$F_a^* = \frac{K_1}{4} \quad \text{for} \quad \alpha_1 = \alpha_2 = \frac{1}{\sqrt{2}}, \ \alpha_3 = 0 \quad ([110]\text{direction}) \ ,$$

$$F_a^* = \frac{K_1}{3} \quad \text{for} \quad \alpha_1 = \alpha_2 = \alpha_3 = \frac{1}{\sqrt{3}} \quad ([111]\text{direction}) \ ,$$

with $K_1 = b'\mathcal{M}^4/2$.

Clearly if K_1 is positive, the equilibrium magnetization is parallel to an edge of the cube and, if K_1 is negative, it is parallel to a diagonal. The sign of K_1 thus determines the easy directions of magnetization. In nickel, $K_1 \simeq -5 \times 10^3$ J m^{-3} while in iron $K_1 \simeq 5 \times 10^4$ J m^{-3}, where the signs are consistent with the observed easy directions of magnetization (Fig. 13.21). In general, terms in $\mathcal{M}^6$ are also included, and the anisotropy energy in a cubic crystal is written in the form

$$F_a^* = K_1 \left(\alpha_1^2 \alpha_2^2 + \alpha_2^2 \alpha_3^2 + \alpha_3^2 \alpha_1^2 \right) + K_2 \alpha_1^2 \alpha_2^2 \alpha_3^2 \tag{13.91}$$

where K_1 and K_2 are anisotropy constants that depend on the temperature.

Measurements of the temperature dependence of K_1 show that anisotropy, which is a very important phenomenon at low temperatures, becomes small in the neighbourhood of the critical point and vanishes at T_C.

13.6.2 Weiss Domains

In zero field, a ferromagnetic sample, even if it is a single crystal, will in general display a magnetic moment that is weaker than its spontaneous value. P. Weiss (1907) explained this observation by assuming that the sample is composed of domains (*Weiss domains*) inside which the magnetization takes its spontaneous value, but its direction varies from one domain to another. This domain structure was confirmed and made visible to the eye by the magnetic powder method developed by F. Bitter.

L. Landau and E.M. Lifschitz (1935) showed that domain structure is a consequence of the fact that at equilibrium the potential of the substance and of the field must both be considered. For example, consider a monocrystalline sample of cubic structure that is cut along the faces of the cube (Fig. 13.22), where the easy direction of magnetization is parallel to the edges. The magnetization can point along six equivalent directions (two directions for each of the three edge axes). Owing to the magnetostatic energy outside the material, it can be shown that the energy of configuration a) in Figure 13.22 is greater than that of configuration b). Configurations such as c) and d) are most favourable since they possess closure domains that prevent any field lines from escaping outside the material. In general, the domain structure is more complex and depends on the shape and orientation of the crystal, as well as on its polycrystallinity, etc.

Domains are separated by transition zones, called Bloch walls, in which the direction of the spins turns gradually. The thickness of these walls is the result of a compromise between the anisotropy energy (the spins no longer lie along the easy

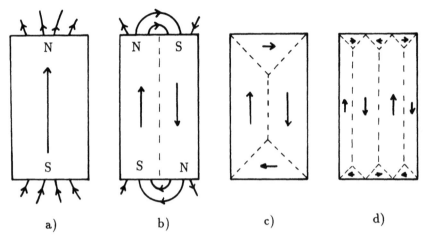

Figure 13.22: Domains in a cubic single crystal cut along the cube faces, where the easy direction of magnetization is parallel to the edges. The arrows indicate the magnetic moments of the domains. a) Single domain: the field lines leave one face and return through the opposite face. b) Two domains of opposite magnetization: most of the lines of force return by the same face. c) Four domains: the two closure domains prevent the field lines from leaving. d) A larger number of domains.

directions of magnetization), which is minimum for a wall of zero thickness, and the increase in exchange energy, which is minimum for an infinite thickness. Since the increase in exchange energy (13.17) between two spins that make an angle ϕ is

$$\Delta\epsilon = 2JS^2(1 - \cos\phi) \simeq JS^2\phi^2 \quad , \tag{13.92}$$

the exchange energy of a wall containing n lattice planes is proportional to $n\phi^2 = \phi_0^2/n$, where $\phi_0 = n\phi$ is the angle between the spins of the two domains. In this way it is found that the thickness of the wall is of the order of several hundred Ångströms ($n \sim 100$), and that the increase in energy due to the wall is of the order of 10^{-3} J m^{-2}. This energy prevents division into increasingly tiny domains. Thus, powder particles of the order of 0.1 μm are composed of only one domain. Note that if a wall touches the surface of a crystal, the gradient of magnetization generates a magnetic field gradient, which the Bitter method reveals by concentrating the grains of powder along the wall.

13.6.3   Hysteresis

Consider a ferromagnetic sample with zero total magnetic moment. It is composed of domains with differently oriented magnetic moments, the resultant of which is zero (Fig. 13.23a). Let us examine the changes in this sample when it is placed in a magnetic field H that increases from zero (first magnetization). In a weak field domains whose magnetic moments are oriented along the field direction grow reversibly at the expense of the others by displacement of their walls (Fig. 13.23b). At higher values of the field, an irreversible rotation of the magnetic moment of whole domains occurs, which orients them along the direction of the field (Fig. 13.23c).

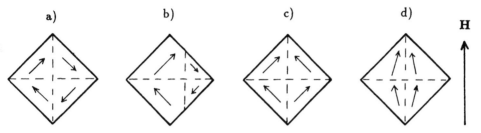

Figure 13.23: Diagram showing the changes in the domains when a field H is applied, increasing from $H = 0$ (a) to a very strong field (d).

These rotations generate sound (Barkhausen effect). Finally, in intense fields, the orientation of the moments, which was previously directed along the easy axes of magnetization of the crystal, changes reversibly inside each domain (Fig. 13.23d). The initial magnetization curve OS in the H, M plane is shown in Figure 13.24. Let us now decrease the field to zero. The substance returns to state c) of Figure 13.23, where it now exhibits a remanent magnetization M_R and its state is different from what it was initially (phenomenon of hysteresis). To reduce the magnetization to zero, a field $-H_C$ must be applied. H_C is called the *coercive field*. As the field H continues to decrease, the curve CS' in Figure 13.24 is traced out. If the magnetic field is increased once again, the curve $S'R'C'S$ is obtained, symmetrical to $SRCS'$ with respect to the origin. The cycle thus described is called the hysteresis cycle. To recover the condition of zero magnetization in zero field, a large number of cycles must be performed with gradually decreasing amplitude.

The shape of the hysteresis cycle depends on the magnitude of the anisotropy energy. If the anisotropy is weak, rotation of the domains is easier and hysteresis is small: the hysteresis cycle is narrow. If the anisotropy is strong, domain rotation occurs at a higher field and the hysteresis cycle is broad. These phenomena condition the methods used to measure spontaneous magnetization of materials. Near the critical point, where anisotropy is weak, the measurements of $M(B)$ are extrapolated to $B = 0$. At low temperatures where anisotropy is strong, the measurements of $M(B)$ are extrapolated to $B \to \infty$.

13.7 ANTIFERROMAGNETISM

13.7.1 Experimental Evidence

Starting in 1932, L. Néel showed that substances containing magnetic ions of a single type, for which the exchange integral is negative, exhibit an ordered anti-ferromagnetic phase below a certain temperature T_N, later to be called the Néel temperature. In this phase at absolute zero, two or more sub-lattices can be distinguished, in each of which the magnetic moments of the ions are parallel, but the total moment is zero. In the simplest cases, for example a simple cubic lattice, there are only two sub-lattices, in which the moments of the ions lie in opposite directions along a preferred axis in the crystal (Fig. 13.25). Above the Néel temperature, the substance is paramagnetic. In both phases, the magnetization in a magnetic field is weak. For this reason, theory has inspired systematic investigations into the

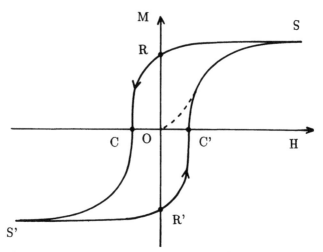

Figure 13.24: Hysteresis cycle in an iron sample. The dotted line is the initial magneti-zation curve. The remanent magnetization M_R and the coercive field H_C are of the order of 10^6 A m^{-1} and 50 A m^{-1}, respectively.

properties of these materials.

The clearest evidence for the phenomenon is found by diffracting neutrons hav-ing the same velocity ($\lambda \sim 1$ Å). In addition to their nuclear interaction with the nuclei, neutrons interact with electronic spins through their magnetic moment. The scattering amplitude depends on the relative orientation of the spins. In the param-agnetic phase, the ion moments are randomly oriented and the scattering, which is incoherent, is not visible. The Bragg diffraction peaks (§8.3) then correspond to the unit cell of the lattice. In the antiferromagnetic phase the scattering is coherent and new diffraction peaks appear. The overall response now corresponds to a double cell. As the intensity of these extra peaks is proportional to the square of the mag-netic moments of the sub-lattices, this method provides an indirect measurement of these moments, whose effect was not observed macroscopically.

13.7.2   Theory of Antiferromagnetism

Molecular field model

With the Heisenberg Hamiltonian (13.18) an interpretation of the properties of an-tiferromagnetic substances can be given. Here we restrict ourselves to the discussion in the framework of the molecular field approximation, which, as we have already seen, accounts for the general properties of ferromagnetic substances.

We suppose that the N magnetic ions are located on two identical sub-lattices A and B, whose moments are respectively $\mathcal{M}_A$ and $\mathcal{M}_B$, with reduced moments $R_A = 2\mathcal{M}_A/\mathcal{M}_S$ and $R_B = 2\mathcal{M}_B/\mathcal{M}_S$. $\mathcal{M}_S = NSg\mu_B$ is the total saturation moment of the crystal. Each magnetic ion of lattice A is subjected to an effective magnetic flux density

$$B_A = B - q_1 R_B - q_2 R_A \tag{13.93a}$$

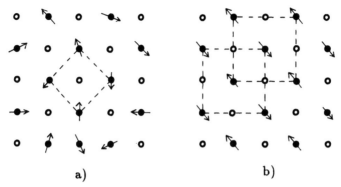

Figure 13.25: Two dimensional diagram of an antiferromagnetic material. a) Paramagnetic phase $T > T_N$, on a simple square lattice. b) Antiferromagnetic phase $T < T_N$, on two simple square sub-lattices.

equal to the sum of the external field B, the molecular field (13.28) $-q_1 R_B$ generated by the ions in the B lattice, and the molecular field $-q_2 R_A$ generated by the other ions of the A lattice. Conversely, the ions of the B sub-lattice are subjected to an effective field

$$B_B = B - q_1 R_A - q_2 R_B \quad . \tag{13.93b}$$

The constants q_1 and q_2 are related to the exchange integrals J_1 (nearest neighbours) and J_2 (second nearest neighbours) through

$$q_i = -4z_i \, \frac{J_i}{g\mu_B} \quad , \tag{13.94}$$

by virtue of equations (13.23b,28b), where z_1 is the number of nearest neighbours and z_2 the number of next nearest neighbours. To produce antiferromagnetism, the constant q_1 must be positive. The constant q_2 is generally smaller in absolute value than q_1.

Magnetic properties

As in (13.34), the reduced magnetization of each lattice is then given in terms of the Brillouin function B_S by

$$
\begin{aligned}
R_A &= B_S\left[\frac{Sg\mu_B}{kT} \, (B - q_1 R_B - q_2 R_A)\right] \quad , \\
R_B &= B_S\left[\frac{Sg\mu_B}{kT} \, (B - q_1 R_A - q_2 R_B)\right] \quad .
\end{aligned}
\tag{13.95}
$$

These two coupled equations in principle allow the moments $\mathcal{M}_A = \mathcal{M}_S R_A/2$ and $\mathcal{M}_B = \mathcal{M}_S R_B/2$ of the two sub-lattices to be calculated as a function of B and T, and hence the total moment $\mathcal{M} = \mathcal{M}_A + \mathcal{M}_B$.

In zero external field $(B = 0)$, equations (13.95) yield $R_B = -R_A = -R°$, with

$$R° = B_S \left[\frac{Sg\mu_B}{kT} (q_1 - q_2)R° \right] \quad . \tag{13.96}$$

This equation is solved as for ferromagnetism (§13.4.2). For temperatures T above the Néel temperature

$$T_N = \frac{(S+1)g\mu_B}{3k} (q_1 - q_2) \quad , \tag{13.97}$$

we get $R° = 0$ and the two sub-lattices display no spontaneous magnetization. For $T < T_N$, the reduced magnetizations of the two sub-lattices are equal and opposite and the total magnetization is still zero.

In weak fields in the paramagnetic phase, we may replace the Brillouin function B_S by its approximate form (4.45). Equations (13.95) then become

$$R_A = \frac{(S+1)g\mu_B}{3kT} (B - q_1 R_B - q_2 R_A)$$

$$R_B = \frac{(S+1)g\mu_B}{3kT} (B - q_1 R_A - q_2 R_B) \quad ,$$

and their solution is

$$R_A = R_B = \frac{(S+1)g\mu_B B}{k(T - \theta)} \quad \text{with} \quad \theta = -\frac{(S+1)g\mu_B}{3k} (q_1 + q_2) \quad . \tag{13.98}$$

The total magnetic moment is then given by

$$M = \frac{M_S}{2} (R_A + R_B) = \frac{NS(S+1)g^2\mu_B^2}{3k} \frac{B}{T - \theta} \quad . \tag{13.99}$$

This is a Curie-Weiss law, similar to equation (4.48) for paramagnetism and (13.37) for ferromagnetism, but with a negative Weiss temperature θ.

In the antiferromagnetic phase for weak fields, two cases arise, depending on whether the applied magnetic field is parallel or perpendicular to the preferred direction of the spontaneous magnetization of the sub-lattices. When the field is applied parallel to the magnetization, the reduced magnetization in each sub-lattice increases by the same amount, i.e.

$$R_A = R° + \delta R \quad \text{and} \quad R_B = -R° + \delta R \quad .$$

The total moment of the substance is $M_\parallel = M_S \delta R$. On substituting these expressions for R_A and R_B into equations (13.95) and making a series expansion around $B = 0$, it is found that the magnetic moment is proportional to the field B as follows,

$$M_\parallel = M_S \delta R = N \frac{S(S+1)g^2\mu_B^2}{3k} \frac{B}{(S+1)/(3S) \times T/B_S'(x_0) - \theta} \tag{13.100}$$

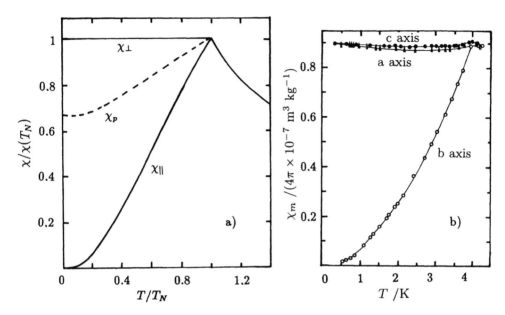

Figure 13.26: Magnetic susceptibility of an antiferromagnetic substance. a) Molecular field theory. The dashed curve shows the susceptibility of a powder. b) Experimental results for a single crystal of gadolinium aluminate $GdAlO_3$, when the field lies along each of the axes of the orthorhombic crystal [J. D. Cashion et al., *Proc. Roy. Soc. A* (London) **318**, 473 (1970)].

where B'_S is the derivative of the Brillouin function B_S and

$$x_0 = \frac{Sg\mu_B}{kT}(q_1 - q_2)R^\circ \ . \tag{13.101}$$

The corresponding susceptibility $\chi_\|$, displayed in Figure 13.26a, is zero at absolute zero, increases with the temperature, and is continuous at the Néel temperature T_N. Since $B'_S(0) = (S+1)/3S$, expressions (13.99 and 100) both take the same value at $T = T_N$,

$$\mathcal{M}(T_N, \ B) = \mathcal{M}_S \frac{B}{2q_1} \ . \tag{13.102}$$

When the applied magnetic field is perpendicular to the direction of spontaneous magnetization of the sub-lattices, the moments $\mathcal{M}_A$ and $\mathcal{M}_B$ rotate, while their modulus remains equal to $\mathcal{M}^\circ = \mathcal{M}_S R^\circ/2$ (Fig. 13.27). The magnetic energy contribution to the free energy is then

$$
\begin{aligned}
E^{(m)} &= -\mathcal{M}_A \cdot \mathbf{B}_A - \mathcal{M}_B \cdot \mathbf{B}_B \\
&= -(\mathcal{M}_A + \mathcal{M}_B) \cdot \mathbf{B} + \frac{1}{2}\left(4q_1 \frac{\mathcal{M}_A \cdot \mathcal{M}_B}{\mathcal{M}_S} + 2q_2 \frac{\mathcal{M}_A^2 + \mathcal{M}_B^2}{\mathcal{M}_S}\right)
\end{aligned}
\tag{13.103}
$$

Figure 13.27: Sub-lattice magnetization in a transverse magnetic field.

where $\mathbf{B}_A$ and $\mathbf{B}_B$ are the effective fields (13.93) and the factor $1/2$ is introduced to avoid counting the interaction energies twice. If the moments rotate by a small angle α, the magnetic energy changes by

$$\begin{aligned} \Delta E^{(m)} &= -2\mathcal{M}^\circ B \sin\alpha - 2q_1 \frac{\mathcal{M}^{\circ 2}}{\mathcal{M}_S} \cos 2\alpha \\ &= -\mathcal{M}_S BR^\circ \sin\alpha - \frac{1}{2}\mathcal{M}_S q_1 R^{\circ 2} \cos 2\alpha \quad . \end{aligned} \tag{13.104}$$

For a given magnetic flux density B, the equilibrium angle α for which $\Delta E^{(m)}$ is a minimum is

$$\sin\alpha = \frac{B}{2q_1 R^\circ} \quad ,$$

and the resulting moment is

$$\mathcal{M}_\perp = 2\mathcal{M}^\circ \sin\alpha = \mathcal{M}_S \frac{B}{2q_1} \quad . \tag{13.105}$$

We thus find that the susceptibility $\chi_\perp$ of the substance is independent of T and continuous at the Néel temperature. When the anisotropy energy (§13.7.4) is explicitly taken into account, a term of the form $K\alpha^2$ is introduced into $\Delta E^{(m)}$. Since this term is in general much smaller than the exchange energy and since it vanishes at the Néel temperature, its effect on the susceptibility is small. If the substance is a powder, the measured susceptibility is, from (4.11)

$$\chi_p = \frac{\chi_\parallel + 2\chi_\perp}{3} \quad . \tag{13.106}$$

In particular, at absolute zero, $\chi_p(0) = 2\chi_p(T_N)/3$.

Magnetic heat capacity

From (13.103), the magnetic energy in zero field is

$$E^{(m)} = -2(q_1 - q_2)\frac{\mathcal{M}^{\circ 2}}{\mathcal{M}_S} = -\frac{1}{2}(q_1 - q_2)\mathcal{M}_S R^{\circ 2} \tag{13.107}$$

since $\mathcal{M}_a = -\mathcal{M}_b = \mathcal{M}^\circ = \mathcal{M}_S R^\circ/2$. The magnetic heat capacity in zero field is therefore

$$C_V^{(m)} = -\frac{1}{2}(q_1 - q_2)\mathcal{M}_S \frac{d}{dT}\left(R^{\circ 2}\right) \quad .$$

Table 13.2: Physical constants of antiferromagnetic materials [A.H. Morrish, *op. cit.* p. 463; Rb Mn F$_3$: L.J. de Jongh and A.R. Miedema, *op. cit.* p. 151; GdAlO$_3$: K.W. Blazey and H. Rohrer, *Phys. Rev.* **173**, 574 (1968)].

Substance	Structure	T_N /K	θ /K	$\chi_p(0)/\chi_p(T_N)$
MnO	f.c.c. (NaCl)	122	−610	0.69
MnF$_2$	tetragonal (rutile)	74	−113	0.75
RbMnF$_3$	s.c. (perovskite)	83	−119	$\simeq 0.7$
FeO	f.c.c. (NaCl)	185	−570	0.77
α Fe$_2$O$_3$	trigonal (Al$_2$O$_3$)	950	−2000	
CoCl$_2$	hexagonal (layers)	25	38	$\simeq 0.6$
GdAlO$_3$	orthorhombic ($\simeq$ perovskite)	3.89	−4.6	0.63

In the paramagnetic phase, the sub-lattices exhibit no spontaneous magnetization ($R° = 0$) and the heat capacity is zero. In the antiferromagnetic phase, $R°$ is non-zero and likewise for the heat capacity. It can be shown, as for ferromagnetism (§13.4.2), that molecular field theory predicts a discontinuity at the Néel temperature given by (13.42). Since the energy $E^{(m)}$ is continuous at this temperature, molecular field theory therefore predicts a second order phase transition.

13.7.3 Experimental Investigations

A great majority of substances containing a single type of magnetic ion exhibit an ordered antiferromagnetic phase at low temperatures. These are generally ionic crystals (oxides, chlorides, fluorides, etc.) in which the magnetic ions are surrounded by anions and interact through a superexchange mechanism. To illustrate this discussion we choose as a typical substance GdAlO$_3$.

Gadolinium ortho-aluminate has an orthorhombic structure that is almost cubic, in which the Gd^{3+} magnetic ions form an approximately simple cubic lattice of side $a = 3.73$ Å. In the antiferromagnetic phase, the ions are located on two face-centred cubic sub-lattices of side $2a$, each ion being surrounded by $z_1 = 6$ nearest neighbours belonging to the other sub-lattice and which form a regular octahedron, and by $z_2 = 12$ second nearest neighbours belonging to the same sub-lattice. The magnetizations of these two sub-lattices are equal and opposite and directed along the diagonal of a face of the cube of Gd^{3+} ions, which coincides with the b axis of the orthorhombic crystal. This configuration is analogous to that of the two-dimensional material in Figure 13.25. We recall (§4.5.3) that the Gd^{3+} ion has zero orbital moment and a spin $S = 7/2$. The Landé factor $g = 2$ is therefore isotropic.

Measurements of the susceptibility of this material are shown in Figure 13.26b. When the magnetic field is applied along the b axis (spontaneous magnetization direction of the sub-lattices), the susceptibility $\chi_\parallel$ behaves as in the model (Fig. 13.26a). When it is applied perpendicular to b, along a or c, the susceptibility is a constant like $\chi_\perp$ in the model. In the paramagnetic phase, at temperatures sufficiently far

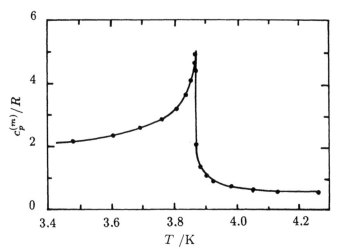

Figure 13.28: Magnetic heat capacity of gadolinium ortho-aluminate GdAlO$_3$. This heat capacity was obtained by subtracting the lattice contribution ($C/R = 2.4 \times 10^{-5}$ [T /K^3]) and that of the Schottky anomaly for the hyperfine structure ($C/R = 5.85$ [T /K^{-2}]) [J.D. Cashion et al., *Proc. Roy. Soc. A* (London) **318**, 473 (1970)].

above the Néel temperature $T_N = 3.89$ K, the substance follows a Curie-Weiss law with $\theta = -4.6$ K and $C_M = 1.00 \times 10^{-4}$ m^3 K mol^{-1}. This value of C_M is in good agreement with the theoretical value (4.50) $C_M = 0.99 \times 10^{-4}$ m^3 K mol^{-1}. The experimental values for T_N and θ , combined with equations (13.97,98) of molecular field theory, yield q_1 and q_2. Hence, from (13.94), we find $J_1/k = -0.068$ K and $J_2/k - 0.003$ K. Clearly, the value of J_1 predominates.

The heat capacity of gadolinium aluminate GdAlO$_3$ is shown in Figure 13.28. It exhibits an increase followed by a sharp drop at the Néel temperature.

As in the case of ferromagnetism, molecular field theory provides a satisfactory description of the experimental data. Its predicted values for the critical exponents are, however, wrong ($\alpha = 0$, $\beta = 1/2$, $\gamma = 1$). The measured values of the critical exponents are close to those in ferromagnetic materials. For example for rubidium manganese fluoride RbMnF$_3$, $\alpha = -0.14 \pm 0.01$ [A. Kornblit and G. Ahlers, *Phys. Rev. B* **8**, 5163 (1973)], $\beta = 0.32 \pm 0.01$ and $\gamma = 1.40 \pm 0.04$ [L. M. Corliss et al., *J. Appl. Phys.* **40**, 1278 (1969)]. In antiferromagnetic substances, the coefficient β defined in (13.9) corresponds to the critical behaviour of the reduced magnetization of each sub-lattice (R°), as measured by the intensity of the neutron diffraction lines.

Similarly, at low temperatures, the experimentally observed variation in T^2 of the magnetic susceptibility $\chi_\parallel$ and that in T^3 of the heat capacity are in disagreement with molecular field theory. Introduction of antiferromagnetic magnons re-establishes agreement. The dispersion relation for these magnons $\epsilon_k \propto k$, differs from that of ferromagnetic magnons (13.65), $\epsilon_k \propto k^2$. The ground state of an antiferromagnetic substance is not an eigenstate of the Heisenberg Hamiltonian and the proof of the relation $\epsilon_k \propto k$ is much more complex. The T^3 dependence follows directly from the linear dispersion relation by a proof similar to that employed for

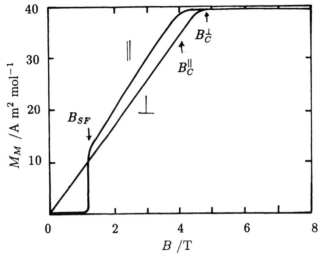

Figure 13.29: Molar magnetic moment of a single crystal of gadolinium aluminate GdAlO₃ at $T = 0.5$ K. The two experimental curves correspond to fields applied along the a and b axes of the orthorhombic crystal, i.e. respectively perpendicular and parallel to the preferred direction (b axis) [J.D. Cashion et al., *Proc. Roy. Soc. A* (London) **318**, 473 (1970)].

phonons (§8.6.2). Finally, we point out that antiferromagnetic substances may also exhibit domains.

13.7.4 Strong Field Behaviour

Perpendicular field

The existence of an anisotropy energy in antiferromagnetic materials endows them with a special behaviour, involving a phase transition that was predicted by Néel (1936). Let us consider a uniaxial substance in which the anisotropy energy, given by the term

$$E_A = K \sin^2 \alpha \ , \tag{13.108}$$

exhibits a minimum at one angle between the sub-lattice magnetization and the preferred direction ($\alpha = 0$), and a maximum in a direction perpendicular to it ($\alpha = \pi/2$). If a magnetic field is applied perpendicular to the preferred direction, a magnetic moment $\mathcal{M}_\perp$ appears. $\mathcal{M}_\perp$ is obtained as in (13.105), where the anisotropy energy is now incorporated in the expression for the energy (13.104). It is given by

$$\mathcal{M}_\perp = \mathcal{M}_S \frac{B}{2q_1 + 2K/\mathcal{M}_S R^{\circ 2}} = 2\mathcal{M}^\circ \frac{B}{2B_W + B_{an}} \ , \tag{13.109}$$

where $B_W = q_1 R^\circ$ is the molecular field from the nearest neighbours and $B_{an} = K/\mathcal{M}^\circ$ is called the anisotropy field. The magnetization therefore increases linearly

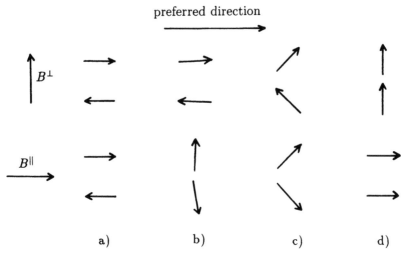

Figure 13.30: Diagram of the orientation in the two sub-lattices when the magnetic field is applied perpendicular or parallel to the preferred direction. The magnetic field progressively increases from a) to d). The critical phase transition occurs between c) and d). The spin-flop occurs between a) and b) for the parallel field.

with the field (Fig. 13.29) up to a critical field

$$B_C^\perp = 2B_W + B_{an} \tag{13.110}$$

at which the magnetic moments of the two sub-lattices are parallel. (Fig. 13.30).

Parallel field

When the magnetic field is applied parallel to the preferred direction, a peculiar phenomenon occurs. For weak fields, the magnetic moment of the substance increases linearly as predicted by (13.100) and its magnetic energy is

$$E^{(m)} \simeq -2q_1 \frac{\mathcal{M}^{\circ 2}}{\mathcal{M}_S} - \mathcal{M}_0 B_W \quad . \tag{13.111}$$

This equation is derived from (13.103), in which the term in B has been neglected and the constant term in q_2 is omitted. For a field equal to a certain value B_{SF}, the magnetizations of the two sub-lattices rotate sharply through an angle close to 90° (Fig. 13.30). The angle α between the moments of the sub-lattices and the preferred direction that minimizes the magnetic energy

$$\begin{aligned}
E^{(m)} &= -2\mathcal{M}^\circ B \cos\alpha + 2q_1 \frac{\mathcal{M}^{\circ 2}}{\mathcal{M}_S} \cos 2\alpha + K \sin^2\alpha \\
&= -2\mathcal{M}^\circ B \cos\alpha + \mathcal{M}^\circ B_W \cos 2\alpha + \mathcal{M}^\circ B_{an} \sin^2\alpha \tag{13.112}
\end{aligned}$$

is given by $\cos\alpha = B/(2B_W - B_A)$. The energy is then equal to

$$E^{(m)} = \mathcal{M}^\circ (B_{an} - B_W) - \mathcal{M}^\circ \frac{B^2}{2B_W - B_{an}} \quad . \tag{13.113}$$

The change in orientation of the moments ("spin-flop") arises when the energy (13.113) is less than (13.111), i.e. for a field

$$B_{SF} = [B_{an}(2B_W - B_{an})]^{1/2} \quad . \tag{13.114}$$

This change is accompanied by a jump in the value of the magnetization of the material (Fig. 13.29). Increasing the field beyond this value causes progressive alignment of the moments of the sub-lattices, and the magnetic moment of the substance is given by

$$\mathcal{M} = 2\mathcal{M}° \cos\alpha = 2\mathcal{M}_0 \frac{B}{2B_W - B_{an}} \quad . \tag{13.115}$$

Finally, when the field reaches the critical value

$$B_C^{\parallel} = 2B_W - B_{an} \tag{13.116}$$

the moments of the two sub-lattices become parallel. This interpretation is in good agreement with the experimental results. The measurements on gadolinium aluminate $GdAlO_3$ at $T = 0.5$ K (Fig. 13.29) are thus consistent with the values $B_W = 2.1$ T and $B_{an} = 0.3$ T. Note that the jump in magnetic moment between the values (13.109) and (13.115) is given by

$$\Delta\mathcal{M} = \frac{4\mathcal{M}° B_{an} B_{SF}}{4B_W^2 - B_{an}^2} \simeq 4\mathcal{M}° \left(\frac{B_{an}}{2B_W}\right)^{3/2} \quad . \tag{13.117}$$

If the anisotropy is very small, as, for instance, in rubidium manganese fluoride $RbMnF_3$, since the spin-flop occurs at very low fields and causes only a small variation in the moment, it cannot be detected, neither can one measure $\chi_{\parallel}$.

Phase diagrams

The above phenomena reveal the existence of different phases and of phase transitions. Thus, when the field is applied perpendicular to the preferred axis and the temperature is below the Néel temperature T_N, each sub-lattice has a component of its moment in a direction perpendicular to the field. This component is used as an order parameter. Its value is non-zero for $B^{\perp} < B_C^{\perp}(T)$ and zero for $B^{\perp} \geq B_C^{\perp}(T)$. The line $B_C^{\perp}(T)$ separates the $(T, B^{\perp})$ plane into two regions corresponding to two phases, called respectively antiferromagnetic (AF) and paramagnetic (P) (Fig. 13.31a). At $B^{\perp} = B_C^{\perp}(T)$ the order parameter vanishes, but remains continuous. The phase transition is therefore a critical transition and the curve $B^{\perp} = B_C^{\perp}(T)$ is a critical line. On this line, the total magnetization is continuous, but its derivative $\partial M/\partial B^{\perp}$ is discontinuous. It is by this property that the critical line is determined experimentally. Note that in the paramagnetic phase the two sub-lattices have identical moments and the notion of sub-lattices loses its meaning.

When the field is applied parallel to the preferred axis, the component of each sub-lattice perpendicular to the field, which is used as the order parameter, is equal

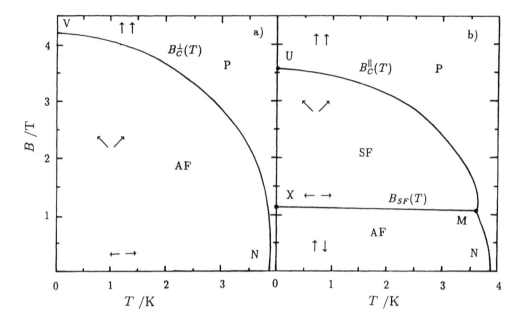

Figure 13.31: Phase diagram of gadolinium aluminate GdAlO$_3$ a) perpendicular field, b) parallel field. The antiferromagnetic (AF), paramagnetic (P) and post spin-flop (SF) domains are distinguished. The arrows indicate the orientation of the moments of the two sub-lattices for a field parallel to the ordinate axis [K.W. Blazey and H. Rohrer, *Phys. Rev.* **173**, 574 (1968)]. As the crystal is not perfectly oriented in the field, the multicritical point M is shifted from its real position $T_M = 3.12$ K and $B_M = 1.25$ T [H. Rohrer and Ch. Gerber, *Phys. Rev. Lett.* **38**, 909 (1977)].

to zero for $B^\parallel < B_{SF}(T)$ (Fig. 13.31b). When the field exceeds this value, this order parameter, as well as the total magnetization, become non-zero. This corresponds to a first order phase transition between the antiferromagnetic (AF) phase and the (SF) phase arising from the spin-flop. When the field reaches the value $B_C^\parallel(T)$, this order parameter vanishes while remaining continuous; a critical transition occurs towards the paramagnetic phase (P). Note that in the first order transition $AF - SF$, the two phases can coexist, and lag phenomena can appear in the phase change. For temperatures lower than T_N but greater than the temperature T_M at which the fields $B_C^\parallel(T)$ and $B_{SF}(T)$ are equal, the system moves directly through a critical transition from the AF phase to the P phase.

When the field is applied in an arbitrary direction, the phase diagram must be considered in the three dimensional space T, $B^\parallel$, $B^\perp$ (Fig. 13.32a). The critical lines $B_C^\parallel(T)$ and $B_C^\perp(T)$ are the intersections of a critical surface $NMUWV$ with the T, $B^\parallel$ and T, $B^\perp$ planes. Outside this surface lies the paramagnetic phase in which the two sub-lattices cannot be distinguished. Inside the surface there is a sheet MXY that cuts the T, $B^\parallel$ plane along the coexistence line XM between the AF and SF phases. This sheet separating the AF phase from the SF phase

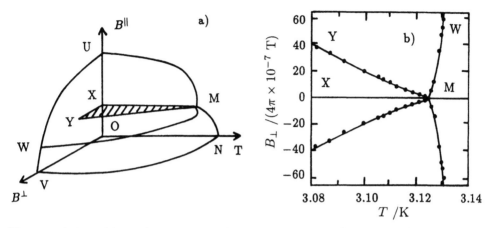

Figure 13.32: Phase diagram of gadolinium aluminate $GdAlO_3$. a) Diagram in T, $B^{\parallel}$, $B^{\perp}$ space. The curves VN and UMN also appear in Figure 13.31. Only one of the four possible octants is shown. b) Section of the previous diagram near the multi-critical point M, which is the junction point of the four critical lines [H. Rohrer and Ch. Gerber, *Phys. Rev. Lett.* **38**, 909 (1977)].

is bounded by a critical line MY. By going around this surface it is possible to move continuously from the AF phase to the SF phase. This is the analogue of the continuous path from the liquid to the vapour phase that goes around the critical point in a fluid. The external critical surface is in fact composed of two critical sheets $NMWV$ and MUW that cut each other along a bicritical line MW and meet perfectly at W. The point M is an umbilical point located at the intersection of several critical lines: it is known as multicritical. In the T, $B^{\parallel}$ plane (Fig 13.31b), this point is located at the intersection of the two critical lines NM and MU: it is then called bicritical. In the plane tangent at M to the sheet MXY (Fig. 13.32b) it lies at the intersection of four critical lines: it is then called tetracritical. The properties of the substance near the multicritical point M are described by critical exponents.

13.7.5 Helimagnetism

In the ordered phases that we have been considering until now, all the spins of the magnetic ions have the same alignment. Many other different arrangements exist, however, one of the most important of which is helimagnetism (Fig 13.1c). The first material in which helimagnetism was discovered was the metallic compound $MnAu_2$. Neutron diffraction showed that in this tetragonal compound, the spins of the manganese ions located in a plane perpendicular to the principal axis are parallel and lie in the plane. The direction of the spins of two consecutive planes is found by rotating through 51°. This substance therefore does not exhibit spontaneous magnetization. Among their various phases, certain rare earth metals (Tb, Dy, Ho) display a helimagnetic phase. Dysprosium, for example, is helimagnetic between 85 and 179 K, ferromagnetic below 85 K and paramagnetic above 179 K.

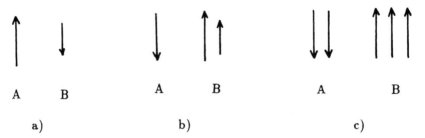

Figure 13.33: Arrangement of ionic moments in the sub-lattices of a ferrimagnetic: a) ions of different types on two sub-lattices; b) model of ferrite (Fe_3O_4); c) model of garnet (YIG = $Y_3Fe_5O_{12}$).

In the rare earth metals, the exchange interaction is not direct but occurs via the conduction electrons. It follows that the exchange constant $J(r)$ is long range and oscillatory (RKKY theory). Let us see how the behaviour of J can generate helimagnetism in the case of the simple model of a row of ions in which the interaction is ferromagnetic ($J > 0$) between nearest neighbours and equal to λJ between second nearest neighbours. The interaction Hamiltonian is then

$$\hat{H} = -2J \sum_i \hat{\mathbf{S}}_i \cdot \hat{\mathbf{S}}_{i+1} - 2\lambda J \sum_i \hat{\mathbf{S}}_i \cdot \hat{\mathbf{S}}_{i+2} \quad .$$

If θ is the angle between neighbouring spins, the interaction energy then has the form

$$E = -2JNS^2(\cos\theta + \lambda \cos 2\theta) \quad .$$

For $\lambda > -1/4$, E is a minimum for $\theta = 0$ (ferromagnetism). For $\lambda < -1/4$, the minimum occurs at $\cos\theta = -1/4\lambda$, i.e. the order is helicoidal, with an angle $\theta < \pi/2$. It can be seen that a sufficiently strong antiferromagnetic interaction between second neighbours leads to helimagnetic ordering.

The RKKY interaction provides an explanation for a large variety of ordered phases, for example, conic phase where the moments rotate about a cone, and phases in which the spin orientation is modulated periodically. Lastly, we point out that anisotropic coupling of the form $\mathbf{D} \cdot (\mathbf{S}_i \times \mathbf{S}_j)$ accounts for the existence of a weak ferromagnetic phase (αFe_2O_3, NiF_2) in which the spins in zero field have an almost antiferromagnetic arrangement with a slight cant in the spins.

13.8   FERRIMAGNETISM

13.8.1   Introduction

The properties of ferrimagnetic materials are similar to those of ferromagnetics. In 1948, L. Néel suggested a model for these substances in which magnetic ions of two different types are distributed in two antiparallel sub-lattices (Fig. 13.33a). Numerous other types of arrangement exist, the most important of which are ferrites and garnets.

Table 13.3: Properties of MO, Fe_2O_3 ferrites. $\overline{m}_S$ and T_N are the magnetic moment for a formula unit and the Néel temperature, respectively [B.I. Bleaney and B. Bleaney, *op. cit.* p.520].

Ion M^{2+}	Spin S	$\overline{m}_S/\mu_B$	T_N/K
Mn^{2+}	5/2	~ 5	573
Fe^{2+}	2	4.1	858
Co^{2+}	3/2	3.7	793
Ni^{2+}	1	2.3	858
Cu^{2+}	1/2	1.3	728

13.8.2 Ferrites

Ferrites (Fig. 13.33b) are oxides of formula MO, Fe_2O_3 (M^{2+} Fe_2^{3+} O_4^{2-}) where M is a divalent metal (M = Fe, Co, Ni, Cu, Mg, Zn, Cd). Magnetite or lodestone, with formula Fe_3O_4, is the most widely known ferrite (M = Fe). Their crystal structure, which is cubic, is that of spinels MgO, Al_2O_3. In this structure, the oxygen ions ($r = 1.32$ Å) form an approximately compact face-centred cubic array, while the metal ions, whose radius is smaller ($r = 0.6 - 0.8$ Å) are housed in the tetrahedral (A sites) and octahedral (B sites) interstices. The (multiple) cubic cell contains the chemical formula 8 times, i.e. 32 O^{--} ions, 8 M^{2+} ions and 16 Fe^{3+} ions. The distribution of these cations among the A and B sites varies, the two limiting cases being

normal structure	8 M^{2+} in A	16 Fe^{3+} in B ,
inverse structure	8 Fe^{3+} in A	8 Fe^{3+}, 8 M^{2+} in B.

The strongest exchange interaction occurring between ions on different sites ($|J_{AB}| \gg |J_{AA}|$, $|J_{BB}|$) is antiferromagnetic ($J_{AB} < 0$). The saturation magnetic moment $\mathcal{M}_S$ for a chemical formula unit is

normal structure	$\overline{m}_S = -\overline{m}_M + 2\overline{m}_{Fe}$
inverse structure	$\overline{m}_S = -\overline{m}_{Fe} + (\overline{m}_{Fe} + \overline{m}_M) = \overline{m}_M$,

where $\overline{m}_M = Sg\mu_B$ is the maximum projection of the magnetic moment of the M^{2+} ion. For the substances we are considering, the experimental measurements are in agreement with an inverse structure, as well as with $g \simeq 2$ (Table 13.3). Note that zinc ferrite has a normal structure. The non magnetic Zn^{2+} ions lie in A sites, and the Fe^{3+} ions, which occupy B sites, have an antiferromagnetic arrangement ($J_{BB} < 0$). The magnetic properties of ferrites can vary continuously if non stoichiometric compounds are considered in which M^{2+} is a mixture of the form X_{1-x}^{2+} Y_x^{2+}.

13.8.3 Ferrimagnetic Garnets

In ferrimagnetic substances of formula $M_3Fe_5O_{12}$ (or $3\,M_2O_3$, $5\,Fe_2O_3$), where M is yttrium or certain rare earths, the metallic ions M and Fe are trivalent. They have the structure of natural garnet, and the cubic cell contains 96 oxygen ions, with the cations occupying the interstitial sites. Three types of site exist, octahedral [a], tetrahedral (d) and dodecahedral {c}. The M^{3+} ions ($r \simeq 1$ Å) occupy the c sites and the Fe^{3+} ions, whose radius is smaller, occupy the a and d sites, giving the notation $\{M_3\}\,[Fe_2]\,(Fe_3)\,O_{12}$.

The most frequently studied of these garnets is yttrium iron garnet $Y_3Fe_5O_{12}$ (called YIG) where the only magnetic ions Fe^{3+} belong to two sub-lattices $A = [a]$ and $B = (d)$. The strongest exchange integral is J_{AB}. As it is negative, it generates an antiparallel arrangement of the $[Fe_2^{3+}]$ and (Fe_3^{3+}) ions, as shown in the diagram of Figure 13.33c. The saturation magnetic moment $\overline{m}_S$ for a formula unit is then

$$\overline{m}_S = -2\overline{m}_{Fe} + 3\overline{m}_{Fe} = \overline{m}_{Fe} = 5\mu_B \quad ,$$

in good agreement with the experimental value 4.96 μ_B.

13.8.4 Néel Model

We consider a Néel model containing two kinds of ion that form two sub-lattices A and B. In the molecular field approximation, and for the sake of simplicity, only AB interactions are included, and the reduced magnetizations $R_A = \mathcal{M}_A/N_A S_A g_A \mu_B$ and $R_B = \mathcal{M}_B/N_B S_B g_B \mu_B$ are connected by two equations of the form (13.95)

$$R_A = B_{S_A}\left[\frac{S_A g_A \mu_B}{kT}\,(B - q_A R_B)\right] \quad \text{and} \quad R_B = B_{S_B}\left[\frac{S_B g_B \mu_B}{kT}\,(B - q_B R_A)\right] \quad ,$$

where q_A and q_B are related by $q_A N_A S_A g_A = q_B N_B S_B g_B$. This ensures that the energy of the interaction AB is equal to that of the interaction BA.

In the paramagnetic region, these equations are linearized by means of the series expansion (4.45) for B_S, yielding

$$R_A = \frac{B}{q_B}\,\frac{T_{NA}\,(T - T_{NB})}{T^2 - T_{NA}T_{NB}}$$

and an equivalent symmetrical formula for R_B. In this formula T_{NA} and T_{NB} are temperatures having the form (13.97) with $q_2 = 0$ and $q_1 = q_A$, q_B respectively. The total magnetic moment $\mathcal{M} = \mathcal{M}_A + \mathcal{M}_B$ is therefore proportional to B, and the molar susceptibility χ_M is given by

$$\frac{1}{\chi_M} = \frac{T - \theta}{C_M} - \frac{C'}{T - \theta'} \quad , \tag{13.118}$$

with

$$\theta = -2\,\frac{T_{NA}\,T_{NB}}{T_{NA} + T_{NB}} \quad , \qquad \theta' = -\theta \, ,$$

$$C_M = \frac{N_A\,C_{MA} + N_B\,C_{MB}}{N_A + N_B} \quad , \qquad C' = \frac{T_{NA}\,T_{NB}}{C_M}\left(\frac{T_{NA} - T_{NB}}{T_{NA} + T_{NB}}\right)^2$$

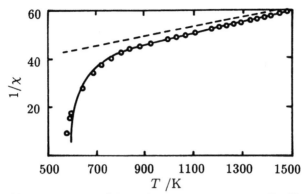

Figure 13.34: Magnetic susceptibility of yttrium iron garnet (YIG). The full curve is equation (13.118) with $\theta = -1525$ K and the dashed curve is the Curie-Weiss law with $C = 50$ K [R. Aléonard and J.C. Barbier, *J. Phys.* (Paris) **20**, 378 (1959)].

where C_{MA} and C_{MB} are the Curie constants (4.50).

When the AA and BB interactions are taken into account, the susceptibility is still given by (13.118), with $\theta < 0$, but θ, $\theta' \neq \theta$ and C' have different definitions. Figure 13.34 shows that the Néel model fits the experimental data satisfactorily.

The susceptibility (13.118) becomes infinite for Néel temperatures T_N below which the sub-lattices exhibit spontaneous magnetization (ferrimagnetic phase). The total spontaneous magnetization of YIG, shown in Figure 13.35, is a decreasing function of T. In more complex cases, the total magnetization may vary in different ways, and, in particular, may vanish at certain temperatures called compensation points. The case of gadolinium iron garnet (GdIG) is shown in Figure 13.35.

The Néel model is confirmed outstandingly well by experiment. At low temperatures it leads to ferrimagnetic magnons that resemble ferromagnetic magnons (§13.4.5). These magnons account for the behaviour of the heat capacity and the magnetization ($T^{3/2}$ power laws) near absolute zero. Lastly, we note the existence of domains resulting from anisotropy.

13.8.5 Applications of Ferrimagnetic Materials

The technological applications of ferrimagnetic materials stem from their ability to display strong spontaneous magnetization while remaining an insulator. For transformer cores, soft ferrites are used (weak coercive field and low hysteresis). For permanent magnets, however, "hard" ferrites are selected with a high coercive field and large anisotropy. An example of this is the highly anisotropic hexagonal ferrite $BaFe_{12}O_{19}$ (BaO, 6 Fe_2O_3), which has a saturation moment per formula unit given by $\overline{m}_S = 20 \ \mu_B$.

13.9 SPIN GLASSES

The low temperature magnetic phase of alloys composed of noble metals and magnetic metals (Fe, Mn) is a spin glass. In these alloys, the magnetic ions, which

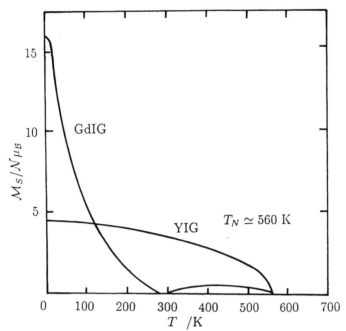

Figure 13.35: Experimental values of spontaneous magnetization $\overline{m}_S$ of a formula unit for the YIG garnets ($Y_3Fe_5O_{12}$) and GdIG ($Gd_3Fe_5O_{12}$) [R. Pauthenet, *Ann. Phys.* (Paris) **3**, 424 (1958)].

are distributed randomly throughout the crystal lattice, interact through a **RKKY** mechanism (§13.3). As the distances between ions are random, the interaction of a given ion with one subset of the other ions favours one orientation, while the other subset favours the opposite orientation. Long distance ferromagnetic or antiferromagnetic order therefore cannot exist. Below a transition temperature T_f, however, a new type of order is established in which the spins are frozen in a variety of directions.

The problem of order in spin glasses has to do with the physics of disordered structures, and it raises such fundamental questions as how to define an order parameter, which statistical methods are appropriate when variables are frozen, or when the system, out of equilibrium, changes very slowly [K. Binder and A.P. Young, *Rev. Mod. Phys.* **58**, 801, (1986)].

13.10 NUCLEAR MAGNETIC ORDER

13.10.1 Interactions

The strong exchange coupling of electrostatic origin that underlies electronic magnetism is absent in nuclear magnetism. The interaction forces are
- magnetostatic dipole-dipole interaction forces, corresponding to the energy (13.13),
- Ruderman-Kittel (RKKY) interactions through intervening electrons.

In the case of ^{3}He, there are in addition exchange interactions involving 3 or 4 nuclei. For insulators of low atomic number (LiF, CaF_2), the only interactions are the ordinary magnetic forces. Predictions can therefore be made with no adjustable parameters.

Since nuclear magnetic moments are 2000 times weaker than electronic moments, the transition to an ordered state occurs at temperatures $10^6 - 10^7$ times lower than for the transitions investigated above, i.e., in the region below a microkelvin. This region is attained by nuclear adiabatic demagnetization (§4.7) but by the nuclear spin system only, since the lattice remains at a temperature of the order of 10 mK. Depending on the spin-lattice coupling, the lifetime of the final state varies from a few minutes (in metals) to several hours (in insulators), which is sufficiently long for observations to be made.

Structures can be investigated by Bragg diffraction of a beam of single energy neutrons, as the neutron-nucleus interaction depends on the relative orientation of the spins (§13.7.1). Nuclear magnetic resonance can also provide relevant structural information.

13.10.2 Experimental Results

For a given crystal, ordered magnetic structures are ferromagnetic or antiferromagnetic depending on the direction of the applied field and the sign of the temperature (§3.7). Antiferromagnetic order is the most common.

In fluorspar CaF_2, the spin of the calcium nuclei is zero, and the fluorine nuclei, which form a simple cubic lattice, consist of the isotope ^{19}F with spin 1/2. Calculation shows that if the magnetic field applied before the adiabatic demagnetization is directed along a [100] axis of the cube or along the [100] diagonal of a face, different antiferromagnetic phases are obtained. When it is directed along a diagonal of the cube [111], the phase is ferromagnetic. Experimental measurements of the magnetic susceptibility are in good agreement with these predictions.

BIBLIOGRAPHY

General References
L.F. Bates, *Modern Magnetism*, Cambridge (1963).
B.I. Bleaney and B. Bleaney, *op. cit.*
S. Chikazumi, *Physics of Magnetism*, Wiley (1964).
J. Crangle, *The Magnetic Properties of Solids*, E. Arnold, London (1977).
D.H. Martin, *Magnetism in Solids*, Iliffe, London (1967).
A.H. Morrish, *Physical Principles of Magnetism*, Wiley (1965).
L.J. de Jongh and A.R. Miedema, *Adv. Phys.* **23**, 1 (1974).
Exchange Interactions
L. Landau and E. Lifschitz, vol. III. *op. cit.* p. 253.
P.W. Anderson, *Solid State Physics* **14**, 99 (1963).
R.J. Elliott, *Magnetism II A*, 406 (1965), edited by G.T. Rado and H. Suhl, Academic Press, New York.
Critical Phenomena
H.E. Stanley, *Introduction to Phase Transitions and Critical Phenomena*, Clarendon Press, Oxford (1971).

M. Ausloos and R.J. Elliott (editors), *Magnetic Phase Transitions, Solid State Sciences* 48, Springer-Verlag (1983).
Spin Waves
F. Keffer, *Handbuch der Physik*, XVIII/2, 1, Springer Verlag (1966).
Nuclear Magnetic Order
A. Abragam and M. Goldman, *Nuclear Magnetism: Order and Disorder*, Clarendon Press, Oxford (1982).
Historical Articles
F. Bloch, *Z. Phys.* **61**, 206 (1930).
W. Heisenberg, *Z. Phys.* **38**, 411 (1926) and **49**, 619 (1928).
E. Ising, *Z. Phys.* **31**, 253 (1925).
L.P. Kadanoff, *Physics* **2**, 263 (1966).
L.D. Landau, *Phys. Z. Sowjet.* **11**, 545 (1937).
L.D. Landau and E.M. Lifschitz, *Phys. Z. Sowjet.* **8**, 153 (1935).
V. Lenz, *Phys. Z.* **21**, 613 (1920).
L. Néel, *Ann. Phys.* (Paris) **17**, 64 (1932); **5**, 232 (1936); **3**, 137 (1948).
L. Onsager, *Phys. Rev.* **65**, 117 (1944).
P. Weiss, *J. Phys.* (Paris) **6**, 667 (1907).
K.G. Wilson, *Phys. Rev. Lett.* **28**, 584 (1972).

COMPREHENSION EXERCISES

13.1 The limiting value of the spontaneous magnetization at absolute zero (saturation magnetization) in nickel is $M_S = 5.10 \times 10^5$ A m^{-1}. Calculate the ferromagnetic effective number of Bohr magnetons n_{eff}. The density and molar mass of nickel are respectively 8.90 g cm^{-3} and 58.71 g mol^{-1} [Ans.: $n_{\text{eff}} = 0.603$].

13.2 The Curie constant per unit mass of nickel in its ferromagnetic phase is equal to $C_m = 6.8 \times 10^{-5}$ m^3 K kg^{-1}. Calculate the effective paramagnetic number of Bohr magnetons p_{eff}. The molar mass of nickel is 58.71 g mol^{-1} [Ans.: $p_{\text{eff}} = 1.59$].

13.3 Repeat the above exercises for europium oxide EuO, where $M_S = 1.88 \times 10^6$ A m^{-1} and $C_m = 5.84 \times 10^{-4}$ m^3 K kg^{-1}. Calculate the effective numbers of Bohr magnetons n_{eff} and p_{eff}. Confirm that these numbers are consistent with a spin $S = 7/2$ and Landé splitting factor $g = 2$. The density and molar mass of europium oxide are 8.21 g cm^{-3} and 168 g mol^{-1} respectively [Ans.: $n_{\text{eff}} = 6.9$; $p_{\text{eff}} = 7.9$].

13.4 The Curie temperatures of nickel and europium oxide are 631 and 69 K respectively. Using the molecular field model, and taking into account only the z nearest neighbours, evaluate the exchange integral J. Take $z = 12$ and the spins respectively equal to $S = 1/2$ and $S = 7/2$ [Ans.: $J = 9.0 \times 10^{-3}$ eV and $J = 4.7 \times 10^{-5}$ eV].

13.5 Estimate the value q of the molecular field B_W at absolute zero in nickel, taking $J = 10^{-2}$ eV, $S = 1/2$, $g = 2$, the number of nearest neighbours being 12 [Ans.: $q \simeq 1000$ T].

13.6 Show that, away from equilibrium, the energy and entropy of a system of spins 1/2 are given by

$$E = -\mathcal{M}B - NwR^2/4$$

$$S = k \ln W = Nk \left[(1 + R) \ln(1 + R) + (1 - R) \ln(1 - R) \right]$$

with $R = M/M_S$. Show that minimizing the thermodynamic potential $F^* = E - TS$ yields the same results as were obtained using the molecular field method (Bragg-Williams method).

13.7 In the above exercise expand the thermodynamic potential F^* to fourth order in R and examine this result in the framework of the Landau theory.

13.8 Show that near the Curie point the molecular field method yields

$$\left(\frac{M}{M_S} \right)^2 = \frac{10}{3} \frac{(S+1)^2}{2S(S+1)+1} \left(\frac{T}{T_C} \right)^2 \frac{T_C - T}{T_C}$$

$$\text{and} \quad \frac{c_V^{(m)}}{R} = \frac{5S(S+1)}{2S(S+1)+1} \left(4 \frac{T}{T_C} - 3 \right) .$$

13.9 Show for the one-dimensional Ising model that the correlation function $\Gamma_1 \equiv \langle \alpha_i \, \alpha_{i+1} \rangle$ is equal to $2kT(\partial \ln Q_N / \partial J)$. Calculate Γ_1 in zero field. Show that in zero field the general expression is

$$\Gamma_p \equiv \langle \alpha_i \, \alpha_{i+p} \rangle = \tanh^p (J/2kT) .$$

13.10 The anisotropy constants of nickel at room temperature are $K_1 = -4.5 \times 10^3$ J m^{-3} and $K_2 = 2.3 \times 10^3$ J m^{-3}. Show that the easy directions of magnetization are [111] plus seven others that are derived from this by symmetry.

13.11 If a spherical particle is uniformly magnetized with magnetization M, then the energy of the magnetic field outside it is $E = 4\pi\mu_0 r^3 M^2/27$ (where r is the radius of the sphere). If the particle consists of two domains, this energy is negligible. Knowing that the surface energy of a Bloch wall in nickel is of the order of 10^{-3} J m^{-2}, show that division into domains is energetically unfavourable for a particle radius smaller than a value r_0 which is to be estimated. Take $M = 5 \times 10^5$ Am^{-1} [Ans.: $r_0 \simeq 2 \times 10^{-8}$ m].

13.12 The saturation moment of one mole of the garnet GdIG (Gd$_3$Fe$_5$O$_{12}$) is equal to $16\mu_B$. Explain this value. The spins of Gd^{3+} and Fe^{3+} are respectively $7/2$ and $5/2$, with $g = 2$ in both cases [Ans.: $16 = 3 \times 7 - (3 - 2) \times 5$].

Chapter 14

Real Gases and Liquids

14.1 INTRODUCTION

In Chapter 5 we considered gases in which interactions between molecules are negligible. This is a good approximation in gases at low densities, and these therefore obey the limiting equation of state for ideal gases. When the density increases moderately (real gases), interactions between molecules begin to have an effect. Two-body, three-body ... interactions must be taken into account, thus forming a basis for series expansions in the thermodynamic functions. At high densities (dense gases and liquids), series expansions no longer suffice and it becomes necessary to revert to special methods.

14.2 REAL GASES

14.2.1 Virial Expansions

At low densities, gases obey the ideal gas law

$$P = \frac{NkT}{V} = \frac{RT}{v} \tag{14.1}$$

where v is the molar volume. As the density increases, the molar volume v decreases and gases deviate from the ideal state. Usually, the equation of state is written as a virial expansion

$$P = \frac{RT}{v}\left[1 + \frac{B_2(T)}{v} + \frac{B_3(T)}{v^2} + \ldots\right] \tag{14.2a}$$

$$\text{or} \qquad v = \frac{RT}{P}\left[1 + B_2'(T)P + B_3'(T)P^2 + \ldots\right] \quad . \tag{14.2b}$$

The virial coefficients B_n and B_n' form two series that are related to each other by

$$B_2' = \frac{B_2}{RT} \quad , \quad B_3' = \frac{B_3 - B_2^2}{(RT)^2} \quad , \quad \ldots \tag{14.3}$$

For convenience, in the following we choose the expansion (14.2a) and denote the

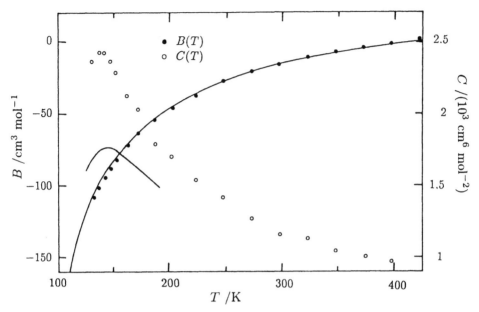

Figure 14.1: Second and third virial coefficients $B(T)$ and $C(T)$ in argon. The curves are the theoretical results (14.20 and 14.25) for a 6-12 Lennard-Jones interaction potential (14.6) with $\sigma = 3.504$ Å and $\epsilon_0/k = 117.7$ K [A. Michels et al., *Physica* **15**, 627 (1949); **24**, 659 (1958) and A.E. Sherwood and J.M. Prausnitz, *J. Chem. Phys.* **11**, 429 (1964)].

second and third virial coefficients by

$$B(T) \equiv B_2(T) \qquad \text{and} \qquad C(T) \equiv B_3(T) \quad . \tag{14.4}$$

The coefficients $B(T)$ and $C(T)$ for argon are shown in Figure 14.1. At room temperature ($T \sim 300$ K), these coefficients are approximately $B = -20$ cm^3 mol^{-1} and $C = 1200$ cm^6 mol^{-2}. At pressure $P = 1$ atm$= 1.013 \times 10^5$ Pa, the molar volume v given by (14.1) is equal to 2.46×10^4 cm^3 mol^{-1} and the two correction terms in (14.2), respectively equal to -8×10^{-4} and 2×10^{-6}, are almost negligible. At $P = 100$ atm, these correction terms become larger, taking the values -0.08 and 0.02. At still higher pressures, higher order coefficients must be included.

The molar free energy

$$f(T, v) = -\int P \, dv = f_0(T) - RT \left[\ln v - \sum_{n=2}^{\infty} \frac{B_n}{n-1} \frac{1}{v^{n-1}} \right]$$

is defined whenever the coefficients $B_n(T)$ and the function $f_0(T)$ are known. In the limit $v \to \infty$, the free energy becomes that of an ideal gas (5.49), i.e.

$$f_{GP}(T, v) = f_0(T) - RT \ln v \quad ,$$

and we have

$$f(T, v) = f_{GP}(T, v) + \sum_{n \geq 2} \frac{B_n(T)}{n-1} \frac{1}{v^{n-1}} \quad . \tag{14.5}$$

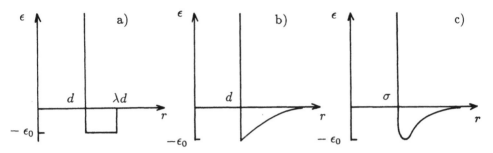

Figure 14.2: Diagram of the interaction potential in the a) square well, b) Sutherland and c) Lennard-Jones 6-12 models.

If the properties of the gas at low pressure (ideal gas) and the virial coefficients are known, expression (14.5) yields all the properties of real gases.

14.2.2 Interactions between Atoms or Molecules

Intermolecular forces are of two types. Long range forces, called van der Waals forces, are attractive. They derive from a potential that varies with distance as $1/r^6$. If the molecules are polar, these forces correspond to interactions between permanent dipoles, statistically averaged over their relative orientations. If the molecules possess no permanent dipole moment, such forces are generated by the interaction between the instantaneous dipole of a molecule and the dipole induced by it in another molecule (London force).

Short range forces appear when the molecules are sufficiently close to each other for their electronic clouds to overlap. They are repulsive and highly directional. The potential from which they derive decreases exponentially, but can also be described by a power law in $1/r^n$ where $n \sim 10 - 15$.

For calculations, simple analytical forms of the interaction potential are used corresponding to the model employed. In particularly simple models, molecules are assumed to have a hard core, i.e. the interaction potential becomes infinite for intermolecular distances less than a certain value d (§6.4.1). The hard sphere model includes no attractive forces (Fig. 6.8b); the square well model has a potential well of width $(\lambda - 1)d$ and depth ϵ_0 (Fig. 14.2a); the Sutherland model contains an attractive part $-\epsilon_0(d/r)^{-n}$ with n equal to about 6 (Fig. 14.2b). In the 6-12 model of Lennard-Jones (Fig. 14.2c), a two-parameter potential is chosen of the form

$$\epsilon(r) = 4\epsilon_0 \left[\left(\frac{\sigma}{r} \right)^{12} - \left(\frac{\sigma}{r} \right)^6 \right] \tag{14.6}$$

which describes the interaction potential between spherical molecules. Other models exist with a larger number of parameters that can describe the interactions between non spherical molecules, either polar or not.

It will be noted that the interaction force between two molecules is modified by the presence of other molecules that polarize them also. As a result, the total interaction energy of N molecules is not the same as the sum of the interaction energies of the molecules taken two by two. Nonetheless, the approximation

$$U_N(\mathbf{r}_1, \ldots \mathbf{r}_N) = \sum_{i<j} \epsilon(r_{ij}) = \frac{1}{2} \sum_{i \neq j} \epsilon(r_{ij}), \tag{14.7}$$

in which only pairs are considered, is often sufficient.

14.2.3 Grand Canonical Partition Function

Grand canonical ensemble methods are the most convenient to get the virial coefficients $B_n(T)$ from the molecular parameters. For simplicity, we consider a monatomic gas. Its energy is

$$E_{r,N} = \sum_{i=1}^{N} \frac{\mathbf{p}_i^2}{2m} + U_N(\mathbf{r}_1, \ldots \mathbf{r}_N) \tag{14.8}$$

when it is in a state (r, N) containing N particles with positions $\mathbf{r}_1, \ldots \mathbf{r}_N$ and momenta $\mathbf{p}_1, \ldots \mathbf{p}_N$. Note that when there is one particle, $U_1 = 0$, and that for two particles, U_2 is the pair interaction potential $\epsilon(r_{12})$.

The grand canonical partition function (12.43)

$$\Xi(T, V, \mu) = \sum_{N=0}^{\infty} e^{\beta N \mu} Q_N(T, V) \tag{14.9}$$

can be written using the canonical partition functions (12.28a)

$$Q_N = \sum_r e^{-\beta E_{r,N}} = \frac{1}{N! \, h^{3N}} \int e^{-\beta E_{r,N}} \, d\mathbf{r}_1 \ldots d\mathbf{r}_N \, d\mathbf{p}_1 \ldots d\mathbf{p}_N \ . \tag{14.10}$$

As the energies $E_{r,N}$ (14.8) are additive for the terms in $\mathbf{p}_i$, the integration can be performed on these $3N$ real variables. From Table 6.1

$$\int_{-\infty}^{+\infty} dp \, e^{-\beta p^2/2m} = (2\pi m k T)^{3/2} \ .$$

The canonical partition functions can therefore be written in the form

$$Q_N = \frac{1}{N!} \left[\frac{V(2\pi m k T)^{3/2}}{h^3} \right]^N Z_N \equiv Q_N^{GP} Z_N \tag{14.11}$$

with $\quad Z_N(T, V) = \frac{1}{V^N} \int e^{-\beta U_N} \, d\mathbf{r}_1 \ldots d\mathbf{r}_N \ . \tag{14.12}$

The grand canonical partition function then becomes

$$\Xi(T, V, \mu) = \sum_{N=0}^{\infty} \frac{1}{N!} z^N Z_N(T, V)$$

where $\quad z(T, V, \mu) = e^{\beta \mu} \frac{V}{h^3} (2\pi m k T)^{3/2} \ . \tag{14.13}$

In the ideal gas limit, the energies U_N are zero and the Z_N are equal to 1. For this particular case we then recover the expressions for Q_N and Ξ in Exercises 12.4 and 12.8.

14.2.4   The Virial Coefficients

General form

The equation of state of a real gas is found by eliminating the chemical potential μ from equations (12.48-49), which can be written here as

$$PV = -\Omega = kT \ln \Xi(T, V, \mu) \tag{14.14}$$

$$N = -\frac{\partial \Omega}{\partial \mu} = kT \frac{\partial \ln \Xi(T, V, \mu)}{\partial \mu} = z \frac{\partial \ln \Xi(T, V, z)}{\partial z} . \tag{14.15}$$

As the partition function Ξ is expressed as a series expansion in integral powers of the variable z, PV and N can also be expanded as integral series of z. This yields

$$PV = kT \sum_{n=1}^{\infty} b_n z^n \qquad \text{and} \qquad \overline{N} \equiv N = \sum_{n=1}^{\infty} n b_n z^n \tag{14.16}$$

where the first coefficients b_n are given by

$$b_1 = Z_1 = 1 \qquad b_2 = \frac{Z_2 - Z_1^2}{2} \qquad b_3 = \frac{Z_3 - 3Z_1 Z_2 + 2Z_1^3}{6} . \tag{14.17}$$

To obtain the equation of state as a virial expansion, z must now be eliminated from the two equations (14.16). Inversion of the series (14.16) for N as a function of z gives

$$z = N - 2b_2 N^2 + \left(8b_2^2 - 3b_3\right) N^3 + \cdots , \tag{14.18}$$

and, substituting into equation (14.16) for PV, we get finally

$$PV = NkT \left[1 - b_2 N + \left(4b_2^2 - 2b_3\right) N^2 + \cdots \right] .$$

On identifying the terms in this expansion with those of (14.2a), we find the following expressions for the virial coefficients

$$B_2 = -\mathcal{N} V b_2 , \qquad B_3 = \mathcal{N}^2 V^2 \left(4b_2^2 - 2b_3\right) , \quad \cdots \tag{14.19}$$

The general expression for the coefficient B_n, given below, was found by H.D. Ursell (1927) using the canonical ensemble. We shall see that these coefficients are independent of the volume V.

Second virial coefficient

From equations (14.12, 17 and 19), the second virial coefficient is

$$B_2 = -\frac{\mathcal{N} V}{2} \left(Z_2 - Z_1^2\right) = -\frac{\mathcal{N}}{2V} \int \left[e^{-\beta U_2} - 1\right] d\mathbf{r}_1 \, d\mathbf{r}_2$$

where $U_2 = \epsilon(r_{12})$. Changing the variable $\mathbf{r}_2 \to \mathbf{r} = \mathbf{r}_2 - \mathbf{r}_1$ and integrating with respect to $\mathbf{r}_1$, gives for the coefficient B_2

$$B_2(T) = -\frac{\mathcal{N}}{2} \int_V \left[e^{-\beta \epsilon(r)} - 1\right] d\mathbf{r} . \tag{14.20}$$

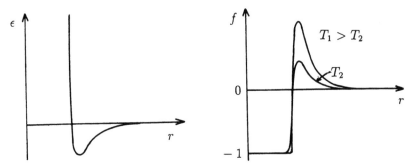

Figure 14.3: Diagram of the interaction potential $\epsilon(r)$ and the function $f(r)$ (14.21) at two temperatures.

Since the interaction potential $\epsilon(r)$ rapidly vanishes as $r \to \infty$, the function

$$f(r) = e^{-\beta\epsilon(r)} - 1 \qquad (14.21)$$

also tends rapidly to zero (Fig. 14.3). The integrand in (14.20) is therefore zero outside the immediate surroundings of a molecule. The coefficient B_2 is thus independent of V.

The integral (14.20) is generally calculated numerically. The result of the integration for a 6-12 Lennard-Jones potential is shown in Figure 14.2. The molecular parameters of the desired potential $\epsilon(r)$ are found by fitting the theoretical curves to the experimental data. Of the two parameter models, the 6-12 Lennard-Jones potential gives the best results (Table 14.1). Potentials with more parameters yield improved fits.

Note that the hard sphere model gives a constant second virial coefficient

$$B_2 = \frac{\mathcal{N}}{2} \int_0^d 4\pi r^2 \, dr = \frac{2\pi\mathcal{N}d^3}{3} \equiv b_0 \ . \qquad (14.22)$$

This coefficient is equal to $4\mathcal{N}$ times the volume $\pi d^3/6$ of one molecule, as each molecule excludes the others within a volume $4\pi d^3/3$ (the volume of the protective sphere). This volume is divided by 2 to avoid double counting. In the square well model, the coefficient B_2 is given by

$$
\begin{aligned}
B_2 &= \frac{\mathcal{N}}{2} \int_0^d 4\pi r^2 \, dr - \frac{1}{2} \int_d^{\lambda d} \left(e^{\beta\epsilon_0} - 1\right) 4\pi r^2 \, dr \\
&= b_0 \left[1 - \left(e^{\beta\epsilon_0} - 1\right)\left(\lambda^3 - 1\right)\right] \ . \qquad (14.23)
\end{aligned}
$$

The square well potential, although unrealistic, has three parameters, and yields a better fit to the data for $B_2 \equiv B(T)$ (Table 14.1) than the 6-12 Lennard-Jones potential.

Higher virial coefficients

To calculate B_n, the integrals Z_N with $N \leq n$ must be calculated from (14.17, 19). For the second coefficient, the calculation of Z_2 requires only $U_2 \equiv \epsilon$. For

Table 14.1: Parameters for the interaction potentials of "spherical" molecules [A.E. Sherwood and J.M. Prausnitz, *J. Chem. Phys.* **11**, 429 (1964)].

gas	Lennard-Jones		square well		
	σ /Å	(ϵ_0/k) /K	d /Å	λ	(ϵ_0/k) /K
A	3.504	117.7	3.067	1.70	93.3
Kr	3.827	164.0	3.278	1.68	136.5
Xe	4.099	222.3	3.593	1.64	198.5
CH_4	3.783	148.9	3.355	1.60	142.5
CF_4	4.744	151.5	4.103	1.48	191.1
C $(CH_3)_4$	7.445	232.5	5.422	1.45	382.6

Z_3, Z_4, ..., the energies U_3, U_4, ... must be considered. Here we assume the additivity approximation (14.7). Z_N is then given by

$$Z_N = \frac{1}{V^N} \int \exp\left[-\beta \sum_{i,j\neq i} \epsilon(r_{ij})\right] d\mathbf{r}_1 \ldots d\mathbf{r}_N .$$

On defining the quantities $f_{ij} = f(r_{ij}) = \exp(-\beta\epsilon_{ij}) - 1$, the integrand in Z_N reduces to a product of $N(N-1)/2$ factors $(1 + f_{ij})$.

It is then straightforward to show that the coefficients b_n (14.17) are given by

$$b_2 = \frac{1}{2V^2} \int f_{12} \, d\mathbf{r}_1 \, d\mathbf{r}_2 ,$$

$$b_3 = \frac{1}{6V^3} \int [f_{12}f_{23}f_{31} + f_{12}f_{23} + f_{23}f_{31} + f_{31}f_{12}] \, d\mathbf{r}_1 \, d\mathbf{r}_2 \, d\mathbf{r}_3, \ldots \quad (14.24)$$

On performing the calculations, we finally obtain

$$B_3(T) = -\frac{\mathcal{N}^2}{3V} \int f_{12}f_{23}f_{31} \, d\mathbf{r}_1 \, d\mathbf{r}_2 \, d\mathbf{r}_3 , \ldots \quad (14.25)$$

It can be seen that the general expression for the b_n and B_n takes the form of an integral in $3n$ dimensions in which the integrand is a sum of products of functions f_{ij}. These products can be symbolized in the following way (Fig 14.4): n circles, numbered from 1 to n, are drawn and each factor f_{ij} is represented by a line connecting the circles i and j. The graphs in Figure 14.4a and b thus correspond respectively to the products f_{12}, $f_{12}f_{23}$, $f_{23}f_{31}$, $f_{31}f_{12}$ and $f_{12}f_{23}f_{31}$. J.E. Mayer (1937) showed that

$$b_n = \frac{1}{n! \, V^n} \int S_n \, d\mathbf{r}_1 \ldots d\mathbf{r}_n \quad \text{and} \quad B_n = -\frac{(n-1)\mathcal{N}^{n-1}}{n!V} \int S'_n \, d\mathbf{r}_1 \ldots d\mathbf{r}_n$$

where S_n is the sum of all connected graphs and S'_n is the sum of all doubly connected graphs. A connected graph is one in which at least one path exists between

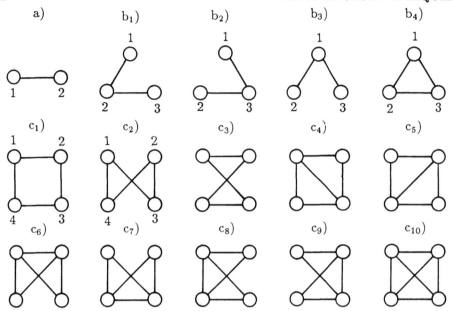

Figure 14.4: Examples of clusters of a) 2, b) 3 and c) 4 particles. The first four graphs are all simply connected for $n \leq 3$. The others are all doubly connected graphs for $n \leq 4$ (there are 28 simply connected graphs $n = 4$). Note that the sets of graphs $(b_1 - b_3)$, $(c_1 - c_3)$ and $(c_4 - c_9)$ are topologically equivalent and yield integrals with the same value.

any two circles. Doubly connected means that at least two paths exist, with no other circles in common (Fig. 14.4).

As the integrand S'_n depends only on the relative distances r_{ij}, the integration gives a factor V which disappears explicitly from B_n. The volume also enters through the limits of the integral, but this makes no contribution since the integrand vanishes by virtue of the f_{ij} as soon as one molecule moves away from the others. It follows that the coefficient B_n is independent of V and is of the order of the volume of $\mathcal{N}$ molecules raised to the power $n - 1$.

The calculation of B_n becomes increasingly complex as n increases and it is performed numerically. The coefficient B_3 for the 6-12 Lennard-Jones potential in shown in Figure 14.1, fitted to the data of $B_2 \equiv B(T)$. The general shape is correct, but there is a quantitative disagreement which does not improve significantly if another form of potential is chosen. Satisfactory agreement with experiment is found only when a non-additive form of the energy $U_3(\mathbf{r}_1, \mathbf{r}_2, \mathbf{r}_3)$ is used. Calculations have rarely been made for $B_4, \ldots$, and anyway the experimental determination of these coefficients becomes increasingly difficult.

Virial coefficients have, however, been calculated for the hard sphere model up to fairly high order. The results are [F.H. Ree and W.G. Hoover, *J. Chem. Phys.* **46**, 4181 (1967)]

$$
\begin{array}{c|c|c|c|c|c|c}
n & 2 & 3 & 4 & 5 & 6 & 7 \\
\hline
B_n/b_0^{n-1} & 1 & 5/8 & 0.2869 & 0.1103 & 0.0386 & 0.0138
\end{array}
\tag{14.26}
$$

14.2.5 Law of Corresponding States

Consider a family of gases whose molecules have a pair-wise interaction with a potential of the form

$$\epsilon(r) = \epsilon_0 \phi \left(\frac{r}{\sigma} \right) \tag{14.27}$$

where ϵ_0 and σ are dimensional parameters that depend on the gas, and ϕ is a fixed function of the reduced distance r/σ. The canonical partition function Q_N of one of the gases is given by (14.11) with

$$
\begin{aligned}
Z_N(T,\,V) &= \frac{1}{V^N} \int \exp\left[-\beta \epsilon_0 \sum_{i<j} \phi\left(\frac{r_{ij}}{\sigma} \right) \right] d\mathbf{r}_1 \ldots d\mathbf{r}_N \\
&= \left(\frac{\sigma^3}{V} \right)^N \int \exp\left[-\beta \epsilon_0 \sum_{i<j} \phi\left(\frac{r_{ij}}{\sigma} \right) \right] \frac{d\mathbf{r}_1}{\sigma^3} \ldots \frac{d\mathbf{r}_N}{\sigma^3} \\
&= \zeta\left(\frac{V}{\sigma^3},\, \frac{kT}{\epsilon_0},\, N \right) .
\end{aligned}
$$

The free energy can then be written in the form

$$F(T,\,V,\,N) = -kT \ln Q_N^{GP} - kT \ln Z_N = F_{GP} - kT \ln \zeta\left(\frac{V}{\sigma^3},\, \frac{kT}{\epsilon_0},\, N \right) ,$$

where F_{GP} is the free energy of the gas at reduced pressure. Since the free energy is an extensive quantity, it follows that

$$F(T,\,V,\,N) = F_{GP}(T,\,V,\,N) - NkT \ln \overline{\zeta}\left(\frac{V}{N\sigma^3},\, \frac{kT}{\epsilon_0} \right)$$

where $\overline{\zeta}$ is a function of the two reduced variables

$$v^* = \frac{V}{N\sigma^3} = \frac{v}{N\sigma^3} \qquad \text{and} \qquad T^* = \frac{kT}{\epsilon_0} . \tag{14.28}$$

The equation of state of the gas is then

$$
\begin{aligned}
P &= -\frac{\partial F}{\partial V} = \frac{NkT}{V} + \frac{kT}{\sigma^3} \frac{\partial \ln \overline{\zeta}}{\partial v^*} (v^*,\, T^*) \\
&= \frac{\epsilon_0}{\sigma^3} \frac{T^*}{v^*} \left[1 + v^* \frac{\partial \ln \overline{\zeta}}{\partial v^*} (v^*,\, T^*) \right] .
\end{aligned}
$$

In terms of the reduced pressure $P^* = P\sigma^3/\epsilon_0$, this equation becomes

$$P^* = \frac{T^*}{v^*} \left[1 + v^* \frac{\partial \ln \overline{\zeta}}{\partial v^*} \right] \equiv P^*(T^*,\, v^*) , \tag{14.29}$$

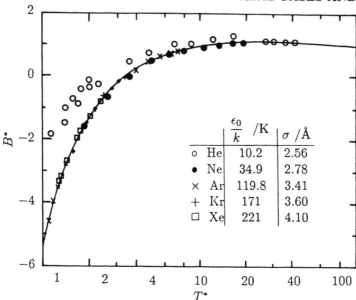

Figure 14.5: Law of corresponding states for the second virial coefficient. The theoretical curve is that of a 6-12 Lennard-Jones potential. Helium deviates from this relation at low temperatures [Helium: W.H. Keesom and W.H. Walstra, *Physica* **13**, 925 (1947), G.P. Nyhoff and W.H. Keesom, *Proc. Kon. Akad. Amsterdam* **31**, 404 (1927) and A. Michels and H. Wouters, *Physica* **8**, 923 (1941); Neon: L. Olborn and J. Otto, *Z. Physik* **38**, 359 (1926); Argon: A. Michels et al., *Physica* **24**, 659 (1958) and E. Whalley and W.G. Schneider, *J. Chem. Phys.* **23**, 1644 (1955); Krypton: J.A. Beattie et al., *J. Chem. Phys.* **20**, 1615 (1952); Xenon: J.A. Beattie et al., *J. Chem. Phys.* **19**, 1219 (1951)].

and we see that all gases belonging to the same family have the same equation of state in reduced variables. This result, called the law of corresponding states, is the explanation of some well known observations. As an illustration, we show the law of corresponding states for the second virial coefficient of the noble gases (Fig. 14.5), namely $B^*(T^*) = B(T)/\mathcal{N}\sigma^3$. It should be pointed out that the third virial coefficient does not obey this law very well, principally on account of the non additivity of the internal energy.

14.2.6 Quantum Effects

Introduction

Hitherto, the quantum character of molecules has entered only through the indistinguishability factor $1/N!$, which turns out to be satisfactory except in the case of the light gases H_2, HD, D_2, ^{3}He and ^{4}He (Fig. 14.5). In our discussion of the second virial coefficient $B_2 \equiv B$ and the canonical partition function Q_2, we shall consider the approximations made on passing from the quantum case

$$Q_2^{qu} = \mathrm{Tr}\, e^{-\beta \hat{H}} = \sum_r < r\, |e^{-\beta \hat{H}}|\, r > \qquad (14.30)$$

to the classical expression

$$Q_2^{cl} = \frac{1}{2!} \int \frac{d\mathbf{r}_1 \, d\mathbf{p}_1}{h^3} \, \frac{d\mathbf{r}_2 \, d\mathbf{p}_2}{h^3} \, e^{-\beta E} \tag{14.31}$$

and give the formulae for various quantum corrections.

To do this, we select as the orthonormal basis space of two-particle states the set $\mid r > \equiv \mid \mathbf{p}_1\mathbf{p}_2 >$. This constitutes an eigenbasis for the momentum operators $\hat{\mathbf{p}}_1$ and $\hat{\mathbf{p}}_2$ and the kinetic energy operator

$$\hat{E}_C = \frac{1}{2m} \left(\hat{\mathbf{p}}_1^2 + \hat{\mathbf{p}}_2^2\right) \ . \tag{14.32}$$

These states are antisymmetric (symmetric) if the particles are fermions (bosons) and correspond to the wave functions

$$\psi_{\mathbf{p}_1,\mathbf{p}_2} (\mathbf{r}_1, \ \mathbf{r}_2) = \frac{1}{h^3} \frac{e^{i(\mathbf{p}_1\mathbf{r}_1+\mathbf{p}_2\mathbf{r}_2)/\hbar} \mp e^{i(\mathbf{p}_2\mathbf{r}_1+\mathbf{p}_1\mathbf{r}_2)/\hbar}}{\sqrt{2}} \ . \tag{14.33}$$

Effect of exchange

In the limit of large volumes, the states $\mid \mathbf{p}_1\mathbf{p}_2 >$ form a continuous set and the quantum expression (14.30) becomes

$$Q_2 = \frac{1}{2!} \int d\mathbf{p}_1 \, d\mathbf{p}_2 \ < \mathbf{p}_1\mathbf{p}_2 \, |e^{-\beta\hat{H}}| \, \mathbf{p}_1\mathbf{p}_2 > \ , \tag{14.34}$$

where the factor $1/2!$ avoids double counting of the identical states $\mid \mathbf{p}_1\mathbf{p}_2 >$ and $\mid \mathbf{p}_2\mathbf{p}_1 >$. This expression, however, does not correctly describe the terms for which $\mathbf{p}_1 = \mathbf{p}_2 \equiv \mathbf{p}$. For fermions, the state $\mid \mathbf{p}\mathbf{p} >$ does not exist (on account of the exclusion principle) but it is included in the integral (14.34). Similarly, for bosons, the state $\mid \mathbf{p}\mathbf{p} >$ corresponding to the wave function

$$\psi_{\mathbf{p}\mathbf{p}} (\mathbf{r}_1, \ \mathbf{r}_2) = \frac{1}{h^3} \, e^{i\mathbf{p}(\mathbf{r}_1+\mathbf{r}_2)/\hbar} \tag{14.35}$$

counts for only half in the integral (14.34). In addition to the integral (14.34), the true partition function therefore contains the expression

$$\mp \frac{1}{2} \int d\mathbf{p} \ < \mathbf{p}\mathbf{p} \, |e^{-\beta\hat{H}}| \, \mathbf{p}\mathbf{p} > \ .$$

The correction term, in which the interactions ($\hat{H} \equiv \hat{E}_C$) can be neglected, then becomes

$$\Delta Q_2^{ex} = \mp \frac{V}{2h^3} \int d\mathbf{p} \, e^{-\beta E_C} \qquad \text{with} \qquad E_C = \frac{\mathbf{p}^2}{m}$$

$$= \mp \frac{V}{2} \frac{(\pi mkT)^{3/2}}{h^3} \ . \tag{14.36}$$

Effect of non-commutativity

The transition from expression (14.34) for Q_2 to its classical form (14.31) would be exact if the operators $\hat{E}_C$ and $\hat{U}_2$ commuted. If this were indeed the case, we would have

$$e^{-\beta\hat{H}} \equiv e^{-\beta(\hat{E}_C + \hat{U}_2)} = e^{-\beta\hat{E}_C} \cdot e^{-\beta\hat{U}_2} \ ,$$

and hence

$$
\begin{aligned}
< \mathbf{p_1 p_2} \, |e^{-\beta\hat{H}}| \, \mathbf{p_1 p_2} > \ &= \ < \mathbf{p_1 p_2} \, |e^{-\beta\hat{E}_C} e^{-\beta\hat{U}_2}| \, \mathbf{p_1 p_2} > \\
&= \ e^{-\beta E_C} < \mathbf{p_1 p_2} \, |e^{-\beta\hat{U}_2}| \, \mathbf{p_1 p_2} > \\
&= \ e^{-\beta E_C} \int d\mathbf{r}_1 \, d\mathbf{r}_2 \, |\psi_{\mathbf{p_1 p_2}}|^2 \, e^{-\beta U_2} \ .
\end{aligned}
$$

For the wave function (14.33), calculation shows that

$$< \mathbf{p_1 p_2} \, |e^{-\beta\hat{H}}| \, \mathbf{p_1 p_2} > = e^{-\beta E_C} \int \frac{d\mathbf{r}_1 \, d\mathbf{r}_2}{h^6} \, e^{-\beta U_2} \ ,$$

which allows us to pass from (14.34) to (14.31).

In fact, the operators $\hat{E}_C$ and $\hat{U}_2 = \epsilon\,(\hat{r}_{12})$ do not commute because the operators $\hat{\mathbf{p}}_i$ and $\hat{\mathbf{r}}_i$ do not commute. It can be shown that a correction to the classical expression (14.31) must be added in the form of a series expansion of even powers of $\hbar$

$$\Delta Q_2^{nc} = -\frac{\hbar^2}{24m} \frac{V(2\pi m)^3}{h^6} \int_0^\infty e^{-\beta\epsilon(r)} \left(\frac{d\epsilon}{dr}\right)^2 4\pi r^2 \, dr + O(\hbar^4) \ . \tag{14.37}$$

Discussion

The correct partition function Q_2 is thus the sum of the classical partition function defined in (14.11,12) and the two correction terms ΔQ_2^{ex} and ΔQ_2^{nc} (14.36,37). The second virial coefficient, calculated as in paragraph 14.2.4, is therefore the sum of the coefficient (14.20) and the correction terms

$$\Delta B_2^{ex} = \pm \frac{\mathcal{N}}{2^{5/2}} \frac{h^3}{(2\pi mkT)^{3/2}} \tag{14.38}$$

and

$$\Delta B_2^{nc} = \frac{\mathcal{N}h^2}{24\pi m(kT)^3} \int_0^\infty e^{-\beta\epsilon(r)} \left(\frac{d\epsilon}{dr}\right)^2 r^2 \, dr + O(\hbar^4) \ . \tag{14.39}$$

The equation for the exchange correction was already obtained in (10.18) and (7.45) in the context of Fermi-Dirac and Bose-Einstein statistics. This correction is small compared with the values of the classical virial coefficient (14.20).

The correction for non-commutativity (14.39), which depends on the interaction potential, is always positive. It becomes large at low temperatures and for light molecules. For a family of gases where the interaction potential is given by (14.27), this correction takes the reduced form

$$\Delta B_2^{*nc} = \frac{\Delta B_2^{nc}}{N\sigma^3} = \frac{\Lambda^{*2}}{12T^{*3}} \int_0^\infty e^{-\phi(x)/T^*} \phi'^2 x^2 \, dx + O\left(\Lambda^{*4}\right) \qquad (14.40)$$

$$\text{with} \qquad \Lambda^* = \frac{h}{\sigma(2\pi m\epsilon_0)^{1/2}} \, . \qquad (14.41)$$

It can be seen that for a given reduced temperature, this correction is proportional to the square of the parameter Λ^*, the de Broglie thermal wavelength at the characteristic temperature ϵ_0/k, reduced to the characteristic length σ. For the rare gas family, this parameter takes the values

	^{3}He	^{4}He	Ne	Ar	Kr	Xe
Λ^*	1.23	1.07	0.24	0.074	0.040	0.025

which explains why the quantum correction is necessary only for helium (Fig. 14.5) and for other light gases (hydrogen and its isotopes). As an illustration, for ^{4}He at 83.5 K,

$$B^{exp} = 10.4 \text{ cm}^3 \text{ mol}^{-1} \qquad \frac{B_2^{cl}}{B^{exp}} = 0.85 \qquad \frac{\Delta B_2^{nc}}{B^{exp}} = 0.16 \qquad \frac{\Delta B_2^{ex}}{B^{exp}} = -0.01 \, ,$$

where 90% of the non-commutativity correction comes from the term in h^2 in the expansion (14.39).

The calculation method for the quantum corrections that we have outlined ceases to be valid at low temperatures. For ^{4}He at 20 K, the term in h^4 of the expansion (14.39) is of the order of that in h^2. In this range of temperature, the partition function

$$Q_2 = \sum_r e^{-\beta E_r}$$

is calculated by examining the bound and the free states in a system of two identical particles that interact with a potential $\epsilon(r)$. The theory is fairly complex and yields good results (Fig. 14.6).

14.3 LIQUIDS

14.3.1 General Considerations

In liquids, the distances between molecules are much smaller than in gases, and the strong interactions can no longer be treated as a series expansion from the limiting case of an ideal gas. The densities of liquids and solids are comparable, and the distances between molecules are also of the same order. Solids and liquids, however, have very different characteristics. In solids, the molecules are ordered and their separations can be described by a discrete series, with variations caused by thermal

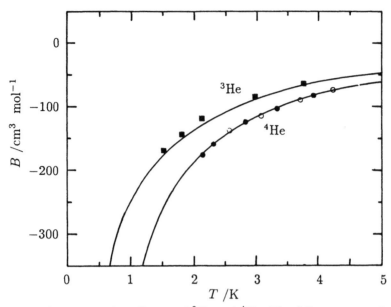

Figure 14.6: Second virial coefficient of ^{3}He and ^{4}He. The full curves are the result of fitting a complete quantum calculation to the data for ^{4}He [W.E. Keller, *Phys. Rev.* **97**, 1 (1955)], and the corresponding prediction for ^{3}He [J.E. Kilpatrick et al., *Phys. Rev.* **97**, 9 (1955)]. The data for ^{3}He were obtained later [W.E. Keller, *Phys. Rev.* **98**, 1571 (1955)].

agitation (Fig. 14.7a). In liquids, this agitation prevents the formation of ordered structures and produces continual rearrangements in the relative positions of the molecules (Fig. 14.7b). Liquids cannot therefore be studied from the starting point of solids. Nonetheless, the distribution of intermolecular distances is reminiscent of that of solids, owing to the impenetrability of the molecules. For this reason, the hard sphere model provides quite a satisfactory approximation (§14.3.5).

The study of liquids is a special subject that relies on the notion of distribution functions, which we shall examine below. It should first be pointed out that a large number of numerical results have been arrived at by computer simulation. Two main methods have been used. In the molecular dynamics method, the movement of N interacting molecules is investigated by step-by-step integration of the equations of motion. The thermodynamic quantities are then calculated in the form of time averages of instantaneous quantities. In the Monte-Carlo method, a large number of configurations of N molecules is randomly generated, and the thermodynamic quantities are calculated in the form of ensemble averages. The Monte-Carlo method is more powerful than that of molecular dynamics, but the advantage of the latter is that it allows the kinetic quantities involved in transport phenomena to be calculated. These numerical methods have certain drawbacks, resulting, for example, from the fact that the number of particles in the systems investigated is limited by the size of the program (10^3 instead of 10^{23}) and that the number of configurations sampled is limited by the calculation time.

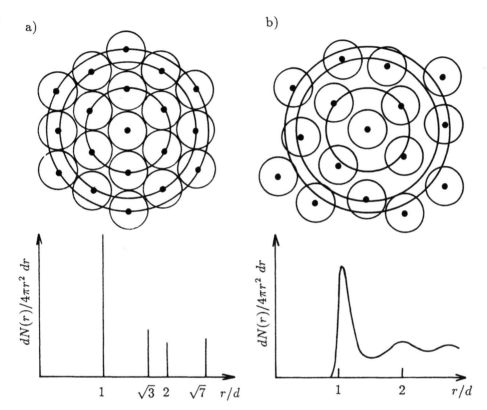

Figure 14.7: Spatial distribution of molecules in a two-dimensional a) solid and b) liquid. The molecules are hard spheres with diameter d. The radii of the concentric circles (relative to d) are 1, $\sqrt{3}$ and 2. The quantity $dN/4\pi r^2\, dr$ is the density of molecules at distance r from a given molecule.

14.3.2 Definition of the Distribution Functions

Consider a system of N point-like particles contained in a volume V at temperature T. The probability density of finding one molecule at $\mathbf{r}_1$, one at $\mathbf{r}_2$, ... one at $\mathbf{r}_N$, irrespective of their momentum, is

$$\rho_N\left(\mathbf{r}_1,\ \ldots\ \mathbf{r}_N\right) = \frac{1}{Q_N}\int e^{-\beta(E_C+U_N)}\frac{d\mathbf{p}_1\ \ldots\ d\mathbf{p}_N}{N!\,h^{3N}} = \frac{e^{-\beta U_N(\mathbf{r}_1,\ \ldots\ \mathbf{r}_N)}}{V^N\,Z_N}\ ,$$

where Z_N is the integral (14.12). This distribution function is normalized to 1 by integrating over all values of $\mathbf{r}_1$, ... $\mathbf{r}_N$, as if the molecules were distinguishable. To take account of their indistinguishability, we write

$$\rho_N\left(\mathbf{r}_1,\ \ldots\ \mathbf{r}_N\right) = N!\,\frac{e^{-\beta U_N(\mathbf{r}_1,\ \ldots\ \mathbf{r}_N)}}{V^N\,Z_N} \tag{14.42}$$

and we take the differential element to be $d\mathbf{r}_1\ \ldots\ d\mathbf{r}_N/N!$.

The probability density of finding n molecules, one at $\mathbf{r}_1$, ... one at $\mathbf{r}_n$, irrespective of their momentum and of the configuration of the other $N - n$, is found by integrating ρ_N with respect to the $N - n$ variables $\mathbf{r}_{n+1}, \ \ldots \ \mathbf{r}_N$. Taking into account the indistinguishability of molecules $n + 1, \ \ldots \ N$, we get

$$
\begin{aligned}
\rho_n\,(\mathbf{r}_1, \ \ldots \ \mathbf{r}_n) \ &= \ \int \rho_N\,(\mathbf{r}_1, \ \ldots \ \mathbf{r}_N) \, \frac{d\mathbf{r}_{n+1} \ \ldots \ d\mathbf{r}_N}{(N-n)!} \\
&= \ \frac{N!}{(N-n)!} \, \frac{\int e^{-\beta U_N(\mathbf{r}_1, \ \cdots \ \mathbf{r}_N)} \, d\mathbf{r}_{n+1} \ \ldots \ d\mathbf{r}_N}{V^N \, Z_N}
\end{aligned}
\tag{14.43}
$$

Note that the integral of ρ_n over all the values $\mathbf{r}_1, \ \ldots \ \mathbf{r}_n$ is not equal to 1, and the same is true if we take the differential element to be $d\mathbf{r}_1 \ \ldots \ d\mathbf{r}_n/n!$. In the latter case we obtain C_N^n, the number of ways of selecting n distinguishable objects out of N.

For an ideal gas in which $U\,(\mathbf{r}_1, \ \ldots \ \mathbf{r}_N) = 0$ and $Z_N = 1$, we have

$$
\rho_n^{GP} \ = \ \frac{N!}{(N-n)!} \, \frac{1}{V^n} \ ,
\tag{14.44}
$$

since the molecules are not correlated. By comparison, this leads us to define the n-particle correlation functions in a system of N molecules as

$$
g_n\,(\mathbf{r}_1, \ \ldots \ \mathbf{r}_n) = \frac{\rho_n(\mathbf{r}_1, \ \ldots \ \mathbf{r}_n)}{\rho_n^{GP}} = \frac{\int e^{-\beta U_N(\mathbf{r}_1, \ \cdots \ \mathbf{r}_N)} \, d\mathbf{r}_{n+1} \ \ldots \ d\mathbf{r}_N}{V^{N-n} \, Z_N} \ ,
\tag{14.45}
$$

in which the integral over $\mathbf{r}_1, \ \ldots \ \mathbf{r}_n$ is V^n.

In the particular case $n = 1$, we find

$$
\rho_1\,(\mathbf{r}_1) = N \, \frac{e^{-\beta U_N(\mathbf{r}_1, \ \cdots \ \mathbf{r}_N)} \, d\mathbf{r}_2 \ \ldots \ d\mathbf{r}_N}{V^N \, Z_N} = \frac{N}{V} \ ,
\tag{14.46}
$$

where the change of variables $\mathbf{r}_i \to \mathbf{r}_{1i} = \mathbf{r}_i - \mathbf{r}_1$ has been applied and use has been made of the fact that U depends only on the intermolecular separation. We see that ρ_1 is the number density of particles and that $g_1\,(\mathbf{r}_1) = 1$.

Radial distribution function

The distribution of molecular distances $dN_p(\mathbf{r})/d\mathbf{r}$ describes the number of pairs of particles i and j for which $\mathbf{r}_{ij} = \mathbf{r}_j - \mathbf{r}_i$ is equal to $\mathbf{r}$ to within $d^3\mathbf{r}$. This distribution depends on the local order and is related to the function

$$
g_2(\mathbf{r}_1, \ \mathbf{r}_2) = \frac{\int e^{-\beta U_N(\mathbf{r}_1, \ \cdots \ \mathbf{r}_N)} \, d\mathbf{r}_3 \ \ldots \ d\mathbf{r}_N}{V^{N-2} \, Z_N}
\tag{14.47}
$$

by $\dfrac{dN_p}{d\mathbf{r}}(\mathbf{r}) = \dfrac{N(N-1)}{2V} \, g_2\,(\mathbf{r}_1, \ \mathbf{r}_2) \simeq \dfrac{N^2}{2V} \, g_2\,(\mathbf{r}_1, \ \mathbf{r}_2)$.   (14.48)

g_2, which depends only on $\mathbf{r} = \mathbf{r}_2 - \mathbf{r}_1$, becomes $g(r)$ in the usual case of isotropic fluids, and is called the radial distribution function. It is normalized to V.

This function can be determined by X-ray or neutron diffraction experiments. Let us consider radiation with wave vector $\mathbf{k}$ impinging on a system of N scattering centres. The amplitude of the radiation scattered at time t along the direction $\mathbf{k}'$ with no change in frequency ($|\mathbf{k}'| = |\mathbf{k}|$) is given by (8.7,8)

$$A\,(\mathbf{k},\ \mathbf{k}') \propto f\,(\mathbf{k},\ \mathbf{k}') \sum_i e^{i(\mathbf{k}'-\mathbf{k})\mathbf{r}_i}$$

where $f(\mathbf{k},\ \mathbf{k}')$ is the structure factor of a scattering centre. The scattered intensity is thus

$$I\,(\mathbf{k},\ \mathbf{k}') \equiv |A^2| \propto |f|^2 \sum_{i,j} e^{i\Delta\mathbf{k}\cdot(\mathbf{r}_i-\mathbf{r}_j)}$$

where $\Delta\mathbf{k} = \mathbf{k}' - \mathbf{k}$. Separating the terms $i = j$ from the terms $i \neq j$ yields

$$I\,(\mathbf{k},\ \Delta\mathbf{k}) \propto |f|^2 \left[N + \sum_{i,j\neq i} e^{i\Delta\mathbf{k}\cdot\mathbf{r}_{ij}} \right] \ .$$

During a measurement, this intensity is integrated over a long time compared to the molecular movements, and the average value of the above expression is given by

$$\begin{aligned}
\bar{I}\,(\mathbf{k},\ \Delta\mathbf{k}) \quad &\propto \quad |f|^2 \left[N + \int dN_p(\mathbf{r})e^{i\Delta\mathbf{k}\cdot\mathbf{r}} \right] \\
&\propto \quad N\,|f|^2 \left[1 + \frac{N}{2V} \int g(r)e^{i\Delta\mathbf{k}\cdot\mathbf{r}}\ d^3\mathbf{r} \right] \ . \qquad (14.49)
\end{aligned}$$

It can thus be seen that the scattered intensity is related to the Fourier transform of the radial distribution function $g(r)$.

Figure 14.8 shows typical experimental results for the scattered intensity and the corresponding radial distribution function. The scattered intensity displays a series of peaks that are broader and less numerous than for crystals, in which long range order prevails. The radial distribution function has several characteristic properties. At large distances, it is equal to 1, demonstrating the absence of long range order in liquids. At short distances it is zero, since the molecules cannot interpenetrate. The first maximum in the distribution corresponds to pairs of molecules in contact. The secondary peaks, corresponding to second, third ... neighbours, become less pronounced as the density decreases or when the temperature increases (Fig. 14.7).

14.3.3   The Thermodynamic Functions

We shall now show that with a knowledge of the radial distribution function $g(r)$ for each temperature T and each density ρ (or molar volume v) all the thermodynamic quantities can be determined, in particular the equation of state and the molar heat capacity c_V.

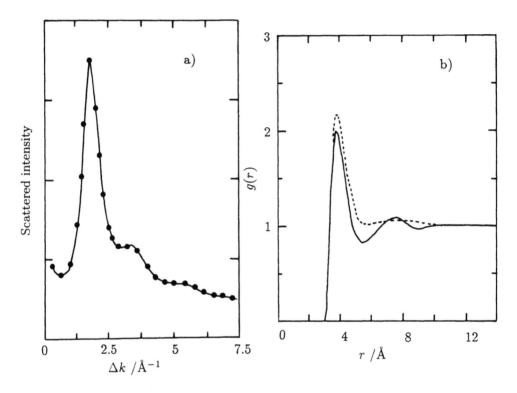

Figure 14.8: a) Scattered intensity and b) radial distribution function $g(r)$ in argon at $T = 148$ K and $\rho = 0.98$ g cm^{-3} ($P = 97.9$ atm). These results were obtained by diffraction of X-rays with wavelength $\lambda = 0.7107$ Å. The dashed curve corresponds to argon at a lower density: $T = 148$ K, $\rho = 0.28$ g cm^{-3} and $P = 42.5$ atm [P.G. Mikolaj and C.J. Pings, *J. Chem. Phys.* **46**, 1401 (1967)].

Virial equation of state

The equation of state can be found from the canonical partition function (14.11) by writing

$$P = -\frac{\partial F}{\partial V} = kT \frac{\partial \ln Q_N}{\partial V} = \frac{NkT}{V} + kT \frac{\partial \ln Z_N}{\partial V} \quad . \tag{14.50}$$

To specify the last derivative, we consider the quantity Z_N defined in (14.12) for a volume $\alpha^3 V$, where α is a real number

$$Z_N(T, \, \alpha^3 V) = \frac{1}{\alpha^{3N} V^N} \int_{\alpha^3 V} e^{-\beta U_N(\mathbf{r}_1, \, \cdots \, \mathbf{r}_N)} \, d\mathbf{r}_1 \, \ldots \, d\mathbf{r}_N \quad .$$

Changing the variables $\mathbf{r}_i \to \alpha\mathbf{r}_i$ yields

$$Z_N(T,\ \alpha^3 V) = \frac{1}{V^N} \int_V e^{-\beta U_N(\alpha\mathbf{r}_1,\ \cdots\ \alpha\mathbf{r}_N)}\ d\mathbf{r}_1\ \cdots\ d\mathbf{r}_N\ .$$

We now differentiate the logarithm of this expression with respect to α

$$\frac{d\ln Z_N(T,\ \alpha^3 V)}{d\alpha} = -\frac{\beta}{Z_N}\frac{1}{V^N}\int \sum_i \mathbf{r}_i \cdot \frac{\partial U_N}{\partial \mathbf{r}_i}(\alpha\mathbf{r}_1,\ \cdots\ \alpha\mathbf{r}_N)\ e^{-\beta U_N}\ d\mathbf{r}_1\ \cdots\ d\mathbf{r}_N\ .$$

This derivative, taken at $(T,\ \alpha^3 V)$, is equal to $3\alpha^2 V \partial\ln Z_N/\partial V$. On choosing $\alpha = 1$, we thus get

$$3V\frac{\partial\ln Z_N}{\partial V} = -\frac{\beta}{Z_N}\frac{1}{V^N}\int \sum_i \mathbf{r}_i \cdot \frac{\partial U_N}{\partial \mathbf{r}_i} e^{-\beta U_N}\ d\mathbf{r}_1\ \cdots\ d\mathbf{r}_N$$

$$= \beta < -\sum_i \mathbf{r}_i \cdot \frac{\partial U_N}{\partial \mathbf{r}_i} >\ .$$

Here we recognize the canonical mean value of what is called the virial of the interaction forces in classical mechanics

$$\mathcal{V} = -\sum_i \mathbf{r}_i \cdot \frac{\partial U_N}{\partial \mathbf{r}_i} = \sum_i \mathbf{r}_i \cdot \mathbf{F}_i\ , \tag{14.51}$$

where $\mathbf{F}_i$ is the force applied to the ith particle. Equation (14.50) then becomes

$$PV = NkT + \frac{1}{3} < \mathcal{V} >\ . \tag{14.52}$$

From now on we assume that interactions between molecules are purely binary, which is a sufficiently good approximation in liquids. The interaction virial is then

$$\mathcal{V} = \sum_i \mathbf{r}_i \cdot \left(\sum_j \mathbf{f}_{ij}\right) = \frac{1}{2}\sum_{i,j\neq i}(\mathbf{r}_i - \mathbf{r}_j)\cdot \mathbf{f}_{ij}$$

where the law of action and reaction $\mathbf{f}_{ji} = -\mathbf{f}_{ij}$ has been used. As the binary interaction force $\mathbf{f}_{ij}$ deriving from the potential $\epsilon(r)$ is directed along $\mathbf{r}_{ij} = \mathbf{r}_j - \mathbf{r}_i$, we finally get

$$\mathcal{V} = -\frac{1}{2}\sum_{i,j\neq i} r_{ij}\epsilon'(r_{ij})\ . \tag{14.53}$$

The canonical mean of $\mathcal{V}$ is found by considering the distribution of distances between pairs, i.e.

$$< \mathcal{V} >= -\int dN_p(\mathbf{r}) \times r\epsilon'(r) = -\frac{N^2}{2V}\int 4\pi r^2\ dr \times g(r)r\epsilon'(r)\ , \tag{14.54}$$

and the equation of state of the gas takes the form

$$P = \frac{NkT}{V} - \frac{2\pi}{3} \frac{N^2}{V^2} \int r^3 \, g(r)\epsilon'(r) \, dr \quad . \tag{14.55}$$

This is called the virial equation of state. The relation between this equation and the virial expansion (14.2) becomes apparent when $g(r)$ is expanded as a function of the variable $1/v$, i.e.

$$g(r \; ; \; T, \; v) = g^{(0)}(r \; ; \; T) + \frac{1}{v} \, g^{(1)}(r \; ; \; T) + \cdots \tag{14.56}$$

On substituting into (14.55), the following expansion is obtained

$$P = \frac{NkT}{V} \left[1 - \frac{2\pi\mathcal{N}}{3kT} \times \frac{1}{v} \sum_{n=0}^{\infty} \frac{1}{v^n} \int r^3 \, g^{(n)}(r)\epsilon'(r) \, dr \right]$$

which can be identified with the virial expansion on setting

$$B_{n+2}(T) = -\frac{2\pi\mathcal{N}}{3kT} \int r^3 \, g^{(n)}(r) \, \epsilon'(r) \, dr \quad . \tag{14.57}$$

On comparing with $B_2(T)$, it can be shown (cf. comprehension exercise 14.3) that the first term in the expansion of g is

$$g^{(0)}(r \; ; \; T) = e^{-\beta\epsilon(r)} \equiv f(r) + 1 \quad . \tag{14.58}$$

This relation explains the analogy between the shapes of the curves in Figures 14.3 and 7.

Heat capacity

We first determine the internal energy of the system. From (12.29), this is

$$U = -\frac{\partial \ln Q_N}{\partial \beta} = \frac{3}{2} \, NkT - \frac{\partial \ln Z_N}{\partial \beta} \quad .$$

To derive equation (14.12) for Z_N, we have

$$\frac{\partial \ln Z_N}{\partial \beta} = -\frac{1}{Z_N V^N} \int U_N \, e^{-\beta U_N} \, d\mathbf{r}_1 \, \ldots \, d\mathbf{r}_N \quad ,$$

which, except for the sign, is the same as the canonical mean of the interaction energy $< U_N >$. The internal energy then becomes

$$U \equiv < E > = \frac{3}{2} \, NkT + < U_N > \quad , \tag{14.59}$$

which shows that the mean kinetic energy of translation of a system is equal to $3NkT/2$ even when interactions are present. This result stems from the fact that we have assumed that the interactions are independent of velocity (or momentum).

Assuming that the interactions are purely binary, we have

$$U_N = \frac{1}{2} \sum_{i,j \neq i} \epsilon(r_{ij})$$

the canonical average of which is found from the distribution of distances

$$< U_N > = \int dN_p(\mathbf{r}) \, \epsilon(r) = \frac{N^2}{2V} \int 4\pi r^2 \, dr \, g(r)\epsilon(r) \ .$$

Finally, the internal energy is

$$U = \frac{3}{2} NkT + 2\pi \frac{N^2}{V} \int r^2 g(r)\epsilon(r) \, dr \qquad (14.60)$$

and the molar heat capacity at constant volume is given by

$$c_V = \frac{3}{2} R + 2\pi \frac{N^2}{v} \int r^2 \frac{dg(r)}{dT} \epsilon(r) \, dr \ . \qquad (14.61)$$

14.3.4 Yvon-Born-Green Equations

Only approximate determinations of the radial distribution function $g(r)$ have been made using (14.47). In one method, we write

$$e^{-\beta U_N} = \exp\left[-\beta \sum_{i,j<i} \epsilon(r_{ij})\right] = \prod_{i,j<i} (1 + f_{ij})$$

where $f_{ij} = f(r_{ij})$ is given by (14.21). Substitution into (14.47) yields a series expansion in N/V for $g_2 \equiv g$

$$g_2(\mathbf{r}_1, \mathbf{r}_2) \equiv g(\mathbf{r} = \mathbf{r}_2 - \mathbf{r}_1) = e^{-\beta\epsilon(r)}\left[1 + \frac{N}{V} \int f_{13}f_{23} \, d\mathbf{r}_3 + \dots\right], \qquad (14.62)$$

which corresponds to the series expansion (14.56). This equation, however, cannot be used at high densities.

Another method that has played a leading role was developed by J. Yvon (1935) and by M. Born and H.S. Green (1946). In this method, expression (14.45) for the correlation function g_n is differentiated with respect to one of its variables, e.g. $\mathbf{r}_1$. Thus

$$\frac{\partial g_n}{\partial \mathbf{r}_1} = -\frac{\beta}{V^{N-n}Z_N} \int \frac{\partial U_N}{\partial \mathbf{r}_1} e^{-\beta U_N} \, d\mathbf{r}_{n+1} \cdots d\mathbf{r}_N \ .$$

Assuming additivity, we have

$$\frac{\partial U_N}{\partial \mathbf{r}_1} = \sum_{i=2}^{N} \frac{\partial \epsilon}{\partial \mathbf{r}_1}(\mathbf{r}_{1i}) \ ,$$

which, on substituting, gives

$$\frac{\partial g_n}{\partial \mathbf{r}_1} = -\frac{\beta}{V^{N-n}Z_N} \left[\sum_{i=2}^{n} \frac{\partial \epsilon}{\partial \mathbf{r}_1}(\mathbf{r}_{1i}) \int e^{-\beta U_N} \, d\mathbf{r}_{n+1} \cdots d\mathbf{r}_N \right.$$

$$\left. + \sum_{i=n+1}^{N} \int \frac{\partial \epsilon}{\partial \mathbf{r}_1}(\mathbf{r}_{1i}) \, e^{-\beta U_N} \, d\mathbf{r}_{n+1} \cdots d\mathbf{r}_N \right] .$$

In the first integral a factor of g_n can be seen. In the second, the $N - n$ terms are equal. Hence

$$\frac{\partial g_n}{\partial \mathbf{r}_1} = -\beta g_n \sum_{i=2}^{n} \frac{\partial \epsilon}{\partial \mathbf{r}_1}(\mathbf{r}_{1i})$$

$$- \frac{(N-n)\beta}{V^{N-n}Z_N} \int \frac{\partial \epsilon}{\partial \mathbf{r}_1}(\mathbf{r}_{1,n+1}) \, d\mathbf{r}_{n+1} \int e^{-\beta U_N} \, d\mathbf{r}_{n+2} \cdots d\mathbf{r}_N$$

$$= -\beta g_n \sum_{i=2}^{n} \frac{\partial \epsilon}{\partial \mathbf{r}_1}(\mathbf{r}_{1i}) - \frac{(N-n)\beta}{V} \int \frac{\partial \epsilon}{\partial \mathbf{r}_1}(\mathbf{r}_{1,n+1}) \, g_{n+1} d\mathbf{r}_{n+1}. \quad (14.63)$$

These differential equations are called Yvon-Born-Green (or BGY) equations. For $n = 1$, the equation becomes

$$\frac{\partial g_1}{\partial \mathbf{r}_1} = -\frac{(N-1)\beta}{V} \int \frac{\partial \epsilon}{\partial \mathbf{r}_1}(\mathbf{r}_{12}) \, g_2(\mathbf{r}_1, \mathbf{r}_2) \, d\mathbf{r}_2 = 0 ,$$

as g_2 does not depend on the direction of $\mathbf{r}_{12} = \mathbf{r}_2 - \mathbf{r}_1$. This is in accord with the result $g_1 = 1$ (§14.3.2). For $n = 2$, we have

$$\frac{\partial g_2}{\partial \mathbf{r}_1} = -\beta g_2 \frac{\partial \epsilon}{\partial \mathbf{r}_1}(\mathbf{r}_{12}) - \frac{(N-2)\beta}{V} \int \frac{\partial \epsilon}{\partial \mathbf{r}_1}(\mathbf{r}_{13}) \, g_3(\mathbf{r}_1, \mathbf{r}_2, \mathbf{r}_3) \, d\mathbf{r}_3 , \quad (14.64)$$

and so on. This series of equations is antirecurrent, since to determine g_n a knowledge of g_{n+1} is required. The equations can therefore be solved only approximately. The simplest method for solving these equations consists in calculating g_2 from (14.64) by adopting for g_3 the superposition approximation due to J.G. Kirkwood (1935)

$$g_3(\mathbf{r}_1, \mathbf{r}_2, \mathbf{r}_3) = g_2(\mathbf{r}_1, \mathbf{r}_2) \, g_2(\mathbf{r}_2, \mathbf{r}_3) \, g_2(\mathbf{r}_3, \mathbf{r}_1) . \quad (14.65)$$

The radial distribution function g_r is then the solution of an integro-differential equation which can be solved numerically. The results obtained are in fairly good agreement with experiment.

Other approximate equations have been suggested to determine $g(r)$, for example the equations of Kirkwood, of Percus and Yevick (PY) and the so-called HNC (hypernetted chain). These are integral equations, which means that $g(r)$ is given as a function of an integral containing g. They are therefore simpler than the BGY equation in the superposition approximation. The PY method gives the best results for temperatures and molar volumes greater than their critical values.

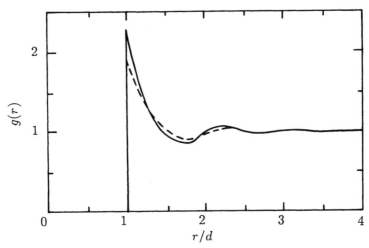

Figure 14.9: Radial distribution function $g(r)$ in the hard sphere model for $\rho/\rho_0 = 0.372$ (full line) and $\rho/\rho_0 = 0.298$ (dashed line). ρ_0 corresponds to compact packing of spheres of diameter d ($\rho_0 = 1.77$ g cm^{-3} and $d = 3.76$ Å for argon [F.H. Ree, Y.T. Lee and T. Ree, *J. Chem. Phys.* **55**, 234 (1971)].

14.3.5 Theoretical Results

The above methods have inspired a large number of scientific investigations. The validity of the various models or approximations is in general tested by comparing with the results of numerical simulations. These allow access to quantities that are of theoretical importance but cannot be obtained directly by experimental means.

We first give the results of numerical simulations for the hard sphere model. They are independent of temperature, since the hard sphere potential has no intrinsic energy scale. Figure 14.9 shows the radial distribution function $g(r)$, whose oscillatory shape is qualitatively similar to that of real fluids (Fig. 14.8b). The oscillations become indistinct as the density decreases. Figure 14.10 shows the equation of state of a substance that obeys this model. Two branches appear in the curve, corresponding to the fluid and solid phases. The limiting molar volume of the latter state occurs for compact packing of spheres of diameter d, i.e. $v_0 = \mathcal{N}d^3/\sqrt{2}$. For the fluid branch, an excellent analytical approximation is at hand [N.F. Karnahan and K.E. Starling, *J. Chem. Phys.* **51**, 635 (1969)] based on the observation that the virial coefficients B_n (for $n \leq 7$) in the hard sphere model (14.26) are approximately given by

$$B_n \simeq (n^2 + n - 2) \times (b/4)^{n-1} \qquad \text{with} \qquad b = \mathcal{N}2\pi d^3/3 \ .$$

Using this expression for all values of n, the virial equation (14.2) can be summed, to yield

$$\frac{Pv}{RT} = \frac{1 + \eta + \eta^2 - \eta^3}{(1 - \eta)^3} \qquad \text{with} \qquad \eta = \mathcal{N}\frac{\pi d^3}{6v} = \frac{\pi}{3\sqrt{2}}\frac{v_0}{v} \ . \qquad (14.66)$$

It can be observed that this equation of state exhibits a singularity for $\eta = 1$ or

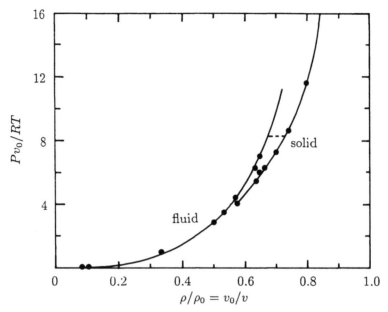

Figure 14.10: Equation of state in the hard sphere model. The points are the results of numerical simulations. The curves represent equations (14.66 and 67). ρ_0 and v_0 correspond to compact packing of spheres ($\rho_0 = 1.77$ g cm^{-3} for argon) [B.J. Alder and T.E. Wainwright, *J. Chem. Phys.* **33**, 1439 (1960)]. The position of the melting plateau (dashes) corresponds to equality of the chemical potentials in the two phases [W.G. Hoover and F.H. Ree, *J. Chem. Phys.* **49**, 3609 (1968)].

$v = \pi v_0/3\sqrt{2} = 0.74\ v_0$, which lies outside the physical range for hard spheres. For the solid branch, the equation of state

$$\frac{Pv}{RT} = \frac{3}{1 - v_0/v}\ ,\tag{14.67}$$

found in certain models, is also satisfactory. Both branches are bounded. At low densities only the fluid phase occurs, and at high densities only the solid phase exists. At intermediate densities both phases occur. This is the zone in which the fluid-solid phase transition must take place.

The hard sphere model qualitatively displays the principal characteristics of liquids. The absence of attraction between molecules, however, leaves it incapable of explaining the existence of two disordered phases, liquid and gas, since the solid phase occurs only as a result of steric hindrance. When a 6-12 Lennard-Jones potential is assumed, the two fluid phases are found, and quantitative agreement with the radial distribution function and with the experimental equation of state becomes satisfactory. This is the case for argon on taking $\sigma = 3.4$ Å and $\epsilon/k = 120$ K [L. Verlet, *Phys. Rev.* **159**, 98 (1967) and **165**, 201 (1968)].

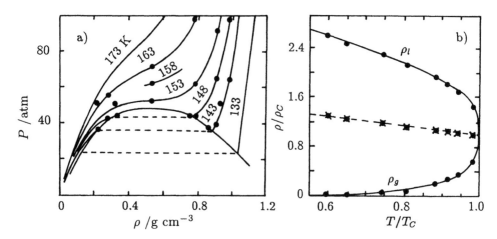

Figure 14.11: Equation of state of argon. a) $(\rho,\ P)$ isotherms. The dashed lines are the liquefaction lines under the saturation curve [P.G. Mikolaj and C.J. Pings, *J. Chem. Phys.* **46**, 1401 (1967)]. b) Densities ρ_l and ρ_g of the liquid and gas phases ($\rho_C = 0.531$ g cm^{-3} and $T_C = 150.7$ K). The curve describes the behaviour of the data for several other gases in reduced co-ordinates. The values of $(\rho_l + \rho_g)/2$ are plotted as crosses and the dashed line is the straight line diameter [E.A. Guggenheim, *J. Chem. Phys.* **13**, 253 (1945)].

14.4 LIQUID-VAPOUR PHASE TRANSITION

14.4.1 Experimental Investigation

Below a certain temperature T_C called the critical temperature, the $P(v,\ T_i)$ or $P(\rho,\ T_i)$ isotherms of a gas, where $\rho = M/v$, exhibit a horizontal region. This is the liquefaction plateau, where two phases coexist, one called liquid, the other vapour (Fig. 14.11a). The ordinate $P_e(T)$ of this plateau is called the saturated vapour pressure and the area of the surface $(\rho,\ P)$ thus created is called the region of coexistence. The limit of this region (saturation curve), gives the densities of the liquid and vapour phases ρ_l and ρ_g. These densities, which are functions of T (or of P_e) only, are shown in Figure 14.11b. The values of $(\rho_l + \rho_g)/2$ for these curves are linear functions of T (law of straight line diameters, L. Cailletet and E. Mathias, 1886). Near the critical temperature, ρ_l and ρ_g tend towards the same value ρ_C, called the critical density, where

$$\rho_l - \rho_g \propto \left(\frac{T_C - T}{T_C}\right)^{\beta}. \tag{14.68}$$

The value of the critical exponent β is close to 0.35 for all gases (Table 14.2).

   At the critical temperature T_C, the isotherm has a point of inflection with a horizontal tangent for $v = v_C$ or $\rho = \rho_C$, such that

$$\frac{\partial P}{\partial v} = 0 \qquad \frac{\partial^2 P}{\partial v^2} = 0 \qquad \text{or} \qquad \frac{\partial P}{\partial \rho} = 0 \qquad \frac{\partial^2 P}{\partial \rho^2} = 0 \ . \tag{14.69}$$

Table 14.2: Physical data at the critical point for certain gases. The critical exponents β and δ are defined in (14.68 and 71) [J.M.H. Levelt Sengers et al., *J. Phys. Chem. Ref. Data* **5**, 1 (1976)].

	P_C /10^6 Pa	ρ_C /kg m^{-3}	T_C / K	$Z_C = P_C v_C / R T_C$	β	δ
^{3}He	0.1168	41.45	3.310	0.307	0.358 ± 0.005	4.26 ± 0.04
^{4}He	0.2274	69.6	5.190	0.303	0.356 ± 0.006	4.34 ± 0.06
Ar	4.865	535	150.7	0.290		
Kr	5.493	908	209.3	0.291		
Xe	5.840	1110	289.7	0.287	0.35 ± 0.04	4.5 ± 0.3
O$_2$	5.043	436.2	154.6	0.288	0.353	4.37
CO$_2$	7.375	467.8	304.1	0.274	0.349 ± 0.005	4.44 ± 0.06
H$_2$O	22.06	322.2	647.1	0.229	0.350 ± 0.013	4.50 ± 0.13

The ordinate P_C at the point of inflection is called the critical pressure, and the point with co-ordinates (ρ_c, P_C) is the critical point. The critical compressibility factor

$$Z_C = \frac{P_C v_C}{RT_C} = \frac{M}{R}\frac{P_C}{\rho_C T_C} \tag{14.70}$$

is a number close to 0.30, even though individual critical co-ordinates vary in a wide range (Table 14.2). This is a manifestation of the law of corresponding states. Near the critical point, the critical isotherm has a power law dependence

$$|P - P_C| \propto |\rho - \rho_C|^\delta \quad , \tag{14.71}$$

where δ is a critical exponent close to 4.4.

Above the critical temperature, there is only one phase that cannot be described either as liquid or as vapour, but which, depending on its density, is closer to a liquid or to a gas. The $P(\rho, T_i)$ isotherms increase monotonically, i.e. the compressibility coefficient

$$\chi_T = -\frac{1}{v}\frac{\partial v}{\partial P} = \frac{1}{\rho}\frac{\partial \rho}{\partial P} \tag{14.72}$$

remains strictly positive. This coefficient diverges at the critical point with

$$\chi_T(T, v_C) \propto \left(\frac{T - T_C}{T_C}\right)^{-\gamma} \tag{14.73}$$

where γ is a critical exponent close to 1.2. The constant volume heat capacity c_V of the substance is a function of T and of P. It diverges at the critical point, and we may write

$$c_V(T, v_C) \propto \left(\frac{T - T_C}{T_C}\right)^{-\alpha} \quad . \tag{14.74}$$

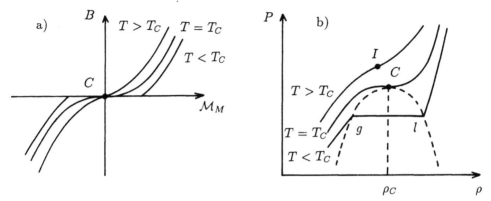

Figure 14.12: Sketch of the isotherms of the equation of state. a) Magnetic substance with variables $(\mathcal{M}_M, B)$. b) Chemical substance with the variables (ρ, P). The point I is the point of inflection of the isotherm. For $T < T_C$, I is chosen such that $\rho_I = \rho_C$.

Measurements of α are difficult and do not allow a distinction to be made between a logarithmic divergence ($\alpha = 0$) and a power law divergence with $\alpha = 0.1$.

14.4.2 Qualitative Analogy Between Chemical and Magnetic Materials

We have just seen that chemical substances have unusual properties in the neighbourhood of a critical point. In Chapter 13 we saw similar behaviour in magnetic materials near the Curie temperature. The analogy between the properties of the two types of substance can be pursued on inspection of Figures 14.12 and 13. It can be seen that the Curie point C with co-ordinates $(0, 0, T_C)$ and the critical point C with co-ordinates (P_C, ρ_C, T_C) play the same role and that there is a correspondence between the variables

$$T \to T \qquad B \to P - P(T, \rho_I) \qquad \mathcal{M}_M \to \rho - \rho_I \ , \qquad (14.75)$$

where I is the point of inflection of the isotherms and where $\rho_I = \rho_C$ for $T < T_C$. The analogy continues in the definition of the magnetic and chemical critical exponents (§13.2.1) and also in their experimental values, which are close to each other. In paragraph 14.4.5 an argument will be set out that provides a theoretical basis for this analogy.

14.4.3 Maxwell Construction

Although the methods described in the above paragraphs are exact, in practice they give rise only to approximate equations of state. As a result, instead of a liquefaction plateau, the isotherms of these equations display undulations where certain portions have a negative compressibility coefficient χ_T (Fig. 14.14). This result is inconsistent with the principles of classical thermodynamics, as well as with a theorem of statistical thermodynamics [L. van Hove, *Physica* **15**, 951 (1949)] which require that χ_T should be a positive quantity or zero.

The liquefaction plateau for these equations is found by applying the Maxwell construction. This method determines the ordinate $P_o(T)$ of the plateau from the

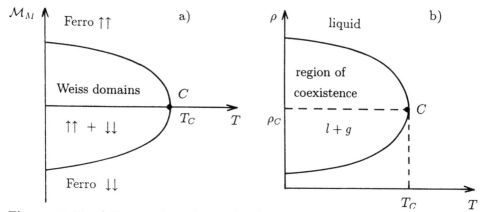

Figure 14.13: a) Region of coexistence for ferromagnetic islets (Weiss domains) with various orientations. b) Region of coexistence of the liquid and vapour phases in a chemical substance.

condition

$$\int_{v_l}^{v_g} P \, dv = P_e(T) \, (v_g - v_l) \quad . \tag{14.76}$$

This equality can be demonstrated in classical thermodynamics by observing that coexistence of phases l and g in a state a can be more stable than the state a' (Fig. 14.14). The curved regions ll' and gg' then correspond to metastable liquid and vapour phases that have been observed. The section $l'g'$ describes an unstable phase, and cannot exist.

Exercise 14.1 *Maxwell construction near the critical point*
   In the neighbourhood of the critical point, many equations of state can be expressed as a series expansion $p = a_0 + a_1\phi + a_2\phi^2 + a_3\phi^3$

$$\text{with} \qquad p = \frac{P - P_C}{P_C} \qquad \text{and} \qquad \phi = \frac{v - v_C}{v_C} \quad , \tag{14.77}$$

where the parameters a_i depend on the temperature. Using the Maxwell construction, determine the values of

$$\xi = \frac{\phi_g + \phi_l}{2} = \frac{v_g + v_l - 2v_C}{2v_C} \qquad \text{and} \qquad \zeta = \frac{\phi_g - \phi_l}{2} = \frac{v_g - v_l}{2v_C} \tag{14.78}$$

as a function of the parameters a_i. (The values of v_l and v_g can then be deduced.)

Solution
   In reduced variables, expression (14.76) becomes

$$\int_{\phi_l}^{\phi_g} p \, d\phi = p_e \, (\phi_l - \phi_g) \quad . \tag{14.79}$$

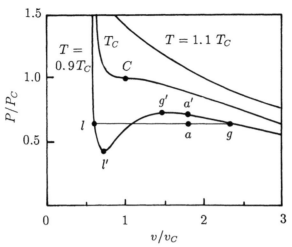

Figure 14.14: Maxwell construction for a van der Waals gas. The plateau between points l and g, found by the Maxwell construction, and the isotherm define two regions of equal area according to (14.76).

This relation, together with the equation of state at l and g

$$p_e = a_0 + a_1\phi_l + a_2\phi_l^2 + a_3\phi_l^3$$
$$p_e = a_0 + a_1\phi_g + a_2\phi_g^2 + a_3\phi_g^3 \qquad (14.80)$$

yields three equations in the three unknowns p_e, ϕ_l and ϕ_g. Upon integrating over ϕ and transforming to the variables ξ and ζ, relation (14.77) becomes

$$p_e = a_0 + \frac{a_1}{2}\,\xi + \frac{a_2}{12}\,(3\xi^2 + \zeta^2) + \frac{a_3}{8}\,(\xi^3 + \xi\zeta^2) \quad . \qquad (14.81)$$

Moreover, on taking the sum and the difference of equations (14.80), we get

$$p_e = a_0 + \frac{a_1}{2}\,\xi + \frac{a_2}{4}\,(\xi^2 + \zeta^2) + \frac{a_3}{8}\,(\xi^3 + 3\xi\zeta^2) \qquad (14.82a)$$
$$0 = a_1 + a_2\xi + \frac{a_3}{4}\,(3\xi^2 + \zeta^2) \quad . \qquad (14.82b)$$

Setting (14.81) and (82a) equal, we find

$$\xi = -\frac{2\,a_2}{3\,a_3} \qquad (14.83a)$$

and substituting into (14.82b), we finally get

$$\zeta^2 = \frac{4}{a_3^2}\left(\frac{a_2^2}{3} - a_1 a_3\right) \quad . \qquad (14.83b)$$

14.4.4 van der Waals Equation of State

J.D. van der Waals (1873) proposed an equation of state interpolating between that of an ideal gas ($Pv = RT$) and an incompressible fluid ($v = b$). This equation was the first that described the liquid-vapour transition using Maxwell's construction and explained the existence of the critical point.

To derive it, interactions between molecules are introduced by empirically changing the canonical partition function of an ideal monatomic gas (12.37)

$$Q_N^{GP} = \frac{Z^N}{N!} \qquad \text{with} \qquad Z = \frac{V}{h^3}\,(2\pi mkT)^{3/2} \ , \tag{14.84}$$

where Z is the single particle partition function. First, the size of the molecules is accounted for by replacing the volume V by

$$V' = V - Nb' \tag{14.85}$$

where b' is a volume of the same order as that of a molecule. Second, it is assumed that the intermolecular interactions are described by a potential well of constant depth ϕ. A Boltzmann factor $e^{-\beta\phi}$ must therefore be included in Z. The modified partition function then becomes

$$Q_N = \frac{(V - Nb')^N}{N!}\frac{(2\pi mkT)^{3N/2}}{h^{3N}}\, e^{-N\beta\phi} = Q_N^{GP}\left(1 - \frac{Nb'}{V}\right)^N e^{-N\beta\phi}$$

and the free energy of the fluid is given by

$$F = -kT\ln Q_N = F_{GP} - NkT\ln\left(1 - \frac{Nb'}{V}\right) + N\phi \ . \tag{14.86}$$

It follows that the equation of state is

$$\begin{aligned}
P &= -\frac{\partial F}{\partial V} = \frac{NkT}{V} + NkT\left(\frac{1}{V - Nb'} - \frac{1}{V}\right) - N\,\frac{\partial\phi}{\partial V} \\
&= \frac{NkT}{V - Nb'} - N\,\frac{\partial\phi}{\partial V} \ ,
\end{aligned} \tag{14.87}$$

where ϕ must be considered to be a function of the molar volume (or the density) of the fluid.

An explicit expression for ϕ can be found by relating the depth of the potential well to the interaction energy between one molecule and the others. Assuming that the other molecules are distributed uniformly with density N/V, we then choose

$$\phi = \frac{1}{2}\int \epsilon(r)\,\frac{N}{V}\,d\mathbf{r} = -\frac{N}{V}\,a' \qquad \text{with} \qquad a' = -\frac{1}{2}\int \epsilon(r)\,d\mathbf{r} \ ,$$

where the factor $1/2$ avoids double counting of the interaction energies. The free energy and the equation of state of the fluid then take the form

$$F(T,\ v,\ N) = F_{G.P.} - NkT\,\ln\left(1 - \frac{b}{v}\right) - \frac{a}{v} \tag{14.88}$$

$$\text{and} \qquad P = \frac{RT}{v - b} - \frac{a}{v^2} \tag{14.89}$$

where

$$a = \mathcal{N}^2 a' = -\frac{\mathcal{N}^2}{2} \int \epsilon(r) \, d\mathbf{r} \qquad \text{and} \qquad b = \mathcal{N} b' \, . \tag{14.90}$$

Equation (14.89) is the van der Waals equation of state. Its isotherms, shown in Figure 14.14, display a single critical point with co-ordinates

$$v_C = 3b \qquad T_C = \frac{8a}{27Rb} \qquad P_C = \frac{a}{27b^2} \tag{14.91}$$

which satisfies equations (14.69). The critical compressibility factor Z_C (14.70) is equal to $3/8 = 0.375$, compared to the value 0.3 found in most non polar fluids (Table 14.2).

Near the critical point, this equation can be expressed as a series expansion $p = a_0 + a_1\phi + a_2\phi^2 + a_3\phi^3$ (Exercise 14.1)

$$\text{with} \qquad a_0 = 4t \qquad a_1 = -6t \qquad a_2 = 9t \qquad a_3 = -\frac{3}{2}\,(1 + 9t) \tag{14.92}$$

where we have set $t = (T - T_C)/T_C$. From (14.79 and 83), Maxwell's construction yields

$$\frac{v_l + v_g}{2} = v_C(1 + \xi) = v_C \frac{1 + 13t}{1 + 9t} \quad \text{and} \quad v_g - v_l = 2v_C\zeta = \frac{4}{3}\,\frac{[-9t(1 + 6t)]^{1/2}}{1 + 9t}\,v_C \, .$$

As these relations are valid near the critical point where $v_l \simeq v_g \simeq v_C$ and $t \ll 1$, they show that the saturation curve has a horizontal straight line diameter (Fig 14.11b), and that

$$\frac{v_g - v_l}{v_C} \simeq \frac{\rho_l - \rho_g}{\rho_C} \simeq 4(-t)^{1/2}$$

i.e. the critical exponent β (14.68) is equal to 1/2 instead of the experimental value 0.35. Furthermore, for the critical isotherm $(t = 0)$ the coefficients a_0, a_1, a_2 are zero, and hence

$$\frac{P - P_C}{P_C} \simeq -\frac{3}{2}\left(\frac{v - v_C}{v_C}\right)^3 \simeq \frac{3}{2}\left(\frac{\rho - \rho_C}{\rho_C}\right)^3 \, .$$

This result indicates that the critical exponent δ (14.71) is equal to 3, in place of the experimental value 4.4. As the inverse of the compressibility is

$$\frac{1}{\chi_T} = -v\,\frac{\partial P}{\partial v} \simeq -P_C\,\frac{\partial p}{\partial \phi} = -P_C\left(a_1 + 2a_2\phi + 3a_3\phi^2\right) \, ,$$

at $v = v_C$ ($\phi = 0$), we have

$$\chi_T(T,\ v_C) = -\frac{1}{P_C a_1} = \frac{1}{6P_C'}\left(\frac{T - T_C}{T_C}\right)^{-1} \ ,$$

which indicates that the critical exponent γ (14.73) is equal to 1 instead of the value found experimentally, 1.2. Finally, from the free energy (14.88), the internal energy and the heat capacity per mole are

$$u = u_{GP} - \frac{a}{v} \qquad \text{and} \qquad c_V(T,\ v) = c_V^{GP} \ . \tag{14.93}$$

Thus, instead of diverging, the heat capacity is continuous at the critical point. This corresponds to a zero value of the critical exponent α (14.74).

The van der Waals model thus provides an interpretation of the liquid and vapour phases that is qualitatively correct, in the same way as the Weiss molecular field model did for the appearance of spontaneous magnetization. It may also be observed that the values of the exponents α, β, γ and δ are the same in both models.

14.4.5 Lattice Model of Gases

Lee and Yang Hamiltonian

The analogy between chemical and magnetic substances was demonstrated by T.D. Lee and C.N. Yang (1952), who introduced a lattice model for gases. In this model, the volume V is divided into N_0 cells of volume b' of the same order as that of a molecule, i.e.

$$N_0 = V/b' \ , \tag{14.94}$$

and it is assumed that at all times each of the N molecules occupies one of these cells ($N \leq N_0$). Consequently, as space is discretized, the Hamiltonian of the fluid has the form

$$H = E_c + \sum_{<ij>} \epsilon\,\lambda_i\lambda_j \tag{14.95}$$

where the subscripts i and $j = 1, \ \dots \ N_0$ denote two neighbouring cells and ϵ is the (negative) interaction energy between molecules in two neighbouring cells. The variables λ_i are numbers that are equal 0 or 1 depending on whether the cell is empty or occupied. The only non vanishing terms in the sum (14.95) then correspond to the $N_0 z/2$ pairs of occupied neighbouring cells, where z is the number of nearest neighbours to a given cell. These variables are subject to the condition

$$N = \sum_{i=1}^{N_0} \lambda_i \ \leq \ N_0 \ . \tag{14.96}$$

The basic entities of the model to which the statistical approach is applied are thus not the molecules, but the cells. The thermodynamic properties of the fluid

Table 14.3: Correspondence between the Ising and lattice gas models

Ising Model	Lattice Gas Model
Number of spins N	Number of cells $N_0 = V/b'$
Magnetic field term $g\mu_B B$	Chemical potential term $\mu - z\epsilon/2$
Exchange integral J	Interaction term $-\epsilon/2$.

are found by calculating the grand canonical partition function Ξ from equation (12.42). Omitting the kinetic energy term, we can write the partition function as

$$\Xi = \text{Tr}\; e^{-\beta \hat{H}'} = \sum_{\lambda_1 = 0,1} \cdots \sum_{\lambda_{N_0} = 0,1} e^{-\beta H'} \tag{14.97}$$

with
$$H' = H - \mu N = \sum_{<ij>} \epsilon\, \lambda_i \lambda_j - \mu \sum_i \lambda_i \; . \tag{14.98}$$

H' is the Lee and Yang Hamiltonian.

Theoretical argument for the analogy between chemical and magnetic materials

The change of variable

$$\lambda_i = \frac{1 - \alpha_i}{2} \qquad \text{with} \qquad \alpha_i = \pm 1 \tag{14.99}$$

yields a new form for H',

$$H' = \frac{\epsilon}{4} \sum_{<ij>} \alpha_i \alpha_j + \left(\frac{\mu}{2} - \frac{z\epsilon}{4}\right) \sum_i \alpha_i - N_0 \left(\frac{\mu}{2} - \frac{z\epsilon}{8}\right) \tag{14.100}$$

which is similar to the Hamiltonian in the Ising model (13.43). All the methods used for the Ising model as well as their results can therefore be transposed to the lattice gas model by applying the correspondences in Table 14.3 that can be found by comparing expressions (13.43) and (14.100). Furthermore, the mean value of α_i, which is the order parameter $R = \mathcal{M}/\mathcal{M}_s$ in the Ising model, is given by

$$\bar{\alpha} = 1 - 2\bar{\lambda} = 1 - \frac{2N}{N_0} \; ;$$

on setting $b = \mathcal{N}b'$, we therefore find

$$\bar{\alpha} = 1 - \frac{2b}{v} = 1 - \frac{2b}{M} \rho \; , \tag{14.101}$$

which shows that the density plays the role of the order parameter.

The equivalence of the two models explains the similarity mentioned above between the properties of chemical and magnetic substances (§14.4.2), namely existence of a critical point, shape of the isotherms, region of coexistence. Note that in the lattice gas model the variable B is made to correspond to the chemical potential μ, but the experimental results correspond to the variable P. However, since the variables μ and P are linearly related in the vicinity of the critical point, they are therefore interchangeable.

Lastly, it should be observed that magnetic systems pertain to the Heisenberg model (spin dimensionality $D = 3$), while chemical systems come under the Ising model ($D = 1$). It follows that there are certain differences between the values of the critical exponents (§13.4.4).

Molecular field approximation

The lattice gas model is particularly useful in interpreting critical phenomena. It can, however, be viewed over the whole of the range of temperature and pressure by adopting the molecular field approximation (§13.4.2).

In this approximation, the Hamiltonian H' (14.98) is linearized by the transformation

$$\lambda_i \lambda_j \;\rightarrow\; \lambda_i \bar{\lambda} + \bar{\lambda}\lambda_j - \bar{\lambda}^2 \;,\tag{14.102}$$

where $\bar{\lambda}$ is the grand canonical mean value of λ_i, i.e.

$$\bar{\lambda} = \frac{N}{N_0} = \frac{Nb'}{V} = \frac{b}{v}\;.\tag{14.103}$$

The Hamiltonian H' then becomes

$$\begin{aligned}
H' &= \bar{\lambda}\epsilon \sum_{<ij>} (\lambda_i + \lambda_j) - \bar{\lambda}^2\epsilon \sum_{<ij>} 1 - \mu \sum_i \lambda_i \\
&= \left(z\bar{\lambda}\epsilon - \mu\right) \sum_i \lambda_i - zN_0\bar{\lambda}^2\epsilon/2 \;,
\end{aligned}\tag{14.104}$$

and, by a calculation similar to that yielding (13.31), the grand canonical partition function is

$$\begin{aligned}
\Xi &= \exp\left(\frac{zN_0\bar{\lambda}^2}{2}\,\beta\epsilon\right)\left[\sum_{\lambda=0,1} \exp\left(\lambda\beta(\mu - z\bar{\lambda}\epsilon)\right)\right]^{N_0} \\
&= \exp\left(\frac{zN_0\bar{\lambda}^2}{2}\,\beta\epsilon\right)\left[1 + \exp\left(\beta(\mu - z\bar{\lambda}\epsilon)\right)\right]^{N_0}\;.
\end{aligned}\tag{14.105}$$

The grand potential $\Omega(T,\,V,\,\mu)$ of the fluid is then given by

$$\Omega = -kT\ln\Xi = -\frac{zN_0\bar{\lambda}^2}{2}\,\epsilon - N_0 kT\ln\left[1 + e^{\beta(\mu - z\bar{\lambda}\epsilon)}\right]\;.\tag{14.106}$$

All quantities in the variables T, V, μ can be found from Ω. To obtain the results in terms of the variables T, V, N, we employ the usual relation

$$N = -\frac{\partial\Omega}{\partial\mu} = \frac{N_0}{e^{\beta(z\bar{\lambda}\epsilon - \mu)} + 1}\;,$$

which, on inversion, gives

$$\mu = z\bar{\lambda}\epsilon - kT \ln \frac{N_0 - N}{N} = z\epsilon \frac{b}{v} - kT \ln \frac{v - b}{b} \ . \tag{14.107}$$

Substituting into (14.106), we finally get

$$
\begin{aligned}
PV \equiv -\Omega \ & = \ \frac{zN_0\bar{\lambda}^2}{2}\epsilon + N_0 kT \ln \frac{N_0}{N_0 - N} \\
& = \ \frac{zN^2 b'}{2V}\epsilon + \frac{VkT}{b'} \ln \frac{v}{v - b} \ .
\end{aligned}
$$

In the molecular field approximation, the equation of state of the fluid in the lattice gas model is thus

$$P \ = \ -\frac{a}{v^2} - \frac{RT}{b} \ln \left(1 - \frac{b}{v}\right) \tag{14.108}$$

$$\text{with} \qquad a \ = \ -z\frac{N^2 b'\epsilon}{2} \qquad \text{and} \qquad b = Nb' \ . \tag{14.109}$$

This equation of state, called the Yang and Lee equation of state, has two parameters like the van der Waals equation, the constants (14.109) being similar in form to those of (14.90). It also interpolates between the equation of state of ideal gases as $v \to \infty$ and that of incompressible fluids at $v = b$. The isotherms are similar in shape to those of van der Waals (Fig. 14.14). The co-ordinates of the critical point are

$$v_C = 2b \qquad T_C = \frac{a}{2Rb} \qquad P_C = \frac{a}{4b^2}(2\ln 2 - 1) \ , \tag{14.110}$$

and the critical compressibility factor Z_C (14.70) is equal to $2\ln 2 - 1 = 0.386$. The values of the critical exponents, obtained as in paragraph 14.4.4, are the same as those of van der Waals, namely

$$\beta = 1/2 \qquad \delta = 3 \qquad \gamma = 1 \qquad \alpha = 0 \ . \tag{14.111}$$

The similarity between the van der Waals and the Yang and Lee isotherms originates in the fact that a mean field approximation is used in both cases. Similar isotherms with the same critical exponents are also obtained for magnetic materials in the Weiss molecular field approximation. This comes from the fact that the mean field and the molecular field approximations are analogous methods applied to analogous phenomena.

14.5 MIXTURES

14.5.1 Molar Fractions - Concentrations

In a mixture containing c components each with N_a, N_b, $\ldots$ N_c molecules, we introduce the molar fractions as variables

$$x_i = \frac{N_i}{N} \qquad \text{with} \qquad N = N_a + N_b + \ldots N_c \ . \tag{14.112}$$

The c molar fractions are connected by the relation

$$\sum x_i = 1 \ . \tag{14.113}$$

In a binary mixture, the two molar fractions

$$x_a = \frac{N_a}{N} \quad \text{and} \quad x_b = \frac{N_b}{N} \qquad \text{with} \qquad N = N_a + N_b \ , \tag{14.114}$$

connected by the relation $x_a + x_b = 1$, are not independent. Only one of them is necessary to define the composition of the mixture.

It is sometimes preferable to use the molecular concentration N_i/V or the molar concentration (molarity)

$$[i] = \frac{n_i}{V} = \frac{N_i}{\mathcal{N}V} \ . \tag{14.115a}$$

This is related to the molar fraction by

$$[i] = \frac{N}{\mathcal{N}V} x_i = \frac{x_i}{v} \ , \tag{14.115b}$$

where v denotes the molar volume of the mixture.

14.5.2   Mixtures of Gases

The methods employed in paragraph 14.2 for pure real gases can be generalized to the case of a gas with several components. We shall consider the case of a mixture of two monatomic gases. The energy is given by

$$E_{r,\,N_a,\,N_b} = \sum_{i=1}^{N_a} \frac{\mathbf{p}_{ai}^2}{2m_a} + \sum_{j=1}^{N_b} \frac{\mathbf{p}_{bj}^2}{2m_b} + U_{N_a,\,N_b} \tag{14.116}$$

when the mixture is in the state r with N_a molecules of gas a and N_b of gas b. The interaction energy $U_{N_a,\,N_b}$ of the mixture includes the interactions of the gas molecules $a - a$, $b - b$ and $a - b$.

The grand canonical partition function of the mixture (12.42),

$$\Xi(T,\,V,\,\mu_a,\,\mu_b) = \sum_{N_a,N_b=0}^{\infty} e^{\beta N_a \mu_a} \, e^{\beta N_b \mu_b} \, Q_{N_a,\,N_b}(T,\,V) \tag{14.117}$$

is found from the canonical partition function $Q_{N_a,\,N_b}$ of the mixture

$$\begin{aligned}
Q_{N_a,\,N_b}(T,\,V) &= \frac{V_a^{N_a}}{N_a!} \frac{V^{N_b}}{N_b!} \frac{(2\pi m_a kT)^{3N_a/2}}{h^{3N_a}} \frac{(2\pi m_b kT)^{3N_b/2}}{h^{3N_b}} Z_{N_a,\,N_b}(T,\,V) \\
&= Q_{N_a,\,N_b}^{GP}(T,\,V) \, Z_{N_a,\,N_b}(T,\,V) \tag{14.118}
\end{aligned}$$

$$\text{with } Z_{N_a,\,N_b} = \frac{1}{V^{N_a+N_b}} \int e^{-\beta U_{N_a,\,N_b}} d\mathbf{r}_{a1} \ldots d\mathbf{r}_{aN_a} \, d\mathbf{r}_{b1} \ldots d\mathbf{r}_{bN_b} \ . \tag{14.119}$$

If we introduce the quantities z_a and z_b defined by

$$z_i(T,\ V,\ \mu_a,\ \mu_b) = e^{\beta\mu_i}\ \frac{V}{h^3}\ (2\pi m_i kT)^{3/2} \qquad (i = a,\ b)\ , \qquad (14.120)$$

the grand canonical partition function becomes

$$\Xi(T,\ V,\ z_a,\ z_b) = \sum_{N_a, N_b = 0}^{\infty} \frac{z_a^{N_a}}{N_a!} \frac{z_b^{N_b}}{N_b!}\ Z_{N_a,\ N_b}(T,\ V)\ . \qquad (14.121)$$

This expression is a generalization of (14.13). As the grand potential is

$$\Omega(T,\ V,\ \mu_a,\ \mu_b) = -kT \ln \Xi$$

the equation of state of the mixture, as in (14.14-15), is given in an implicit form by the three equations

$$
\begin{aligned}
PV &= -\Omega = kT \ln \Xi\,(T,\ V,\ z_a,\ z_b) \\
N_a &= -\frac{\partial \Omega}{\partial \mu_a} = z_a\ \frac{\partial \ln \Xi}{\partial z_a}\ (T,\ V,\ z_a,\ z_b) \\
N_b &= -\frac{\partial \Omega}{\partial \mu_b} = z_b\ \frac{\partial \ln \Xi}{\partial z_b}\ (T,\ V,\ z_a,\ z_b)\ .
\end{aligned}
\qquad (14.122)
$$

Elimination of the intermediate variables z_a and z_b yields the equation of state $P = P(T,\ V,\ N_a,\ N_b)$ in the form of a virial expansion of the type

$$
\begin{aligned}
P &= RT \sum_{n_a + n_b = 1}^{\infty} \frac{x_a^{n_a}\ x_b^{n_b}}{v^{n_a + n_b}}\ B_{n_a n_b}(T) \\
&= RT \sum_{n_a + n_b = 1}^{\infty} [a]^{n_a}\ [b]^{n_b}\ B_{n_a n_b}(T)
\end{aligned}
\qquad (14.123)
$$

where x_a and x_b are the molar fractions (14.112), $[a]$ and $[b]$ are the molarities (14.115) and $B_{n_a n_b}$ are the virial coefficients such that $B_{10} = B_{01} = 1$.

Let us now determine the first virial coefficients B_{20}, B_{11} and B_{02}. To do this, we expand expression (14.121) for Ξ to order 2,

$$\Xi = 1 + z_a Z_{10} + z_b Z_{01} + \frac{z_a^2}{2}\ Z_{20} + z_a z_b Z_{11} + \frac{z_b^2}{2}\ Z_{02}\ ,$$

in which $Z_{10} = Z_{01} = 1$. In an expansion to the same order, its logarithm becomes

$$\ln \Xi = z_a + z_b + z_a^2 b_{20} + z_a z_b b_{11} + z_b^2 b_{02}$$

with $\quad b_{20} = \dfrac{Z_{20} - Z_{10}^2}{2}\ , \qquad b_{11} = Z_{11} - Z_{10} Z_{01} \quad$ and $\quad b_{02} = \dfrac{Z_{02} - Z_{01}^2}{2}\ .$

The three equations (14.122) can then be written

$$
\begin{aligned}
PV &= kT \left[z_a + z_b + z_a^2 b_{20} + z_a z_b b_{11} + z_b^2 b_{02} \right] \\
N_a &= z_a + 2z_a^2 b_{20} + z_a z_b b_{11} \\
N_b &= z_b + 2z_b^2 b_{02} + z_a z_b b_{11}\ ,
\end{aligned}
$$

the last two of which, by inversion, yield z_a and z_b to the same order, as functions of N_a and N_b, i.e.

$$
\begin{aligned}
z_a &= N_a - 2N_a^2 b_{20} - N_a N_b b_{11} \\
z_b &= N_b - 2N_b^2 b_{02} - N_a N_b b_{11} \ .
\end{aligned}
$$

On substituting into the equation for PV, we finally get

$$
PV = kT \left[N_a + N_b - N_a^2 b_{20} - N_a N_b b_{11} - N_b^2 b_{02} \right] \ ,
$$

which is of the same form as (14.123), with

$$
B_{20} = -\mathcal{N} V b_{20} \qquad B_{11} = -\mathcal{N} V b_{11} \qquad \text{and} \qquad B_{02} = -\mathcal{N} V b_{02} \ . \tag{14.124}
$$

It can be observed that in first order we recover Dalton's law (5.111).

The succeeding virial coefficients are found by taking higher terms in the series expansions. The coefficients $B_{n_a 0}$ and $B_{0 n_b}$ are identified in general with the virial coefficients of the pure a and b gases. In particular, it can be verified that B_{20} and B_{02} are the second virial coefficients (14.19) of the pure gases. The first correction to the equation of state, due to interactions between a and b molecules, enters through the coefficient B_{11}. This is given by

$$
\begin{aligned}
B_{11} &= -\mathcal{N} V \left(Z_{11} - Z_{10} Z_{01} \right) = -\frac{\mathcal{N}}{V} \int \left[e^{-\beta \epsilon_{ab}} - 1 \right] \, d\mathbf{r}_a \, d\mathbf{r}_b \\
&= -\mathcal{N} \int \left[e^{-\beta \epsilon_{ab}(r)} - 1 \right] \, d\mathbf{r} \ , \tag{14.125}
\end{aligned}
$$

in which $\epsilon_{ab}(r)$ is the interaction energy of the a and b molecules when their separation is r. This expression for B_{11} is similar to that for the virial coefficient B_2 of a pure gas (14.20). Measurement of the coefficient B_{11} as a function of temperature gives information about the form of the a b interaction energy.

14.5.3   Dilute Solutions

In spite of the fact that the theory of liquids is complicated, dilute solutions can be handled in a fairly straightforward way. We shall consider the case of a binary mixture a b in which substance b (solute) is at a very low concentration in substance a (solvent).

The grand canonical partition function (14.117) of the mixture can be written in the form

$$
\Xi (T, \ V, \ \mu_a, \ \mu_b) = \sum_{N_b=0}^{\infty} e^{\beta N_b \mu_b} \ \Xi_{N_b} (T, \ V, \ \mu_a)
$$

$$
\text{with} \qquad \Xi_{N_b} = \sum_{N_a=0}^{\infty} e^{\beta N_a \mu_a} \ Q_{N_a, \, N_b}(T, \ V) \ . \tag{14.126}
$$

The function $\Xi_{N_b}(T, \ V, \ \mu_a)$ is a "semi-grand canonical" partition function corresponding to an open system for the a molecules and a closed system for the b

molecules. The function Ξ_0, corresponding to the case $N_b = 0$, is identified with the grand canonical partition function Ξ_a of the pure solvent a.

On writing the first terms of Ξ,

$$\Xi = \Xi_a \left[1 + e^{\beta \mu_b} \frac{\Xi_1}{\Xi_a} + e^{2\beta \mu_b} \frac{\Xi_2}{\Xi_a} + \cdots \right] , \qquad (14.127)$$

we find a series expansion in the solvent concentration. To first order, the grand potential of the mixture is given by

$$\begin{aligned}
\Omega \left(T, \, V, \, \mu_a, \, \mu_b\right) &= -kT \ln \Xi = -kT \ln \Xi_a - kT e^{\beta \mu_b} \frac{\Xi_1}{\Xi_a} \\
&= \Omega_a \left(T, \, V, \, \mu_a\right) - kT e^{\beta \mu_b} \frac{\Xi_1(T, \, V, \, \mu_a)}{\Xi_a(T, \, V, \, \mu_a)} \quad (14.128)
\end{aligned}$$

where Ω_a is the grand potential of the pure solvent a and the second term is due to the presence of solute in the solvent. At equilibrium, the number of solute molecules is given by

$$N_b = -\frac{\partial \Omega}{\partial \mu_b} = e^{\beta \mu_b} \frac{\Xi_1}{\Xi_a}$$

and the chemical potential μ_b is likewise given by

$$\mu_b = kT \ln N_b \frac{\Xi_a}{\Xi_1} = \psi_b + kT \ln[b] \qquad (14.129a)$$

$$\text{with} \quad \psi_b = kT \ln V \frac{\Xi_a}{\Xi_1} . \qquad (14.129b)$$

In classical thermodynamics, equation (14.129a) can be obtained directly without having to define ψ_b. The calculation of ψ_b from (14.129b) is, however, fairly complicated.

At higher concentrations, the succeeding terms in the series (14.127) must be considered and the expression for μ_b is modified. Usually these modifications are accounted for by introducing an activity coefficient γ_b defined for all concentrations by

$$\mu_b = \psi_b + kT \ln \gamma_b [b] \qquad (14.130)$$

where ψ_b is still given by (14.129b).

14.5.4 Concentrated Solutions

A concentrated solution consists of a mixture of several components whose molar fractions are of the same order of magnitude. For such systems we must generalize the methods used for pure liquids. We shall restrict ourselves to a simple model that allows us to recover the principal characteristics of a mixture of two liquids.

Consider two liquids whose molecules a and b have similar sizes, giving rise to a so-called regular solution. Taken separately, their respective molecular interaction energies can be written

$$\overline{U}_{N_a} = \frac{1}{2}\, N_a w_{aa} \qquad \text{and} \qquad \overline{U}_{N_b} = \frac{1}{2}\, N_b w_{bb}$$

where N_a (N_b) is the number of a (b) molecules and w_{aa} (w_{bb}) is the (negative) mean energy of interaction of an a (b) molecule with all the others. When they are mixed, each a or b molecule interacts with the fraction x_a of a molecules and x_b of b molecules. In addition to the mean energies w_{aa} and w_{bb}, we must now consider a mean interaction energy w_{ab} between a and b molecules such that the interaction energy of the mixture is given by

$$\overline{U}_{N_a,\, N_b} = \frac{1}{2}\, N_a\,(x_a w_{aa} + x_b w_{ab}) + \frac{1}{2}\, N_b\,(x_a w_{ab} + x_b w_{bb}) \quad .$$

The change in interaction energy produced by mixing is then

$$\Delta U \;=\; \overline{U}_{N_a,\, N_b} - \overline{U}_{N_a} - \overline{U}_{N_b} = -\frac{N}{2}\, x_a x_b\,(w_{aa} + w_{bb} - 2w_{ab})$$

i.e. $\qquad \Delta U \;=\; -\frac{N}{2}\, x_a x_b w \qquad \text{with} \qquad w = w_{aa} + w_{bb} - 2w_{ab} \quad . \quad (14.131)$

To evaluate the change in entropy due to mixing, we apply the Boltzmann relation with the assumption that the a and b molecules are randomly distributed throughout the positions occupied by the $N_a + N_b$ molecules of the mixture, that is

$$\Delta S = k \ln \frac{(N_a + N_b)!}{N_a!\, N_b!} = -Nk\,(x_a \ln x_a + x_b \ln x_b) \quad . \tag{14.132}$$

The free energy of mixing is thus

$$\Delta F = \Delta U - T\Delta S = -\frac{N}{2}\, x_a x_b w + NkT\,(x_a \ln x_a + x_b \ln x_b) \qquad (14.133)$$

where $x_a + x_b = 1$. This confirms that for $x_a = 0$ or $x_b = 0$ the mixing contribution is zero and that ΔF is symmetrical in x_a and x_b.

If w is positive, i.e. if the a b interaction is more attractive than the average of the a a and b b interactions, then the dependence of the free energy of mixing upon $x \equiv x_b$ has the form shown in Figure 14.15a for any temperature. If w is negative, the behaviour of the solution depends on the temperature (Fig. 14.15b). For $T > |w|/4k$ the curve $\Delta F(x)$ has a minimum at $x = 1/2$. For $T < |w|/4k$, the curve passes through a maximum at $x = 1/2$ and two symmetric minima occur at $x = \alpha(T)$ and $x = \beta(T)$, where $\alpha + \beta = 1$. Solutions with composition x intermediate between α and β are then unstable and separate into two phases with respective compositions α and β of lower free energy. This is called phase separation. The phase diagram corresponding to this phenomenon (Fig. 14.16) is observed experimentally. The temperature

$$T_c = |w|/4k \quad , \tag{14.134}$$

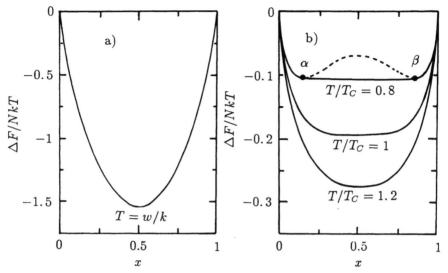

Figure 14.15: Free energy of mixing for a regular solution as a function of the molar fraction x. a) Positive $w = w_{aa} + w_{bb} - 2w_{ab}$. b) Negative w for 3 temperatures around $T_c = |w|/4k$. Below T_c, the mixture separates into two phases (phase separation).

below which phase separation appears is a critical temperature, and the phase separation is a critical phenomenon, the order parameter, $R(T)$, being given by

$$\alpha = \frac{1-R}{2} \qquad \text{and} \qquad \beta = \frac{1+R}{2} . \qquad (14.135)$$

The analogy between the paramagnetic-ferromagnetic and the liquid-vapour transitions becomes apparent on comparing Figures 14.13 and 16. Critical exponents are defined for phase separation that are similar to those that we have already encountered.

14.6 ELECTROLYTES

14.6.1 Introduction

When ionic compounds are placed in solution, they dissociate into two or more ions. This dissociation may be complete, as with sodium chloride NaCl in water, such substances being called strong electrolytes. It may however be partial, as with acetic acid CH_3COOH in water. Such substances are called weak electrolytes. The degree of dissociation in fact depends on the solvent. It is stronger in polar solvents, which are characterized by a high relative permittivity ($\epsilon_r = 78.54$ in water at $t = 25\,°C$), than in non polar solvents ($\epsilon_r = 2.015$ for benzene at $t = 25\,°C$.)

The difficulty encountered in treating electrolyte systems comes from the fact that virial series expansion techniques (§14.2.4 and 14.5.2) no longer apply. The virial coefficients (14.20 and 125) in fact diverge at $r = \infty$ for a Coulomb potential in r^{-2}. Special techniques must therefore be used that take into account the long range of the Coulomb potential. The first satisfactory theory of electrolytes is due to

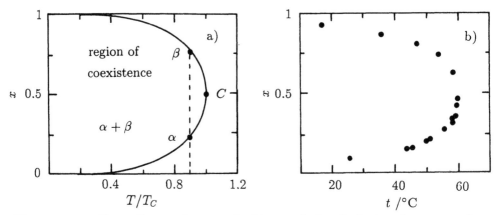

Figure 14.16: Phase diagram for a regular binary solution displaying phase separation. a) Theoretical curve. b) Solution of aniline $C_6H_5NH_2$ and n-hexane C_6H_{14}. x is the mass fraction of hexane [D.B. Keyes and J.H. Hildebrand, *J. Am. Chem. Soc.* **39**, 2126 (1917)].

P. Debye and E. Hückel (1923) (§14.6.2). It has been subsequently shown that this theory is applicable to aqueous solutions of concentrations C less than 0.01 mol L^{-1}. In 1937, J.E. Mayer developed a perturbation method, the lowest order of which corresponds to the Debye-Hückel theory, and which is useful for concentrations greater than 0.1 mol L^{-1}. More recent theories, employing distribution functions, provide agreement with experiment for concentrations in excess of 1 mol L^{-1}.

14.6.2   Debye-Hückel Theory

Model of electrolytes

In models of electrolytes, it is supposed that the solvent behaves as a continuous medium with a relative permittivity ϵ_r that is independent of T. The electrolyte, which for simplicity is taken to be of the form A^+B^-, dissociates into ions that are assumed to be hard spheres of diameter d, in which all the charge $\pm e$ is concentrated at the centre. Let us calculate the contribution F_e to the free energy F of the Coulomb interaction of the ions. This contribution vanishes if we take $e \to 0$ (e = charge of the proton). It is obtained from the Helmholtz relation $U = -T^2 \partial(F/T)/\partial(T)$ in classical thermodynamics (see Eq. 3.9), by considering the contribution of the Coulomb interaction

$$U_e = \frac{N}{2} \ (e\psi_+ - e\psi_-) \tag{14.136}$$

to the internal energy of the system of $2N$ ions. In this expression, $\psi_\pm$ is the mean electrostatic potential to which an ion of charge $\pm e$ is subjected by all the other ions. The factor $1/2$ is to avoid double counting.

Basic hypothesis

In the Debye-Hückel theory, $\psi_\pm$ is calculated by making the approximation (14.58) in which the radial distribution functions of the ions with the same charge and with opposite charges are respectively

$$g_+(r) = \exp\left[-\beta e\phi(r)\right] \qquad \text{and} \qquad g_-(r) = \exp\left[\beta e\phi(r)\right]. \qquad (14.137)$$

Here, $\phi(r)$ is the mean electrostatic potential prevailing at a distance r from a positive ion and, by symmetry, $-\phi(r)$ is that for a negative ion. With these functions we can write that the mean number of positive and negative ions per unit volume at a distance r from a given positive ion are respectively

$$n_+(r) = \frac{N}{V} g_+(r) \qquad \text{and} \qquad n_-(r) = \frac{N}{V} g_-(r) . \qquad (14.138)$$

The mean electrostatic charge density around the "central" positive ion

$$\rho(r) = e\left(n_+ - n_-\right) = -\frac{2Ne}{V} \sinh\ \left[\beta e\phi(r)\right] \qquad (14.139)$$

is directly related to the potential $\phi(r)$ by Poisson's equation

$$\Delta\phi = -\frac{\rho}{\epsilon_0\epsilon_r} = \frac{2Ne}{V\epsilon_0\epsilon_r} \sinh\ \left[\beta e\phi(r)\right], \qquad (14.140)$$

which, written in this way, allows us to calculate ϕ. Note that the potential ϕ is the sum of the potential $e/4\pi\epsilon_0 r$ generated by the central ion and the one of interest for the calculation of $\psi_\pm$, namely that generated by the other ions.

14.6.3 Approximate Determination of ϕ

Since equation (14.140) has no simple solution, we use the approximation whereby the function sinh can be replaced by its argument. This then gives

$$\Delta\phi \equiv \frac{1}{r} \frac{\partial^2 (r\phi)}{\partial r^2} \simeq \frac{2N\beta e^2}{V\epsilon_0\epsilon_r} \phi$$

$$\text{or} \qquad \frac{\partial^2 (r\phi)}{\partial r^2} = \kappa^2 r\phi \qquad \text{with} \qquad \kappa^2 = \frac{2e^2}{\epsilon_0\epsilon_r} \frac{N}{V} \beta . \qquad (14.141)$$

The general solution of this equation is

$$\phi = \frac{A}{r} e^{-\kappa r} + \frac{A'}{r} e^{\kappa r} , \qquad (14.142)$$

where the constant A' must vanish so that ϕ can remain finite at $r = \infty$. To determine A, we must examine the expression for ϕ in the neighbourhood of the central ion.

Equation (14.142) for ϕ is in fact valid only for $r > d$ since the hard spheres cannot interpenetrate. For $r < d$, ρ is zero, and Poisson's equation becomes $\Delta\phi = 0$. The general solution of this equation is

$$\phi = B + \frac{B'}{r} , \qquad (14.143)$$

where, on considering the limit $r \to 0$, the constant B' can be identified with $e/4\pi\epsilon_0\epsilon_r$. The two constants A and B are then found by writing the continuity equation for ϕ and for its derivative at $r = d$,

$$B + \frac{e}{4\pi\epsilon_0\epsilon_r\,d} = \frac{A}{d}\,e^{-\kappa d} \qquad \text{and} \qquad -\frac{e}{4\pi\epsilon_0\epsilon_r d^2} = -\frac{A\,e^{-\kappa d}}{d^2}\,(1 + \kappa d)\ .$$

Hence

$$A = \frac{e}{4\pi\epsilon_0\epsilon_r}\,\frac{e^{\kappa d}}{1 + \kappa d} \qquad \text{and} \qquad B = -\frac{e}{4\pi\epsilon_0\epsilon_r}\,\frac{\kappa}{1 + \kappa d}\ . \qquad (14.144)$$

The potential ϕ is then given by

$$\phi(r) = \frac{e}{4\pi\epsilon_0\epsilon_r}\,\frac{e^{\kappa(d-r)}}{1 + \kappa d} \qquad\qquad\qquad r > d \qquad (14.145)$$

$$\phi(r) = -\frac{e}{4\pi\epsilon_0\epsilon_r}\,\frac{\kappa}{1 + \kappa d} + \frac{e}{4\pi\epsilon_0\epsilon_r r} \qquad r < d\ .$$

Free energy

To find ψ_+, the mean electrostatic field due to the other ions and to which the central positive ion is subjected, the potential $e/4\pi\epsilon_0\epsilon_r r$ generated by the central ion is subtracted from ϕ, and we set $r = 0$. In the Debye-Hückel theory we thus get

$$\psi_+ = -\frac{e}{4\pi\epsilon_0\epsilon_r}\,\frac{\kappa}{1 + \kappa d}\ . \qquad (14.146)$$

Similarly, for a negative central ion, we would get $\psi_- = -\psi_+$.

Substitution of these expressions for $\psi_\pm$ into equation (14.136) for the internal energy gives

$$U_e = -N\,\frac{e^2}{4\pi\epsilon_0\epsilon_r}\,\frac{\kappa}{1 + \kappa d}\ , \qquad (14.147)$$

where κ is the parameter that is homogeneous with the inverse length defined in (14.141). With the help of the Helmholtz equation, the contribution F_e to the free energy is found to be

$$F_e = kT \int U_e\,d\beta\ .$$

Since U_e depends on β only through κ, we can apply the change of variable $\beta \to \kappa$, and hence

$$F_e = \frac{\epsilon_0\epsilon_r}{e^2}\,\frac{V}{N}\,kT \int U_e\,\kappa\,d\kappa = -\frac{VkT}{4\pi} \int \frac{\kappa^2}{1 + \kappa d}\,d\kappa\ . \qquad (14.148)$$

The integration constant is determined by the fact that the Coulomb contribution F_e vanishes when the charge e tends to zero ($\kappa \to 0$). Thus

$$F_e = -\frac{VkT}{12\pi}\,\kappa^3\tau(\kappa d) \quad \text{with} \quad \tau(x) = \frac{3}{x^3}\left[\ln(1 + x) - x + \frac{x^2}{2}\right]\ . \qquad (14.149)$$

The function τ tends to 1 as $x \to 0$.

From the above expression for F_e we can therefore determine the contribution of the Coulomb interaction to the various thermodynamic quantities. In particular, the contribution to the chemical potential is given by

$$\mu_e = \frac{\partial F}{\partial N} = \frac{\partial F}{\partial \kappa} \frac{\partial \kappa}{\partial N} = -\frac{VkT}{4\pi} \frac{\kappa^2}{1+\kappa d} \frac{\kappa}{2N} = -\frac{e^2}{4\pi\epsilon_0\epsilon_r} \frac{\kappa}{1+\kappa d} \, . \tag{14.150}$$

The above results were obtained for a 1-1 A^+B^- electrolyte . For an electrolyte that dissociates into ions A_1, A_2, $\ldots$, having numbers ν_1, ν_2, $\ldots$, and charges $z_1 e$, $z_2 e$, $\ldots$ respectively, the chemical potential (14.150) can be generalized to

$$\mu_e = -\frac{e^2}{8\pi\epsilon_0\epsilon_r} \frac{\kappa}{1+\kappa d} \sum_i \nu_i z_i^2 \tag{14.151}$$

$$\text{with} \qquad \kappa^2 = \frac{e^2}{\epsilon_0\epsilon_r kT} \sum_i z_i^2 \frac{N_i}{V} = \frac{\mathcal{N}e^2}{\epsilon_0\epsilon_r kT} \sum_i z_i^2 [A_i] \, . \tag{14.152}$$

Mean activity coefficient

Adopting the language of chemistry, we define the ionic strength as

$$I = \frac{1}{2} \sum_i z_i^2 [A_i] \tag{14.153}$$

which is the same as the concentration in moles per litre for a 1-1 electrolyte, and we use the activity coefficient γ defined in (14.130). Equations (14.151 and 152) then become

$$\ln \gamma = \frac{\mu_e}{kT} = -\frac{e^2}{8\pi\epsilon_0\epsilon_r kT} \frac{\kappa}{1+\kappa d} \sum_i \nu_i z_i^2 \tag{14.154}$$

$$\text{and} \qquad \kappa^2 = \frac{2\mathcal{N}e^2}{\epsilon_0\epsilon_r kT} I \, . \tag{14.155}$$

The ionic strength is calculated directly from the concentrations, and the activity coefficient can be measured by various methods, e.g. electromotive force of batteries, vapour pressure of the solvent, diffusion coefficient, ... In fact, since the solute molecules are dissociated, we should consider an activity coefficient γ_i for each ionic species created. However, these individual quantities cannot be measured separately, and the results are therefore given in the form of an average

$$\ln \gamma_\pm = \frac{\sum_i \nu_i \ln \gamma_i}{\sum_i \nu_i} = \frac{\ln \gamma}{\sum_i \nu_i} \, . \tag{14.156}$$

For binary electrolytes, for example,

$$\ln \gamma_\pm = -\frac{e^2}{8\pi\epsilon_0\epsilon_r kT} \frac{\kappa}{1+\kappa d} \frac{\nu_+ z_+^2 + \nu_- z_-^2}{\nu_+ + \nu_-} \, , \tag{14.157}$$

where the last factor can be written as $|z_+ z_-|$ owing to the electrical neutrality of the solution ($\nu_+ z_+ + \nu_- z_- = 0$).

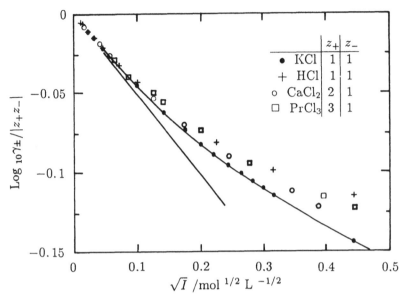

Figure 14.17: Mean activity coefficient $\gamma_\pm$ for aqueous solutions of various binary salts at 25 °C as a function of the ionic strength. The straight line is the limiting case (14.159) of Debye-Hückel. The curve shows equation (14.157) for $d = 4$ Å[KCl and CaCl$_2$: G.N. Lewis and M. Randall, *op. cit.*, p. 643 and 654; HCl: H.S. Harned and B.B. Owen, *op. cit.* p. 547; PrCl$_3$: F.H. Spedding, P.E. Porter and J.M. Wright. *J. Am. Chem. Soc.* **74**, 2781 (1952)].

Comparison with experiment

The parameter κ is fundamental in the Debye-Hückel theory. According to (14.155), it takes the form

$$\kappa /\text{Å}^{-1} = 50.3 \left[\frac{I \,/(\text{mol L}^{-1})}{\epsilon_r \, T \,/\text{K}} \right]^{1/2}$$

$$= 0.329 \left[I \,/(\text{mol L}^{-1}) \right]^{1/2} \quad \text{(water at 25 °C) ;} \quad (14.158)$$

the latter numerical coefficient, calculated for water at 25 °C , varies little with temperature between 0 °C and 100 °C, because the product $\epsilon_r T$ decreases only by 16% in this range. At low ionic strengths ($I < 10^{-3}$ mol L^{-1}), i.e. at low concentrations, the parameter κ^{-1} is in excess of 30 Å and is therefore much larger than the diameters of the ions, and consequently the product κd is much smaller than 1. According to (14.157, 158) therefore, the Debye-Hückel theory predicts that dilute solutions are governed by the limiting relation

$$\ln {}_{10}\gamma_\pm = -1.83 \times 10^6 \, |z_+ z_-| \frac{\left[I \,/(\text{mol L}^{-1}) \right]^{1/2}}{[\epsilon_r \, T \,/\text{K}]^{3/2}}$$

$$= -0.51 \, |z_+ z_-| \left[I \,(\text{mol L}^{-1}) \right]^{1/2} \quad \text{(water at 25 °C) . (14.159)}$$

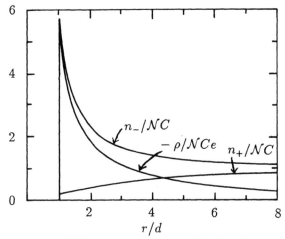

Figure 14.18: Distribution of charges around a positive ion in an aqueous solution of KCl at 25 °C and at concentration 10^{-3} mol L^{-1}.

Figure 14.17 shows that the experimental values confirm this limiting law up to ionic strengths of the order of 10^{-3} mol L^{-1}, but deviate above this. The data can be simulated over a wider range, however, by using the more complete formula (14.157), which differs from the limiting case (14.159) in having a denominator $1 + \kappa d$, where d is taken to be an adjustable parameter of the same order of magnitude as the diameter of the ions. The data for KCl can be described in this way up to 1 mol L^{-1}, while for HCl this relationship is valid only up to 0.3 mol L^{-1}. For electrolytes such as sulfates, negative values of d must be taken, which has no physical meaning. These limitations in the Debye-Hückel theory stem from the approximations made in (14.137 and 141), and from the fact that at higher concentrations short-range interactions become significant. These depend on the ion species.

Exercise 14.2 *Distribution of charges around an ion*

A solution of potassium chloride KCl in water at 25 °C ($\epsilon_r = 78.54$) is prepared at concentration $C = 10^{-3}$ mol L^{-1}.

1 Plot the number densities n_+ and n_- (14.138) as a function of r, as well as the charge density ρ (14.139). Take $d = 4$ Å.

2 Under what condition can equation (14.140) be linearized? Is this condition fulfilled for the present solution?

3 Assuming that the condition is satisfied, derive an approximate analytical expression for $n_\pm(r)$ and $\rho(r)$. Calculate the total charge outside the central ion. Comment on the result.

Solution

1 The quantities are shown in Figure 14.18. The central ion is surrounded by an excess of opposite charges (ionic atmosphere) in a region extending out to approximately $10d$. At this concentration, the mean distance between two ions, if they had no charge, would be $D \simeq (V/N)^{1/3} = 118$ Å$\simeq 30d$.

2 The condition is $\beta e\phi(r) \ll 1$ for $r > d$. As the potential ϕ (14.145) is a decreasing function of r, we must have

$$\beta e\phi(d) = \beta \, \frac{e^2}{4\pi\epsilon_0\epsilon_r d} \, \frac{1}{1+\kappa d} \ll 1 \ . \tag{14.160}$$

Under the prevailing experimental conditions, $\beta e\phi(d) = 1.71$ and the condition is not satisfied. However, Gronwall et al. (1928), using a series expansion of the function sinh in (14.140), showed that the above results are valid at low concentrations, even though condition (14.160) is not fulfilled.

3 We have

$$n_\pm \ \simeq \ \frac{N}{V} \, [1 \mp \beta e\phi(r)] = \frac{N}{V} \mp \frac{\kappa^2}{8\pi} \, \frac{e^{\kappa(d-r)}}{r(1+\kappa d)}$$

and $\qquad \rho \ = \ (n_+ - n_-)\,e \simeq -\frac{\kappa^2 e}{4\pi} \, \frac{e^{\kappa(d-r)}}{r(1+\kappa d)} \ . \tag{14.161}$

The total charge outside the positive central ion is

$$q \ = \ \int_d^\infty 4\pi r^2 \rho \, dr = -\kappa^2 \, e \, \frac{e^{\kappa d}}{1+\kappa d} \int_d^\infty re^{-\kappa r} \, dr$$

$$= \ -\kappa^2 e \, \frac{e^{\kappa d}}{1+\kappa d} \times \frac{1}{\kappa^2} \, \left[-(1+x)e^{-x}\right]_{\kappa d}^\infty = -e \ .$$

This result, which is what is expected in a solution with no net charge, shows that the Debye-Hückel approximation is also satisfactory here.

Exercise 14.3 *Calculation of the free energy by a charging process*

Another way of calculating the free energy F_e consists in calculating the work required to increase the charge of all the ions progressively from 0 to $\pm e$ at constant temperature T and volume V. At a given moment in the charging process, the charge of the ions is $\pm\lambda e$ ($0 \leq \lambda \leq 1$).

Find an expression for the increment of work required to increase the ionic charge by $\pm e \, d\lambda$. Hence derive F_e and compare with equations (14.148,149).

Solution

On incrementing the charge, we have $dW_e = \sum V_i \, dq_i$, which in this case is

$$dW_e = N\psi_+(\lambda)e \, d\lambda - N\psi_-(\lambda)e \, d\lambda = 2Ne\psi_+(\lambda) \, d\lambda,$$

where $\psi_\pm(\lambda)$ is the potential (14.146) when the charge is $\pm\lambda e$. The free energy F_e then becomes

$$F_e = 2Ne \int_0^1 \psi_+(\lambda) \, d\lambda = -\frac{2Ne^2}{4\pi\epsilon_0\epsilon_r} \int_0^1 \lambda \, \frac{\kappa(\lambda)}{1+\kappa(\lambda)d} \, d\lambda$$

where, from (14.141), $\kappa(\lambda)$ is equal to $\kappa\lambda$. Then, setting $x = \kappa\lambda$, we find

$$F_e = -\frac{VkT}{4\pi} \int_0^\kappa \frac{x^2 \, dx}{1+xd} \ , \tag{14.162}$$

which is identical to (14.148). The progressive charge method and the Helmholtz equation produce the same result. In situations where the relative permittivity ϵ_r of the solvent depends on temperature, the Helmholtz equation, when applied to expression (14.147) for the internal energy U_e, yields a result for F_e that is different from (14.162). N. Bjerrum (1926) explained this apparent discrepancy by observing that equation (14.136) for U_e neglects the effect of the ions on the orientation of the surrounding solvent molecules when these have a permanent dipole moment. The progressive charge method at constant temperature gives the result directly, and both equation (14.149) for F_e and the results that follow from it are valid.

BIBLIOGRAPHY

C. Croxton, *Introduction to Liquid State Physics*, Wiley (1975).
R.H. Fowler and E.A. Guggenheim, *op. cit.* p. 255.
J.P. Hansen and I.R. McDonald, *Theory of Simple Liquids*, Acad. Press (1976).
H.S. Harned and B.B. Owen, *The Physical Chemistry of Electrolytic Solutions*, Reinhold, New York (1950).
J.H. Hildebrand and R.L. Scott, *The Solubility of Non Electrolytes*, Dover, N.Y. (1964).
T.L. Hill, *op. cit.*, p. 261.
J.O. Hirschfelder, C.F. Curtiss and R.B. Byrd, *Molecular Theory of Gases and Liquids*, Wiley (1967).
K. Huang, *op. cit.*, p. 313.
G.N. Lewis and M. Randall, *Thermodynamics*, McGraw Hill (1965), p. 184.
D.A. McQuarrie, *op. cit.* p. 222.
A. Münster, *op. cit.*, p. 317.
Historical References
N. Bjerrum, *Z. Physik. Chem.* **119**, 145 (1926).
M. Born and H.S. Green, *Proc. Roy. Soc. A* (London) **188**, 10 (1946).
L. Cailletet and E. Mathias, *C. R. Acad. Sci.*, Paris **102**, 1202 (1886).
P. Debye and E. Hückel, *Physik. Z.* **24**, 185 (1923).
T.H. Gronwall, V.K. La Mer and K. Sandved, *Physik. Z.* **29**, 358 (1928).
J.G. Kirkwood, *J. Chem. Phys.* **3**, 300 (1935).
T.D. Lee and C.N. Yang, *Phys. Rev.* **87**, 410 (1952).
J.E. Mayer, *J. Chem. Phys.* **5**, 67 (1937).
H.D. Ursell, *Proc. Cambridge Phil. Soc.* **23**, 685 (1927).
J.D. van der Waals, *Over de Continuiteit van den Gas en Vloeistoftoestand*, Thesis, Leiden (1873).
J. Yvon, *Actualités scientifiques et industrielles*, n° 203, Hermann, Paris (1935).

COMPREHENSION EXERCISES

14.1 Find the second and third virial coefficients for the equation of state
$\left(P + A(T)/v^2\right)(v - b) = RT$ [Ans.: $B(T) = b - A(T)/RT$, $C(T) = b^2$].

14.2 Show that the first correction term in the molar heat capacity c_V and in the chemical potential μ of a real gas are respectively $-2RT\left(TB_2'' + 2B_2'\right)/v^3$ and

$2kT B_2(T)/v$.

14.3 By integrating equation (14.20) by parts, show that the second virial coefficient can be written

$$B_2 = -\frac{2\pi}{3} \frac{\mathcal{N}}{kT} \int_0^\infty \frac{d\epsilon}{dr} e^{-\beta\epsilon(r)} r^3 \, dr \quad .$$

Hence derive equation (14.58).

14.4 For the Sutherland interaction potential (§14.2.2), show that the second virial coefficient B_2 at high temperature ($\epsilon_0 \ll kT$) is

$$B_2 = \frac{2\pi}{3} \mathcal{N} d^3 \left[1 - \frac{3}{n-3} \frac{\epsilon_0}{kT} + O\left(\frac{\epsilon_0}{kT}\right)^2 \right] \quad .$$

14.5 Show that the virial coefficient B_2 is not meaningful for a potential $\epsilon(r) \propto r^n$ if $n < -3$.

14.6 Show that the molar free energy of a fluid of hard spheres having Eq. (14.66) as equation of state is

$$f(T, v) = f^{GP}(T, v) + \frac{\eta(4 - 3\eta)}{(1 - \eta)^2} \quad .$$

Note that when $v \to \infty$ ($\eta \to 0$), then $f \to f^{GP}$.

14.7 For the hard sphere model, equation (14.55) takes the particular form

$$\frac{Pv}{RT} = 1 + \frac{2\pi}{3} \frac{\mathcal{N} d^3}{v} g(d)$$

where d is the diameter of the spheres and $g(r)$ the radial distribution function. From Figure 14.9 read the values of $g(d)$, and confirm that they yield the values of P in Figure 14.10.

14.8 Determine the co-ordinates of the critical point for the van der Waals (14.89) and the Yang and Lee (14.108) equations. [Ans.: (14.91 and 110)].

14.9 Write the equation of state of Yang and Lee (14.108) in the form (14.77) and hence deduce that the critical exponents β, γ and δ take the values (14.111) [Ans.: $a_0 = 2\ln(2t)/Z_C$; $a_1 = -2t/Z_C$; $a_3 = 3t/Z_C$; $a_4 = -2(1 + 7t)/3Z_C$ with $Z_C = 2\ln 2 - 1$].

14.10 Show that below the critical temperature T_C(14.134), a regular binary mixture separates into two phases α and β with compositions $(1 \pm R)/2$, where the order parameter R is given by $R = \tanh (RT_C/T)$.

14.11 Determine the proportions of the α and β phases into which a solution of composition x separates. [Ans.: $(\beta - x)/(\beta - \alpha)$ and $(x - \alpha)/(\beta - \alpha)$].

14.12 Calculate the ionic strength I of the electrolytes 1-1 (KCl), 2-1 (CaCl$_2$), 3-1 (PrCl$_3$) as a function of the concentration C [Ans.: C, $3C$, $6C$].

14.13 In the Debye-Hückel theory, rederive expression (14.146) for the electrostatic potential ψ_+ felt by a positive ion, starting from the charge density (14.161) around the ion.

14.14 In the Debye-Hückel theory, the contribution from the electrostatic interaction of the ions to the free energy F_e is given in (14.148-149). If ϵ_r depends on T, show that the internal energy U_e is the same as that given in (14.147), multiplied by a factor $1 + T\epsilon'_r/\epsilon_r$. Discuss the case $\epsilon_r =$ constant (non polar molecules) and $\epsilon_r = b/T$ (rigid polar molecules, cf. exercise 5.4).

Chapter 15

Superconductivity

15.1 INTRODUCTION

While investigating the limiting value of electrical resistivity in very pure metals at absolute zero, H. Kamerlingh Onnes (1911) discovered that below a certain critical temperature $T_C = 4.2$ K, the resistivity of mercury suddenly vanishes. This phenomenon, called superconductivity, occurs in many substances. He later discovered (1914) that superconductivity is destroyed by magnetic fields greater than a critical value of the flux density $B_C(T)$ that is quite well described empirically by the so-called Tuyn relation (Fig. 15.1)

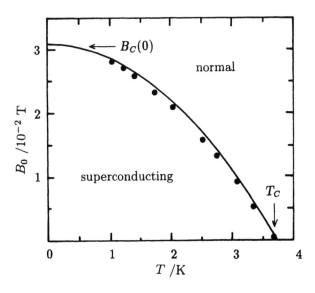

Figure 15.1: Phase diagram of tin. The experimental points are the applied magnetic flux densities $B_C(T)$ above which superconductivity vanishes. The curve shows the empirical relation (15.1) with $T_C = 3.72$ K, $B_C(0) = 3.09 \times 10^{-2}$ T [R.W. Shaw, D.E. Mapother and D.C. Hopkins, *Phys. Rev.* **120**, 88 (1960)].

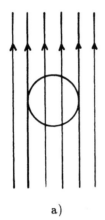   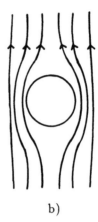

a)   b)

Figure 15.2: Meissner effect. a) In the normal state the substance is weakly magnetic and the field lines are hardly deflected. b) In the superconducting state the field lines are expelled from the material. Inside the sample the flux density is zero.

$$\frac{B_C(T)}{B_C(0)} = 1 - \left(\frac{T}{T_C}\right)^2 \; . \tag{15.1}$$

In 1933, W. Meissner and R. Ochsenfeld discovered a new property of super-conductors, called the Meissner effect: the magnetic field is always zero inside a superconductor (Fig. 15.2).

Following these discoveries, F. and H. London (1935) gave a quantum mechanical explanation for the magnetic properties of these materials. They assumed that the wave function of the superconducting electrons is sufficiently "rigid" not to be influenced by moderate magnetic fields. They postulated that the ground state of the superconducting electrons is separated from the excited states by a "gap". The existence of this gap, of the order of kT_C, was demonstrated in the 1950s by absorption of electromagnetic radiation in the centimetre wavelength range, as well as by other methods.

In 1950, the isotope effect was independently discovered by E. Maxwell and C.A. Reynolds et al. They found that the critical temperature T_C varies as $M^{-1/2}$, where M is the molar mass of an isotope of a given superconducting element. This result indicated that superconductivity is related to electron-phonon interactions. In 1950, H. Fröhlich, taking these interactions into account, independently predicted the existence of an isotope effect, and in 1954, showed that a gap appears for a one dimensional model. The idea then arose that pairs of electrons can behave like bosons, and L. N. Cooper (1956) showed this to be the case whenever there is an overall attraction, however weak, between electrons. This work culminated in the BCS theory (J. Bardeen, L.N. Cooper and J.R. Schrieffer (1957)), which provides a general explanation of all the properties of superconductors, in particular the electromagnetic properties, which had been treated on a purely phenomenological basis by F. and H. London and by A.B. Pippard (1953).

Table 15.1: Data for some superconducting elements. T_C critical temperature, α isotope effect constant (15.2), $B_C(0)$ and $Eg(0)$ critical flux density (in 10^{-4} T) and gap at absolute zero, Θ_D Debye temperature, C_s and C_n molar heat capacities of the superconducting and normal phases at the critical temperature [N.W. Ashcroft and N.D. Mermin, *op. cit.* p. 729, 745 and 747, C. Kittel, *op. cit.* p.368].

Element	T_C /K	α	Type	$B_C(0)$	$\dfrac{Eg(0)}{kT_C}$	$\dfrac{C_s-C_n}{\gamma T_C}$	Θ_D /K
Al	1.196		I	99	3.4	1.4	394
Zn	0.875	0.45	I	53	3.2	1.3	234
Ga	1.091		I	51		1.4	240
Cd	0.56	0.32	I	30	3.2	1.4	120
In	3.40		I	293	3.6	1.7	129
Sn	3.72	0.47	I	305·	3.5	1.6	170
Hg(α)	4.15	0.50	I	411	4.6	2.4	100
Hg(β)	3.95		I	339			
Tl	2.39	0.61	I	171	3.6	1.5	96
Pb	7.19	0.49	I	803	4.3	2.7	88
V	5.30		II	1 020	3.4	1.5	390
Nb	9.26		II	1 980	3.8	1.9	275
Ta	4.48		I	830	3.6	1.6	225

The phenomenon of superconductivity is of special importance. For example, it can reveal wave functions that are coherent over macroscopic distances. It also has many applications, such as loss-free transport of electric current, powerful electromagnets, Josephson junctions.

Until the middle of the 80s, the superconductors in use were metals, metallic compounds and alloys. The highest critical temperature ($T_C = 23.2$ K) was that of the compound Nb_3Ge. In 1986, a new class of materials was discovered possessing a higher critical temperature, even exceeding 77 K, the normal boiling point of nitrogen. These were ceramics containing copper oxide. Their existence opens new perspectives for the development of applications of superconductivity.

15.2   EXPERIMENTAL PROPERTIES

15.2.1   Critical Temperature

At present a large number of metallic substances are known that are superconductors at low temperatures. The list includes elements (Table 15.1) as well as compounds (Table 15.2) or alloys.

Superconducting elements can be divided into two groups. The first, composed of nine metals listed at the beginning of Table 15.1, belongs to the family of simple metals on the right hand side of the periodic table of the elements. The second

Table 15.2: Critical temperatures of some compounds [J.S. Blakemore, *op. cit.* p. 268].

Substance	$Nb_3\,Ge$	$Nb_3\,Ga$	$Nb_3\,Sn$	Nb N	Ni Bi
T_C /K	23.2	20.3	18.05	16.0	4.25

group comprises the following seventeen transition metals.

    3d : Ti, V

    4d : Zr, Nb, Mo, Tc, Ru   4f : La(α and β)

    5d : Hf, Ta, W, Re, Os, Ir   5f : Th, Pa, U(α and γ) .

No alkali or alkali-earth or noble metal has been found to be superconducting. To become superconducting a substance must display strong electron-phonon coupling, which implies poor conductivity in the normal phase. Similarly no ferromagnetic element is a superconductor. Finally, thousands of compounds or alloys are known to be superconducting even though one or several (or even all) of the constituents are not.

The critical temperatures T_C of metallic substances are all lower than 25 K. For pure elements, these critical temperatures vary with the mass M of the isotope (isotope effect) as follows

$$T_C \propto M^{-\alpha} \ , \tag{15.2}$$

where α is a constant close to 0.5 (Table 15.1).

15.2.2 Critical Field

At sufficiently strong magnetic flux densities B_C the superconductivity in a substance is destroyed. The limiting flux density $B_C(T)$ decreases from its maximum value at absolute zero to zero at the critical temperature T_C (Fig. 15.1).

Values of $B_C(0)$ are fairly small, less than 0.2 T (Table 15.1). The function $B_C(T)$ is approximately consistent with the Tuyn equation (15.1). Deviations from this relation, defined by

$$D(T) = \frac{B_C(T)}{B_C(0)} - \left[1 - \left(\frac{T}{T_C}\right)^2\right] \ , \tag{15.3}$$

are quite small (Fig. 15.3). In the T, B_0 plane the $B_C(T)$ curve bounds the region of existence of two phases, one superconducting, the other normal. At absolute zero, the slope of its tangent is zero, in conformity with the Third Law, and the slope is finite at the critical point.

15.2.3 Meissner Effect

Consider a long thin cylindrical rod (Fig. 15.4) with its axis parallel to the direction of the applied magnetic field (other configurations will be discussed in §15.3), in zero

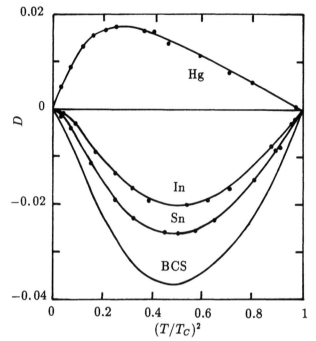

Figure 15.3: Deviation from the Tuyn equation (15.3): experimental data for tin, indium and mercury [D.K. Finnemore and D.E. Mapother, *Phys. Rev. A* **140**, 507 (1965)]; BCS theory [B. Mühlschlegel, *Z. Physik* **155**, 313 (1959)].

field and at a temperature above T_C (state a). When its temperature is reduced below T_C, the rod becomes superconducting (state b). If a weak external field of flux density $B_0 = \mu_0 H_0$ is applied, Foucault currents are generated in the surface that oppose penetration of the field, and which persist owing to the lack of electrical resistance (state c). These currents, whose density is equal to $j = H_0$, compensate the external field and endow the substance with a total magnetic moment $\mathcal{M} = Sjl$, where S and l are the cross-sectional area and length of the cylinder respectively. This yields the following result, which is valid for all configurations,

$$\mathcal{M} = -V\mathbf{H}_0 = -V\frac{\mathbf{B}_0}{\mu_0} \ . \tag{15.4}$$

If, starting once again from the normal state a, the field is increased to its value B_0 (state d), the field then pervades the metal with a value $B = B_0$, since superconductors are not magnetic. If the temperature is now reduced, it might be thought that a superconducting state c' $\neq$ c would be reached in which the magnetic field on the inside is still $B = B_0$. This does not occur, since at the transition the material expels the magnetic field by generating the same surface currents as before. The system thus returns to state c again. This property of superconductors whereby the flux density is zero ($\mathbf{B} = 0$) in the bulk material is called the Meissner effect. It is not a consequence of the fact that the resistivity is zero. It is another aspect of the phenomenon acting on a molecular level.

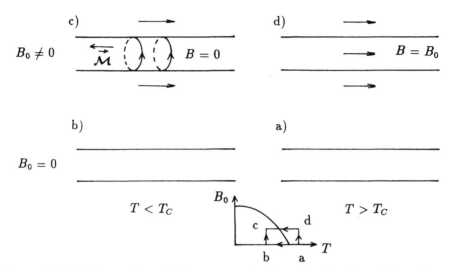

Figure 15.4: Meissner effect. In the superconducting state, $B = 0$ at all times (see text).

It is worth remarking that authors frequently express equation (15.4) in the form

$$\mathbf{M} = \frac{\mathcal{M}}{V} = -\mathbf{H}_0 , \quad \text{i.e.} \quad \chi = -1 ,$$

and assimilate superconductors with ideal diamagnetic substances. This analogy may simplify certain discussions of the overall behaviour, but it is incorrect locally, since $\mathbf{M} = 0$ everywhere. The magnetic moment is the consequence of the surface currents.

15.2.4   Heat Capacity

In zero field, the heat capacity of superconducting materials exhibits a discontinuity at the critical temperature, but remains finite (Fig. 15.5a). Also, as no latent heat is observed, the transition is second order. For magnetic flux densities B_0 smaller than the critical field $B_C(0)$, the heat capacity also displays a discontinuity at the transition temperature defined by $B_C(T) = B_0$. Latent heat is, however, observed, and the phase transition is first order. The two phases can then coexist (§15.4). For fields greater than the critical value $B_C(0)$ no transition occurs and the heat capacity is continuous (Fig. 15.5a).

Since the substance is non magnetic in its normal state, its properties are independent of B_0. The free energy is then a function $F_n(T)$ of the temperature alone (the volume is not involved at low temperatures), as is also the heat capacity $C_n(T)$. In the superconducting phase the substance has a magnetic moment given by (15.4) and its free energy, whose differential is

$$dF = -S\,dT - \mathcal{M}\,dB_0 = -S\,dT + \frac{V}{\mu_0}\,B_0\,dB_0$$

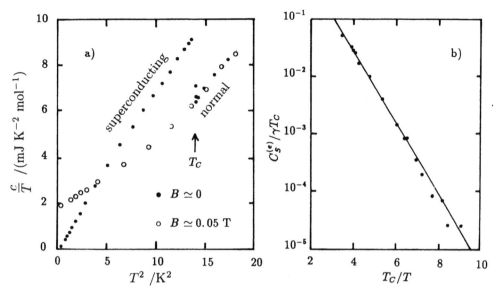

Figure 15.5: Molar heat capacities of tin at low temperature. a) Total heat capacity c in zero field (superconducting and normal states) and in fields greater than the critical field $B_C(0) = 0.03$T (normal state). The critical temperature $T_C = 3.72$ K is indicated [C.A. Bryant and P.H. Keesom, *Phys. Rev.* **123**, 491 (1961)]. b) Electronic heat capacity at very low temperatures. The data are satisfactorily described by equation (15.8) with $a = 7.85$ and $b = 1.42$. γ is taken to be 1.78 mJ K^{-2} mol^{-1} [H.R. O'Neal and N.E. Phillips, *Phys. Rev. A* **137**, 748 (1965)].

is given by

$$F(T,\ B_0) = F_s(T) + \frac{V}{2\mu_0}\ B_0^2 \ . \tag{15.5}$$

It follows that, in the superconducting state as well, the entropy and the heat capacity are independent of B_0. The latter quantity is denoted $C_s(T)$.

At the transition between the normal and the superconducting state, the free energy $F(T,\ B_0)$ is continuous, i.e.

$$F_n(T) = F_s(T) + \frac{V}{2\mu_0}\ B_C^2(T) \ . \tag{15.6}$$

This equation yields $B_C(T)$. On differentiating, we get successively

$$S_n(T) - S_s(T) = -\frac{V}{2\mu_0}\ \frac{d}{dT}\left[B_C^2(T)\right]$$

$$\text{and} \qquad C_n(T) - C_s(T) = -\frac{VT}{2\mu_0}\ \frac{d^2}{dT^2}\left[B_C^2(T)\right] \ . \tag{15.7}$$

The last equation relates the heat capacities to the critical field, thereby reducing the number of measurements that need to be made.

For each of the two phases the heat capacity contains a term $C^{(r)}$ coming from the lattice vibrations and a term $C^{(e)}$ from the electrons. Since no change in the lattice occurs at the transition, it may be assumed that $C^{(r)}$ is the same function in both phases. Thus

$$C_n = C^{(r)} + C_n^{(e)}$$
$$C_s = C^{(r)} + C_s^{(e)} \ .$$

Since the electronic heat capacity of the normal phase has the form γT (10.37), the electronic heat capacity of the superconducting phase can be found from

$$C_s^{(e)} = C_s - C^{(r)} = C_s - (C_n - \gamma T) \ ,$$

where γ is obtained by extrapolating C_n (§10.3.4) to absolute zero. For $T \ll T_C$, the heat capacity found in this way (Fig. 15.5b) varies exponentially with $1/T$ as

$$\frac{C_s^{(e)}}{\gamma T_C} = a \, \exp\left(-b \, \frac{T_C}{T}\right) \ . \tag{15.8}$$

The values of a and b are generally close to 8 and 1.4 respectively. At the critical temperature, the electronic heat capacity exhibits a discontinuity, the relative value of which is (Table 15.1)

$$\left|\frac{C_s^{(e)} - C_n^{(e)}}{C_n^{(e)}}\right|_{T_C} = \frac{|C_s - C_n|_{T_C}}{\gamma T_C} \simeq 1.4 \ . \tag{15.9}$$

15.2.5   Gap

The behaviour of the heat capacity $C_s^{(e)}$ is typical of a two-level system, which implies the existence of a gap in the electronic energy spectrum. Several methods can be used to observe and measure this gap. When a far infrared electromagnetic wave ($\sigma = 1/\lambda \sim 10 \text{ cm}^{-1}$) impinges on a superconducting film, absorption occurs only above a certain cut-off frequency, while a film in the normal state absorbs at all frequencies. The value of the gap E_g is found from the energy of the photons at the cut-off frequency. Other methods of measurement are available, such as the tunnel effect (§15.5), or acoustic wave attenuation.

At absolute zero, the value of the gap $E_g(0)$ is of the order of 10^{-3} eV and is such that the ratio $E_g(0)/kT_C$ is close to 3.5 (Table 15.1). It is almost constant at low temperature, then decreases (Fig. 15.6) and, close to T_C, varies as

$$E_g(T) \propto \left(1 - \frac{T}{T_C}\right)^{1/2} \ . \tag{15.10}$$

It can be seen that $E_g(T)$ vanishes at the critical temperature. This shows that superconductivity is directly related to the existence of the gap. The behaviour of the gap (15.10) makes it an appropriate order parameter for the normal $\rightarrow$ superconducting transition. The critical exponent $1/2$ and the jump in heat capacity are indications of critical behaviour similar to that found in molecular field theory (§13.4.2).

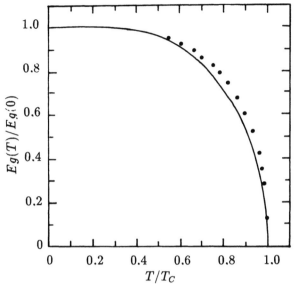

Figure 15.6: Gap $Eg(T)$ measured by tunnel effect in tin. $T_C = 3.8$ K and $E_g(0) = 1.15 \times 10^{-3}$ eV [P. Townsend and J. Sutton, *Phys. Rev.* **128**, 591 (1962)]. The curve is the prediction (15.30) of the BCS theory for $\Delta(T)/\Delta_0$.

15.2.6 Heat Conduction

Normal metals are both good electrical and thermal conductors. Superconductors, however, which are perfect conductors, have a lower heat conductivity K_s in the superconducting phase than in the normal metal K_n (Fig. 15.7). This has to do with the fact that as the temperature decreases the number of "normal" electrons, which are the only heat carriers, decreases. At the same time the number of "superconducting" electrons is different from zero, and the electrical conductivity is therefore infinite.

This property is used at very low temperatures to construct heat switches. The sample under investigation is connected to the heat source by a superconducting rod (Fig. 4.13). In the absence of a magnetic field the sample and source are isolated, but thermal contact is established when the current in a coil generates a magnetic field and returns the rod to its normal state.

15.3 BCS THEORY OF SUPERCONDUCTIVITY

15.3.1 Electron - Phonon Interaction

The principal cause of electrical resistance in metals is the electron-phonon interaction. In this interaction, an electron colliding with an ion creates or removes an oscillation that propagates through the whole lattice. In other words, the electron has created or annihilated a phonon. This interaction is reflected in the Hamiltonian of the electron gas through terms of the form

$$v(\mathbf{k}, \ \mathbf{q}) \ c^+_{-\mathbf{q}} \ a^+_{\mathbf{k}+\mathbf{q}} \ a_{\mathbf{k}} \qquad \text{or} \qquad v(\mathbf{k}, \ \mathbf{q}) \ c_{\mathbf{q}} \ a^+_{\mathbf{k}+\mathbf{q}} \ a_{\mathbf{k}}. \qquad (15.11)$$

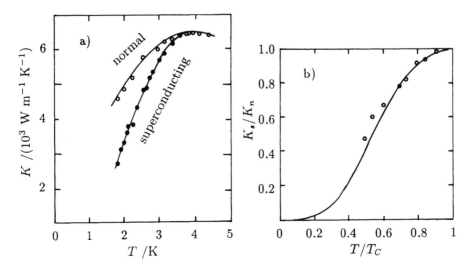

Figure 15.7: Thermal conductivity of tin in the normal phase, K_n, and the supercon-ducting phase, K_s [J.K. Hulm, *Proc. Roy. Soc. A* (London) **204**, 98 (1950)]. In a) the upper curve shows $K_n^{-1} = \alpha T^2 + \beta T^{-1}$, where the two terms correspond to scattering of electrons by phonons and by impurities respectively. In b) the curve is the prediction of the BCS theory [J. Bardeen, G. Rickayzen and L. Tewordt, *Phys. Rev.* **113**, 982 (1959)].

These terms mean that an electron in the state $\mathbf{k}$ has been annihilated (operator $a_{\mathbf{k}}$) and an electron in the state $\mathbf{k}+\mathbf{q}$ has been created (operator $a_{\mathbf{k}+\mathbf{q}}^{+}$). Simultaneously, a phonon is created in the state $-\mathbf{q}$ (operator $c_{-\mathbf{q}}^{+}$) or annihilated in the state $\mathbf{q}$ (operator $c_{\mathbf{q}}$). The energy $v(\mathbf{k}, \mathbf{q})$ characterizing the probability amplitude of the interaction is practically independent of $\mathbf{k}$. These terms can be represented by the graphs

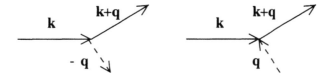

Since the energy of the phonons is less than or of the order of $k\Theta_D \sim 10^{-2}$ eV (where Θ_D is the Debye temperature), and that of the electrons is of the order of the Fermi energy $\epsilon_F \sim 5$ eV, these collisions are almost elastic for the electrons. Owing to the Pauli exclusion principle the electrons inside the Fermi sphere cannot be scattered since the neighbouring energy states are already completely occupied. Only the electrons close to the Fermi surface are therefore subject to collisions.

15.3.2 Electron - Electron Interaction

The scattering of two electrons, initially in the states $\mathbf{k}_1$ and $\mathbf{k}_2$, towards the states $\mathbf{k}_1 - \mathbf{q}$ and $\mathbf{k}_2 + \mathbf{q}$, is reflected in the Hamiltonian by the presence of a term

$$V(\mathbf{k}_1, \mathbf{k}_2, \mathbf{q}) \; a_{\mathbf{k}_1-\mathbf{q}}^+ \, a_{\mathbf{k}_2+\mathbf{q}}^+ \, a_{\mathbf{k}_2} \, a_{\mathbf{k}_1}, \tag{15.12}$$

and can be represented by the graph

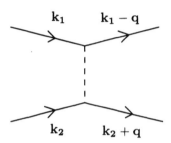

By virtue of the Pauli exclusion principle, the only electrons still involved are those lying close to the Fermi surface. The strongest interaction is the repulsive Coulomb interaction ($V > 0$). This, however, is screened by the ions of the lattice at large distances, i.e. for small values of $\mathbf{q}$. The energy V therefore never exceeds a certain maximum value.

A different type of interaction between electrons forms the basis of the microscopic theory of superconductivity. When an electron moves in the lattice it attracts the positive ions and thus perturbs their movement. This lattice perturbation can act in turn on a second electron. In terms of phonons, the first electron emits a phonon with wave vector $\mathbf{q}$ which is absorbed by the second electron. This phonon is virtual and its energy

$$\epsilon_{\mathbf{k}_1} - \epsilon_{\mathbf{k}_1-\mathbf{q}} = \epsilon_{\mathbf{k}_2+\mathbf{q}} - \epsilon_{\mathbf{k}_2}$$

may take a value different from $\hbar\omega_{\mathbf{q}}$, the energy that a real phonon of wave vector $\mathbf{q}$ would have. The calculation shows that for this type of interaction the energy V is given by

$$V = \frac{\hbar\omega_{\mathbf{q}} \, |v(\mathbf{q})|^2}{(\epsilon_{\mathbf{k}} - \epsilon_{\mathbf{k}-\mathbf{q}})^2 - (\hbar\omega_{\mathbf{q}})^2} \; , \tag{15.13}$$

where v is the energy introduced in (15.11). V can be either positive (repulsion), or negative (attraction). Thus elastic scattering ($\epsilon_{\mathbf{k}} = \epsilon_{\mathbf{k}-\mathbf{q}}$) corresponds to an attraction. Under certain conditions the total interaction, obtained by including the Coulomb interaction, can turn out to be attractive. This is the case for electrons located in a shell of thickness of the order of $k\Theta_D$ at the Fermi surface, since for them the condition

$$\epsilon_{\mathbf{k}} - \epsilon_{\mathbf{k}-\mathbf{q}} < \hbar\omega_{\mathbf{q}} \simeq k\Theta_D \tag{15.14}$$

is satisfied.

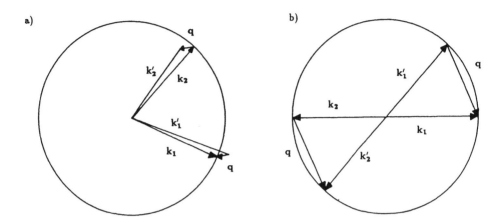

Figure 15.8: Scattering of two electrons from the states k_1 and k_2 to the states $k'_1 = k_1 - q$ and $k'_2 = k_2 + q$. a) arbitrary k_1 and k_2; only small values of q are allowed. b) $k_2 = -k_1$; many values of q are possible.

15.3.3   Cooper Pairs

When two electrons with arbitrary momenta k_1 and k_2 scatter, the constraints of conservation of energy and momentum, in addition to the Pauli exclusion principle, limit q to small values (Fig. 15.8a). The probability of interaction is therefore small. When, however, k_1 and k_2 are equal and opposite in value (Fig. 15.8b), many configurations are available after scattering. The probability of interaction between electrons with momenta k and $-k$ is therefore very large, and these electrons will be especially correlated (Cooper pairs).

Until now we have not considered the electron spin. For each electron pair there are four spin configurations making up a singlet and a triplet. It can be shown that the singlet state favours correlation between pairs. As a result, only pairs of states $|k \uparrow, -k \downarrow >$ are considered in the following.

15.3.4   BCS Hamiltonian

The Hamiltonian of the electron system is composed, firstly, of the kinetic term

$$\sum_{k,\lambda} \epsilon_k n_{k\lambda} \ ,$$

where $n = a^+ a$ is the electron number operator in the state k with spin $\lambda = \pm 1/2$, and, secondly, of the interaction term, which is a sum of terms of the form (15.12). In the BCS theory, the fact that the pairs $|k \uparrow, -k \downarrow>$ play a fundamental role is taken into account by introducing the pair annihilation and creation operators

$$b_k = a_{k\uparrow} a_{-k\downarrow} \qquad \text{and} \qquad b_k^+ = a_{-k\downarrow}^+ a_{k\uparrow}^+, \tag{15.15}$$

and by considering only those terms (15.12) for which $\mathbf{k}_2 = -\mathbf{k}_1$. The interaction Hamiltonian then becomes

$$\sum_{\mathbf{k},\mathbf{k}'} V(\mathbf{k},\ \mathbf{k}')\ b_{\mathbf{k}'}^+\ b_{\mathbf{k}}$$

where $V(\mathbf{k},\ \mathbf{k}')$ is a symmetrized energy in $\mathbf{k}$ and $\mathbf{k}'$ and which is negative when $\mathbf{k}$ and $\mathbf{k}'$ are close to the Fermi surface.

As we already saw in paragraph 12.6.4, use of the annihilation and creation operators calls for the grand-canonical formalism. For the BCS Hamiltonian we therefore write

$$\hat{H}' = \hat{H} - \mu\hat{N} = \sum_{\mathbf{k},\lambda}(\epsilon_{\mathbf{k}} - \mu)\,n_{\mathbf{k}\lambda} + \sum_{\mathbf{k},\mathbf{k}'} V(\mathbf{k},\ \mathbf{k}')\ b_{\mathbf{k}'}^+\ b_{\mathbf{k}} \qquad (15.16)$$

where μ is the Fermi energy ϵ_F. This Hamiltonian is difficult to use on account of the quadratic term in the operators b. To alleviate this difficulty, the following linearization is used

$$b_{\mathbf{k}'}^+\ b_{\mathbf{k}} \;\to\; b_{\mathbf{k}'}^+\ \bar{b}_{\mathbf{k}} + \bar{b}_{\mathbf{k}'}^+\ b_{\mathbf{k}} - \bar{b}_{\mathbf{k}'}^+\ \bar{b}_{\mathbf{k}}$$

in which $\bar{b}_{\mathbf{k}}$ is the statistical average of the operator $b_{\mathbf{k}}$ to be calculated in paragraph 15.3.6. This linearization is exact in the limit $N \to \infty$. In fact, it is similar to the molecular field approximation (13.25), which improves as the number of nearest neighbours z increases. Here, N plays the same role. The Hamiltonian (15.16) then becomes

$$\hat{H}' \;=\; \sum_{\mathbf{k},\lambda}\epsilon_{\mathbf{k}}' n_{\mathbf{k}\lambda} + \sum_{\mathbf{k}}\left(\Delta_{\mathbf{k}} b_{\mathbf{k}}^+ + \Delta_{\mathbf{k}}^* b_{\mathbf{k}}\right) - C \qquad (15.17)$$

$$\text{with} \quad \epsilon_{\mathbf{k}}' \;=\; \epsilon_{\mathbf{k}} - \mu = \epsilon_{\mathbf{k}} - \epsilon_F \qquad (15.18)$$

$$\Delta_{\mathbf{k}} \;=\; \sum_{\mathbf{k}'} V(\mathbf{k},\ \mathbf{k}')\ \bar{b}_{\mathbf{k}'}$$

$$\text{and} \quad C \;=\; \sum_{\mathbf{k},\mathbf{k}'} V(\mathbf{k},\ \mathbf{k}')\ \bar{b}_{\mathbf{k}'}^+\ \bar{b}_{\mathbf{k}} = \sum_{\mathbf{k}} \Delta_{\mathbf{k}}^* \bar{b}_{\mathbf{k}} \;. \qquad (15.19)$$

Collecting the terms for a pair of states $|\mathbf{k}\uparrow,\ -\mathbf{k}\downarrow>$, we get finally

$$\hat{H}' \;=\; \sum_{\mathbf{k}}\hat{h}_{\mathbf{k}}' - C \qquad (15.20)$$

$$\text{with} \quad \hat{h}_{\mathbf{k}}' \;=\; \epsilon_{\mathbf{k}}'\left(n_{\mathbf{k}\uparrow} + n_{-\mathbf{k}\downarrow}\right) + \Delta_{\mathbf{k}} b_{\mathbf{k}}^+ + \Delta_{\mathbf{k}}^* b_{\mathbf{k}} \;.$$

In this form, to within the energy $-C$, the Hamiltonian $\hat{H}'$ can be decomposed into a sum of Hamiltonians $\hat{h}_{\mathbf{k}}'$ of pairs of independent states.

15.3.5 Eigenstates of the BCS Hamiltonian

The eigenstates of the Hamiltonian $\hat{H}'$ are found from the eigenstates of each of the Hamiltonians $\hat{h}_{\mathbf{k}}'$. For the electrons, there are four ways of occupying the pair of states $|\mathbf{k} \uparrow, -\mathbf{k} \downarrow>$: the pair of states can be empty (labelled $\| \, 0 \, 0 >$), occupied by one electron (two states labelled $\| \, 0 \, 1 >$ and $\| \, 1 \, 0 >$), or occupied by two electrons (labelled $\| \, 1 \, 1 >$). These four states form a basis for the space in which $\hat{h}'_{\mathbf{k}}$ operates. The operators n, b and b^{+} act on the states in this basis as follows

$$
\begin{array}{ll}
n_{\mathbf{k}\uparrow} \| \, 0 \, 0 > \, = \quad 0 & \qquad n_{-\mathbf{k}\downarrow} \| \, 0 \, 0 > \, = \quad 0 \\
n_{\mathbf{k}\uparrow} \| \, 0 \, 1 > \, = \quad 0 & \qquad n_{-\mathbf{k}\downarrow} \| \, 0 \, 1 > \, = \quad \| \, 0 \, 1 > \\
n_{\mathbf{k}\uparrow} \| \, 1 \, 0 > \, = \quad \| \, 1 \, 0 > & \qquad n_{-\mathbf{k}\downarrow} \| \, 1 \, 0 > \, = \quad 0 \\
n_{\mathbf{k}\uparrow} \| \, 1 \, 1 > \, = \quad \| \, 1 \, 1 > & \qquad n_{-\mathbf{k}\downarrow} \| \, 1 \, 1 > \, = \quad \| \, 1 \, 1 >
\end{array}
$$

$$
\begin{array}{ll}
b_{\mathbf{k}} \| \, 0 \, 0 > \, = \quad 0 & \qquad b_{\mathbf{k}}^{+} \| \, 0 \, 0 > \, = \quad \| \, 1 \, 1 > \\
b_{\mathbf{k}} \| \, 0 \, 1 > \, = \quad 0 & \qquad b_{\mathbf{k}}^{+} \| \, 0 \, 1 > \, = \quad 0 \\
b_{\mathbf{k}} \| \, 1 \, 0 > \, = \quad 0 & \qquad b_{\mathbf{k}}^{+} \| \, 1 \, 0 > = \quad 0 \\
b_{\mathbf{k}} \| \, 1 \, 1 > \, = \quad \| \, 0 \, 0 > & \qquad b_{\mathbf{k}}^{+} \| \, 1 \, 1 > \, = \quad 0
\end{array}
$$

The operator $\hat{h}'_{\mathbf{k}}$ can then be represented by the following 4×4 matrix

$$
\begin{pmatrix}
0 & 0 & 0 & \Delta_{\mathbf{k}}^{*} \\
0 & \epsilon'_{\mathbf{k}} & 0 & 0 \\
0 & 0 & \epsilon'_{\mathbf{k}} & 0 \\
\Delta_{\mathbf{k}} & 0 & 0 & 2\epsilon'_{\mathbf{k}}
\end{pmatrix}
$$

We see that the $\| \, 0 \, 1 >$ and $\| \, 1 \, 0 >$ states, in which the electron is single and does not interact, are eigenstates of the operator $h'_{\mathbf{k}}$, the eigenvalue being $\epsilon'_{\mathbf{k}}$. In the sub-space generated by the other two states $\| \, 0 \, 0 >$ and $\| \, 1 \, 1 >$, the operator $h'_{\mathbf{k}}$ is represented by the 2×2 matrix

$$
\begin{pmatrix}
0 & \Delta_{\mathbf{k}}^{*} \\
\Delta_{\mathbf{k}} & 2\epsilon'_{\mathbf{k}}
\end{pmatrix} \, ,
$$

where the eigenvalues

$$
\lambda_{\mathbf{k}\pm} = \epsilon'_{\mathbf{k}} \pm E_{\mathbf{k}} \qquad \text{with} \qquad E_{\mathbf{k}} = \sqrt{\epsilon'^{2}_{\mathbf{k}} + |\Delta_{\mathbf{k}}|^{2}} \tag{15.21}
$$

are associated with the eigenvectors

$$
\| \, \psi_{\mathbf{k}\pm} \, >= \frac{-(\mp\epsilon'_{\mathbf{k}} + E_{\mathbf{k}}) \| \, 0 \, 0 > \mp \Delta_{\mathbf{k}} \| \, 1 \, 1 >}{[2E_{\mathbf{k}} (\epsilon'_{\mathbf{k}} + E_{\mathbf{k}})]^{1/2}} \, . \tag{15.22}
$$

We now discuss these results. When $\Delta_{\mathbf{k}}$, defined in (15.18), is zero, the energies corresponding to the eigenvalues of $h'_{\mathbf{k}}$ are 0, $\epsilon'_{\mathbf{k}}$ (twice) and $2\epsilon'_{\mathbf{k}}$, and are respectively associated with the pair states $\| \, 0 \, 0 >$, $\| \, 0 \, 1 >$, $\| \, 1 \, 0 >$ and $\| \, 1 \, 1 >$. The minimum energy, equal to zero, is then that of the $\| \, 1 \, 1 >$ state if $\epsilon'_{\mathbf{k}} < 0$ ($\epsilon_{\mathbf{k}} < \epsilon_{F}$) and that of $\| \, 0 \, 0 >$ if $\epsilon'_{\mathbf{k}} > 0$ ($\epsilon_{\mathbf{k}} > \epsilon_{F}$). If $\Delta_{\mathbf{k}}$ is zero for all $\mathbf{k}$, which would be the case if V were zero (free electrons), we recover the well known property of normal metals at absolute zero, in which all electronic states below the Fermi level are occupied, and the others are empty.

When $\Delta_{\mathbf{k}}$ is non zero, the minimum energy is $\lambda_{\mathbf{k}-}$ for all $\mathbf{k}$. Since its value is negative, the ground state $\| \psi_{\mathbf{k}-} >$ is therefore more stable than for a normal metal. This is the state that is referred to when we speak of Cooper pairs. It should be emphasized that this is a superposition of the $\| 0\ 0 >$ and $\| 1\ 1 >$ states, and that it is occupied by one pair of electrons $\mathbf{k} \uparrow, -\mathbf{k} \downarrow$ only, with a probability

$$P_{\mathbf{k}} = \frac{|\Delta_{\mathbf{k}}|^2}{2E_{\mathbf{k}}\,(\epsilon'_{\mathbf{k}} + E_{\mathbf{k}})} = \frac{1}{2}\left(1 - \frac{\epsilon'_{\mathbf{k}}}{E_{\mathbf{k}}}\right). \tag{15.23}$$

$\| 0\ 1 >$ and $\| 1\ 0 >$ are excited states situated at an energy $E_{\mathbf{k}}$ above $\| \psi_{\mathbf{k}-} >$. However, as these excited states do not have the same number of electrons as the ground state, only the excitation leading to $\| \psi_{\mathbf{k}+} >$ and requiring an additional energy $\lambda_{\mathbf{k}+} - \lambda_{\mathbf{k}-} = 2E_{\mathbf{k}}$ need be considered. Since $E_{\mathbf{k}} > |\Delta_{\mathbf{k}}|$, a minimum energy $2\,|\Delta_{\mathbf{k}}|$ must be supplied to form an excited state. This gives us a glimpse of how a microscopic explanation can be given for the experimentally observed gap.

15.3.6 Occurrence of a Phase Transition

The grand-canonical partition function Ξ in the BCS theory is found from the Hamiltonian (15.20) in the form

$$\Xi = \mathrm{Tr}\ e^{-\beta \hat{H}'} = e^{\beta C} \prod_{\mathbf{k}} \xi_{\mathbf{k}} \qquad \text{where} \qquad \xi_{\mathbf{k}} = \mathrm{Tr}\ e^{-\beta \hat{h}'_{\mathbf{k}}}. \tag{15.24}$$

This trace can be calculated directly from the eigenvalues of $\hat{h}'_{\mathbf{k}}$. We have

$$
\begin{aligned}
\xi_{\mathbf{k}} &= 2 \exp\left(-\beta \epsilon'_{\mathbf{k}}\right) + \exp\left(-\beta \lambda_{\mathbf{k}-}\right) + \exp\left(-\beta \lambda_{\mathbf{k}+}\right) \\
&= 2 \exp\left(-\beta \epsilon'_{\mathbf{k}}\right) \left[1 + \cosh \beta\, E_{\mathbf{k}}\right] \\
&= 4 \exp\left(-\beta \epsilon'_{\mathbf{k}}\right) \cosh^2 \frac{\beta E_{\mathbf{k}}}{2}.
\end{aligned}
\tag{15.25}
$$

To find $\xi_{\mathbf{k}}$ and then Ξ, we must first determine $\Delta_{\mathbf{k}}$, which is defined in (15.18) and enters the equation (15.21) for $E_{\mathbf{k}}$. We have therefore to calculate the mean values $\bar{b}_{\mathbf{k}}$. Applying (12.16 and 42), we have

$$\bar{b}_{\mathbf{k}} = \frac{1}{\Xi}\ \mathrm{Tr}\ b_{\mathbf{k}}\ e^{-\beta \hat{H}'} = -kT\ \frac{\partial \ln \Xi}{\partial \Delta^*_{\mathbf{k}}} = -kT\ \frac{\partial \ln \xi_{\mathbf{k}}}{\partial \Delta^*_{\mathbf{k}}}.$$

Substitution of (15.25) for $\xi_{\mathbf{k}}$ yields

$$\bar{b}_{\mathbf{k}} = -2kT\ \frac{\partial}{\partial \Delta^*_{\mathbf{k}}}\ \ln \cosh \frac{\beta E_{\mathbf{k}}}{2} = -\frac{\Delta_{\mathbf{k}}}{2E_{\mathbf{k}}}\ \tanh \frac{\beta E_{\mathbf{k}}}{2}. \tag{15.26}$$

This equation implicitly determines $\bar{b}_{\mathbf{k}}$. Substituting this equation into (15.18) gives in turn the implicit equation defining $\Delta_{\mathbf{k}}$

$$\Delta_{\mathbf{k}} = -\sum_{k'} V(\mathbf{k},\ \mathbf{k}')\ \frac{\Delta_{\mathbf{k}'}}{2E_{\mathbf{k}'}}\ \tanh \frac{\beta E_{\mathbf{k}'}}{2}. \tag{15.27}$$

This equation can be solved only when the exact form of $V(\mathbf{k}, \mathbf{k}')$ is known. However, all the main characteristics of the result that would thus be obtained can be found by assuming that $V(\mathbf{k}, \mathbf{k}')$ is zero if $\mathbf{k}$ or $\mathbf{k}'$ lie outside a layer of thickness $k\Theta_D$ on either side of the Fermi surface while remaining constant inside. Thus

$$V(\mathbf{k}, \mathbf{k}') = -V_0 \, \theta \, (k\Theta_D - |\epsilon'_\mathbf{k}|) \, \theta \, (k\Theta_D - |\epsilon'_{\mathbf{k}'}|) \quad , \tag{15.28}$$

where the positive constant V_0 represents the attractive interaction between electrons due to exchange of virtual phonons. The Heaviside function θ is equal to 0 for negative arguments and 1 for positive arguments. On substituting into (15.27), it can be seen that $\Delta_\mathbf{k}$ is zero if $\mathbf{k}$ lies outside the layer of thickness $k\Theta_D$ and has a constant value Δ inside this layer, i.e.

$$\Delta_\mathbf{k} = \Delta \times \theta \, (k\Theta_D - |\epsilon'_\mathbf{k}|) \qquad \Delta = V_0 \sum_{\mathbf{k}'} \frac{\Delta}{2E_{\mathbf{k}'}} \, \tanh(\beta E_{\mathbf{k}'}/2) \, . \tag{15.29}$$

It can be seen that this equation, and more generally (15.27), has a non zero solution for Δ only if $V(\mathbf{k}, \mathbf{k}')$ is negative ($V_0 > 0$). On replacing the sum over $\mathbf{k}'$ by an integral and substituting the formula for $E_{\mathbf{k}'}$, we find that Δ is given by the implicit equation

$$1 = \frac{V_0}{2} \int_{\epsilon_F - k\Theta_D}^{\epsilon_F + k\Theta_D} \frac{g(\epsilon)}{2} \, d\epsilon \, \frac{\tanh(\beta\sqrt{\epsilon'^2 + \Delta^2}/2)}{\sqrt{\epsilon'^2 + \Delta^2}} \quad ,$$

where $g(\epsilon)$ is the density of electronic states taking into account the spin degeneracy (10.43). Neglecting variations in $g(\epsilon)$ close to the Fermi surface, it can be seen that Δ is defined by

$$\int_{-k\Theta_D}^{+k\Theta_D} \frac{\tanh(\beta\sqrt{\epsilon'^2 + \Delta^2}/2)}{\sqrt{\epsilon'^2 + \Delta^2}} \, d\epsilon' = \frac{4}{V_0 \, g(\epsilon_F)} = \frac{2}{g} \tag{15.30}$$

and that its value depends on the temperature. The parameter

$$g = \frac{V_0 \, g(\epsilon_F)}{2} \tag{15.31}$$

combines the two electronic parameters V_0 and $g(\epsilon_F)$ of the theory into a single dimensionless quantity. Making use of an identity for the function tanh, and given that

$$\int_{-k\Theta_D}^{k\Theta_D} \frac{d\epsilon'}{\sqrt{\epsilon'^2 + \Delta^2}} = 2\ln\left(k\Theta_D + \sqrt{\Delta^2 + k\Theta_D^2}\right) \simeq 2\ln(2k\Theta_D/\Delta) \quad ,$$

equation (15.30) can also be written

$$\ln \frac{2k\theta_D}{\Delta} = \int \frac{d\epsilon'}{E} \frac{1}{e^{\beta E} + 1} + \frac{1}{g} \, . \tag{15.32}$$

The function $\Delta(T)$, found by numerically solving (15.30 or 31), is shown in Figure 15.6. Its value Δ_0 at absolute zero is found by observing that the integral in (15.32) is zero. Thus

$$\Delta(0) \equiv \Delta_0 = 2k\Theta_D e^{-1/g} . \tag{15.33}$$

Close to absolute zero, it can be shown that Δ takes the form

$$\frac{\Delta(T)}{\Delta_0} = 1 - \sqrt{\frac{2\pi kT}{\Delta_0}} \, e^{-\Delta_0/kT} , \quad (kT \ll \Delta) \tag{15.34}$$

and decreases, vanishing at a temperature T_C given by

$$\int_{-k\Theta_D}^{k\Theta_D} \frac{\tanh(\beta_C \epsilon'/2)}{|\epsilon'|} \, d\epsilon' = \frac{2}{g} .$$

Analytical calculation yields

$$T_C = 2 \frac{\exp \gamma_E}{\pi} \Theta_D \exp(-1/g) = 1.13 \, \Theta_D \exp(-1/g) \tag{15.35}$$

where $\gamma_E = 0.577$ is Euler's constant. Close to T_C $(T < T_C)$, a series expansion shows that

$$\Delta(T) = kT_C \left[\frac{8\pi^2}{7\zeta(3)} \left(1 - \frac{T}{T_C} \right) \right]^{1/2} = 3.06 \, kT_C \left(1 - \frac{T}{T_C} \right)^{1/2} \tag{15.36}$$

where $\zeta(3) = 1.202$ is the value of the Riemann function $\zeta(z)$ at $z = 3$. For $T > T_C$, equation (15.30) no longer has a solution and the only solution of the complete equation (15.29) defining Δ is zero. These variations of $\Delta_k(T)$ change very little if other forms for $V(\mathbf{k}, \mathbf{k}')$ are chosen, as long as they are negative close to the Fermi surface.

We have just seen for $T > T_C$ that $\Delta_k = 0$ for all $\mathbf{k}$; the metal then behaves normally. For $T < T_C$, however, Δ_k differs from zero for certain values of $\mathbf{k}$ and the metal behaves as a superconductor, as we shall show below.

The parameter g (15.31) can be evaluated from the transition temperature T_C given in (15.35). In the case of tin for which $\Theta_D \simeq 200$ K and $T_C \simeq 3.7$ K, this parameter is given by $g = 0.24$. If we had chosen a layer thickness other than $\pm k\Theta_D$, only a small variation in the value of g would have resulted $(g = 0.21$ for a thickness $\pm k\Theta_D/2)$. This is due to the fact that g enters through the factor $\exp(-1/g)$. Moreover, when the isotope is varied, the electronic properties remain unchanged but the Debye temperature varies. The vibration frequencies of the lattice depend on the mass of the atoms. For example, in the acoustic branch these frequencies vary as $M^{-1/2}$ (8.30). The BCS theory thus provides an explanation of the isotope effect (§15.2.1).

15.3.7 Two-Fluid Model

The grand potential $\Omega(T, V, \mu)$ is calculated from the grand-canonical partition function Ξ (15.24),

$$\Omega = kT \ln \Xi = -C - kT \sum_k \ln \xi_k$$

$$= -C + \sum_{\mathbf{k}} \left[\epsilon'_{\mathbf{k}} - 2kT \ln \left(e^{\beta E_{\mathbf{k}}/2} + e^{-\beta E_{\mathbf{k}}/2} \right) \right] \quad .$$

The constant C, defined in (15.19), reduces to $-\Delta^2/V_0$ when $\bar{b}_{\mathbf{k}}$ is expressed in terms of (15.26), and equation (15.27) is used to define Δ. Transforming to a continuous integral, we thus find

$$\Omega = \frac{\Delta^2}{V_0} + \int \frac{g(\epsilon)}{2} \, d\epsilon \, \left[\epsilon' - E - 2kT \ln \left(1 - e^{-\beta E} \right) \right] \quad . \tag{15.37}$$

Note that the grand potential depends explicitly on T, but also implicitly through Δ and $E = \left(\epsilon'^2 + \Delta^2 \right)^{1/2}$.

We now calculate this grand potential at absolute zero. We have

$$\Omega_0 = \frac{\Delta_0^2}{V_0} + \int \frac{g(\epsilon)}{2} \, [\epsilon' - E] \, d\epsilon \quad .$$

The range of integration can be divided into 3. For $\epsilon' \equiv \epsilon - \epsilon_F$ greater than $k\Theta_D$, $\Delta_{\mathbf{k}} = 0$ and therefore $E = \epsilon'$; the corresponding contribution is zero. For ϵ' less than $-k\Theta_D$, once again $\Delta_{\mathbf{k}} = 0$ and $E = |\epsilon'| = -\epsilon'$. We therefore have

$$\Omega_0 = \frac{\Delta_0^2}{V_0} + \int_0^{\epsilon_F - k\Theta_D} g(\epsilon)\epsilon' \, d\epsilon + \frac{g(\epsilon_F)}{2} \int_{\epsilon_F - k\Theta_D}^{\epsilon_F + k\Theta_D} (\epsilon' - E) \, d\epsilon$$

$$= \frac{\Delta_0^2}{V_0} + \int_0^{\epsilon_F - k\Theta_D} g(\epsilon)\epsilon' \, d\epsilon - g(\epsilon_F) \int_0^{k\Theta_D} E \, d\epsilon' \quad .$$

Invoking the grand potential of a free electron gas at absolute zero,

$$\Omega_{e0} = U_{e0} - N\mu = \int_0^{\epsilon_F} g(\epsilon)\epsilon' \, d\epsilon,$$

for which the formula is given in paragraph 10.3.3, we may write

$$\Omega_0 = \Omega_{e0} + \frac{\Delta_0^2}{V_0} - g(\epsilon_F) \int_0^{k\Theta_D} (E - \epsilon') \, d\epsilon' \quad .$$

Remarking that

$$2 \int \sqrt{x^2 + 1} \, dx = x\sqrt{x^2 + 1} + \ln \left(x + \sqrt{x^2 + 1} \right) \simeq x^2 + \ln 2x + \frac{1}{2} \qquad (x \gg 1) \quad ,$$

and making use of expression (15.33), we finally get

$$\Omega_0 = \Omega_{e0} + \frac{\Delta_0^2}{V_0} - \Delta_0^2 \frac{g(\epsilon_F)}{2} \left(\ln \frac{2k\Theta_D}{\Delta_0} + \frac{1}{2} \right)$$

$$= \Omega_{e0} - \frac{\Delta_0^2}{4} g(\epsilon_F) \quad . \tag{15.38}$$

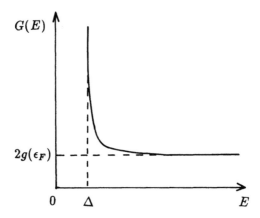

Figure 15.9: Density of states of excitations in a superconducting electron gas.

This result shows that the superconducting state is more stable than the normal state at absolute zero. However, taking $\Delta_0 \sim kT_C \sim 10^{-3}$ eV and $\epsilon_F \sim 5$ eV, and calculating the difference $\Omega_0 - \Omega_{e0}$ normalized to one electron, we find

$$\frac{\Delta_0^2}{4}\frac{g(\epsilon_F)}{N} \sim 10^{-7} \text{ eV} \quad,$$

where $g(\epsilon_F)$ is calculated from (10.49). Clearly this difference is very small compared to $\Omega_0/N \sim \epsilon_F$, which explains why superconductivity vanishes when the perturbations applied to the system exceed a certain threshold.

For arbitrary temperatures, an identical calculation gives the grand potential in the form $\Omega = \Omega_{e0} + \Omega_s + \Omega_n$ where

$$
\begin{aligned}
\Omega_s &= -\Delta^2 \frac{g(\epsilon_F)}{2} \left(\frac{2}{V_0\, g(\epsilon_F)} - \ln \frac{2k\Theta_D}{\Delta} - \frac{1}{2} \right) \\
&= -\Delta^2 \frac{g(\epsilon_F)}{2} \left(\ln \frac{\Delta_0}{\Delta} + \frac{1}{2} \right) \quad (15.39)
\end{aligned}
$$

and
$$
\begin{aligned}
\Omega_n &= -kT \int_0^\infty g(\epsilon) \ln\left(1 + e^{-\beta E}\right)\, d\epsilon \\
&= -2kT \int_\Delta^\infty g(\epsilon) \ln\left(1 + e^{-\beta E}\right) \frac{E\, dE}{\sqrt{E^2 - \Delta^2}} \quad . \quad (15.40)
\end{aligned}
$$

The term Ω_n corresponds to the third term of the integral in (15.37). It can be taken as the grand potential (2.34) of a gas of an indeterminate number of excitations obeying Fermi-Dirac statistics, whose energy is E and whose density of states $G(E)$ is shown in Figure 15.9. $G(E)$ is given by

$$G(E) = 2g(\epsilon) \frac{E}{\sqrt{E^2 - \Delta^2}} \simeq 2g(\epsilon_F) \frac{E}{\sqrt{E^2 - \Delta^2}} \quad . \quad (15.41)$$

These excitations, whose contribution is essentially at $E \simeq \Delta$, in fact correspond to the electrons of Cooper pairs that are excited into the state $\mid \psi_{+\mathbf{k}} >$.

By analogy with the two fluid model for liquid helium (§9.3), the electron gas can be considered as containing a "superconducting" component with grand potential Ω_s and a "normal" component with grand potential Ω_n. At absolute zero, Ω_n (15.40) vanishes and Ω_s (15.39) is equal to $-\Delta_0^2\, g(\epsilon_F)/4$, i.e.

$$\Omega = \Omega_{e0} + \Omega_s + \Omega_n = \Omega_{e0} - \frac{\Delta_0^2}{4}\, g(\epsilon_F) \ ,$$

a result already found in (15.38). At temperatures greater than or equal to T_C, it is Ω_s that vanishes, and since then $E = |\epsilon'|$, the total grand potential $\Omega = \Omega_{e0} + \Omega_n$ is identified with the grand potential of a free electron gas.

At a given temperature the minimum energy required to create an additional excitation is $\Delta(T)$. Since these excitations are of fermion type, conservation of the total angular momentum requires them to be created in pairs. It follows that the minimum energy required for the excitation in the metal is equal to $2\Delta(T)$. This energy corresponds to the width $E_g(T)$ of the gap (§15.2.5). Figure 15.6 shows that the BCS theory satisfactorily simulates the experimental data for the gap in tin, in spite of the simplified form (15.28) used for the interaction energy $V(\mathbf{k},\ \mathbf{k}')$. Lastly, from (15.33 and 35) it can be seen that

$$\frac{E_g(0)}{kT_C} = 2\,\frac{\Delta_0}{kT_C} = 2\pi \exp\left(-\gamma_E\right) = 3.53 \tag{15.42}$$

in very good agreement with the experimental results (Table 15.1).

15.3.8   Thermodynamic Functions

To find the free energy of the electron gas from the grand potential Ω, ϵ_F must first be determined as a function of N and of the other variables, then $F = \Omega + N\epsilon_F$ must be calculated. We have seen that in fact the attractive interaction of the electrons has little effect on the value of the grand potential. It follows that ϵ_F is not very different from the Fermi energy of a gas of free electrons at absolute zero (10.24). Hence, from (15.39), we can write directly

$$\begin{aligned} F(T,\ V,\ N) \ = \ & F_{e0}(V,\ N) - \frac{g(\epsilon_F)}{2}\, \Delta^2 \left(\ln\frac{\Delta_0}{\Delta} + \frac{1}{2}\right) \\ & - \ kT \int_\Delta^\infty G(E) \ln\left(1 + e^{-\beta E}\right)\, dE \ , \end{aligned} \tag{15.43}$$

where F_{e0} is the free energy of a free electron gas at absolute zero.

We can then derive, for example, the internal energy

$$\begin{aligned} U \ = \ & -T^2\, \frac{\partial}{\partial T}\left(\frac{F}{T}\right) \\ = \ & F_{e0} - \frac{g(\epsilon_F)}{2}\, \Delta^2 \left(\ln\frac{\Delta_0}{\Delta} + \frac{1}{2}\right) + \int_\Delta^\infty E\,\frac{G(E)\, dE}{e^{\beta E} + 1} \ . \end{aligned} \tag{15.44}$$

Here, the derivative $\Delta'(T)$ does not enter as it is multiplied by the factor

$$\ln\frac{\Delta_0}{\Delta} - \int \frac{d\epsilon'}{E}\,\frac{1}{e^{\beta E} + 1} = 0 \ , \tag{15.45}$$

which is zero by virtue of equations (15.32-33). We thus see that the energy can be decomposed into three terms, namely a constant term F_{e0}, the internal energy of the superconducting component, and the internal energy of the normal component. Note that the internal energy of the superconducting component is identical to its free energy, which shows that its entropy is zero and that it is perfectly ordered. This result is analogous to a similar property of the superfluid component in helium (§9.2).

Using relation (15.45), we find for the heat capacity

$$
\begin{aligned}
C_V &= \frac{\partial U}{\partial T} = \int_\Delta^\infty G(E)\, dE\, \frac{e^{\beta E}}{(e^{\beta E}+1)^2}\, \frac{\left(E^2 - \Delta\Delta'T\right)}{kT^2} \\
&= \int_0^\infty g(\epsilon)\, d\epsilon\, \frac{e^{\beta E}}{(e^{\beta E}+1)^2}\, \frac{\left(E^2 - \Delta\Delta'T\right)}{kT^2} \quad .
\end{aligned}
\tag{15.46}
$$

Note that the heat capacity of a free electron gas would be given by the same expression with $\Delta = 0$, and hence $E \to |\epsilon'| = |\epsilon - \mu|$.

Close to absolute zero, Δ' tends to zero exponentially and the corresponding term vanishes. Moreover, since the main contribution to the integral (15.46) comes from the region $E \simeq \Delta_0$ or $\epsilon = \epsilon_F$, we have

$$
C_V = \frac{g(\epsilon_F)}{kT^2} \int_0^\infty d\epsilon\, e^{-\beta E} E^2 \simeq \frac{g(\epsilon_F)\, \Delta_0^2}{kT^2} \int_0^\infty d\epsilon\, e^{-\beta E} \quad .
$$

On replacing E by its approximate form

$$
E = \sqrt{\Delta_0^2 + \epsilon'^2} \simeq \Delta_0 + \frac{\epsilon'^2}{2\Delta_0} \quad ,
$$

we finally get

$$
C_V = \sqrt{2\pi}\, g(\epsilon_F)\Delta_0 k \left(\frac{\Delta_0}{kT}\right)^{3/2} e^{-\beta\Delta_0} \quad .
\tag{15.47}
$$

This equation reveals the exponential behaviour of the heat capacity observed close to absolute zero. Using expression (10.48) for the heat capacity constant γ and equation (15.42), which relates Δ_0 to kT_C, we get

$$
\frac{C_V}{\gamma T_C} = 3.15 \left(\frac{T_C}{T}\right)^{3/2} \exp\left(-1.76\, \frac{T_C}{T}\right) \quad .
\tag{15.48}
$$

When T/T_C is close to 0.2, $(T_C/T)^{3/2}$ can be replaced by the numerical approximation $2.50 \exp(-T_C/T)$, and hence

$$
\frac{C_V}{\gamma T_C} = 7.9\, \exp\left(-1.46\, \frac{T_C}{T}\right) \qquad (T_C/T \simeq 5)
$$

in good agreement with the observed behaviour (15.8).

On passing through the critical point, the product $\Delta\Delta' = d\left(\Delta^2/2\right)/dT$ exhibits a discontinuity, which, according to (15.36), is equal to $-4\pi^2 k^2 T_C/7\zeta(3)$. It follows that the heat capacity (15.46) at $T = T_C$ displays a jump

$$\Delta C_V = \frac{4\pi^2}{7\zeta(3)} \; k \; \int_0^\infty g(\epsilon) \, d\epsilon \; \frac{e^{\beta c \epsilon'}}{\left(e^{\beta c \epsilon'} + 1\right)^2} \; .$$

Since the integrand differs significantly from zero only close to ϵ_F we find

$$\Delta C_V \;\; = \;\; \frac{4\pi^2}{7\zeta(3)} \; k^2 T_C \; g(\epsilon_F) \; , \qquad (15.49a)$$

$$\text{or} \quad \frac{\Delta C_V}{\gamma T_C} \;\; = \;\; \frac{12}{7\zeta(3)} = 1.43 \; . \qquad (15.49b)$$

This value is close to that found experimentally in the majority of superconductors (Table 15.1).

In paragraph 15.2.4 we saw that the critical field $B_C(T)$ is related by equation (15.6) to the free energies of the normal and superconducting phases in zero field. Since the superconducting free energy is given by (15.43) and the normal energy by the same expression in which $\Delta = 0$, the value of B_C at any temperature can be found by numerical calculation. The calculation can be performed analytically at absolute zero, giving

$$\frac{V}{2\mu_0} \; B_C^2(0) = \frac{g(\epsilon_F) \, \Delta_0^2}{4} \; , \qquad \text{i.e.} \qquad B_C(0) = \sqrt{\frac{\mu_0}{2} \; \frac{g(\epsilon_F)}{V}} \; \Delta_0 \; . \qquad (15.50)$$

This relation is generally found to be valid. For tin, with $B_C(0) = 0.03$ T and $g(\epsilon_F)/V = 1.7 \times 10^{47}$ J^{-1} m^{-3} (found from the constant γ of the heat capacity), we obtain $\Delta_0/k = 5.7$ K, in agreement with the other prediction (15.42) of the BCS theory that gives $\Delta_0/k = 1.76\,T_C = 6.6$. The temperature dependence of the critical field is very close to Tuyn's law (15.1). The deviation (15.3) from this relation is shown in Figure 15.3. For most superconductors the agreement is satisfactory. We draw attention here, however, to the fact that lead and mercury, as also in most of their other properties (Table 15.1), deviate slightly from the results of the BCS theory. This has to do with the fact that the value of g is approximately 0.35 while it about 0.2 for other superconductors.

15.3.9 Electromagnetic Properties

London equation

In quantum mechanics, each particle of charge e and wave function ψ is associated with a current density

$$\mathbf{j}_0(\mathbf{r}) = -\frac{ie\hbar}{2m} \; (\psi_0^* \, \boldsymbol{\nabla} \, \psi_0 - \psi_0 \, \boldsymbol{\nabla} \, \psi_0^*) \; . \qquad (15.51)$$

In an electromagnetic field of vector potential $\mathbf{A}(\mathbf{r})$, the equation for the current is found by replacing $-i\hbar \, \boldsymbol{\nabla}$ by $-i\hbar \, \boldsymbol{\nabla} - e\mathbf{A}$ in (15.51), thus

$$\mathbf{j}(\mathbf{r}) = -\frac{ie\hbar}{2m} \; (\psi^* \, \boldsymbol{\nabla} \, \psi - \psi \, \boldsymbol{\nabla} \, \psi^*) - \frac{e^2}{m} \; \psi^* \psi \mathbf{A} \; , \qquad (15.52)$$

where ψ is the wave function in the presence of a field. These formulae can be generalized to an electron gas, where ψ stands for the wave function of the N electrons.

In a normal metal, $\mathbf{j}_0$ is zero, and in a magnetic field the two terms of $\mathbf{j}$ cancel, even though they are large, and only the current associated with electronic diamagnetism survives (§10.4.2). In a superconducting metal, $\mathbf{j}_0$ is also zero in zero magnetic field. When a field is present, however, the existence of a gap in the energy spectrum means that the state of the electrons is almost unaffected and that the wave function ψ remains very close to ψ_0. The current density correspondingly takes the form

$$\mathbf{j}(\mathbf{r}) = -\frac{e^2}{m}\,\psi_0^*\psi_0\mathbf{A} = -\frac{e^2}{m}\,n(\mathbf{r})\mathbf{A}(\mathbf{r}) \ , \tag{15.53}$$

where $n(\mathbf{r})$ is the local electron density. This equation, called the London equation, was proposed by F. and H. London (1935) before it was demonstrated in the context of the BCS theory.

We now consider the case of a static magnetic field, for which Maxwell's equations give

$$\mathbf{rot}\ \mathbf{B} = \mu_0\mathbf{j} \qquad \text{and} \qquad \mathbf{B} = \mathbf{rot}\ \mathbf{A}\ . \tag{15.54}$$

On eliminating first $\mathbf{j}$ and then $\mathbf{A}$ from these relations and using the London equation, we find successively

$$\mathbf{rot}\ \mathbf{B} = -\frac{\mu_0 e^2 n}{m}\,\mathbf{A}$$

$$\text{then} \qquad \Delta\mathbf{B} = \frac{1}{\lambda_L^2}\,\mathbf{B} \qquad \text{where} \qquad \lambda_L = \sqrt{\frac{m}{\mu_0 e^2 n}}\ . \tag{15.55}$$

The solution of this equation near a flat surface of separation has the form

$$\mathbf{B} = \mathbf{B}_0\ e^{-z/\lambda_L} \tag{15.56}$$

where z is the co-ordinate perpendicular to the surface, and B_0 is the flux density of the applied field. It follows that the magnetic field vanishes inside the superconductor within a characteristic distance λ_L called the London penetration depth. As the penetration depth is very small (140 Å for tin), this explains the Meissner effect which states that the magnetic field is zero inside a superconductor.

The above results are valid at absolute zero where the N electrons are in the ground state. At other temperatures the contribution to the current $\mathbf{j}$ comes only from the superconducting component and n should be replaced by n_s, the density of superconducting electrons. In particular at the critical temperature, for which $n_s = 0$, the penetration depth (15.55) becomes infinite and the field penetrates the material completely.

Coherence length

The fact that magnetic fields penetrate the surface of superconductors has been demonstrated for example in thin films, which offer a means of measuring the penetration depth λ. The experimental temperature dependence of λ can be described by

$$\lambda = \frac{\lambda(0)}{[1 - (T/T_C)^4]^{1/2}} \cdot \tag{15.57}$$

The measured values of $\lambda(0)$ are, however, generally larger than the theoretical value λ_L (15.55) found from the London equation. $\lambda(0)$ increases with increasing sample purity.

These observations can be understood when it is remarked that London's equation (15.53) relies on the assumption that ψ can be replaced by ψ_0, which is independent of the field. In fact, if the small variation in ψ is taken into account, the London equation (15.53) becomes

$$\mathbf{j}(\mathbf{r}) = - \int K(\mathbf{r} - \mathbf{r}') \, \mathbf{A}(\mathbf{r}') \, d\mathbf{r}'. \tag{15.58}$$

$K(\mathbf{r} - \mathbf{r}')$ is a function defined in the BCS theory with a maximum at zero and a width

$$\xi_0 = \frac{\hbar v_F}{\pi \Delta_0} \tag{15.59}$$

called the coherence length (v_F is the velocity of the electrons on the Fermi surface). This non local equation was introduced on a phenomenological basis by A.B. Pippard.

In a situation in which $\mathbf{A}$ varies slowly over the distance ξ_0, equation (15.58) reduces to

$$\mathbf{j}(\mathbf{r}) = - \frac{1}{\mu_0 \lambda^2} \, \mathbf{A}(\mathbf{r}) \ , \qquad \text{where} \quad \lambda^2 = \lambda_L^2 \, \frac{d(\beta \Delta)}{\Delta \, d\beta} \ . \tag{15.60}$$

This equation is similar to that of London (15.53) but yields a penetration depth λ and not λ_L. The limiting forms (15.34) and (15.36) for Δ show that $\lambda \to \lambda_L$ at absolute zero and that $\lambda \to \infty$ at the critical point, corresponding to complete penetration of the magnetic field.

The above expressions are valid only if the penetration depth λ characterizing the magnetic field variation is large compared to the coherence length ξ_0. In metals, however, the latter is of the order of 5×10^{-7} m and is very much larger than λ_L. This means that the approximation (15.59) can no longer be used. On taking the limit of equation (15.58) for the case in which variations in K are neglected ($\xi_0 \gg \lambda$), the penetration depth is given by

$$\frac{\lambda(T)}{\lambda(0)} = \left[\frac{\Delta(T)}{\Delta_0} \tanh \left[\frac{\beta \Delta(T)}{2} \right] \right]^{-1/3} , \tag{15.61a}$$

$$\text{with} \qquad \frac{\lambda(0)}{\lambda_L} = \left[\frac{\sqrt{3}}{2\pi} \frac{\xi_0}{\lambda_L} \right]^{1/3} . \tag{15.61b}$$

The above relation is very close to the empirical form (15.57) and the value of $\lambda(0)$ is significantly larger than λ_L, in agreement with experiment.

For metals containing impurities or defects, Pippard suggested replacing ξ_0 by a coherence length ξ defined by

$$\frac{1}{\xi} = \frac{1}{\xi_0} + \frac{1}{l} \ , \tag{15.62}$$

where l is the mean free path of the electrons in the normal phase. In a metal containing a large number of impurities or defects, we thus have $\xi \simeq l \ll \xi_0$. For these impure metals the condition $\xi \ll \lambda$ applies, and Pippard suggested retaining equation (15.60), but with the substitution

$$\lambda \to \lambda \left(\frac{\xi_0}{\xi}\right)^{1/2} \tag{15.63}$$

Note that in the BCS theory the coherence length is interpreted as the spatial range of the Cooper pair. As this size is of the order of 10^3 Å, different Cooper pairs overlap and are coherent.

Persistent currents

In the same way as equation (15.55) was derived for static magnetic fields, elimination of **B** and **A** from equations (15.53,54) yields

$$\Delta \mathbf{j} = \frac{1}{\lambda_L^2} \ \mathbf{j} \ .$$

This equation shows that a persistent current can exist in the absence of an electric field. Note that this current penetrates the superconductor in the same way as the magnetic field.

The general proof of the absence of electrical resistance in the BCS theory is complex. Suffice it to say that from the equations of Maxwell and of London, the following relation can be obtained for a static electric field

$$\frac{d\mathbf{j}}{dt} = \frac{ne^2}{m} \ \mathbf{E},$$

which corresponds to the fact that each electron is accelerated without resistance in accordance with

$$\frac{d\mathbf{v}}{dt} = -\frac{e}{m} \ \mathbf{E} \ .$$

The following qualitative explanation can be given for the absence of resistance. When the potential difference is removed in a normal conductor, the current stops because the electrons experience successive collisions which decelerate them. In a superconductor, however, the electrons move as coherent Cooper pairs. It follows that the current can be decreased only by simultaneous collision of all the Cooper pairs, which is highly improbable.

At temperatures other than absolute zero, both superfluid and normal electron components must be taken into account. Since the electrical conductivity of the

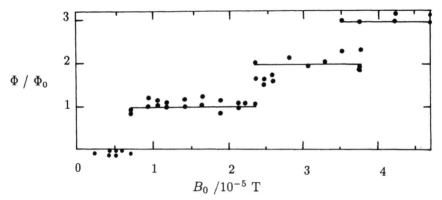

Figure 15.10: Experimental confirmation of magnetic flux quantization in a superconducting cylinder. The flux density of the applied field is B_0; the quantum of magnetic flux Φ trapped inside the cylinder is $\Phi_0 = h/2e$ [B.S. Deaver Jr. and W.M. Fairbank, *Phys. Rev. Lett.* **7**, 43 (1961)].

former is infinite, it follows that the conductivity of the whole of the electron gas is infinite. As for the thermal conductivity, for each of the components it is proportional to the corresponding heat capacity (§6.5.2). Since the heat capacity of the superconducting component tends to zero exponentially, this component does not conduct heat, which explains the behaviour of the thermal conductivity (§15.2.6).

15.3.10   Flux Quantization

When a particle is in a certain quantum state, the probability density of finding it at a point $\mathbf{r}$ is given by the square of the modulus of its wave function, i.e. $|\psi(\mathbf{r})|^2$. If a large number N of particles (bosons) are in the same quantum state, $\rho(\mathbf{r}) = N\,|\psi(\mathbf{r})|^2$ can be taken as the particle density at $\mathbf{r}$. The function $\Psi = \sqrt{N}\,\psi$ is then a macroscopic wave function obeying the same Schrödinger equation as ψ. Such a function, introduced by Ginzburg and Landau (1950), can describe the superconducting state, where the bosons are the Cooper pairs. It is defined by its modulus $\sqrt{\rho}$ and its phase θ, the coherence of which extends to macroscopic distances.

The electric current density associated with the wave function Ψ is given by the general formula (15.52), where $e \to 2e$ and $m \to 2m$. It is given by

$$\mathbf{j}(\mathbf{r}) = \frac{e\hbar}{m}\left(\boldsymbol{\nabla}\,\theta - \frac{2e}{\hbar}\,\mathbf{A}\right)\rho \ . \tag{15.64}$$

In the bulk superconductor where there is no current, we then have

$$\boldsymbol{\nabla}\,\theta = \frac{2e}{\hbar}\,\mathbf{A}.$$

On calculating the circulation of both sides of this equation around a closed circuit, we find

$$\oint \boldsymbol{\nabla}\,\theta \cdot d\mathbf{r} = \frac{2e}{\hbar}\oint \mathbf{A}\cdot d\mathbf{r} = \frac{2e}{\hbar}\,\Phi \ ,$$

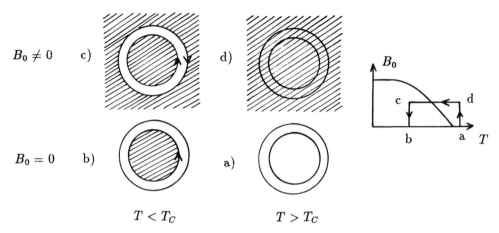

Figure 15.11: Meissner effect for a hollow cylinder seen in section (circuit adcb). The region pervaded by the magnetic field is shaded. The surface currents are indicated by arrows and correspond to an applied field coming out of the page. The state of the cylinder in b) is metastable.

where Φ is the magnetic flux crossing the circuit. When the sample is simply connected the circulation of $\nabla \theta$ is zero for all closed paths and we recover the result that the field is zero everywhere in the bulk material.

In samples that are multiply connected (hollow cylinder or ring, for example), for a circuit around the hole the circulation of the phase must be equal to $2n\pi$ (n integer) for Ψ to be single valued. It follows that the flux Φ is quantized as follows

$$\Phi = \int \mathbf{B} \cdot \mathbf{dS} = n\,\frac{h}{2e} = n\Phi_0 \qquad \left(\text{where} \quad \Phi_0 = \frac{h}{2e}\right) \quad . \tag{15.65}$$

This quantization had been suggested by F. London already in 1950, but without the factor 2 due to the Cooper pairs. The value of the flux quantum $h/2e$ is very small, $\Phi_0 = 2.07 \times 10^{-15}$ Wb, which is, for example, equivalent to a flux density of 2.5×10^{-5} T (0.25 gauss) crossing a 10 μm diameter circle. The first observation of this quantization, which requires extremely high sensitivity, was made only in 1961 (Fig. 15.10).

For topologies such as these, the Meissner effect shows up in a special way. For this, we consider a hollow cylinder (Fig. 15.11) that is placed successively in the states a, d, c, b indicated in the phase diagram of Figure 15.4. In state d the cylinder is not superconducting and the field inside the cylinder is equal to the applied flux density B_0. Cooling to the state c calls into existence permanent surface currents. The currents on the outer surface of the cylinder generate a field inside the surface that exactly compensates the external field. To create this field, however, requires an expenditure of energy that is proportional to the volume. This energy is minimized by generating currents on the inner surface of the cylinder which maintain the applied field in the cavity. The value of this field is, however, slightly different from the applied field, on account of quantization (15.65).

When the applied field is removed (state b) the outer currents vanish, but the inner currents remain. The cylinder is then in a metastable state, where the stable

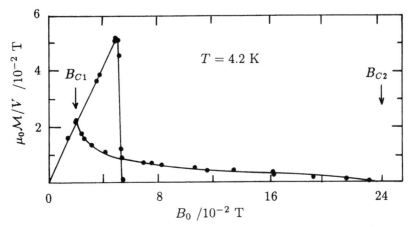

Figure 15.12: Magnetic moment $\mathcal{M}$ of superconductors of type I (lead) and type II (8.23 % by weight lead-indium alloy). The flux densities B_{C1} and B_{C2} for this alloy are indicated [J.D. Livingston, *Phys. Rev.* **129**, 1943 (1962)].

state corresponds to the inner field being zero. As the inner field is quantized through its flux, it cannot be decreased continuously. Fluctuations are not sufficient to generate the required jumps.

If we consider a circuit that is too close to the surface (in particular in a thin ribbon where **j** may be different from zero), it is not the magnetic flux that is quantized but a quantity called the "fluxoid", which reduces to ϕ when $\mathbf{j} = 0$.

If the reverse sequence of transformations is followed (trajectory a, b, c, d), the states a and d are identical. In b, however, the cylinder is in its stable state (inner field and current zero). It follows that in c (applied field B_0), the field inside the cylinder remains equal to zero because of its flux quantization. This state is then metastable, and the stable state corresponds to a field being present inside the cylinder.

15.4   TYPE II SUPERCONDUCTORS

15.4.1   Experimental Investigation

The behaviour described in paragraph 15.2 is found only in certain superconductors (tin, lead, etc.) called type I, or soft superconductors (on account of their plastic properties). Other substances (transition elements, compounds and alloys) have more complex behaviou, and are called type II or hard superconductors. They are distinguished by the fact that the Meissner effect operates completely if the applied field is lower than a critical value $B_{C1}(T)$, and imperfectly between this value and the critical flux density $B_{C2}(T)$, at which the substance becomes normal again. This property manifests itself by the fact that the magnetic moment $\mathcal{M}$ is given by equation (15.4) for $B_0 < B_{C1}$, then gradually decreases above B_{C1}, vanishing at B_{C2} (Fig. 15.12). Below $B_{C1}(T)$ the substance behaves like a type I superconductor, while above B_{C2} it is normal. Between B_{C1} and B_{C2} the superconductor is in a

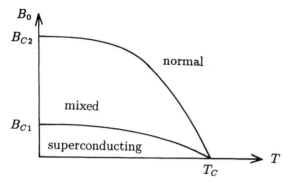

Figure 15.13: Phase diagram of a type II superconductor.

so-called mixed state (Fig. 15.13).

The interest of type II superconductors lies in the fact that, since B_{C2} can adopt very large values (several Teslas), they have zero resistance in high magnetic fields. They are therefore choice materials for constructing superconducting magnets.

15.4.2 The Mixed State

When a cylinder of type II superconducting material enters the mixed state as the applied field B_0 is increased, filaments of normal phase are created in which the magnetic field is non zero. These filaments are caused by the fact that the penetration depth λ of the magnetic field is longer than the coherence length. As we shall see, this yields a negative surface energy when the two phases are separated, thus favouring the creation of filaments.

Let us consider the spatial variation of the electronic and magnetic free energy densities, f_e and f_m, near the interface between the normal n and superconducting s phases (Fig. 15.14). As a result of pair creation, the electronic free energy in the superconducting phase is lower than that of the normal phase (§15.3.7), where the change begins at the interface and extends over a distance of the order of ξ characterizing the Cooper pairs. The intrinsic magnetic free energy of the substance is zero in the normal phase and is equal to $B_0^2/2\mu_0$ in the superconducting phase, since the surface currents give rise to a flux density $-\mathbf{B}_0$ in the superconductor. This energy is established within a distance λ characteristic of the thickness of the current layer. Depending on the relative values of λ and ξ, two cases arise.

When $\lambda \ll \xi$ (Fig. 15.14a), the free energy density displays an excess close to the surface. This is unfavourable to the coexistence of the two phases. This situation corresponds to type I superconductors. However, if $\lambda \gg \xi$ (Fig. 15.14b), the free energy is deficient close to the surface and the two phases can coexist. This is the case for type II superconductors. Filaments then arise with their axis parallel to the field B_0. Inside these filaments the density of Cooper pairs is zero on the axis but returns to the equilibrium value at a distance of the order of ξ. The field generated by a vortex of superconducting currents of density j (Fig. 15.15a) has a flux density B_0 on the axis and vanishes at a distance of the order of λ.

These filaments and their spatial arrangement (Fig. 15.15b) have been observed

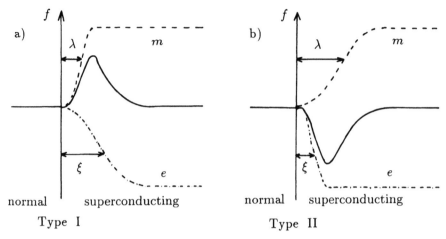

Figure 15.14: Free energy density f near an interface between normal and superconducting phases in a type I (a) or type II (b) substance. The dash-dot, dashed and full
lines correspond respectively to the electronic, magnetic and total energy. λ and ξ are
respectively the penetration depth of the field and the coherence length.

experimentally (magnetic powder method, neutron diffraction). Measurement of
their density confirms that each vortex corresponds to a flux quantum $h/2e$. If
the sample contains defects, the vortices anchor themselves to the impurities or
imperfections. As the external field increases above B_{C1}, new vortices are generated.
At B_{C2}, the density of the lattice of vortices becomes so large that the size of the
superconducting interstices shrinks to the order of ξ, so that the superconducting
phase vanishes and the whole cylinder is in the normal state.

    Note that the analogies between these phenomena and those occurring in liquid
helium(§9.6) have stimulated progress in both these subjects simultaneously.

15.4.3 Superconducting Magnets

Intense magnetic fields are of enormous interest in physics. Up to 1 T conventional
electromagnets are used, made from copper wires wound around an iron core. To
obtain higher fields, solenoids are constructed out of stacks of copper disks with a
hole in the centre (Bitter magnets). These provide fields up to 30 T in volumes of
a few tens of cubic centimetres. Generating such fields requires currents of several
thousand ampères, producing some 10^6 W through the Joule effect. The factor
that limits the maximum field is the release of heat, which is removed by a flow of
cooling water. It may be remarked that these techniques can yield fields in excess
of 50 T, for periods of 10 ms (pulsed fields).

    The situation described above explains the interest of using superconductors to
produce electromagnets that are free of the Joule effect. However, the fields that are
reached (~ 20 T) are limited by the critical field. In practice superconducting alloys
(type II) are used in the form of wires composed of very fine strands ($d \sim 30$ μm)
of niobium alloys embedded in a copper or aluminium matrix. This configuration
is dictated by the fact that a superconducting current generating a field of the

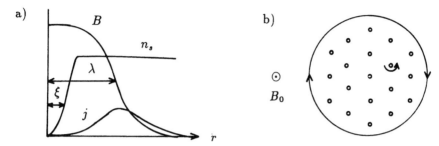

Figure 15.15: Mixed state of a type II superconductor. a) Structure of a vortex: density of superconducting electrons n_s, flux density B and current density $j = \partial B/\partial r$. b) Regular arrangement of the vortices. The arrows indicate the direction of the currents at the surface of the cylinder and in the vortices.

order of 10 T corresponds to a stored magnetic energy density of 40 J cm^{-3}. Any local and temporary transit to the normal state would liberate energy through the Joule effect, which, by a chain reaction, would seriously damage the device. The bulk of copper, which is more conducting than the superconductor in its normal state, limits the consequences of the Joule effect, and the use of fine wires greatly reduces the incidence of local resistance. This resistance appears in thicker wires if the vortices jump from one trapping position to another on being subjected to a magnetic force.

The strongest continuous magnetic fields (~ 40 T) are produced by hybrid electromagnets. These are composed of a superconducting solenoid (~ 10 T) inside which is placed a Bitter magnet ($\simeq 30$ T).

15.4.4   High Temperature Superconductivity

The technological interest of superconductors is obvious. Their use, however, is limited by the fact that they require considerable cryogenic resources. For this reason there is a constant search for materials having as high a critical temperature as possible. Since Kamerlingh Onnes's discovery ($T_C = 4.15$ K for mercury), the highest observed critical temperature increased constantly until 1973 ($T_C = 23.2$ K for the alloy Nb$_3$ Ge). At that time investigations were directed towards metallic compounds, and the temperature reached was close to a theoretical maximum estimated to be 25-30 K.

Beginning in the 70s superconductivity was being sought in conducting organic compounds. In these molecular compounds, the delocalized electrons in planar organic molecules propagate along their stacking axis. The critical temperature reached in these compounds did not exceed 10 K ($T_C = 8.1$ K in 1985).

At the same time superconductivity was being investigated in perovskite double oxides (e.g. Sr Ti O$_{3-x}$ with $T_C = 0.3$ K in 1967). In 1986 [J.G. Bednorz and K.A. Müller, *Z. Phys.* **B 64**, 189 (1986)], a record critical temperature in excess of 30 K was obtained with the double oxide La$_2$ Cu O$_4$ doped with Ba^{2+} ions replacing the La^{3+} ions. Then in 1987, a critical temperature $T_C = 92$ K was reached, for the first time above the boiling point of nitrogen, with the compound Y Ba$_2$ Cu$_3$

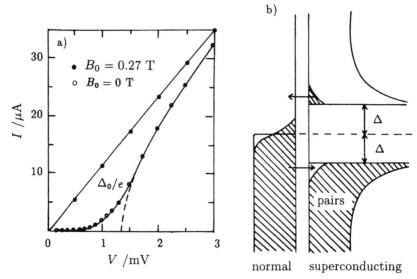

Figure 15.16: a) Tunnel effect at $T = 1.6$ K between normal aluminium and lead (normal for $B_0 = 0.27$ T and superconducting for $B_0 = 0$ T). The dashed curve shows a limiting tunnel effect at 0 K. $2\Delta_0$ is the gap at absolute zero [J. Giaever, *Phys. Rev. Lett.* **5**, 147 (1960)]. b) Diagram of the densities of states and occupation for a normal metal-oxide-superconductor junction. The arrows indicate the displacement of the electrons. Applying a potential difference V increases the energy of the electrons in the normal metal and induces a flow of electrons towards the superconductor. The increase in current becomes rapid when eV reaches a value close to Δ.

O_{7-x} (denoted by Y-Ba-Cu-O). In these compounds the conduction electrons move in the flat lattices of Cu and O ions, giving rise to Cu^{2+}-Cu^{3+} valency changes. It has been confirmed that these compounds exhibit the same properties as ordinary superconductors, i.e., the Meissner effect, flux quantization (15.65), and gap. These type II superconductors have the additional advantage of having critical fields B_{C2} of the order of several tens of Teslas. They are ceramics, however, and are difficult to work mechanically, in particular to produce wires or ribbons.

15.5   TUNNEL EFFECTS

15.5.1   Electronic Tunnel Effect

If two conductors are separated by a very thin layer of non conducting material (oxide of one of the metals), the electrons can cross the insulating layer by tunnel effect, and the Fermi levels of the two metals become equal. Applying a potential difference increases the chemical potential of one metal with respect to the other and generates a current that obeys Ohm's law (Fig. 15.16a).

If one of the metals is superconducting and the temperature is far below T_C, no current flows as long as the potential difference V is smaller than a value given by $eV = \Delta(T) \simeq \Delta_0$, where 2Δ is the gap of the superconductor. At higher temperatures the current starts more gradually (Fig. 15.16a). This is explained by

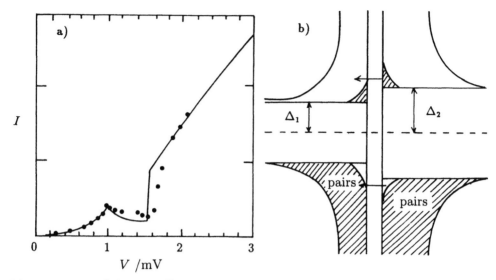

Figure 15.17: a) Tunnel effect between superconducting tin and lead at $T = 3.4$ K. The curve corresponds to the BCS theory [S. Shapiro et al., *IBM J. Res. Dev.* **6**, 34 (1962)]. b) Diagram of the densities of states and of the occupation. The arrows indicate the displacements of the electrons and the holes. The discontinuities in $I(V)$ at $e|V| = |\Delta_1 - \Delta_2|$ and $\Delta_1 + \Delta_2$ correspond to equal energies for infinite densities of states on either side of the junction.

the fact that the electronic levels of the two metals have the configuration shown in Figure 15.16b. This so-called normal-superconducting tunnel effect provides a method for determining the gap experimentally.

When both metals are superconducting and the temperature is much lower than the critical temperatures, no current passes as long as the potential difference V is smaller than a value given by $eV = \Delta_1 + \Delta_2$ (Fig. 15.17a). At higher temperatures the phenomena are more complex and a discontinuity arises at $eV = |\Delta_1 - \Delta_2|$, with a reversal of slope in the characteristic curve $I(V)$ (negative resistance) (Fig. 15.17b). The shape of the characteristic curve is due to a superposition of several mechanisms: the tunnel effect of electrons and holes, high densities of states at certain energy values. These effects are satisfactorily explained by the BCS theory .

15.5.2 Josephson Effects

d.c. Josephson effect

B.D. Josephson (1962) showed that when a superconducting loop is obstructed by a very thin insulating layer (~ 10 Å), a persistent current of Cooper pairs can propagate by tunnelling through the junction, with no potential difference between either side of the junction. This current, however, cannot exceed a certain limiting value I_0 that is less than a milliampere. At absolute zero I_0 is given by $I_0 = \pi\Delta_0/2eR$, where R is the resistance of the junction to a normal current. Using the

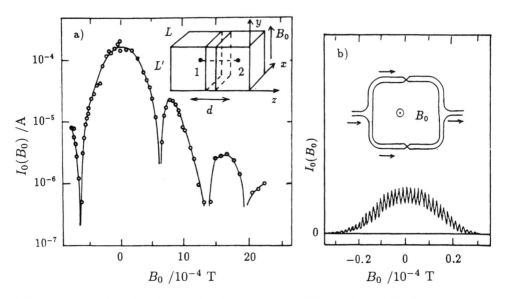

Figure 15.18: d.c. Josephson effect in a magnetic field. a) A Pb-PbO-Pb junction at 1.3 K with $L = 40$ μm and $d = 780$ Å [J.M. Rowell, *Phys. Rev. Lett.* **5**, 200 (1963)]. b) Two Sn-SnO-Sn junctions at 2 K. The period of the flux Φ through a junction is $B_0 = 0.35 \times 10^{-4}$ T and that of the flux Φ_T crossing the zone between the two junctions is $B_0 = 16 \times 10^{-7}$ T. The maximum current is of the order of 0.5 mA [R.C. Jaklevic et al., *Phys. Rev. A* **140**, 1628 (1965)].

Ginzburg-Landau theory (§15.3.10) it can be shown that the current and its density are related to the difference in phase $\phi \equiv \Delta\theta$ of the wave Ψ of the Cooper pairs at the contacts of the junction through

$$I = I_0 \, \sin\phi \qquad \text{and} \qquad j = j_0 \, \sin\phi \ . \tag{15.66}$$

The phase ϕ adapts to the value of the applied current.

When a magnetic field is applied in the vicinity of the junction, the phase difference ϕ between two points 1 and 2 on either side of the junction depends on the position of these points. Thus

$$\phi = \phi_0 - \frac{2e}{\hbar} \int_1^2 \mathbf{A} \cdot d\mathbf{l} \ , \tag{15.67}$$

where $\mathbf{A} = \mathbf{B}_0 \times \mathbf{r}/2$ is the vector potential. It follows that the current density given by (15.66) varies from one point to another in the junction, and can even change sign. For a junction with a rectangular cross-section (Fig. 15.18a) of sides L (Ox) and L' (Oy) and thickness d (Oz) (taking into account the penetration depth on either side of the junction), and a flux density B_0 directed along Oy, the current density depends only on the abscissa x and is equal to

$$j(x) = \frac{I_0}{LL'} \, \sin\left[\phi - \frac{2ed}{\hbar} \, B_0 x\right] \ .$$

The total current is found by integrating $j(x)$. Its limiting value is

$$I_0(B_0) = I_0 \left| \frac{\sin(\pi\Phi / \Phi_0)}{\pi\Phi / \Phi_0} \right| ,$$

where Φ_0 is the flux quantum (15.65) and $\phi = Ld\, B_0$. This effect, which is similar to that of diffraction by a slit, is shown in Figure 15.18a.

When the superconducting loop has two identical junctions in parallel, the limiting total current caused by interference between the two derived currents is

$$I_0(B_0) = I_0 \left| \frac{\sin(\pi\Phi / \Phi_0)}{\pi\Phi / \Phi_0} \right| |\cos(\pi\Phi_T / \Phi_0)| , \tag{15.68}$$

where Φ_T is the magnetic flux crossing the zone between the two junctions (Fig. 15.18b). This maximum current exhibits a double periodicity in B_0, where the period of Φ_T can be made much smaller than that of Φ. A device of this kind can detect changes in flux density of the order of 10^{-10} T.

a.c. Josephson effect

Josephson also predicted that applying a d.c. potential difference to the contacts of a junction generates an alternating pair current of frequency

$$\nu = \frac{2e}{h} V \tag{15.69}$$

and amplitude I_0.

If a small alternating voltage of frequency ν_0 is added to V, the frequency ν becomes modulated, giving rise to numerous additional frequencies $\nu \pm n\nu_0$. It follows that when V takes the values $V_n = nh\nu_0/2e$, a continuous (d.c.) current appears. The characteristic curve $I(V)$ then displays a series of steps (or levels) of width $\Delta V = h\nu_0/2e$ (Fig. 15.19). In general, the a.c. voltage is produced by irradiating with microwaves ($\nu_0 \sim 10$ GHz), which corresponds to quanta of approximately 20 μV.

Applications

The Josephson effects are the basis of a large number of metrological devices, on account of their extremely high precision. These are used in metrology and in laboratories.

SQUIDs (superconducting quantum interference devices) make use of the very high sensitivity to magnetic flux density B_0 of the current $I_0(B_0)$ (15.68) in the two-junction d.c. Josephson effect. They measure extremely small changes in flux ($\delta\phi_T \sim 10^{-18}$ Wb, corresponding to $\delta B \sim 10^{-11}$ T for an area of 0.1 cm^2) between junctions. Even higher precision ($\delta\phi_T \sim 10^{-21}$ Wb) can be obtained with a single junction, by coupling with resonating circuits (a.c. SQUID).

The phenomenon of current steps in the characteristic curve $I(V)$ of a junction in the presence of electromagnetic waves of frequency ν_0 is used to make extremely

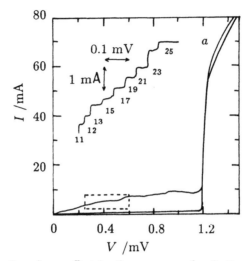

Figure 15.19: a.c. Josephson effect in the presence of radiation ($V_0 \simeq 10$ GHz) for the Sn-SnO-Sn junction at 1.2 K. A magnification of the dotted region is shown together with the level number. In the absence of radiation (curve a), the current is practically zero below $V = 1.2$ mV. Above this voltage, an electronic tunnel effect appears [B.N. Taylor, *J. Appl. Phys.* **39**, 2490 (1968)].

precise measurements of potential differences through the relation $V_n = nh\nu_0/2e$. This is limited by the precision in the knowledge of the ratio e/h ($\sim 6 \times 10^{-6}$). Conversely, standard batteries can be used to measure the ratio e/h.

The Josephson effects are also used to measure small voltages (10^{-15} V), to detect microwaves (down to 10^{-10} W in the range 5 to 1 000 GHz), as memories for computers (switching in $\sim 10^{-10}$ s and consuming $\sim 10^{-18}$ J).

BIBLIOGRAPHY

N.W. Ashcroft and N.D. Mermin, *op. cit.* p. 725.

J.S. Blakemore, *op. cit.* p. 266.

B.I. Bleaney and B. Bleaney, *op. cit.*, p. 397.

T. van Duzer and C.W. Turner, *Principles of Superconductive Devices and Circuits*, Elsevier (1981).

R.J. Elliott and A.F. Gibson, *op. cit.*, p. 343.

C. Kittel, *op. cit.* p. 355.

C.G. Kuper, *Theory of Superconductivity*, Clarendon Press, Oxford (1968).

L.D. Landau and E.M. Lifschitz, *op. cit.* p. 301.

E.A. Lynton, *La supraconductivité*, Dunod (1964).

R.D. Parks (Ed.), *Superconductivity*, M. Dekker, N. Y. (1969).

G. Rickayzen, *Theory of Superconductivity*, Wiley (1965).

H.M. Rosenberg, *op. cit.*

M. Tinkham, *Introduction to Superconductivity*, McGraw Hill. (1975).

Historical References

J. Bardeen, L.N. Cooper and J.R. Schrieffer, *Phys. Rev.* **108**, 1175 (1957).

L.N. Cooper, *Phys. Rev.* **104**, 1189 (1956).

H. Fröhlich, *Phys. Rev.* **79**, 845 (1950); *Proc. Roy. Soc. A* (London) **223**, 296 (1954).

V.L. Ginzburg and L.D. Landau, *Zh. Eksp. Teor. Fiz.* **20**, 1064 (1950).

B.D. Josephson, *Phys. Lett.* **1**, 251 (1962).

H. Kamerlingh Onnes, *Comm. Phys. Lab. Univ. Leiden*, n° 119 b, 120, 122 b, 124 c (1911); 139 f (1914).

F. London, *Superfluids* Vol. I, Wiley (1950).

F. London and H. London, *Proc. Roy. Soc. A* (London) **149**, 71 (1935).

E. Maxwell, *Phys. Rev.* **78**, 477 (1950).

W. Meissner and R. Ochsenfeld *Natürwiss.* **21**, 787 (1933).

A.B. Pippard, *Proc. Roy. Soc. A* (London) **216**, 547 (1953).

C.A. Reynolds et al., *Phys. Rev.* **78**, 487 (1950).

APPENDIX A

Physical Constants and Common Units

Table A1 Physical constants [*Physics Letters*, B239, III.1 (1990)]. The numbers in brackets are the standard deviation of the uncertainty in the last figures. Values stated without error are exact.

velocity of light	c	2.99792458×10^8 m s^{-1}
Planck's constant	h	$6.6260755(40) \times 10^{-34}$ J s
reduced Planck's constant	$\hbar$	$1.05457266(63) \times 10^{-34}$ J s
gravitional constant	G	$6.67259(85) \times 10^{-11}$ m^3 kg^{-1} s^{-2}
electron charge	e	$1.60217733(49) \times 10^{-19}$ C
Boltzmann constant	k	$1.3806513(25) \times 10^{-23}$ J K^{-1}
Avogadro's number	N	$6.0221367(36) \times 10^{23}$ mol^{-1}
Gas constant	$R = Nk$	$8.314471(15)$ J K^{-1} mol^{-1}
permeability of free space	μ_0	$4\pi 10^{-7}$ N A^{-2}
permittivity of free space	$\epsilon_0 = (\mu_0 c^2)^{-1}$	$8.854187817\ldots \times 10^{-12}$ F m^{-1}
electron rest mass	m_e	$9.1093897(54) \times 10^{-31}$ kg
proton rest mass	m_p	$1.6726231(10) \times 10^{-27}$ kg
Bohr magneton	$\mu_B = e\hbar/2m_e$	$9.2740154(31) \times 10^{-24}$ J T^{-1}
nuclear magneton	$\mu_N = e\hbar/2m_p$	$5.0507866(17) \times 10^{-27}$ J T^{-1}

Table A2 Common Physical Units. The sign $\equiv$ means equivalent to. For magnetic susceptibilities see § 4.2.3.

1 Å$= 10^{-10}$ m
1 atm $= 1.01325 \times 10^5$ Pa
1 torr $= 1/760$ atm   1 torr $\equiv 1$ mm Hg
1 eV $= 1.60217733(49) \times 10^{-19}$ J
1 K $\equiv 8.617347 \times 10^{-5}$ eV   1 cm$^{-1} \equiv 1.2398 \times 10^{-4}$ eV
1 cal $= 4.184$ J   1 kcal mol$^{-1} = 0.04336$ eV molecule^{-1}
1 debye $= 3.336 \times 10^{-10}$ C m
1 gauss $= 10^{-4}$ T   1 oersted $= 1/(4\pi \times 10^3)$ A m^{-1}

APPENDIX B

The Elements

Table B1 Correspondence between the elements and their atomic number

89	Actinium	63	Europium	60	Neodymium	21	Scandium
13	Aluminium	9	Fluorine	93	Neon	34	Selenium
95	Americium	87	Francium	10	Neptunium	14	Silicon
51	Antimony	64	Gadolinium	28	Nickel	47	Silver
18	Argon	31	Gallium	41	Niobium	11	Sodium
33	Arsenic	32	Germanium	7	Nitrogen	38	Strontium
85	Astatine	79	Gold	76	Osmium	16	Sulfur
56	Barium	72	Hafnium	8	Oxygen	73	Tantalum
4	Beryllium	2	Helium	46	Palladium	43	Technetium
83	Bismuth	67	Holmium	15	Phosphorous	52	Tellurium
5	Boron	1	Hydrogen	78	Platinum	65	Terbium
35	Bromine	49	Indium	94	Plutonium	81	Thallium
48	Cadmium	53	Iodine	84	Polonium	90	Thorium
20	Calcium	77	Iridium	19	Potassium	50	Thulium
6	Carbon	26	Iron	59	Praseodymium	22	Tin
58	Cerium	36	Krypton	61	Promethium	69	Titanium
55	Cesium	57	Lanthanum	91	Protoactinium	74	Tungsten
17	Chlorine	82	Lead	88	Radium	92	Uranium
24	Chromium	3	Lithium	86	Radon	23	Vanadium
27	Cobalt	71	Lutecium	75	Rhenium	39	Xenon
29	Copper	12	Magnesium	45	Rhodium	54	Ytterbium
96	Curium	25	Manganese	37	Rubidium	70	Yttrium
66	Dysprosium	80	Mercury	44	Ruthenium	30	Zinc
68	Erbium	42	Molybdenum	62	Samarium	40	Zirconium

Table B2 Periodic table of the elements

H^1																	He^2
Li^3	Be^4											B^5	C^6	N^7	O^8	F^9	Ne^{10}
Na^{11}	Mg^{12}											Al^{13}	Si^{14}	P^{15}	S^{16}	Cl^{17}	Ar^{18}
K^{19}	Ca^{20}	Sc^{21}	Ti^{22}	V^{23}	Cr^{24}	Mn^{25}	Fe^{26}	Co^{27}	Ni^{28}	Cu^{29}	Zn^{30}	Ga^{31}	Ge^{32}	As^{33}	Se^{34}	Br^{35}	Kr^{36}
Rb^{37}	Sr^{38}	Y^{39}	Zr^{40}	Nb^{41}	Mo^{42}	Tc^{43}	Ru^{44}	Rh^{45}	Pd^{46}	Ag^{47}	Cd^{48}	In^{49}	Sn^{50}	Sb^{51}	Te^{52}	I^{53}	Xe^{54}
Cs^{55}	Ba^{56}	La^{57}	Hf^{72}	Ta^{73}	W^{74}	Re^{75}	Os^{76}	Ir^{77}	Pt^{78}	Au^{79}	Hg^{80}	Tl^{81}	Pb^{82}	Bi^{83}	Po^{84}	At^{85}	Rn^{86}
Fr^{87}	Ra^{88}	Ac^{89}															

Ce^{58}	Pr^{59}	Nd^{60}	Pm^{61}	Sm^{62}	Eu^{63}	Gd^{64}	Tb^{65}	Dy^{66}	Ho^{67}	Er^{68}	Tm^{69}	Yb^{70}	Lu^{71}
Th^{90}	Pa^{91}	U^{92}	Np^{93}	Pu^{94}	Am^{95}	Cm^{96}							

Table B3 Physical properties of the elements. The symbols in the table below have the following meanings :

- Z : atomic number.

- 1 : chemical symbol.

- M /(g mol^{-1}) : atomic mass. Values in brackets indicate the mass of the most stable isotope.

- t_B and t_M/°C : boiling point and melting point at atmospheric pressure; add 273.15 to transform to kelvins. The sign - indicates absence of phase change and "s" indicates sublimation.

- ΔH_V and ΔH_M /(kcal mol^{-1}) : latent heat of vaporization and melting at atmospheric pressure ; 1 kcal mol^{-1} = 4.184 kJ mol^{-1}. The sign - indicates absence of phase change.

- 2 : state of the element at normal temperature and pressure. This column lists the molecule of the gas or liquid, or the Bravais lattice of the solid. In the latter case the nomenclature of Table 8.2 and Figure 8.3 is used; *bcc* stands for body centred cubic, *fcc* for face centred cubic, *hcp* for a hexagonal compact structure, and diam denotes diamond structure.

- ρ /(g cm^{-3}) : density at normal temperature and pressure ; for gases this value is omitted.

- $\Delta H°$ /(kcal mol^{-1}) : standard enthalpy under normal conditions ; 1 kcal mol^{-1} = 4.184 kJ mol^{-1}.

- 3 : electronic configuration of the neutral atom.

- 4 : most common ionization states.

- 5 : ground state(s) corresponding to the most common ionization state(s).

- I/eV : ionization energy of the most common state.

Z	1	M	t_M	ΔH_M	t_B	ΔH_V	2	ρ	ΔH^o	3	4	5	I
1	H	1.008	−259.9	0.014	−252.9	0.109	gas H_2		52	1s			
2	He	4.003	-	-	−268.9	0.020	gas		0	$1s^2$	0	$^1S_0$	1.28
3	Li	6.941	180.5	0.72	1347	32	bcc	0.53	38	2s	1	$^1S_0$	0.68
4	Be	9.012	1280	3.52	2970	70	hcp	1.84	78	$2s^2$	2	$^1S_0$	0.35
5	B	10.81	2300		2550		rhomb	2.46	130	$2s^2$ 2p	3	$^1S_0$	0.23
6	C	12.01	3550		4827	170	hex	0.81	170	$2s^2$ $2p^2$			
7	N	14.01	−209.9	0.086	−195.8	0.67	gas N_2		113	$2s^2$ $2p^3$			
8	O	16.00	−218.4	0.053	−183.0	0.81	gas O_2		60	$2s^2$ $2p^4$	−2	$^1S_0$	1.40
9	F	19.00	−219.6	0.061	−188.1	0.76	gas F_2		19	$2s^2$ $2p^5$	−1	$^1S_0$	1.36
10	Ne	20.18	−248.7	0.080	−246.0	0.42	gas		0	$2s^2$ $2p^6$	0	$^1S_0$	0
11	Na	22.9	97.81	0.62	882.9	23	bcc	0.97	26	3s	1	$^1S_0$	0.98
12	Mg	24.31	649	2.14	1090	32	hcp	1.74	36	$3s^2$	2	$^1S_0$	0.66
13	Al	26.98	660.4	2.56	2467	68	fcc	2.70	77	$3s^2$ 3p	3	$^1S_0$	0.51
14	Si	28.09	1410	12.0	2355		diam	2.33	105	$3s^2$ $3p^2$			
15	P	30.98	44.1	0.15	280	3.0		1.80	75	$3s^2$ $3p^3$			
16	S	32.06	112.8	0.335	444.7		orth F	2.08	66	$3s^2$ $3p^4$			
17	Cl	35.45	−101.0	0.77	−34.6	2.44	gas Cl_2		29	$3s^2$ $3p^5$	−1	$^1S_0$	1.81
18	Ar	39.95	−189.2	0.28	−185.7	1.57	gas		0	$3s^2$ $3p^6$	0	$^1S_0$	0
19	K	39.10	63.65	0.56	774	19	bcc	0.86	21	4s	1	$^1S_0$	1.33
20	Ca	40.08	839	2.07	1484	36	fcc	1.53	42	$4s^2$	2	$^1S_0$	0.99
21	Sc	44.96	1539	3.70	2832		hcp	2.99	80	3d $4s^2$	3	$^1S_0$	0.81
22	Ti	47.90	1660		3287		hcp	4.51	112	$3d^2$ $4s^2$	2	$^3F_2$	0.94
23	V	50.94	1890		3380		bcc	6.09	123	$3d^3$ $4s^2$	2	$^4F_{3/2}$	0.88
24	Cr	52.00	1860	3.47	2672		bcc	7.19	95	$3d^5$ 4s	2	$^5D_0$	0.84
25	Mn	54.94	1244	3.50	1962	53	bcc	10.2	67	$3d^5$ $4s^2$	2	$^6S_{5/2}$	0.80
26	Fe	55.85	1535	3.67	2750	84	bcc	7.87	100	$3d^6$ $4s^2$	2	$^5D_4$	0.74
											3	$^6D_{5/2}$	0.64
27	Co	58.93	1495	3.70	2870	92	hcp	8.8	102	$3d^7$ $4s^2$	2	$^4F_{9/2}$	0.72
28	Ni	58.70	1453	4.21	2732	90	fcc	8.90	102	$3d^8$ $4s^2$	2	$^3F_4$	0.69
29	Cu	63.55	1083	3.12	2567	73	fcc	8.93	81	$3d^{10}$ 4s	1	$^1S_0$	0.96
											2	$^2D_{5/2}$	0.72
30	Zn	65.38	419.6	1.77	907	27	hcp	7.13	31	$3d^{10}$ $4s^2$	2	$^1S_0$	0.74
31	Ga	69.72	29.78	1.34	2403	60	orth C	5.91	65	$4s^2$ 4p	3	$^1S_0$	0.62
32	Ge	72.59	937.4	7.6	2830		diam	5.33	90	$4s^2$ $4p^2$	4	$^1S_0$	0.53
33	As	74.92	−	−	613(s)	7.55	rhomb	5.78		$4s^2$ $4p^3$	3	$^1S_0$	0.58
34	Se	78.96	217	1.30	685		hex	4.81	49	$4s^2$ $4p^4$			
35	Br	79.90	−7.2	1.26	58.78	3.58	liq Br_2	3.12	27	$4s^2$ $4p^5$	−1	$^1S_0$	
36	Kr	83.80	−156.6	0.39	152.3	2.16	gas		0	$4s^2$ $4p^6$	0	$^1S_0$	0

Z	1	M	t_M	ΔH_M	t_B	ΔH_V	2	ρ	ΔH^o	3	4	5	I
37	Rb	85.47	38.89	0.56	688	18	bcc	1.53	20	$5s$	1	$^1S_0$	1.47
38	Sr	87.62	769		1384	34	fcc	2.58	39	$5s^2$	2	$^1S_0$	1.12
39	Y	88.91	1520	2.73	3337		hcp	4.47	98	$4d\ 5s^2$	3	$^1S_0$	0.89
40	Zr	91.22	1852		4377		hcp	6.56	146	$4d^2\ 5s^2$	4	$^1S_0$	0.99
41	Nb	92.91	2470		4742		bcc	8.58	175	$4d^4\ 5s$	5	$^1S_0$	0.69
42	Mo	95.94	2617	6.66	4612		bcc	10.2	157	$4d^5\ 5s$	4	$^1S_0$	0.70
											6	$^1S_0$	0.62
43	Tc	(97)											
44	Ru	101.1	2310		3900		hcp	12.4	154	$4d^7\ 5s$	4		0.67
45	Rh	102.9	1966		3700		fcc	12.4	133	$4d^8\ 5s$	3		0.68
46	Pd	106.4	1552	4.10	3140	90	fcc	12.0	90	$4d^{10}$	2	$^3F_4$	0.80
47	Ag	107.9	961.9	2.78	2212	61	fcc	10.5	68	$4d^{10}\ 5s$	1	$^1S_0$	1.26
											2	$^2D_{5/2}$	0.89
48	Cd	112.4	320.9	1.48	765	24	hcp	8.64	27	$4d^{10}\ 5s^2$	2	$^1S_0$	0.97
49	In	114.8	156.6	0.78	2080	54	tetrag I	7.29	57	$5s^2\ 5p$	3	$^1S_0$	0.81
50	Sn	118.7	232.0	1.71	2270	70	tetrag I	7.29	72	$5s^2\ 5p^2$	4	$^1S_0$	0.71
51	Sb	121.8	630.7	4.74	1750	46	rhomb	6.69	63	$5s^2\ 5p^3$	3	$^1S_0$	0.76
52	Te	127.6	449.5	4.18	989	12	hex	6.24	46	$5s^2\ 5p^4$			
53	I	126.9	113.5	1.87	184.4	5.2	orth C	4.93	26	$5s^2\ 5p^5$	-1	$^1S_0$	2.16
54	Xe	131.3	-111.9	0.55	-107.1	3	gas		0	$5s^2\ 5p^6$	0	$^1S_0$	2.0
55	Cs	132.9	28.40	0.51	678.4	16	bcc	1.90	19	$6s$	1	$^1S_0$	1.67
56	Ba	137.3	725	1.83	1640	36	bcc	3.6	42	$6s^2$	2	$^1S_0$	1.34
57	La	138.9	920	1.48	3454		hex	6.17	102	$5d\ 6s^2$	3	$^1S_0$	1.02
58	Ce	140.1	798	1.24	3257		fcc	6.77	98	$4f^2\ 6s^2$	3	$^2F_{5/2}$	1.03
59	Pr	140.9	930	1.65	3212		hex	6.77	86	$4f^3\ 6s^2$	3	$^3H_4$	1.01
60	Nd	144.2	1010	1.71	3127		hex	7.00	76	$4f^4\ 6s^2$	3	$^4I_9$	1.00
61	Pm	(145)											
62	Sm	150.4	1070	2.06	1778		rhomb	7.54	50	$4f^6\ 6s^2$	3	$^6H_{5/2}$	0.97
63	Eu	152.0	820	2.20	1597		bcc	5.25	43	$4f^7\ 6s^2$	2	$^8S_{7/2}$	1.09
64	Gd	157.3	1311	2.44	3233		hcp	7.87	82	$4f^7\ 5d\ 6s^2$	3	$^8S_{7/2}$	0.94
65	Tb	158.9	1360	2.46	3041		hcp	8.27		$4f^8\ 5d\ 6s^2$	3	$^7F_6$	0.92
66	Dy	162.5	1409		2335		hcp	8.53		$4f^{10}\ 6s^2$	3	$^6H_{15/2}$	0.91
67	Ho	164.9	1470		2720		hcp	8.80	70	$4f^{11}\ 6s^2$	3	$^5I_8$	0.89
68	Er	167.3	1522		2510		hcp	9.04		$4f^{12}\ 6s^2$	3	$^4I_{15/2}$	0.88
69	Tm	168.9	1550		1727		hcp	9.33	58	$4f^{13}\ 6s^2$	3	$^3H_6$	0.87
70	Yb	173.0	824	1.83	1193		fcc	6.97	40	$4f^{14}\ 6s^2$	3	$^2F_{7/2}$	0.86
71	Lu	175.0	1655	2.9	3315		hcp	9.74	99	$5d\ 6s^2$	3	$^1S_0$	0.85
72	Hf	178.5	2230		4602		hcp	13.2		$5d^2\ 6s^2$	4	$^1S_0$	0.78
73	Ta	180.9	2996		5400		bcc	16.6	187	$5d^3\ 6s^2$	5	$^1S_0$	0.68
74	W	183.9	3410		5660	185	bcc	19.3	200	$5d^4\ 6s^2$	4		0.72
75	Re	186.2	3180				hcp	21.0	186	$5d^5\ 6s^2$	4		0.72
76	Os	190.2	3045		5000		hcp	22.6		$5d^6\ 6s^2$	4		0.69
77	Ir	192.2	2410		4130		fcc	22.7	159	$5d^8\ 6s$	4		0.68
78	Pt	195.1	1772		3800	122	fcc	21.5	135	$5d^9\ 6s$	2		0.80

Z	1	M	t_M	ΔH_M	t_B	ΔH_V	2	ρ	ΔH^o	3	4	5	I
79	Au	197.0	1064	2.96	2807	80	fcc	19.3	88	$5d^{10}\ 6s$	1	$^1S_0$	1.37
											3		0.85
80	Hg	200.6	−38.87	0.549	356.6	14	liq	13.6	15	$5d^{10}\ 6s^2$	2	$^3D_3$	1.10
81	Tl	204.4	393.5	1.02	1460	39	hcp	11.9	43	$6s^2\ 6p$	3	$^3D_3$	0.95
82	Pb	207.2	327.5	1.14	1740	42	fcc	11.3	47	$6s^2\ 6p^2$	4		0.84
83	Bi	209.0	271.3	2.60	1560	40	rhomb	9.81	50	$6s^2\ 6p^3$	3	$^1S_0$	0.96
84	Po	(209)											
85	At	(210)											
86	Rn	(222)											
87	Fr	(223)											
88	Ra	226.0	700		1140		bcc	6.0		$7s^2$	2		1.43
89	Ac	227.0	1050		3200		fcc	10.1		$6d\ 7s^2$	3		1.18
90	Th	232.0	1750		4790		fcc	11.7		$6d^2\ 7s^2$	4		1.02
91	Pa	231.0					tetrag I	15.4		$5f^2\ 6d\ 7s^2$			
92	U	238.0	1132		3818		orth C	19.0	125	$5f^3\ 6d\ 7s^2$	4		0.97
93	Np	237.0	640		3902		orth P	20.5		$5f^4\ 7s^2$			
94	Pu	(244)											
95	Am	(243)											
96	Cm	(247)											

Subject Index

D

d.c. Josephson effect 569-571

Dalton's law 159, 522

Daunt and Mendelssohn experiment 300, 302

de Broglie, thermal wavelength 129, 165, 497

de Haas-van Alphen effect 333, 342, 376

Debye, P. 526
  dispersion relation 293
  function 269ff, 287, 292
  model 263, 266, 268, 270-272, 275, 286-292, 294, 304
    density of states 268
    improvements 286
  temperature 269, 272, 277ff, 284ff, 294, 325, 349, 546, 553
    metals 329
    superconductors 539

Debye-Hückel theory 526, 528, 530, 534

degeneracy 2, 7-9, 37, 40, 53, 56, 58ff, 62ff, 65ff, 70, 85, 96, 127
  translational energy levels 2

de Haas-van Alphen effect 342, 353

demagnetizing field 80, 428

density matrix 397, 406

density operator 397, 420
  statistical 399

detailed balance 229ff

deuterium 174

deuterium D_2 148-150, 152

diamagnetism 77, 81
  Landau 341
  Larmor 89

diameters, molecular 206

diamond 243, 244

diamond
  dispersion curve 265
  lattice 274

dielectric susceptibility of HCl 142

diffraction of waves by crystals 251
  cell structure factor 253
  condition 254
  scattering factor 252

diffusion 179, 195, 200, 209ff
  coefficient 199-201, 210
    ethyl alcohol 202
    gases 210
  equation 202

dilute solutions 522

dilution refrigerator 315

diode 391

dipolar interaction energy 70

Dirac δ function 330, 370

direct rotations 247

dispersion relation
  elastic waves 258
  electronic 330, 369
  liquid helium 303
  phonons 261-264, 279
  photons 218
  spin waves 452

distance between neighbouring molecules 192ff

distinguishable particles 9, 38, 43, 53, 58

distribution function 498ff

distribution relation
  Bose-Einstein 44, 218, 231
  corrected Maxwell-Boltzmann 49, 179
  Fermi-Dirac 43, 319
  free paths 196
  Maxwell 181, 184, 186, 188, 189
  Maxwell-Boltzmann 39
  rotational levels 141
  speeds 181
  velocity components 184
  velocity in a beam 186

donor, level 382ff, 389

doping
  n-type 382
  p-type 382

Doppler shift 210

Drude P. 345
  model of conduction in metals 345

Dulong and Petit law 63-65, 70-72, 75, 267, 270ff, 282, 289, 292

dynamic viscosity 205

dysprosium, helimagnetism 475

Author Index